Hudson River Fishes and their Environment

American Fisheries Society Symposium 51

American Fisheries Society
Bethesda, Maryland
2006

Hudson River Fishes and their Environment

John R. Waldman, Karin E. Limburg,
and David L. Strayer

American Fisheries Society Symposium 51

Proceedings of the Conference
Hudson River Fishes and their Environment
held in Poughkeepsie, New York
20–21 March 2003

American Fisheries Society
Bethesda, Maryland
2006

Suggested citation formats are

Entire Book

Waldman, J. R., K. E. Limburg, and D. L. Strayer, editors. 2006. Hudson River fishes and their environment. American Fisheries Society, Symposium 51, Bethesda, Maryland.

Chapter in book

Blumberg, A. F., and F. L. Hellweger. 2006. Hydrodynamics of the Hudson River Estuary. Pages 9–28 *in* J. R. Waldman, K. E. Limburg, and D. L. Strayer, editors. Hudson River fishes and their environment. American Fisheries Society, Symposium 51, Bethesda, Maryland.

Front cover photographs: Atlantic sturgeon capture, photograph by John Waldman; young-of-the-year river herring and gizzard shad courtesy of Karin Limburg; aerial view of Bear Mountain Bridge; striped bass from Upper New York Bay courtesy of Daniel Shaw.

Printed in the United States of America on acid-free paper.

Library of Congress Control Number 2006936580

ISBN 1-888569-82-4
ISSN 0892-2284

American Fisheries Society Web site address: *www.fisheries.org*

American Fisheries Society
5410 Grosvenor Lane, Suite 110
Bethesda, Maryland 20814-2199
USA

Contents

Reviewer Acknowledgments

The editors thank the following individuals who assisted with reviewing the manuscripts:

Kenneth Able
Thomas Axenrot
John Boreman
Tommy Brown
Douglas Carlson
Thomas Cronin
Robert Daniels
Joseph DiLorenzo
Edward Durbin
Barbara Entwisle
Stuart Findlay
Dan Gefell
Rockwell Geyer
Edward Houde
Mark Mattson
Kim McKown
Sean Moser
Frank Nitsche
Michael Pace
Christian Reiss
Lars Rudstam
Eric Schultz
Charles Simenstad
Ron Sloan
Dennis Suszkowski
Dennis Swaney
John Thedinga
Gary Wall
Thad Wasklewicz
Cathleen Wigand
Wil Wollheim
David Yozzo

Symbols and Abbreviations

The following symbols and abbreviations may be found in this book without definition. Also undefined are standard mathematical and statistical symbols given in most dictionaries.

A	ampere	eq	equivalent
AC	alternating current	et al.	(et alii) and others
Bq	becquerel	etc.	et cetera
C	coulomb	eV	electron volt
°C	degrees Celsius	F	filial generation; Farad
cal	calorie	°F	degrees Fahrenheit
cd	candela	fc	footcandle (0.0929 lx)
cm	centimeter	ft	foot (30.5 cm)
Co.	Company	ft^3/s	cubic feet per second ($0.0283\ m^3/s$)
Corp.	Corporation	g	gram
cov	covariance	G	giga (10^9, as a prefix)
DC	direct current; District of Columbia	gal	gallon (3.79 L)
D	dextro (as a prefix)	Gy	gray
d	day	h	hour
d	dextrorotatory	ha	hectare (2.47 acres)
df	degrees of freedom	hp	horsepower (746 W)
dL	deciliter	Hz	hertz
E	east	in	inch (2.54 cm)
E	expected value	Inc.	Incorporated
e	base of natural logarithm (2.71828...)	i.e.	(id est) that is
		IU	international unit
e.g.	(exempli gratia) for example	J	joule

K	Kelvin (degrees above absolute zero)
k	kilo (10^{3}, as a prefix)
kg	kilogram
km	kilometer
l	levorotatory
L	levo (as a prefix)
L	liter (0.264 gal, 1.06 qt)
lb	pound (0.454 kg, 454g)
lm	lumen
log	logarithm
Ltd.	Limited
M	mega (10^{6}, as a prefix); molar (as a suffix or by itself)
m	meter (as a suffix or by itself); milli (10^{23}, as a prefix)
mi	mile (1.61 km)
min	minute
mol	mole
N	normal (for chemistry); north (for geography); newton
N	sample size
NS	not significant
n	ploidy; nanno (10^{29}, as a prefix)
o	ortho (as a chemical prefix)
oz	ounce (28.4 g)
P	probability
p	para (as a chemical prefix)
p	pico (10^{212}, as a prefix)
Pa	pascal
pH	negative log of hydrogen ion activity
ppm	parts per million
qt	quart (0.946 L)
R	multiple correlation or regression coefficient
r	simple correlation or regression coefficient
rad	radian
S	siemens (for electrical conductance); south (for geography)
SD	standard deviation
SE	standard error
s	second
T	tesla
tris	tris(hydroxymethyl)-aminomethane (a buffer)
UK	United Kingdom
U.S.	United States (adjective)
USA	United States of America (noun)
V	volt
V, Var	variance (population)
var	variance (sample)
W	watt (for power); west (for geography)
Wb	weber
yd	yard (0.914 m, 91.4 cm)
α	probability of type I error (false rejection of null hypothesis)
β	probability of type II error (false acceptance of null hypothesis)
Ω	ohm
μ	micro (10^{26}, as a prefix)
′	minute (angular)
″	second (angular)
°	degree (temperature as a prefix, angular as a suffix)
%	per cent (per hundred)
‰	per mille (per thousand)

HUDSON RIVER ESTUARY

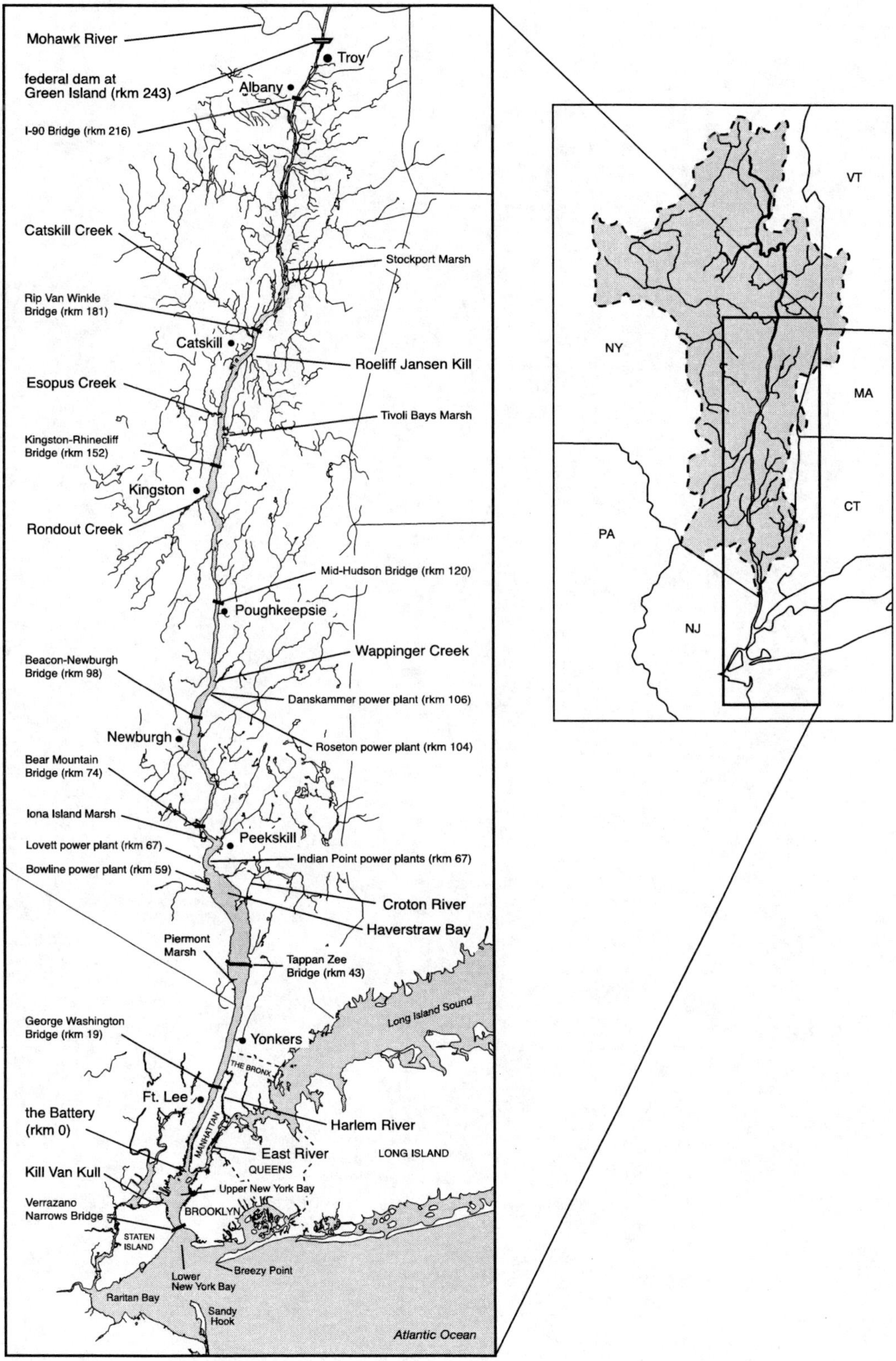

American Fisheries Society Symposium 51:1–7, 2006

The Hudson River Environment and its Dynamic Fish Community

JOHN R. WALDMAN
Queens College, Biology Department, 65-30 Kissena Boulevard
Flushing, New York 11367 USA
john.waldman@qc.cuny.edu

KARIN E. LIMBURG
State University of New York College of Environmental Science and Forestry
Syracuse, New York 13210 USA
KLimburg@esf.edu

DAVID L. STRAYER
Institute of Ecosystem Studies, 65 Sharon Turnpike, Box AB
Millbrook, New York 12545-0129 USA
strayerd@ecostudies.org

Introduction

For its modest size, the Hudson is a well-known river, an outcome of prolonged and intense postcolonial human use and abuse. Indeed, the Hudson is one of the most scientifically-studied rivers in the world (e.g., Levinton and Waldman 2006). Why has the river received so much attention?

Part of its fame is due to its important historical roles as an early inland route for colonization, a Revolutionary War battleground, and the setting for the development of an esthetic movement known as the Hudson River School of painting. Later, it achieved infamy as a waterway that received untold quantities of contaminants: a remarkably extensive suite of industrial wastes and, for many decades, the raw sewage of hundreds of thousands and, eventually millions of inhabitants along its shores. After, as water quality improved due to government programs, other environmental issues emerged, some of them leading to heated conflicts, such as the controversies over the effects of electric-generating facilities on fish populations, the potential harm to the river's striped bass *Morone saxatilis* population from the large area of landfill that would have lined one of its shores had a Manhattan highway project nicknamed "the Westway" been approved, and over the desirability of dredging to remove polychlorinated biphenyls (PCBs) from Hudson River sediments.

Science on the Hudson has evolved in keeping with this history (i.e., much of what is known about the river has emerged through applied studies funded to answer practical and often nagging questions). But this history of sometimes painful conflict has also yielded fruits not often found on other estuaries. One is the launching in 1973 and continuation to the present of an annual physical and biological monitoring program funded by the electric utilities that is among the most comprehensive anywhere. Another is the creation in 1981 of the Hudson River Foundation, that, through financial support and otherwise, fosters not only applied but also fundamental studies of how the river functions (Suszkowski and D'Elia 2006). These efforts, together with a host of other city, state, and federal programs, plus the increasing interest of academic researchers and amateur volunteers and observers, have led to a golden age of Hudson River research, much of it centered around its fishes. This volume is one product of this golden age and reflects just how much is known about Hudson River fishes and their environment.

Hudson River Fishes

The contributions in this volume show a detailed, long-term understanding of fish populations in the Hudson. Waldman et al. (2006, this volume) review the composition of the fauna, describe its complex zoogeographic origins, and discuss major compositional changes over historical times. Several chapters examine key populations in detail. A fish species which was found to be heavily impacted by power plants on the Hudson (Boreman and Goodyear 1988) but which had received little directed study on its fundamental biology in the system is bay anchovy *Anchoa mitchilli.* Four papers address these issues: Schultz et al. (2006, this volume) examine fine-scale movements and transport, fecundity, and fluctuations in abundance; Dunning et al. (2006a, this volume) consider densities of early life stages of bay anchovy in the lower river in comparison with surrounding waterways and the implications of population discreteness for the Hudson to population mortality estimates caused by Hudson River power plants; the question of population discreteness is further examined by Tipton and Hartman (2006a, this volume) and, in another paper, Tipton and Hartman (2006b, this volume) use hydroacoustics to evaluate the distribution of fishes (primarily bay anchovy) near two power plants in relation to light, depth, and distance from the facilities.

A major component of the Hudson's ichthyofauna is its approximately one-dozen diadromous fishes (Waldman 2006). Because they inhabit not only the river but also marine waters, their environment extends at least as far as their ranges outside the river. Moreover, their use of nearby coastal waters may occur as adults but also, for some species, at early life stages, which raises questions about estuarine transport processes. Dunning et al. (2006b, this volume) examine processes that could account for postyolk-sac larvae of striped bass outside the mainstem Hudson River in the adjoining East River tidal strait. Within-population competition is another aspect of a fish's environment. In another paper by Dunning et al. (2006c, this volume) the relationship is explored between cohort abundance and stock size and dispersal of age-2+ striped bass outside the Hudson.

Like all river systems, the Hudson consists of a complex mosaic of interconnected habitats. As shown by the chapters just discussed, many of the Hudson's fishes use multiple habitats. Understanding movements of fish populations among habitats is an active area of research and preserving corridors for fish movement is an important challenge for conservation and restoration. In addition to movements within the estuary and between the estuary and nearby coastal waters, movement into the Hudson's tributaries, many of which have been blocked by dams, is important for some fish species. Schmidt et al. (2006, this volume) examine the importance of tributaries for the Hudson's fishes, focusing on American eel *Anguilla rostrata*.

Habitats also are important for their resident species. Knowledge of the fish communities of difficult-to-sample habitats such as wetlands and beds of vegetation has lagged behind that of open-water fishes. Osgood et al. (2006, this volume) show that the marshes of the invasive common reed *Phragmites australis* and narrow-leaf cattail *Typha angustifolia* each support distinctive fish communities. Sampling difficulties also have limited our knowledge of habitat use during the winter. However, Hartman and Nagy (2006, this volume) used hydroacoustics to provide new insight into fish use of the lower river in relation to ambient environmental factors.

Environmental Influences

A theme in many of the chapters is how dynamic the Hudson's fish populations are, from week to week, year to year, and decade to decade. We have made significant progress in understanding how environmental factors, including the direct and indirect effects of human activities, affect the Hudson's fishes.

Interannual variation in weather, especially the amount and timing of freshwater that runs off the watershed, may be substantial (Abood et al. 2006; this volume; Blumberg and Hellweger 2006; this volume), and should affect the numbers, location, and condition of fishes in the Hudson. Caraco and Cole (2006, this volume) show how variations in freshwater inputs affect the allochthonous and autochthonous inputs that are at the base of the Hudson's food web. Over the longer term, recent climatic warming, resulting in part from human activities, may be responsible for the disappearance of the rainbow smelt *Osmerus mordax* and declines in populations of tomcod *Microgadus tomcod*, both of which reach their southern range limits in the Hudson (Daniels et al. 2005). Longer-term changes in climate and sea level certainly have also affected the ecological character of the Hudson (Peteet et al. 2006; this volume), even if explicit links to the fish community have not yet been demonstrated.

As mentioned earlier, point sources of pollution—chiefly municipal sewage and industrial waste—have been of enormous importance in the Hudson. Control of these inputs has led to vastly increased water quality in the estuary, especially near New York City and Albany (Abood et al. 2006; Brosnan et al. 2006). There is at least some evidence (Strayer et al. 2004) that this improved water quality has led in turn to a redistribution of fish populations in the Hudson, with fish moving into formerly degraded reaches. Despite reductions in inputs, persistent pollutants such as PCBs, metals, and PAHs still contaminate the Hudson and its fishes, with marked effects on fish health and fisheries (Wirgin and Chambers 2006; this volume) and, perhaps, human health (Gobeille et al. 2006).

Land use in the Hudson watershed has changed greatly since first European settlement, and continues to change today. Swaney et al. (2006, this volume) summarize major trends over the past 200-plus years in land use, population density, and damming within the entire Hudson watershed, and discuss likely consequences for the ecological function of the system. This land-use change affects the timing and amount of inputs of water, sediment, nutrients, and toxins into the Hudson, which presumably have pervasive effects on the Hudson's fishes. So far, only a few links between land use and the Hudson's fishes have been demonstrated (e.g., Limburg and Schmidt 1990); understanding such links is an important challenge for research and management.

Humans have transformed the physical condition of the Hudson and its tributaries to make them more suitable for navigation, power production, and other human uses. The result has been loss of thousands of hectares of shallow-water habitats, especially in the upper estuary (Miller et al. 2006; this volume), and the blocking of many tributaries against passage of the Hudson's many migratory (anadromous, catadromous, potamodromous) and residential lotic fishes. Schmidt et al. (2006) discuss some of the consequences of tributary blockage. The loss of shallow-water habitats must have had large effects on populations of fishes and other animals, but these effects have not yet been estimated in the Hudson.

Human activities have also brought hundreds of species of nonnative plants and animals into the Hudson and its tributaries. Several of these species—most notoriously the zebra mussel *Dreissena polymorpha* and the water chestnut *Trapa natans*—have been shown to have major effects on habitat quality, food webs, and fish populations (e.g., Strayer et al. 1999; Caraco and Cole 2002; Osgood et al. 2006; Strayer 2006).

Finally, harvest of fishes, whether for recreational or commercial purposes, may affect fish populations. Although historically important, commercial fisheries within the Hudson have dropped to nearly nothing as a result of PCB contamination and declining stocks (MacKenzie 1992; Limburg et al. 2006). However, offshore fisheries for diadromous species may have large effects on the Hudson's fish stocks (Limburg et al. 2006). The chapter by Euston et al. (2006, this volume) documents the extent and characteristics of the current recreational fishery, and the surprising extent to which health advisories about PCB contamination are ignored by anglers.

Mysteries

Despite the enormous amounts of data and the impressive advances in understanding Hudson River fishes, much remains unknown. Typically, quantitative analyses of fish populations and communities in the river report large amounts of unexplained variance (e.g., Hurst et al. 2004; Strayer et al. 2004). Even some major events in the Hudson, such as the near-disappearance of the four-spine stickleback *Apeltes quadracus* over the past two decades and the recent irruption of the gizzard shad *Dorosoma cepedianum* population (Daniels et al. 2005), are poorly understood. These mysteries reflect gaps in our basic knowledge of the natural history of fishes in the Hudson, including inadequate exploration of some habitats, as well as the need for data and integrative tools that bear on the problems of understanding multiple controlling factors and interactions among multiple habitats in a complex estuary and its neighborhood. We have come a long way towards understanding the dynamic and complex fish community of the Hudson, but much work remains to be done.

Related Literature

Current knowledge of Hudson River fishes and related topics has occasionally been assembled from diverse sources through books, conference proceedings, and monographs. A seminal work for the lay public was *The Hudson River: A Natural and Unnatural History* by Boyle (1969). Early among conference proceedings was the *Hudson River Ecology* series, published by the Hudson Valley Commission in 1966, the New York State Department of Environmental Conservation in 1969, and the Hudson River Environmental Society (HRES) in 1973 and 1976. The proceedings of two later HRES conferences were edited by C. L. Smith and published in book form (Smith 1988, 1992). The vast amount of information generated by the electric utilities controversy, and the controversy itself was summarized in Barnthouse et al. (1988); Limburg et al. (1986) also digested elements of this conflict and other critical Hudson River issues. Fisheries in the lower portion of the estuary were reviewed by MacKenzie (1992). The proceedings of a Hudson River session at the 1993 Biennial International Estuarine Research Federation Conference were published in a special issue of *Estuaries* (volume 19: issue 4). Recently, Levinton and Waldman (2006) edited a broad volume on the Hudson River estuary. Important early works on New York fishes, including those of the Hudson estuary, are cited in Waldman et al. (2006).

The present book links ecosystem function, including legacies from human use of the watershed, to fisheries in a way that perhaps has not been done for the Hudson before. We hope that future work continues to examine habitat and ecosystem connections to fisheries there.

Acknowledgments

This volume is based largely on a conference titled *Hudson River Fishes and their Environment* held in Poughkeepsie, New York, on 20–21 March 2003 and organized by the Hudson River Environmental Society (HRES). HRES Director Stephen Wilson, the HRES Board of Directors, and various members and volunteers deserve credit for conceptualizing and organizing the conference. HRES Board Chairman Mark Mattson was particularly instrumental in bringing this book to fruition. One of us (JRW) thanks the Hudson River Foundation for support while planning the volume and editing its earlier stages. The many external reviewers greatly improved the final manuscripts. We are also deeply appreciative to Megan Hubbard of the American Fisheries Society for her editing and for easing this volume through the entire publication process.

References

Abood, K. A., T. L. Englert, S. G. Metzger, C.V. Beckers, Jr., T. J. Groninger, and S. Mallavaram. 2006. Current and evolving physical and chemical conditions in the Hudson River estuary. Pages 39–62 *in* J. R. Waldman, K. E. Limburg, and D. L. Strayer, editors. Hudson River fishes and their environment. American Fisheries Society, Symposium 51, Bethesda, Maryland.

Barnthouse, L. W., R. J. Klauda, D. S. Vaughan, and R. L. Kendall, editors. 1988. Science, law, and Hudson River power plants: a case study in environmental impact assessment. American Fisheries Society Monograph 4.

Blumberg, A. F., and F. L. Hellweger. 2006. Hydrodynamics of the Hudson River estuary. Pages 9–28 *in* J. R. Waldman, K. E. Limburg, and D. L. Strayer, editors.. Hudson River fishes and their environment. American Fisheries Society, Symposium 51, Bethesda, Maryland.

Boreman, J., and C. P. Goodyear. 1988. Estimates of entrainment mortality for striped bass and other fish species inhabiting the Hudson River estuary. American Fisheries Society Monograph 4:152–160.

Boyle, R. H. 1969. The Hudson River: a natural and unnatural history. Norton, New York.

Brosnan, T. M., A. Stoddard, and L. J. Hettling. 2006. Hudson River sewage inputs and impacts: past and present. Pages 335–348 *in* J. S. Levinton and J. R. Waldman, editors. The Hudson River estuary. Cambridge University Press, New York.

Caraco, N. F., and J. J. Cole. 2002. Contrasting impacts of a native and alien macrophyte on dissolved oxygen in a large river. Ecological Applications 12:1496–1509.

Caraco, N. F., and J. J. Cole. 2006. Hydrologic control of external carbon loads and primary production in the tidal freshwater

Hudson. Pages 63–74 *in* J. R. Waldman, K. E. Limburg, and D. L. Strayer, editors. Hudson River fishes and their environment. American Fisheries Society, Symposium 51, Bethesda, Maryland.

Daniels, R. A., K. E. Limburg, R. E. Schmidt, D. L. Strayer, and C. Chambers. 2005. Changes in fish assemblages in the tidal Hudson River, New York. Pages 471–503 *in* J. N. Rinne, R. M. Hughes, and B. Calamusso, editors. Historical changes in large river fish assemblages of America. American Fisheries Society Symposium 45.

Dunning, D. J., Q. E. Ross, A. F. Blumberg, and D. G. Heimbuch. 2006b. Transport of striped bass larvae out of the lower Hudson River estuary. Pages 273–286 *in* J. R. Waldman, K. E. Limburg, and D. L. Strayer, editors. Hudson River fishes and their environment. American Fisheries Society, Symposium 51, Bethesda, Maryland.

Dunning, D. J., Q. E. Ross, M. T. Mattson, and D. G. Heimbuch. 2006a. Distribution and abundance of bay anchovy eggs and larvae in the Hudson River and nearby waterways. Pages 215–226 *in* J. R. Waldman, K. E. Limburg, and D. L. Strayer, editors. Hudson River fishes and their environment. American Fisheries Society, Symposium 51, Bethesda, Maryland.

Dunning, D. J., J. R. Waldman, Q. E. Ross, and M. T. Mattson. 2006c. Dispersal of age-2+ striped bass out of the Hudson River. Pages 287–294 *in* J. R. Waldman, K. E. Limburg, and D. L. Strayer, editors. Hudson River fishes and their environment. American Fisheries Society, Symposium 51, Bethesda, Maryland.

Euston, E. T., S. A. Haney, K. A. Hattala, and A.W. Kahnle. 2006. Overview of Hudson River recreational fisheries, with an emphasis on striped bass. Pages 295–315 *in* J. Waldman, K. Limburg, and D. Strayer, editors. Hudson River fishes and their environment. American Fisheries Society, Symposium 51, Bethesda, Maryland.

Gobeille, A. K., K. B. Morland, R. F. Bopp, J. H. Godbold, and P. J. Landrigan. 2006. Body burdens of mercury in lower Hudson River area anglers. Environmental Research 101:205–212.

Hartman, K. J., and B. W. Nagy. 2006. Winter distribution and abundance of Hudson River fishes using hydroacoustics. Pages 175–196 *in* J. R. Waldman, K. E. Limburg, and D. L. Strayer, editors. Hudson River fishes and their environment. American Fisheries Society, Symposium 51, Bethesda, Maryland.

Hurst, T. P., K. A. McKown, and D. O. Conover. 2004. Interannual and long-term variation in the nearshore fish community of the mesohaline Hudson River estuary. Estuaries 27:659–669.

Levinton, J. S., and J. R. Waldman, editors. 2006. The Hudson River estuary. Cambridge University Press, New York.

Limburg, K. E., K. A. Hattala, A. W. Kahnle, and J. R. Waldman. 2006. Fisheries of the Hudson River estuary. Pages 189–204 *in* J. S. Levinton and J. R. Waldman, editors. The Hudson River estuary. Cambridge University Press, New York.

Limburg, K. E., M. A. Moran, and W. H. McDowell, editors. 1986. The Hudson River ecosystem. Springer-Verlag, New York.

Limburg, K. E., and R. E. Schmidt. 1990. Patterns of fish spawning in Hudson River tributaries: response to an urban gradient? Ecology 71:1238–1245.

MacKenzie, C. L., Jr. 1992. The fisheries of Raritan Bay. Rutgers University Press, New Brunswick.

Miller, D., J. Ladd, and W. C. Nieder. 2006. Channel morphology in the Hudson River estuary: Historical changes and opportunities for restoration. Pages 29–37 *in* J. R. Waldman, K. E. Limburg, and D. L. Strayer, editors. Hudson River fishes and their environment. American Fisheries Society, Symposium 51, Bethesda, Maryland.

Osgood, D. T., D. J. Yozzo, R. M. Chambers, S. Pianka, and J. Lewis. 2006. Patterns of habitat utilization by resident nekton in *Phragmites* and *Typha* marshes of the Hudson River Estuary. Pages 151–173 *in* J. Waldman, K. Limburg, and D. Strayer, editors. Hudson River fishes and their environment. American Fisheries Society, Symposium 51, Bethesda, Maryland.

Peteet, D. M., D. C. Pederson, D. Kurdyla, and T. Guilderson. 2006. Hudson River paleoecology from marshes: environmental change and its implications for fisheries.

Pages 113–128 *in* J. R. Waldman, K. E. Limburg, and D. L. Strayer, editors. Hudson River fishes and their environment. American Fisheries Society, Symposium 51, Bethesda, Maryland.

Schmidt, R. E., T. R. Lake, and R. Petersson. 2006. Hudson River tributaries in the lives of fishes with emphasis on the American eel. Pages 317–330 *in* J. R. Waldman, K. E. Limburg, and D. L. Strayer, editors. Hudson River fishes and their environment. American Fisheries Society, Symposium 51, Bethesda, Maryland.

Schultz, E. T., K. M. M. Lwiza, J. R. Young, K. J. Hartman, and R. C. Tipton. 2006. The dynamics of bay anchovy in the Hudson River estuary: process-oriented studies and long-term changes. Pages 197–213 *in* J. Waldman, K. Limburg, and D. Strayer, editors. Hudson River fishes and their environment. American Fisheries Society, Symposium 51, Bethesda, Maryland.

Smith, C. L., editor. 1988. Fisheries research in the Hudson River. State University of New York Press, Albany.

Smith, C. L., editor. 1992. Estuarine research in the 1980s. State University of New York Press, Albany.

Strayer, D. L. 2006. Alien species in the Hudson River. Pages 296–310 *in* J. S. Levinton and J. R. Waldman, editors. The Hudson River estuary. Cambridge University Press, New York.

Strayer, D. L., N. F. Caraco, J. J. Cole, S. Findlay, and M. L. Pace. 1999. Transformation of freshwater ecosystems by bivalves: a case study of zebra mussels in the Hudson River. BioScience 49:19–27.

Strayer, D. L., K. Hattala, and A. Kahnle. 2004. Effects of an invasive bivalve (*Dreissena polymorpha*) on fish populations in the Hudson River estuary. Canadian Journal of Fisheries and Aquatic Sciences 61:924–941.

Suszkowski, D. J., and C. F. D'Elia. 2006. The history and science of managing the Hudson River. Pages 313–334 *in* J. S. Levinton and J. R. Waldman, editors. The Hudson River estuary. Cambridge University Press, New York.

Swaney, D. P., K. E. Limburg, and K. Stainbrook. 2006. Some historical changes in the patterns of population and land use in the Hudson River watershed. Pages 75–112 *in* J. Waldman, K. Limburg, and D. Strayer, editors. Hudson River fishes and their environment. American Fisheries Society, Symposium 51, Bethesda, Maryland.

Tipton, C., and K. J. Hartman. 2006a. Distributional ecology of bay anchovy *Anchoa mitchilli* in the Hudson River estuary. Pages 227–249 *in* J. R. Waldman, K. E. Limburg, and D. L. Strayer, editors. Hudson River fishes and their environment. American Fisheries Society, Symposium 51, Bethesda, Maryland.

Tipton, C., and K. J. Hartman. 2006b. Fine-scale distribution and abundance of pelagic fish near two Hudson River power plants. Pages 251–272 *in* J. R. Waldman, K. E. Limburg, and D. L. Strayer, editors.. Hudson River fishes and their environment. American Fisheries Society, Symposium 51, Bethesda, Maryland.

Waldman, J. R. 2006. The diadromous fish fauna of the Hudson River: life histories, conservation concerns, and research avenues. Pages 171–188 *in* J. S. Levinton and J. R. Waldman, editors. The Hudson River estuary. Cambridge University Press, New York.

Waldman, J. R., T. R. Lake, and R. E. Schmidt. 2006. Biodiversity and zoogeography of the fishes of the Hudson River watershed and estuary. Pages 129–150 *in* J. R. Waldman, K. E. Limburg, and D. L. Strayer, editors. Hudson River fishes and their environment. American Fisheries Society, Symposium 51, Bethesda, Maryland.

Wirgin, I., and R. C. Chambers. 2006. Atlantic tomcod *Microgadus tomcod*: a model species for the responses of Hudson River fish to toxicants. Pages 331–364 *in* J. R. Waldman, K. E. Limburg, and D. L. Strayer, editors. Hudson River fishes and their environment. American Fisheries Society, Symposium 51, Bethesda, Maryland.

American Fisheries Society Symposium 51:9–28, 2006

Hydrodynamics of the Hudson River Estuary

Alan F. Blumberg

Stevens Institute of Technology, Department of Civil, Environmental, and Ocean Engineering Castle Point on Hudson, Hoboken, New Jersey 07030 USA.
ablumberg@stevens.edu

Ferdi L. Hellweger

Northeastern University, Department of Civil and Environmental Engineering 360 Huntington Avenue, 400 SN, Boston, Massachusetts 02115 USA.
ferdi@coe.neu.edu

Abstract.—The Hudson River Estuary can be classified as a drowned river valley, partially mixed, tidally dominated estuary. Originally, it had a fjord-like morphology as a result of glacial scour which filled in over the past 3,000 years with river sediments. The hydrodynamics of the estuary are best described by the drivers of circulation, including the upstream river inflows, the oceanographic conditions at the downstream end, and meteorological conditions at the water surface and the response of the waters to these drivers in terms of tides and surges, currents, temperature, and salinity. Freshwater inflow is predominantly from the Mohawk and Upper Hudson rivers at Troy (average flow = 400 m^3/s, highest in April, lowest in August). At the downstream end at the Battery the dominant tidal constituent is the semidiurnal, principal lunar constituent (the M_2 tide), with an evident spring/neap cycle. The amplitude of the tide is highest at the Battery (67 cm), lower at West Point (38 cm), and higher again at Albany (69 cm), a function of friction, geometry, and wave reflection. Meteorological events can also raise the water surface elevation at the downstream end and propagate into the estuary. Freshwater pulses can raise the water level at the upstream end and propagate downstream. Tidal flows are typically about an order of magnitude greater than net flows. The typical tidal excursion in the Hudson River Estuary is 5–10 km, but it can be as high as 20 km. Temperature varies seasonally in response to atmospheric heating and cooling with a typical August maximum of 25°C and January-February minimum of 1°C. Power plants cause local heating. The salinity intrusion varies with the tide and amount of upstream freshwater input. The location of the salt front is between Yonkers and Tappan Zee in the spring and just south of Poughkeepsie in the summer. Vertical salinity stratification exists in the area of salt intrusion setting up an estuarine circulation. The effect of wind is limited due to a prevailing wind direction perpendicular to the main axis and the presence of cliffs and hills. Dispersive processes include shear dispersion and tidal trapping, resulting in an overall longitudinal dispersion coefficient of 3–270 m^2/s. The residence or flushing time in the freshwater reach is less than 40 d in the spring and about 200 d in the summer. In the area of salt intrusion, it is about 8 d.

Introduction

The Hudson River Estuary extends from Troy to the Battery at the southern tip of Manhattan Island, as shown in Figure 1. Strictly speaking, in order to meet the definition of an "estuary" a water body has to contain seawater "measurably diluted with freshwater derived from land drainage" (Pritchard 1967). The upper reaches of this estuary (from Troy to Poughkeepsie) are always fresh and therefore that part of the system is technically not an "estuary" but a "tidal river." However, in this paper we concern ourselves little with that distinction and simply refer to the whole water body as the Hudson River Estuary.

The Hudson River Estuary was created about 6,000 years ago when rising sea level flooded the lower portions of the Hudson River with ocean water. Originally, the estuary had a fjord-like morphology as a result of glacial scour. About 18,000 years ago, the Laurentide glacier retreated northward leaving behind a deep gouge in the bedrock that was filled with melt water. Substantial quantities of river sediments were brought into the gouge over the next 3,000 years, altering the morphology so that today the Hudson River Estuary can be classified as a drowned river valley (McHugh et al. 2004). Along its 247-km length (from Troy to the Battery), the geometry is extremely variable reflecting its geological past. The river contains wide shallow bays (e.g., Newburgh Bay), narrow deep channels (e.g., World's End), islands (e.g., Esopus Island), peninsulas (e.g., Croton Point), coves (e.g., Foundry Cove) and numerous other features (tidal flats, shoals, and rock outcroppings) that affect how water moves through the estuary. The Hudson River Estuary is a partially mixed estuary, where salt water and freshwater mix, resulting in a significant vertical density gradient. The currents in the estuary are predominantly driven by the tide. Meteorological events and freshwater inflows also play an important role in affecting the circulation.

The hydrodynamics of the Hudson River Estuary are of major importance to fish and fishery studies for many reasons. Most notably, in their early life stages, fish are planktonic and their movements are controlled by the ambient currents. Hydrodynamic processes lead to transport, dispersion, mixing, and flushing, important elements to consider when undertaking studies of the life histories of fishes and other aquatic organisms.

This paper describes the hydrodynamics of the Hudson River Estuary. It is written for the nonhydrodynamic professional and therefore attempts to describe many of the phenomena in nontechnical terms. The vocabulary of hydrodynamics is used where appropriate as an instructional aid. The paper starts with a description of the external drivers of the circulation at the upstream and downstream ends of the estuary and at the water surface. Then the responses of the estuary to these drivers are discussed, including tides and surges, currents, and temperature and salinity distributions. Finally, the processes that transport and mix salt, heat and any introduced pollutants are described.

Drivers of the Hydrodynamics

The hydrodynamics of the Hudson River Estuary are primarily driven by upstream inflows (mostly at Troy), oceanographic conditions at the downstream end (at the Battery), and meteorological conditions at the water surface. Following is a description of each of these drivers.

Upstream End

Upstream inflows are important mainly for

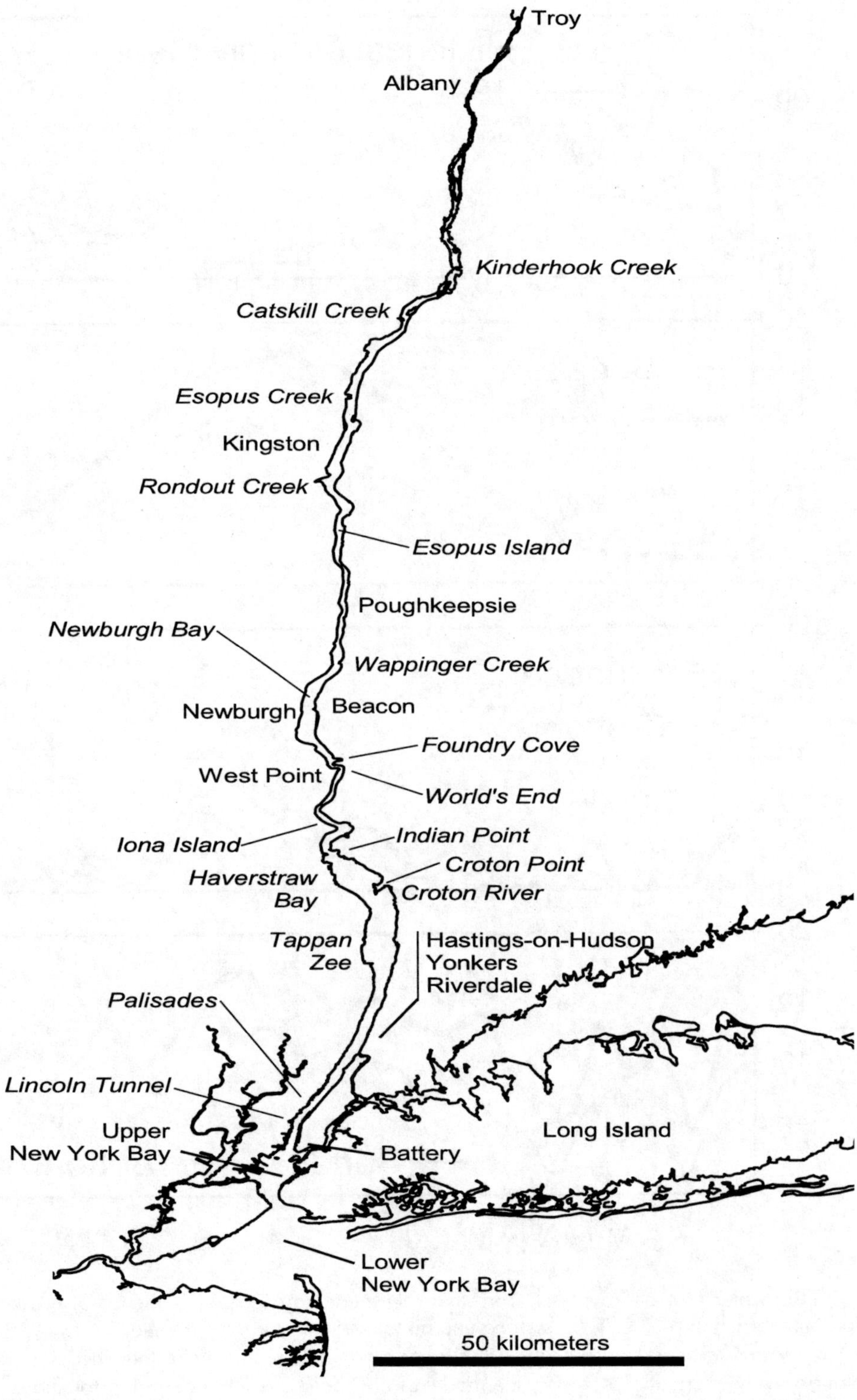

Figure 1. Overview map of the Hudson River Estuary.

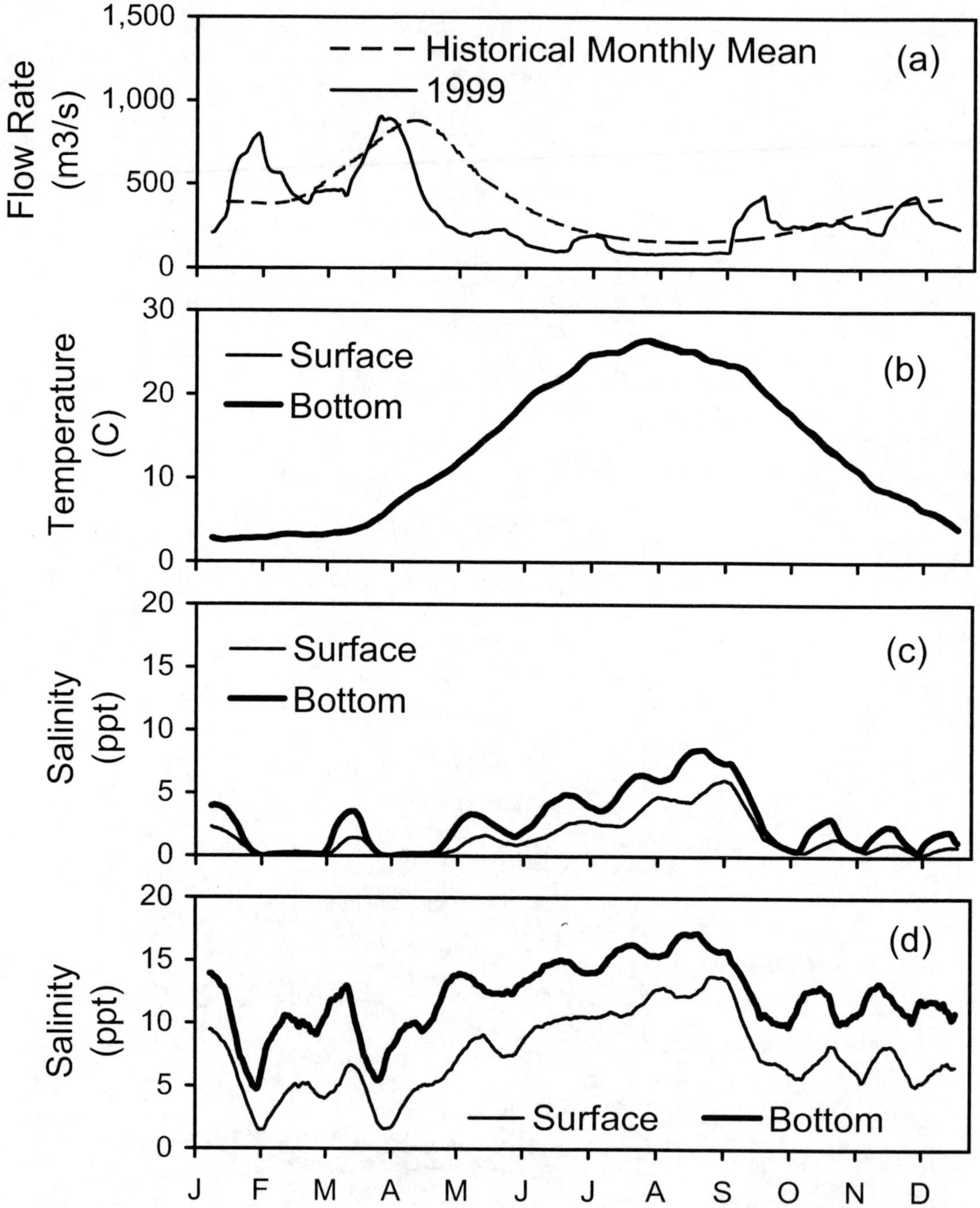

Figure 2. (a) Freshwater flow at Troy, New York (monthly means are based on the USGS Green Island gauge, period of record: 1946–2002), (b) surface and bottom temperature at Croton-on-Hudson, (c) surface and bottom salinity at Croton-on-Hudson, and (d) surface and bottom salinity at the Lincoln Tunnel. The temperature and salinity information are from the model of Blumberg et al. (1999) using forcing data from 1999. All data are "filtered" to remove oscillations occurring on time scales of less than 15 d.

two reasons. First, the added water increases the water surface elevation at the upstream end. This rise in elevation pushes water downstream (via the pressure gradient force). Second, the added water is fresh and the continuous input of freshwater prevents the salt water from flowing all the way up the estuary. Freshwater inflow is predominantly from the Mohawk and Upper Hudson rivers, which jointly enter the estuary from the north at Troy (average inflow of about 400 m^3/s). Other tributaries (see Figure 1 for locations of the inflow tributaries) include Rondout Creek (50 m^3/s), Kinderhook Creek (20 m^3/s), Esopus Creek (20 m^3/s), Catskill Creek (20 m^3/s), Croton River (10 m^3/s), and Wappinger Creek (10 m^3/s). Figure 2a shows the historical monthly average flows and the flows from 1999 at Troy (the other panels in the figure will be discussed subsequently). The inflow has a strong seasonal signal with highest flows in April and lowest flows in August. In any given year, the flow can vary significantly from the long-term mean. For example, in 1999, there were two peaks in the spring, one in late January and one in late March/early April. The summertime low flows in 1999 were considerably less than the long-term average. The freshwater inflows, the reader will learn later using Figures 2c and 2d, have a significant impact on the salinity distribution in the estuary. For a more detailed discussion on freshwater inputs to the Hudson River Estuary refer to Abood et al. (1992) and Wells and Young (1992).

Downstream End

The downstream end of the estuary is at the Battery where the Hudson River meets the upper part of the New York/New Jersey Harbor Estuary. Changes in the water surface elevation at this location are the most important drivers of water movement in the Hudson River Estuary. When the water surface elevation at the downstream end rises (e.g., high tide), water is pushed upstream into the estuary. Conversely, when it falls (e.g., low tide), water is pulled out of the estuary. The mechanism is the same as that responsible for the net downstream flow due to freshwater inputs at the upstream end (pressure gradient force).

On a global scale, the ocean's surface elevation oscillates as a result of the balance between gravitational attraction and centrifugal forces on the ocean water in the Earth, Moon, and Sun system. There are 399 "tidal constituents" (individual components that make up the overall tide) that are used to describe the tides on earth (Doodson 1922). Each constituent represents a periodic change or variation in the relative positions of the Earth, Moon and Sun and each has a unique period. The amplitude and phase associated which each constituent varies from place to place, however. Most constituents have very small amplitudes, and the observed tide is dominated by only a few of them.

The dominant tidal constituent in the Hudson River Estuary is the semidiurnal, principal lunar constituent (the M_2 tide), which has a period of 12.42 h. It therefore produces approximately two "high tides" and two "low tides" per day. The semidiurnal, principal solar component (the S_2 tide) also produces two "high tides" and two "low tides" per day and has a period of 12 h. This constituent is smaller than the lunar component. It works to amplify or reduce the amplitude of the lunar component, depending on whether it is in or out of phase with it. The result is an oscillation in the observed tidal amplitude with a period of 14.8 d. So, about every 15 d, the lunar and solar components are in phase and the observed tidal amplitude is highest. This is called the spring tide. And about every 15 d, the two components are out of phase and the observed tidal amplitude is lowest, which is called the neap tide. Therefore, this oscillation is commonly referred to as the spring/neap (spring here has no relation to the season) tidal cycle. In the Hudson River Estuary, the N_2 constituent (larger lunar elliptic) with a period of 12.66 h is also important. The M_2 - N_2 combination produces an oscillation that repeats every 27.6 d. This combination modulates the amplitude of the spring/neap cycle, causing it to be larger and smaller. Also observed in the Hudson River Estuary, although of lesser importance, are the K_1 (solilunar) and O_1 (diurnal lunar) constituents. The water surface elevation at the Battery during the spring of 1998 is shown in Figure 3e. The twice daily oscillation is due to the M_2 tide. The spring/neap component of the tidal system is also evident in the water surface oscillations. The amplitude of water level oscillation is lowest during the neap tide of 4 April 1998 and is highest during spring tide about 7.5 d (about midway in the spring/neap cycle) earlier, on 27 March 1998.

Besides the astronomical tidal forcing, meteorological events, such as storm surges due to strong persistent onshore wind, often raise the water surface elevation at the downstream boundary. The pronounced increase in water surface elevation on 21 March 1998 was caused by a coastal low pressure system. Strong winds from the East (Figure 3b) produced a surge of water that propagated into lower and upper New York Bay and raised the water surface elevation for several days at the Battery (Figure 3e), and further upstream into the estuary as will be discussed later.

The salinity at the Battery depends on the amount of freshwater input upstream. The salinity reflects the mixture of water from upstream freshwater inflows (mostly at Troy, 0 parts of salt per thousand parts of water [ppt]) and the offshore saltier waters of the New York Bight (about 33 ppt). The saline waters near the Battery are the primary source of salt to the estuary. During low flow conditions, surface and bottom salinities can be as high as 20 and 28 ppt, respectively. During periods of high freshwater inputs, the surface and bottom salinities are much lower, about 5 and 20 ppt, respectively.

Water Surface

The water surface itself is an important boundary of the estuary because it is where atmospheric heating (e.g., solar radiation) and cooling (e.g., latent heat of evaporation) occur. In addition, wind over the water surface can drag water and modify the currents. There are two factors that limit the effect of wind on the Hudson River Estuary, compared to other estuaries. First, the prevailing wind direction is from the west, whereas the estuary is oriented north-south. This implies that the wind does not blow along the main axis, which would lead to a maximum effect of the wind. Second, the estuary is relatively narrow, and cliffs (e.g., Palisades) and hills often shelter and reduce the wind at the water surface. However, winds can have a strong impact on limited straight sections of the river, if they blow in the right direction. Hunkins (1981) showed a weak effect of wind on surface currents at Yonkers. In an isolated event (16 February 1967) Busby and Darmer (1970) found that strong northerly (along the main axis) winds with gusts up to 55 mi/h prevented the occurrence of high tide at Poughkeepsie. The effect of wind is expected to be largest in the open, shallow areas of the estuary (Tappan Zee/Haverstraw Bay, Newburgh Bay).

Water Level

The drivers, discussed above, affect the hydrodynamics of the estuary. How and why

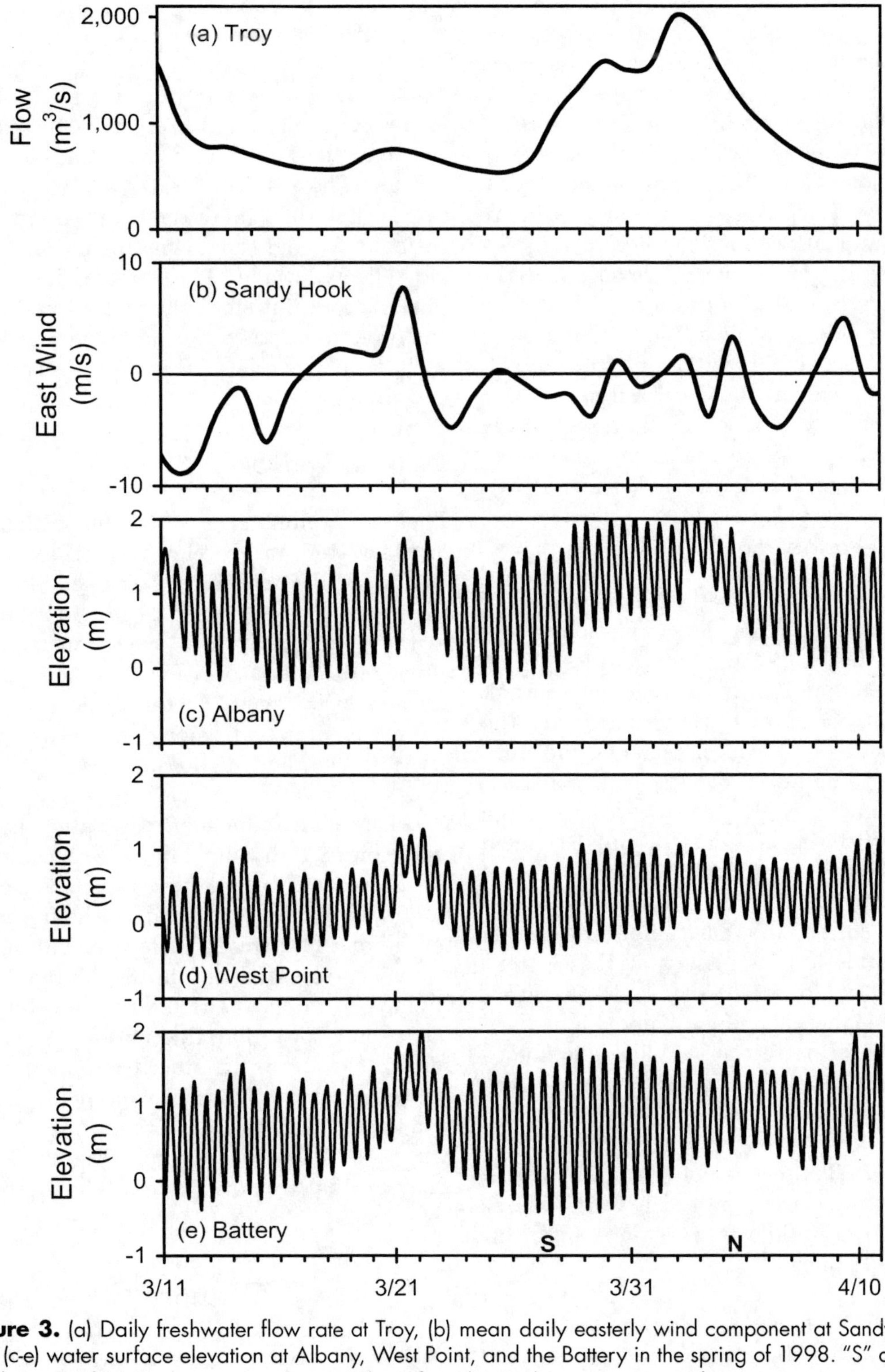

Figure 3. (a) Daily freshwater flow rate at Troy, (b) mean daily easterly wind component at Sandy Hook, and (c-e) water surface elevation at Albany, West Point, and the Battery in the spring of 1998. "S" and "N" labels in panel (e) mark spring and neap tides referred to in the text, respectively.

the hydrodynamics respond to the drivers is the subject of the remainder of this paper, starting in this section with the water levels. The most obvious feature of the water surface elevation of the Hudson River Estuary is its periodic oscillations, due to tides (solely due to the moon and sun). Water levels also respond to surges (due to winds and atmospheric pressure effects). Changes in water surface elevation at the downstream end propagate all the way up to Troy. The tidal peak moves up river at about 25–30 km/h with high tide at Albany occurring about 9–10 h after high tide at the Battery. Freshwater input, especially at the most upstream end, can also affect the water surface elevation (Darmer 1970).

Spatial Variability

While the M_2 tide is the major source of energy which drives the circulation in the region, the N_2, S_2, K_1, and O_1 tides are all significant and contribute to the diurnal, fortnightly, and monthly variations in the magnitude of the tides, currents and mixing. The amplitudes of these tidal constituents (M_2, N_2, S_2, K_1, and O_1) vary, at the Battery (67 cm, 16 cm, 13 cm, 10 cm, 5 cm), West Point (38 cm, 10 cm, 4 cm, 8 cm, 4 cm) and Albany (69 cm, 11 cm, 10 cm, 13 cm, 7 cm). The amplitude of the overall tide is thus highest at the Battery (Figure 3e), lower at West Point (Figure 3 d), and higher again at Albany (Figure 3c). This spatial pattern in amplitude is due to the interplay of three main factors, including friction, geometry, and wave reflection.

Friction.—The energy of the tidal wave is dissipated by friction, which works to reduce the amplitude with distance upstream into the estuary.

Geometry.—The cross sectional area decreases with distance upstream. To conserve energy, the amplitude of the wave increases. This tends to cause an increase in amplitude with distance upstream.

Wave reflection.—The tidal wave traveling upstream from the Battery is reflected at the dam at Troy and travels back downstream. The resulting water surface elevation is then the sum of the upstream propagating wave and the downstream propagating reflected wave. These waves have the same period, but since they travel in different directions, they can amplify or reduce each other's signals (this varies in space and time).

Temporal Variability

There are a number of variations in the observed water surface elevations that occur on time scales longer than the semidiurnal and diurnal tides. These can be attributed to the spring/neap tidal cycle and surges, already discussed previously. Fluctuations in water surface elevation due to storm surges can easily propagate along the entire estuary, just like the tide does.

Another factor influencing the water surface elevation is freshwater inflow. Especially at the upstream end, the water surface elevation is significantly affected by the freshwater input. The water surface elevation at Albany is elevated around 31 March 1998 (Figure 3c), due to the high freshwater input (Figure 3a). The fluctuations in water surface elevation introduced by changes in inflow propagate in the downstream direction (opposite to tide and storm surges) and can be seen at West Point (Figure 3 d). They are hardly noticeable at the Battery, however.

Currents

Tides and currents are intimately connected. Spatial gradients in water surface elevations cause currents (water flows from higher to lower elevation with some modification due to the earth's rotation) and spatially nonuniform currents cause changes in water surface elevation (water "piles up" where currents converge). Therefore, much of the oscillatory nature of the tides is also reflected in the currents.

Timing of Water Level and Current Fluctuations

In the vicinity of the Battery, the tide is produced by a progressive wave which propagates in from the Atlantic Ocean. The time of maximum flood currents occurs at the same time as high tide and the time of maximum ebb currents occurs at the same time as low water. The times of the two slack (time when the current is not moving) waters of a tidal cycle occur at the times that the water surface is at its tidal mean level. This situation changes markedly as you move upstream. For example, near the George Washington Bridge, maximum flood occurs about 30 min before high tide and maximum ebb occurs about 30 min before low tide. This time shift increases even more farther upstream. As Albany is approached, the tide wave takes on the characteristics of a standing wave. The time of slack water occurs much closer to high water and low water. Maximum ebb currents begin to occur nearly 3 h before low water and maximum flood currents begin to occur about 3 h before high water.

Temporal Pattern

The current variability within the tidal cycle near Indian Point is shown in Figure 4. The tidal cycle reversals are most obvious. Bottom currents are generally slower than the surface currents, due to the effect of friction acting of the water column at the bottom. The only exception occurs when strong winds oppose the direction of flow, causing the surface currents to slow down. Then the highest currents are located at middepth. As with the water surface elevations, the spring/neap cycle manifests itself by varying the amplitude of the oscillations (Figure 4). The spring tide on 27 March 1998 is accompanied by the largest currents, while the neap tide on 4 April 1998 has the lowest currents, within the time period shown.

Tidal versus Net Flow

The previous section illustrated that the observed flow direction oscillates from upstream to downstream in response to the tidal forcing. Since the flow is driven primarily by the tidal forcing, it is called "tidal flow." The "net flow" is defined as the long-term average flow at a given point. Practically, the net flow is difficult to measure because it involves a very large fluctuating component and a small mean value. The net flow rate has to be in the downstream direction and equal to the magnitude of freshwater flow entering the estuary upstream of that point (as well as precipitation minus evaporation and any other inputs), which is on the order of 400 m^3/s (mean flow at Troy). It should be noted that this only has to be true for the total net flow rate over the entire cross section. As discussed below, net bottom currents are typically in the upstream direction in estuaries.

The time series of currents at Indian Point (Figure 4) demonstrates that the tidal flows are significantly larger than the net flow. At that location (average depth = 12 m, width = 1,300 m), a current of 50 cm/s corresponds to a flow rate of 7,700 m^3/s. Busby and Darmer (1970) measured tidal flows of 4,800 m^3/s at Poughkeepsie. de Vries and Weiss (1999) measured tidal flows of 11,000 m^3/s in Haverstraw Bay. Thus, it is evident that tidal flows in the Hudson River Estu-

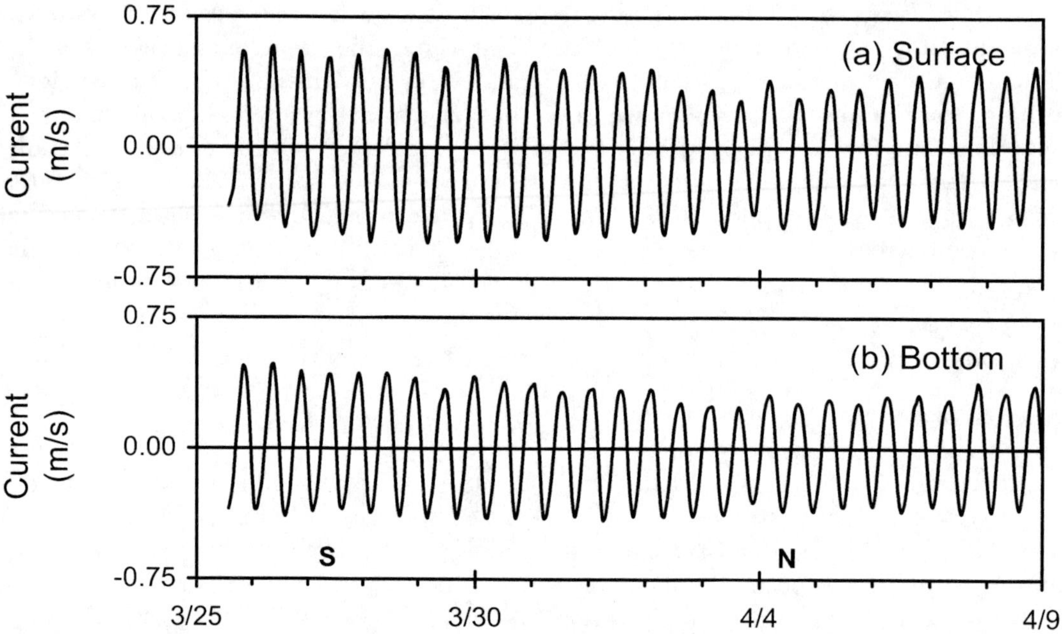

Figure 4. Time series of alongshore currents in the middle of the estuary near Indian Point during 1998 at (a) surface and (b) bottom. Positive is upstream. "S" and "N" labels in panel (b) mark spring/neap tides referred to in the text, respectively.

ary are typically about an order of magnitude greater than the net flows. Since the net flow is small compared to the tidal flow, low frequency variations in the tidal flow (e.g., storm surges) can lead to temporary reversal of net flow. Busby and Darmer (1970) found net upstream movement for several days at a time at Poughkeepsie.

Spatial pattern

Spatially, the currents vary across and along the estuary due to geometry effects. Typically currents are higher over the deeper center channel and lower on the shallower side banks. However, such an idealized picture is rarely observed in the Hudson River Estuary because the geometry can be very complex. Figure 5 shows lateral profiles of near surface currents by Indian Point at four different times in the tidal cycle. During times of strong ebb (Figure 5a, b), flow is in the downstream direction with typical currents of 50–100 cm/s. During the slack before flood period (Figure 5c), the flow is in the process of switching from downstream to upstream. At that time, the current is actually directed in opposite directions at several cross sections. The current is upstream on the west side of the river and downstream on the east side. Two hours later (Figure 5 d), the flood currents are upstream throughout the Indian Point area, with typical currents of less than 50 cm/s.

The fact that the tide generates currents that flow around bending regions of the Hudson River Estuary complicates the circulation even further. Georgas and Blumberg (2004) demonstrate that water level is slightly higher on a bend's outer bank during both flood and ebb than it is on the bend's inner bank. A transverse circulation cell is setup that is directed towards the outer bank at the surface and toward the inner bank at the bottom. In the presence of stratification, the transverse circulation tends to produce upwelling of salt at the inner part of

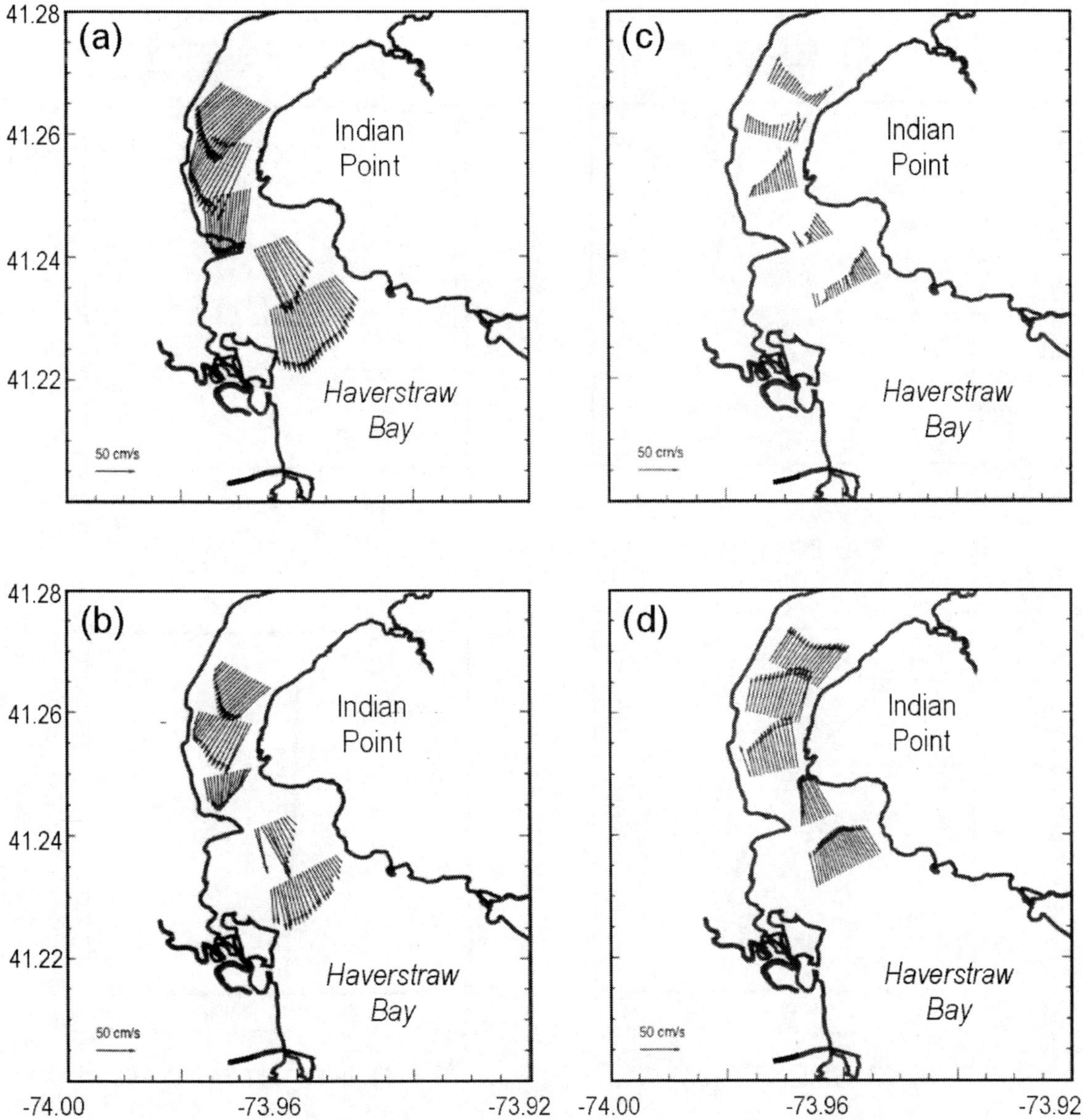

Figure 5. Observed near-surface currents in the region of the river near Indian Point at four points in the tidal cycle on 2 April 1998 (from HydroQual 1999).

the bend leading to stronger cross estuary density gradients (Chant and Wilson 1997).

Tidal Excursion

An important transport concept in estuaries is the tidal excursion, which characterizes the distance a water parcel travels as a result of tidal currents. The tidal excursion is the distance between the most upstream and downstream locations occupied by a water parcel during one tidal cycle. If a parcel is released at high tide, the ebb tide will carry it downstream a distance equal to the tidal excursion. Of course, the tidal excursion varies in time (freshwater flow, spring/

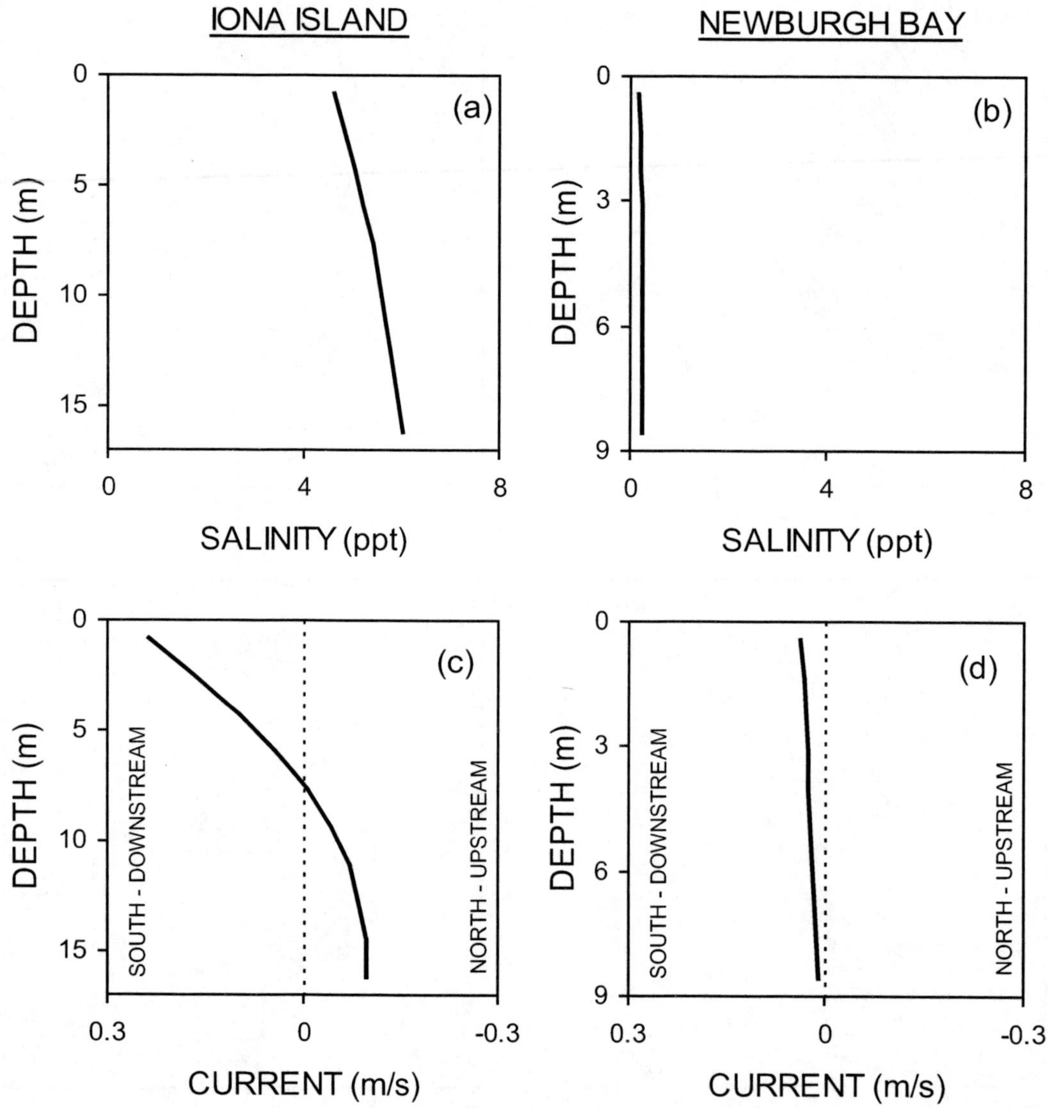

Figure 6. Vertical profiles of modeled (a, b) salinity and (c, d) net currents (a, c) at Iona Island and (b, d) in northern Newburgh Bay for 25 July 2000–8 August 2000 (from Hellweger et al. 2004).

neap) and space (location and depth). The typical tidal excursion in the Hudson River Estuary is 5–10 km, but it can be as high as 20 km.

Estuarine Circulation

Salt adds mass to water, and as a result, salt water has a higher density than freshwater. The density difference between fresh and salt water is small (3.5% or less) but sufficient to significantly affect the circulation in estuaries. Denser salt water enters the estuary in the deeper parts of the water column and lighter freshwater floats as a lens on top of the salt water. The result is net upstream currents in the bottom layer and net downstream currents in the surface layer, in locations where salt is present. This is commonly referred to as the "estuarine circulation." An example of this is presented in Figure 6, which shows vertical profiles of salinity and net currents at two locations. In northern Newburgh Bay, the salinity is very low, and as a result the net currents are in the downstream direction at all depths. At Iona Island, there is significant salinity, which causes an estuarine circulation. The net currents are in the upstream direction in the bottom layer and in the downstream direction in the surface layer. Steward (1958) measured weak upstream bottom currents at Riverdale. Further upstream, at West Point, upstream directed bottom currents were only present on 1 out of 4 d. At Yonkers, Hunkins (1981) measured net downstream surface and upstream directed bottom currents of 13 and 2.3 cm/s, respectively.

There are large temporal variations in the estuarine circulation patterns described above due to the spring-neap changes in the tidal forcing. The intense currents that occur during spring tides and the much smaller currents of the neap tide lead to a dramatic variation of vertical mixing intensity and to changes in vertical salinity stratification. The greater the stratification, the stronger the estuarine circulation becomes. Bowen and Geyer (2003) show an order of magnitude change in the horizontal salt transport as a result of spring-neap cycle changes in the estuarine circulation. The effect of the spring-neap cycle on salinity will be discussed in more detail later.

Temperature

The water temperature in the Hudson River Estuary varies seasonally primarily in response to changes in atmospheric heating and cooling. Tributary temperature and ocean temperature play a lesser although not insignificant role. Spatially, the temperature is influenced in part by power plant cooling water discharges.

Temporal Variability

Typical maximum (August) temperatures are 25°C and minimum (January and February) temperatures are 1.0°C (Poughkeepsie Water Works; period of record: 1951–1987; Wells and Young 1992). This seasonality is also shown in Figure 2b. There is little surface to bottom temperature difference because the water column is relatively well mixed in the vertical. Ice shells are observed floating on the water surface in the winter. Further downstream in the area of saltwater intrusion, the temperature of the waters of the Hudson River Estuary is similar to that of the Atlantic Ocean. Sometimes on a hot summer day there can be pockets of warm water confined to the near surface layers. They are typically short-lived and disappear as a result of mixing by winds or tidal currents. For a more detailed discussion on the seasonality of temperature in the Hudson River Estuary, see Wells and Young (1992) and Mancroni et al. (1992).

Spatial Pattern

Spatial differences in temperature occur because some areas (e.g., shallow banks) tend to respond faster to changes in the forcing functions (i.e., water surface heat flux). Also, the shallower tributaries tend to warm and cool faster than the deeper estuary and therefore can be a source of warm water in the spring and cool water in the fall. Maximum summer temperatures can easily exceed 30°C in these shallow regions. In the summer, the temperature is coolest in the downstream portions of the estuary. It reaches a maximum near the region where several power plants are located and then decreases slightly until the upper reaches of the estuary where the temperature again increases. There are five major power plants located along the

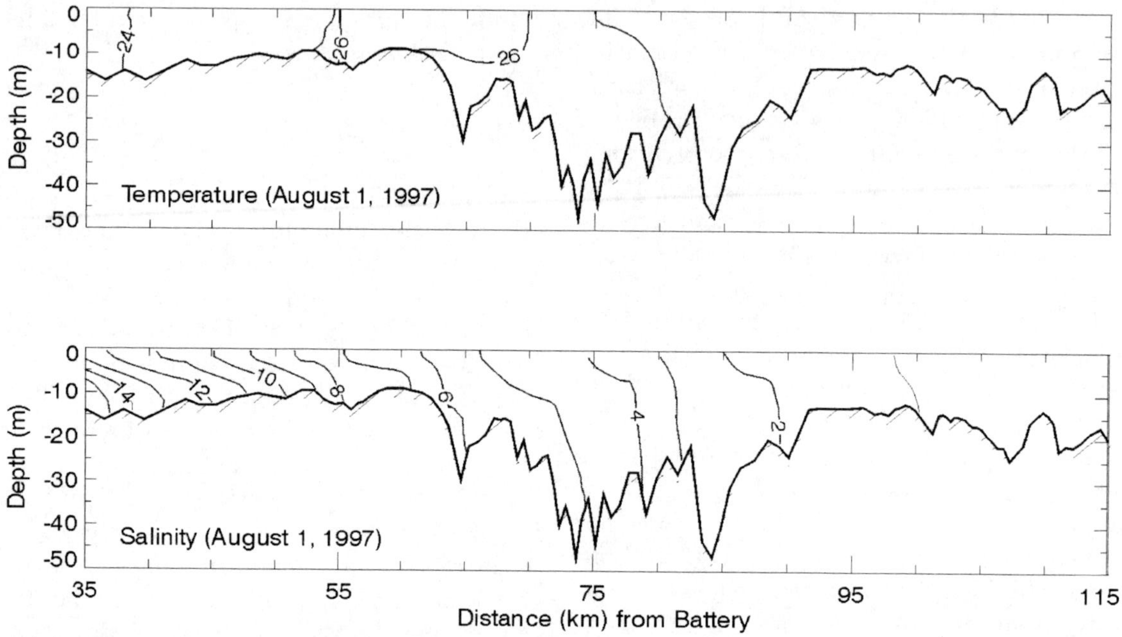

Figure 7. Longitudinal profile of observed temperature (a, top) and salinity (b, bottom) on 1 August 1997. Distance is measured from the Battery (from HydroQual, 1999). Indian Point is located at the 55 km mark.

river and they are an important factor influencing the temperature in the Hudson River Estuary (e.g., Wrobel 1974; HydroQual 1999). These power plants withdraw large volumes of water for cooling and discharge back to the river at an elevated temperature. The maximum cooling water flow from the Indian Point nuclear power plant is 110 m^3/s (Hutchison 1988), which is comparable in magnitude to the summer low flow of the largest natural freshwater input into the estuary (Green Island, August, 160 m^3/s). Withdrawal and discharge locations for power plants are typically located close enough so that their effect on currents is localized and constrained to the immediate area of the power plant. However, the increase in temperature can be evident over a larger area. The maximum temperature increase for the Indian Point nuclear power plant is between 8°C and 9°C in the vicinity of the discharge (Hutchison 1988) and about 1°C or so mid-river (HydroQual 1999). The effect of the plant discharge can be seen in the spatial temperature profile shown in Figure 7a.

Salinity

Freshwater enters the estuary at the upstream end, and salt water mixes upstream from the ocean. The result is a mixture of fresh and salt water throughout much of the estuary. The horizontal and vertical distribution of salt varies dynamically at various time scales in response to changes at the upstream and downstream boundaries.

Spatial and Temporal Pattern

The salinity in the Hudson River Estuary varies along the length of the estuary, as illustrated in the longitudinal profile presented in Figure 7b. The Figure shows that salinity is higher at the downstream end and lower at the upstream end. Also, the salinity tends to be higher in the bottom layer than at the surface. The "S" shaped salinity contours are quite typical of estuarine environments (vertical stratification is discussed in the next section).

The salinity distribution varies temporally. At the tidal time scale, the salinity can change rapidly at a certain location due to the tidal water movement, especially in the area of strong longitudinal salinity gradient (salt front, see below). At longer time scales, salinity intrudes further into the estuary during neap tides and retreats during spring tides (Bowen and Geyer 2003). At even longer time scales, the salinity responds to the freshwater input, which varies at seasonal and shorter time scales. Short time scale pulses of freshwater are known as freshets. Freshwater tends to push the salt water out of the estuary during high flows and permits salt water to intrude during low flow periods. The freshwater flow variability is evident in both the surface and bottom salinity at Croton-on-Hudson and at the Lincoln Tunnel as shown in Figure 2c, d. The salinity is generally higher during the low-flow summer and lower during the high-flow spring. Also, the effect of each of the freshets (Figure 2a) is evident in the salinity (Figure 2c, d). During the spring high-flow period, the salt front (the upper limit of saltwater intrusion, defined here as a salinity of 0.1 ppt) is located between Yonkers and Tappan Zee and during the summer low-flow period it moves north and is typically located south of Poughkeepsie, just south of the City of Poughkeepsie drinking water intake (Wells and Young 1992). For a more detailed discussion of the estuarine circulation and the movement of the salt front in the Hudson River Estuary, the reader is referred to Geyer et al. (2000) and de Vries and Weiss (2000).

Vertical Stratification

The saltiest water resides at the bottom with fresher water at the surface, and hence there is a vertical salinity gradient. This vertical salinity stratification exists throughout the estuary (in the presence of salt) and can be seen in the longitudinal salinity profile shown in Figure 7b. The vertical distribution of salinity is characterized by an upper layer of low salinity, which very slowly increases with depth, an intermediate layer of more rapid salinity increase, called the halocline, and a deep layer in which the salinity increase with depth is small. This "S" shape salinity distribution is the result of the interplay of factors that "want" to keep the water column stratified and those that "want" to mix it. The vertical salinity gradient serves to stabilize the water column and inhibit vertical mixing. A lighter water parcel from the fresh surface layer will resist mixing into the heavier salty bottom layer. This "stability" tends to keep the water column stratified. The water column can become destratified through two mechanisms, tidal straining and turbulence mixing. Both tend to mix the water column and erase the vertical stratification.

Tidal straining (Simpson et al. 1990; Nepf and Geyer 1996) arises from the vertical variation in tidal currents in the presence of a longitudinal salinity gradient. The maximum stratification typically occurs at low water after the ebb flow has moved fresher water in the upper layers seaward over the saltier water in the deeper layers. During the flood, this process is reversed with tidal straining acting to reduce the stability of the water column, which results in a mixed water column close to high water.

Turbulence is produced at the water surface due to wind stirring and at the bottom, where tidal currents move back and forth over the sediment bed. The turbulent energy is highest during the spring tide, and it is then when the vertical stratification often breaks down. This is evident in the

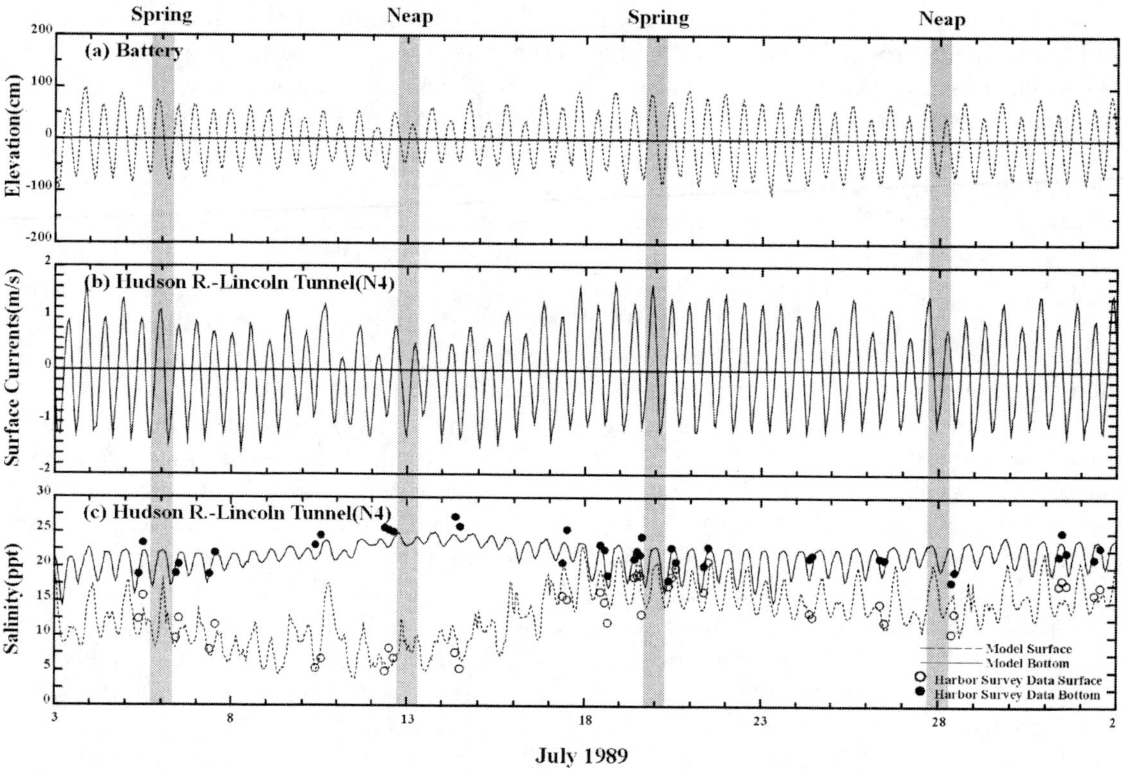

Figure 8. A 30 day time series beginning July 3, 1989 of (a) water surface elevation at the Battery, (b) currents at the nearby Lincoln Tunnel (N4) station and (c) surface and bottom salinity at the nearby Lincoln Tunnel station (from Blumberg et al. 1999).

salinity at the Lincoln Tunnel (Figure 8c). Throughout most of the period shown, the bottom salinity is significantly higher than at the surface. However, during spring tide (e.g., Day 292, see tidal amplitudes of Figure 8a and currents of Figure 8b), the energy is sufficient to overcome the stabilization and the water column mixes vertically causing the surface and bottom salinities to become the same. In general, however, the vertical stratification, which is observed in the water column, is the result of the interplay between both the straining and mixing, both of which vary in time.

Dispersive Processes

Various processes, operating at different spatial scales, contribute to the horizontal spreading of salinity and other constituents that have been introduced into the water column. The main processes are shear dispersion and tidal trapping, both of which will be described below.

Shear Dispersion

The currents in the Hudson River Estuary have significant lateral (Figure 5) and vertical (Figure 6) structure. This structure in currents spreads out pollutants that are in the water column by a process termed shear dispersion (Pritchard 1954). Pritchard considered the vertical current structure of the estuarine circulation, in what turns out to be the dominant, but not only mechanism for shear dispersion in the Hudson River Estuary. The following examples illustrate the mechanisms. An initial vertically mixed instantaneous release ("slug release") of a substance (e.g., tracer) near Iona Island will travel upstream and downstream with the tide, but the difference in net currents between the surface and bottom layers is 20 cm/s (see Figure 6), which means that after 1 d, the slug at the bottom is almost 20 km further upstream than the slug at the surface. The estuarine circulation will have created this slug dispersion. The effect of this net estuarine circulation is evident in the longitudinal tracer concentration profile presented in Figure 10b. The tracer plume shows much higher spreading downstream of the 90 km location, which is the area where the salinity is high enough to cause an estuarine circulation.

Consider a second example where lateral variations in currents exist and a hypothetical instantaneous spill (slug) release by Indian Point at a time when the current is switching from ebb to flood (Figure 5c). At this time, water flows upstream and downstream on the west and east sides of the river, respectively. These currents would spread out the slug by stretching it in the upstream and downstream directions. This variation in currents is extreme and specific to this time in the tidal cycle only. Most of the time water either flows upstream or downstream throughout the cross section. However, the upstream component is stronger on the west side and the downstream component is stronger on the east side, and this structure causes dispersion as well. For example, the strong flood currents shown on Figure 5 d have significant lateral (across river) structure and would also lead to significant longitudinal spreading of a slug in the upstream direction.

Tidal Trapping

The geometry of the Hudson River Estuary can be very complex, containing many irregularities (coves and inlets). The interplay of the tidal flow with these geometric

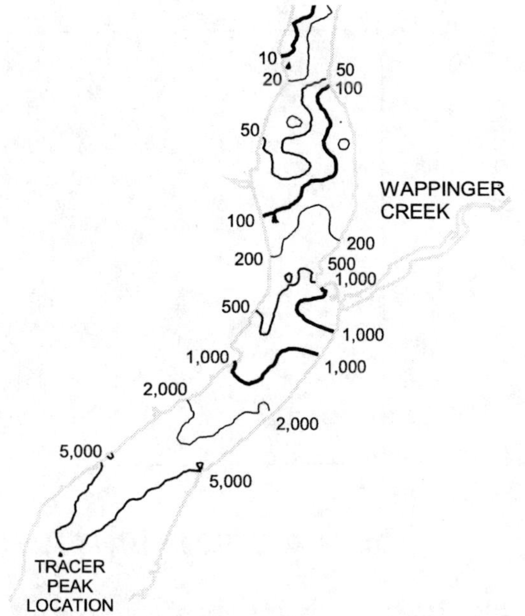

Figure 9. Horizontal distribution of modeled SF_6 tracer concentration in fmol/L in northern Newburgh Bay by Wappinger Creek 2 d after a slug release (from Hellweger et al. 2004).

irregularities enhances longitudinal dispersion by a process called "tidal trapping" (Okubo 1973). The basic concept of tidal trapping is that geometric irregularities can temporarily trap a water parcel as it passes by and then release it at some later time. This effectively removes a small amount of water from the original main channel water mass and then adds it back later to a new main channel water mass. For example, a water parcel with low salinity is removed from its original low salinity main channel water mass and then added later to a new main channel water mass with higher salinity.

The process of tidal trapping is illustrated in Figure 9, which presents tracer concentrations in northern Newburgh Bay near Wappinger Creek. In that experiment, a slug of tracer was released further south in Newburgh Bay. The Figure shows the hori-

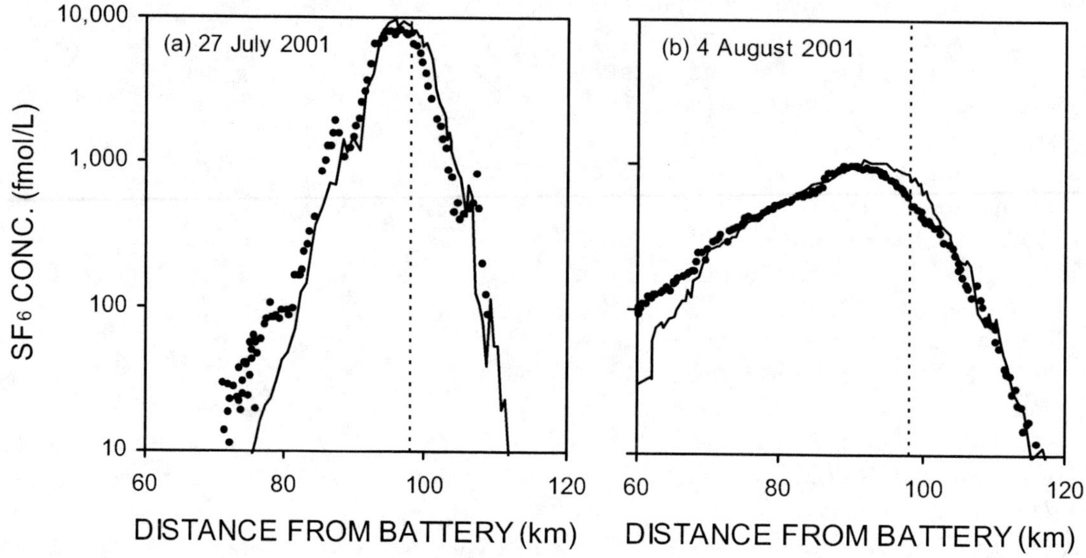

Figure 10. Longitudinal profiles of tracer concentration 2 and 10 d after a slug release in Newburgh Bay. Symbols are data and lines are model (from Hellweger et al. 2004).

zontal tracer distribution 2 d after the release. The concentration at the mouth of Wappinger Creek is higher compared to that in the main channel (1,000 versus 500). A water mass with a high tracer concentration became trapped there. The tidal trapping by Wappinger Creek is also evident in longitudinal profiles of tracer concentration, presented in Figure 10a (Wappinger Creek is located at the 105 km point in that Figure). At the upstream and downstream end of the plume, there are smaller scale "secondary peaks," which are also the result of tidal trapping (located at the 90, 105, and 110 km positions).

Longitudinal Dispersion Coefficient

The combined effect of all the dispersive processes is to cause a spreading of constituents, which can be characterized using a longitudinal dispersion coefficient. The longitudinal dispersion coefficient has been estimated using tracer studies, with values ranging from 3 to 270 m^2/s at various locations from Troy to Newburgh (Hohman and Parke 1969; Clark et al. 1996, 1997; Ho et al. 2002). Longitudinal dispersion coefficients are useful for characterizing the dispersion in an estuary, but it is important to realize that the dispersion changes in space and time and that there is not one coefficient that applies to all locations of the estuary or even one location at all times. This variability is evident in the estimated dispersion coefficients for the Hudson River Estuary given above.

Residence Time

Another important concept related to currents and dispersion is residence time or flushing time. This concept is defined as the average time a water parcel spends in the estuary or a certain part of the estuary (e.g., freshwater reach). For the freshwater region of the Hudson, the flushing time is simply the volume divided by the upstream inflow. During the springtime high inflow periods, the flushing time in the estuary is less than 40 d. However, it is on the order of 200 d during the summer low flow periods. In the estuarine portions of the Hudson River, the flushing time is defined as the average volume of estuary divided by the seaward rate of outflow. For a salinity intrusion of 80 km and a typical estuarine surface current of 10 cm/s (see Figure 6), the residence time is about 8 d. This is obviously much shorter than the residence times in the freshwater regions, again demonstrating the impressive dispersive characteristics of the estuarine circulation.

Acknowledgments

This paper builds on research by the Hydrodynamics Groups at HydroQual, Inc. and Stevens Institute of Technology and the Environmental Tracer Group at Columbia University. Graphics and data analysis support was provided by HydroQual's Honghai Li and Heidi Salerno. The authors want to acknowledge the excellent comments received from Richard Hires of Stevens and two anonymous reviewers. AFB was funded under ONR grant N00014-03-1-0633 and FLH was funded in part by HydroQual, Inc.

References

Abood, K. A., G. A. Apicella, and A. W. Wells. 1992. General evaluation of Hudson River freshwater flow trends. Pages 3–28 in C. L. Smith, editor. Estuarine research in the 1980s. Hudson River Environmental Society, 7th Symposium on Hudson River Ecology. State University of New York Press, Albany, New York.

Blumberg, A. F., L. A. Khan, and J. P. St. John. 1999. Three-dimensional hydrodynamic model of New York harbor region. Journal of Hydraulic Engineering 125:799–816.

Bowen, M. M., and W. R. Geyer. 2003. Salt transport and the time-dependent salt balance of a partially stratified estuary. Journal of Geophysical Research 108(C5): DOI: 10.1029/2001JC001231.

Busby, M. W., and K. I. Darmer. 1970. A look at the Hudson River Estuary. Water Resources Bulletin 6:802–812.

Chant, R. J., and R. E. Wilson. 1997. Secondary circulation in a highly stratified estuary, Journal of Geophysical Research 102: 23207–23215.

Clark, J. F., P. Schlosser, M. Stute, and H. J. Simpson. 1996. SF6–3He tracer release experiment: a new method of determining longitudinal dispersion coefficients in large rivers. Environmental Science and Technology 30:1527–1532.

Clark, J. F., P. Schlosser, M. Stute, and H. J. Simpson. 1997. SF6–3He tracer release experiment: A new method of determining longitudinal dispersion coefficients in large rivers. (Correction). Environmental Science and Technology 31:308.

Darmer, K. I. 1970. Hydrologic characteristics of the Hudson River Estuary. Pages 40–55 *in* G. P. Howells and G. J. Lauer, editors. Hudson River ecology – proceeding of a symposium. New York State Department of Environmental Conservation, Albany.

de Vries, M. P., and L. A. Weiss. 1999. Salt-front movement in the Hudson River Estuary, New York-simulations by one-dimensional flow and solute-transport models. United States Geological Survey, Water-Resources Investigations Report 99–4024, Troy, New York.

Doodson, A. T. 1922. The harmonic development of the tide-generating potential. Proceedings of the Royal Society, *A*: 100–305.

Georgas, N. and A. F. Blumberg. 2004. The influence of centrifugal and coriolis forces on the circulation in a curving estuary. Estuarine and Coastal Modeling, ASCE: 541–558.

Geyer, R. W., J. H. Trowbridge, and M. M. Bowen. 2000. The dynamics of a partially stratified estuary. Journal of Physical Oceanography 30:2035–2048.

Hellweger, F. L., A. F. Blumberg, P. Schlosser, D. T. Ho, T. Caplow, U. Lall, and H. Li. 2004. Transport in the Hudson Estuary: a modeling study of estuarine circulation and tidal trapping. Estuaries 27:527–538.

Ho, D. T., P. Schlosser, and T. Caplow. 2002. Determination of longitudinal dispersion coefficient and net advection in the tidal Hudson River with a large-scale, high resolution SF6 tracer release experiment. Environmental Science and Technology 36:3234–3241.

Hohman, M. S., and D. B. Parke. 1969. Hudson River dye studies, 150 miles of red water. Pages 60–81 *in* Hudson River ecology. Hudson River Valley Commission of New York, Albany, New York.

Hunkins, K. 1981. Salt dispersion in the Hudson Estuary. Journal of Physical Oceanography 11:729–738.

Hutchison, J. B., Jr. 1988. Technical descriptions of Hudson River electricity generating stations. Pages 113–120 *in* L. W. Barnthouse, R. J. Klauda, D. S. Vaughn, and R. L. Kendall, editors. Science, law, and Hudson River power plants: a case study in environmental impact assessment. American Fisheries Society, Monograph 4, Bethesda, Maryland.

HydroQual. 1999. Modeling the entrainment of passive particles into Hudson River power plants. Draft report to Hudson River Utilities, No. OARU0010. HydroQual, Mahwah, New Jersey.

Mancroni, W. J., M. W. Daley, and W. Dey. 1992. Recent trends and patterns in selected water quality parameters in the mid-Hudson River Estuary. Pages 59–81 *in* C. L. Smith, editor. Estuarine research in the 1980s. Hudson River Environmental Society, 7th Symposium on Hudson River Ecology. State University of New York Press, Albany, New York.

McHugh, C. M. G., S. F. Pekar, N. Christie-Blick, W.B.F. Ryan, S. Carbotte and R. Bell. 2004. Spatial variations in a condensed interval between estuarine and open-marine settings: Holocene Hudson River Estuary and adjacent continental shelf. Geology 32 (2):169–172.

Nepf, H. M., and W. R. Geyer. 1996. Intratidal variations in stratification and mixing in the Hudson Estuary. Journal of Geophysical Research 101(C5):12,079–12086.

Okubo, A. 1973. Effect of shoreline irregularities on streamwise dispersion in estuaries and other embayments. Netherlands Journal of Sea Research 6:213–224.

Posmentier, E. S., and J. W. Rachlin. 1976. Distribution of salinity and temperature in the Hudson Estuary. Journal of Physical Oceanography 6:775–777.

Pritchard, D. W. 1954. A study of salt balance in a coastal plain estuary. Journal of Marine Research 13:133–144.

Pritchard, D. W. 1967. Observations of circulation in coastal plain estuaries. Pages 37–44 *in* G. H. Lauff, editor. Estuaries. AAAS Publ. No. 83, Washington, D.C.

Pritchard, D. W., A. Okubo, and E. Mehr. 1962. A study of the movement and diffusion of an introduced contaminant in New York Harbor waters. Chesapeake Bay Institute, Technical Report 31, Johns Hopkins, Baltimore, Maryland.

Simpson, J. H., J. Brown, J. P. Matthews, and G. Allen. 1990. Tidal straining, density currents, and stirring in the control of Estuarine Stratification. Estuaries 12:129-132.

Steward, Jr., H. B. 1958. Upstream bottom currents in New York Harbor. Science 127:1113–1115.

Wells. 1992. General evaluation of Hudson River freshwater flow trends. Pages 3-28 *in* C. L. Smith, editor. Estuarine research in the 1980s. Hudson River Environmental Society, 7th Symposium on Hudson River Ecology. State University of New York Press, Albany, New York.

Wells, A. W., and J. R. Young. 1992. Long-term variability and predictability of Hudson River physical and chemical characteristics. Pages 29–58 *in* C. L. Smith, editor. Estuarine research in the 1980s. Hudson River Environmental Society, 7th Symposium on Hudson River Ecology, State University of New York Press, Albany, New York.

Wrobel, W. E. 1974. Thermal balance in the Hudson Estuary. Annals of the New York Academy of Sciences 250:157–168.

American Fisheries Society Symposium 51:29–37, 2006

Channel Morphology in the Hudson River Estuary: Historical Changes and Opportunities for Restoration

Daniel Miller

Hudson River Estuary Program, Hudson River National Estuarine Research Reserve Bard College Field Station, Annandale, New York 12504 USA demiller@gw.dec.state.ny.us

John Ladd

Hudson River Estuary Program, Hudson River National Estuarine Research Reserve 43 Hudson Watch Drive, Ossining, New York 10562 USA

William C. Nieder

New York Department of Environmental Conservation, Hudson River National Estuarine Research Reserve, Bard College Field Station, Annandale, New York 12504 USA

Abstract.—Dredge and fill activities associated with navigational channel maintenance have occurred in the northern third of the Hudson River estuary since the late 18th century and have caused significant losses of intertidal and shallow-water acreage—areas likely important as nursery habitats for fish. The habitat loss suggests restoration efforts in the estuary should focus on replacing these habitats with functional equivalents. Replacing these lost habitats will involve tradeoffs between the lost habitats and those habitats that currently occupy this section of the river. An evaluation of these tradeoffs will require more complete mapping of existing upland and aquatic habitat in the Hudson River estuary watershed and evaluation of the functions of these various habitats.

Introduction

The Hudson River has a long history of channel modifications that have altered ecosystem structure and function. Physical modification of the river morphology and hydrology fall into three broad categories: dredging and filling, shoreline hardening, and construction of dams on tributaries to the river. The degree of alteration is best assessed by developing reference conditions from historic channel maps, which predate the human-induced channel modifications. We use charts from the early 1800s to assess the morphology of the Hudson River channel before navigation channel dredging. Historic and current charts are used to identify shoreline location, channel depths, and spatial arrangement of shallow and intertidal areas to determine the loss or decline of physical habitats. The results from these analyses, along with recent studies identifying the important role of Hudson River freshwater shallow and intertidal habitats for fish reproduction, are then used to identify opportunities for restoring ecosystem functions to the Hudson River estuary.

History

Large-scale modification of Hudson River channel morphology began in the early to mid 1800s and continues today. With the completion of the Erie Canal in 1825, the city of Albany became a transportation center for the northeastern and midwestern United States for goods arriving by ocean-going ships via the Hudson River (Stanne et al. 1996). Much of the southern portion of the estuary is naturally deep enough to accommodate deep-draft vessels. However, near Catskill (river kilometer [rkm] 185), river morphology changed to a shallow water, braided channel system, requiring major modifications to allow modern oceangoing vessels to pass. Early attempts to create and maintain a navigation channel included construction of longitudinal dikes designed to focus river flow through a self-maintaining channel. The method worked, but maintained a limited channel depth. In 1925, to accommodate deeper-draft vessels capable of transporting more materials, the U.S. Congress authorized deepening the original 3.5 m channel to 8.2 m. In 1932, the federal navigation channel was again deepened to 9.7 m below mean low water where it is maintained to this day (USACE 1995).

Dredging the navigation channel had numerous impacts on the tidal Hudson River. The shallows and intertidal habitats in the upper Hudson River estuary (rkm 185–241) were modified in two ways. First, shallows and intertidal habitats were converted to deep-water habitat (navigation channel) through dredging. Second, dredge spoil was placed and retained on adjacent shallows and intertidal and backwater habitats, converting them to uplands. The resulting dredge spoil islands were subsequently colonized by upland vegetation and, through the process of succession, developed into mature upland habitats, often unrecognizable as dredge spoil.

Recent efforts to assess the need and feasibility of restoration of the upper Hudson River estuary have focused on existing habitat function. There remains a need to identify historic ecosystem structure and quantify the loss of habitat function as a result of navigation channel dredging and spoil deposition. Unfortunately, there is little record or knowledge of historic ecosystem structure and function that predates significant modification of the river.

Methods

Identifying loss of function

Identifying loss of historic river ecosystem structure and function is a five-step process:

1) Identify functions of habitat types that remain in existence today;

2) Identify current distribution of existing habitat types;

3) Identify historic distribution of these habitat types based on available information;

4) Project current habitat functions to historic distributions in an attempt to quantify historic ecosystem function; and

5) Estimate loss of ecosystem function as a result of altered ecosystem structure.

Historic charts contain limited information about habitat types and distribution. Reliable information is usually limited to depth soundings and delineation of intertidal areas and shorelines. The information avail-

Table 1. Historic maps of the Hudson River estuary. Sources marked with * were used for our analysis. USACE= Army Corps of Engineers.

Source	Date	Scale	Range
Randel (NYS)*	**1820**	**1:15,000**	**Troy-Hudson**
US Coast Survey*	**1863**	**1:40,000**	**Troy-Poughkeepsie**
USACE	1867	1:5,000	Troy-New Baltimore
USACE*	**1890**	**1:20,000**	Troy-Hudson
Hudson River Atlas	1891	1:14,000	Troy-New York City
USACE-Turn-of-the-century	1897-1909	1:5,000	Albany-Hudson
Modern USACE	Present	1:5,000	Green Island-Germantown
NYS DOT Quads*	**Present**	**1:24,000**	**All**
NOAA Charts	Present	1:40,000	All

able in numerous historic charts allowed us to define and map historic and current distributions of three habitat types based on depth and tidal inundation. These broad habitat descriptions are defined as follows.

1. Intertidal: areas of the river that are exposed at low tide and submerged at high tide. This habitat consists of mud flats, sand, broadleaf marsh, and emergent intertidal vegetation (Reschke 1990).

2. Shallows: areas of the river less than 2-m deep at mean low tide. This habitat typically supports all the submerged aquatic vegetation (SAV) in the river and is considered one of the most productive habitats in the estuary and of great ecological importance (Reschke 1990).

3. Deepwater: areas greater than 2-m deep at low tide. Historically, this depth was the limit of light penetration. Rooted aquatic vegetation is not found at this depth (Reschke 1990).

Shallows and intertidal habitats have vital ecological roles in the Hudson River estuary. Submerged aquatic vegetation and emergent aquatic vegetation occurs only in the shallow subtidal and lower intertidal habitats of the estuary (Reschke 1990). Submerged aquatic vegetation has significant effects on pore-water chemistry, sediment nutrient cycling, and oxygen production (Wigand et al. 2001). Many animals depend on the aquatic plants, plankton, nekton, and benthic invertebrates that create a complex food web in shallow-water and intertidal habitats (Limburg et al. 1986; Strayer et al. 2003). Plants provide habitat for production of forage species and spawning and nursery habitat for many larger anadromous and resident fishes. Additionally, their importance to the entire tidal Hudson River ecosystem may be growing as a location for primary production as the traditional source, phytoplankton, has been drastically reduced by the invasion of zebra mussels *Dreissena polymorpha* (Strayer et al. 1999; Caraco et al. 2000).

Identifying Loss of Ecosystem Structure

In 1995, staff of the Hudson River National Estuarine Research Reserve (HRNERR) conducted a preliminary analysis to quantify change in Hudson River estuary channel morphology as a result of navigation channel dredging. The results, which appear in the Hudson River Habitat Restoration, Hudson River Basin, New York, Reconnaissance Report (USACE 1995) combined several sets of historic charts of the Hudson

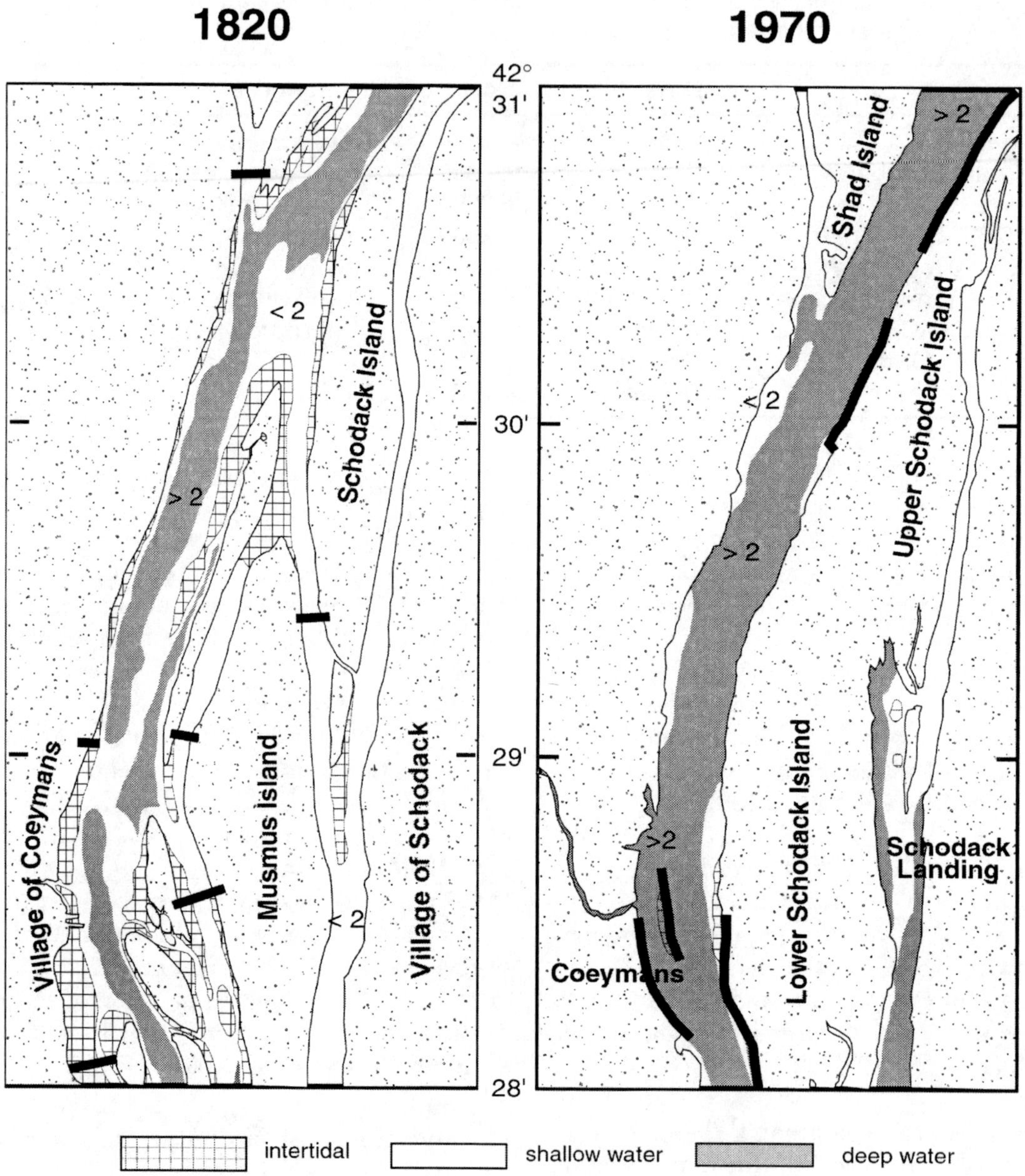

Figure 1. A comparison of a 5-kilometer stretch of the Hudson River estuary near Coeymans as it appeared in 1820 as indicated by the Randel map with how it appeared in 1970 as indicated by NYS DOT quadrangle maps.

River (Table 1) that predate navigation channel dredging to map deepwater, shallows, and intertidal habitats as defined above. The maps used in this analysis indicated water depths in shallow areas. Intertidal areas were inferred from map symbology. Areas of present and historic deep and shallow water were determined by dividing the maps of Table 1 into a series of polygons, bounding segments of the estuary and shallow and deep water within these segments. A mechanical planimeter was used to measure the length of boundaries and the areas of polygons. To increase accuracy of measurement, each polygon was measured twice and the average of the two measurements was used in further calculations. In most cases the historical maps did not show the extent of tributaries to the Hudson including the Mohawk River, so our estimates of historical area should be taken as a minimum. Shoreline length and numbers of islands were also noted. This information was then compared to information gathered from modern New York State Department of Transportation Quad maps using similar methods. We divided the river into 8-kilometer segments and the gain or loss of each type of habitat was calculated in square meters. The comparison identified substantially altered river morphology with potentially massive ecological effects (Figures 1–3).

Some of the charts indicated the presence of vegetation. However, these indications cannot be considered a quantitative assessment. To identify historic areas that potentially supported submerged and emergent vegetation we combined historic channel morphology data with present-day understanding of factors limiting the extent of vegetation communities on the Hudson. Reschke (1990) identified shallow-water and intertidal habitats as the only vegetated habitats within the estuarine system. This is likely due to light limitations. With this understanding, we assume that historic shallow-water and intertidal habitats were potentially vegetated habitat.

Results

The Hudson River channel has been physically altered as a result of channel dredging and spoil disposal (Figure 1). This system, formerly dominated by shallows that potentially supported aquatic vegetation, now is dominated by deepwater and upland habitat. Data from the analysis indicate that approximately 57% (1,740 ha) of the historic shallow-water and intertidal habitats from km 185–241 were obliterated by dredging and filling. Of total surface area, 30% (1,335 ha) of the river surface was filled. Figure 2 shows the loss of intertidal and shallow-water habitat for km 185–241. The result of these actions was a dramatic shift in the relative amounts of habitat types (and their functions) in a large portion of the estuary (Figure 3).

Opportunities for Restoration

Restoration of Hudson River estuary shallow-water and intertidal main-channel habitats for fish and other organisms faces many challenges. Identifying the need for and goals of restoration efforts depends on understanding the historic structure and function of the Hudson River estuary prior to navigation channel dredging. However, while restoring to predisturbance function may be ideal, it is often not a viable option (Brookes 1995). Ecosystems are dynamic systems that continually respond to internal and external pressures. The balances that ecosystems achieve often change along a trajectory over time and are irreversible. Additionally, continued presence of humankind as an inseparable component of the ecosystem often precludes restoration of ecosystems to predisturbance form.

In addition to identifying lost historic ecosystem structure, mapping efforts aid in identifying areas of former wetland that were filled and converted to upland habitat. Filled areas can then be evaluated as potential restoration sites. However, simply identifying areas as dredge spoil is not the only factor in determining restoration feasibility. Additional complicating factors include restoration dredge–spoil disposal, habitat trading issues, alien species, contaminants, and

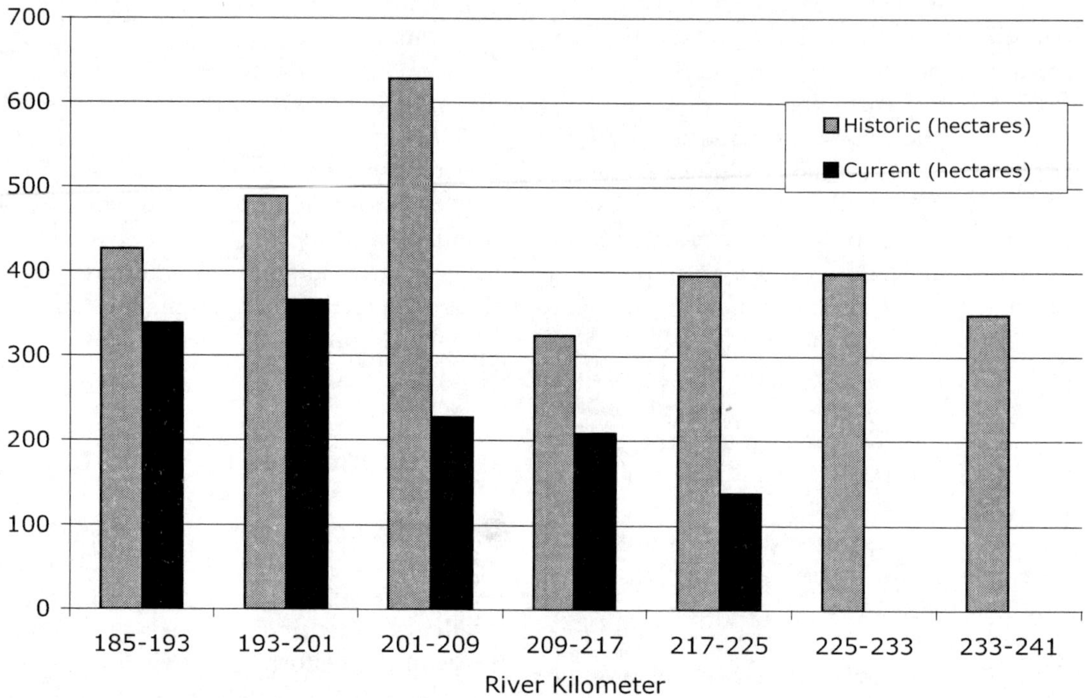

Figure 2. Changes in amount of shallow-water and intertidal habitat between 1820 and 1970 in 8-km segments from river kilometer 185–241.

property ownership.

Spoil Disposal

Restoration often involves restoring the ecosystem structure in an attempt to recreate conditions conducive to reestablishing historic plant and animal communities (NRC 1992). Taking this approach, restoration of the Hudson River estuary would require restoring shallow-water and intertidal habitats by removing dredge spoil and placing it back into the dredged channel. This option, however, is precluded by the continued need to maintain a navigation channel. Therefore, an important constraint to restoring Hudson River shallow-water habitats is finding an appropriate way to dispose or reuse the material that is generated by restoration. During this process one needs to recognize that destroying habitat while disposing of material minimizes the overall benefit of the restoration project. Given the massive amount of material that could be generated by shallow-water and intertidal habitat restoration on the Hudson River, where and how to dispose of dredge material is one of the most significant issues for determining feasibility of such an effort.

Habitat Trades

Because navigation channel dredging occurred nearly a century ago, many of the dredge spoil islands have established functioning upland habitats that host a variety of species including some that are rare and endangered. Proposals to eliminate the created upland habitat in favor of restoring shallow-water habitat may make sense in a historic context, but current issues of relative values of habitats being traded have to be addressed by a variety of stakeholders, each with unique priorities. While evaluating habitat trades, restoration practi-

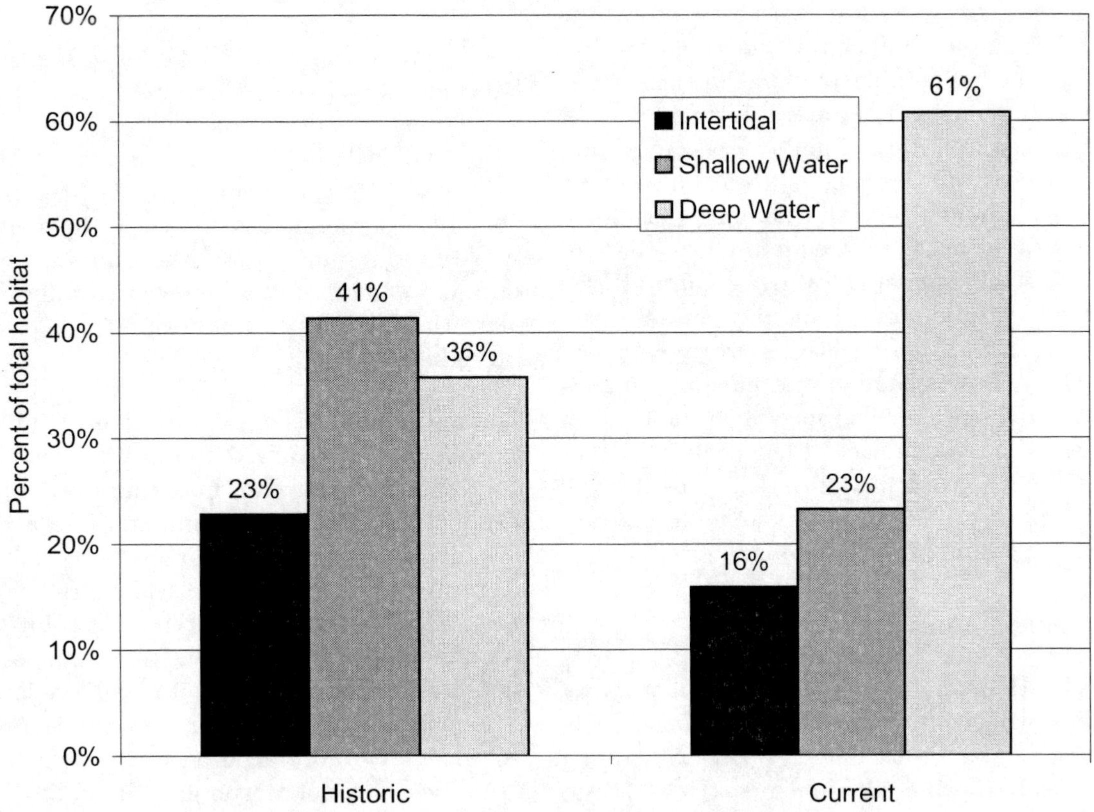

Figure 3. Historic and current proportions of intertidal, shallow-, and deepwater habitat in the upper Hudson River estuary from river kilometer 185–241.

tioners should consider the ecosystem functions provided by current conditions and proposed restored conditions. A functional assessment of Hudson River fresh-tidal marshes using a modified hydrogeomorphic approach was produced in 2003 (Mihocko et al. 2003). The assessment identifies several functional attributes of Hudson River fresh-tidal marshes and indicators of function that will be useful in evaluating the potential gains from fresh-tidal wetland restoration. Development of a compatible assessment and inventory of upland resources is needed to understand the net value of restoration efforts that will include loss of upland resources.

Alien Species

The presence of alien species can complicate main-channel restoration efforts (Strayer et al. 2005). Water-chestnut *Trapa natans*, zebra mussel *Dreissena polymorpha*, and common reed *Phragmites australis* are among several alien species that will continue to alter the way the Hudson River estuary functions. Elimination of alien species often involves costly, long-term efforts with uncertain outcomes. Understanding the emerging role of these alien species and their qualities (negative or positive) is an important step in defining achievable goals for restoration. Restoration initiatives need to account for the continued presence of these species and the role they play in the future of the estuary with or without restoration. For example, shallow-water restoration sites may be vulnerable to invasion by water-chestnut instead of native, historic vegetation communities. Common reed may provide ecosystem services such as nutrient removal, an attribute that may be considered desirable as watershed land use changes and zebra mussels have drastically altered food web dynamics that may influence future restoration priorities (Strayer et al. 2005). Assessment of the current and future influence of alien species on the Hudson River ecosystem needs to continue. With or without control measures, their presence will continue to be a formative issue in restoration planning and feasibility.

Contaminants

The Hudson River has a long history of industrial contamination. The General Electric Company discharged up to 590,000 kg of polychlorinated biphenyls (PCBs) into the upper Hudson River from 1947 to 1977 (EPA 2005). As a result of these and other actions, PCBs are found in the water column, biota, and sediments in the upper river and throughout the estuary. Logistical challenges to deposition or beneficial reuse of sediments produced by restoration will be complicated by the need to address contamination issues. However, much of the dredging and creation of dredge spoil islands was performed prior to the introduction of PCBs into the Hudson River. It may be possible that existing dredge spoil that was never exposed to waterborne PCBs is "clean" and available for beneficial reuse applications. Because dredge disposal options, which depend on contamination levels, will be a critical issue in determining the feasibility of Hudson River restoration, a definitive analysis of PCB contamination of existing spoils should be conducted early in the feasibility process.

Conclusions

Preliminary analysis of historic Hudson River estuary channel morphology has identified massive alteration to the river and loss of critical shallow-water and intertidal habitats. This loss, combined with the important functional role of these habitats, suggests a dramatic loss of ecosystem function and demonstrates the need to fully explore the feasibility of restoring these critical habitats.

The initial analysis of historic river channel morphology by HRNERR staff was preliminary. Since this work, mapping technologies including remote sensing and GIS utilities have created the opportunity for more accurate use of these historical maps. New York State has conducted high resolution mapping of modern Hudson River morphology (Ladd et al. 2002), habitat, and vegetation communities. There are several current initiatives to more accurately identify historic river channel morphology including an analysis of historic wetland distribution and change for the entire estuary being conducted by the New York State Department of Environmental Conservation. These efforts will allow a more accurate depiction of channel morphology change over the past two centuries. This information, combined with continued assessment of habitat distribution and function, will allow quantitative assessment of lost environmental function as a result of a variety of human activities including construction of the navigation channel. It is important to maintain an understanding of the effects we have had on our environment as we consider the substantial task of restoring the Hudson River estuary. Identifying historic conditions contributes to this effort by preserving an understanding of the structure and function of the historic Hudson River and identifying changes and lost resources that have occurred as a result of our actions.

References

Brookes, A. 1995. River channel restoration: theory and practice. Changing River Channels. Wiley, New York.

Caraco, N. F., J. J. Cole, S. E. G. Findlay, D. T. Fischer, G. G. Lampman, M. L. Pace, and D. L. Strayer. 2000. Dissolved oxygen declines in the Hudson River associated with the invasion of the zebra mussel (Dreissena polymorpha). Environmental Science and Technology 34:1204–1210.

Ladd, J. W., R. E. Bell, E. A. Blair, H. Bokuniewicz, S. Carbotte, R. M. Cerrato, S. Chillrud, V. L. Ferrini, R. D. Flood, N. P. Maher, C. M. G. McHugh, F. O. Nitsche, W. B. F. Ryan, D. L. Strayer, J. Thissen, and R. Versteeg. 2002. Mapping the Hudson estuary's submerged lands. Clearwaters 32:5–7. New York Water Environment Association, Syracuse.

Limburg, K. E., M. A. Moran, and W. H. McDowell. 1986. The Hudson River ecosystem. Springer-Verlag, New York.

Mihocko, G. E. Kiviat, R. E. Schmidt, S. E. G. Findlay, W. C. Nieder, and E. A. Blair. 2003. Assessing ecological functions of Hudson River fresh-tidal marshes: reference data and a modified hydrogeomorphic (HGM) approach. Hudsonia, Annandale, New York.

National Research Council. 1992. Restoration of aquatic ecosystems. National Academy Press, Washington, D. C.

Reschke, C. 1990. Ecological communities of New York State. New York State Department of Environmental Conservation, New York Natural Heritage Program, Albany.

Stanne, S. P., R. G. Panetta, and B. E. Forist. 1996. The Hudson: an illustrated guide to the living river. Rutgers University Press, New Brunswick, New Jersey.

Strayer, D. L., E. A. Blair, N. F. Caraco, J. J. Cole, S. Findlay, W. C. Nieder, and M. L. Pace. 2005. Interactions between alien species and restoration of large-river ecosystems. Archiv für Hydrobiologie Supplementband 155:133–145.

Strayer, D. L., N. F. Caraco, J. J. Cole, S. Findlay, and M. L Pace. 1999. Transformation of freshwater ecosystems by bivalves: a case study of zebra mussels in the Hudson River. BioScience 49:19–27.

Strayer, D. L., C. Lutz, H. M. Malcom, K. Munger, and W. H. Shaw. 2003. Invertebrate communities associated with a native (*Vallisneria americana*) and an alien (*Trapa natans*) macrophyte in a large river. Freshwater Biology 48:1938–1949.

United States Army Corps of Engineers. 1995. Hudson River habitat restoration Hudson River basin, New York. Reconnaissance Report, New York.

United States Environmental Protection Agency. 2005. Available: http://www.epa.gov/hudson/. (April 2005).

Wigand, C., M. Finn, S. Findlay, and D. Fischer. 2001. Submersed macrophyte effects on nutrient cycling in riverine sediments: contribution of "new" inputs. Estuaries 24:398–406.

American Fisheries Society Symposium 51:39–61, 2006

Current and Evolving Physical and Chemical Conditions in the Hudson River Estuary

Karim A. Abood, Thomas L. Englert, Susan G. Metzger, Charles V. Beckers, Jr., Timothy J. Groninger, and Sumant Mallavaram

HDR-LMS

One Blue Hill Plaza, Pearl River, New York 10965 USA

Karim.Abood@hdrinc.com

Abstract.—The physical and chemical environment of the Hudson River Estuary is characterized from data collected during the last 6 decades, with a focus on hydrology (primarily freshwater flow) and water quality (mainly salinity, temperature, and dissolved oxygen [DO]). The most remarkable change during this period is the substantial improvement in dissolved oxygen levels, particularly in the vicinity of New York City and around Albany. A primary reason behind this improvement is wastewater treatment and minimization of point source discharges (raw sewage and industrial effluents). During the summer of 1973, water quality was poor (DO of zero), with no fish life present in the Albany region. It is a given that water quality is but one element affecting biota and ecosystem functions; however, current levels no longer support the septic, polluted label given to the River in times past. The recent DO improvement is one measure of the resiliency of the Hudson River and its recovery from abuse. Also, the last 2 decades experienced greater than average freshwater flows. The data suggest a slight increase in water temperature in certain segments and an apparent, yet to be confirmed, small increase in salinity intrusion for a given flow.

Introduction

The Hudson River watershed and its estuary is a moderate-sized, but regionally and nationally important, mid-Atlantic system which has received substantial study because of management issues with its fishes (Daniels et al. 2005). To provide a physical and chemical context for the Hudson's fishes, we present a description of the Estuary and its basin, with emphasis on freshwater flow patterns and recent and evolving water quality characteristics, primarily water temperature, salinity, and dissolved oxygen (DO). Freshwater flow observations are from the Green Island U.S. Geological Survey station (the most downstream gauging station above tidewater). Semiempirical methods to quantify the contribution of the basin area below Green Island at Poughkeepsie and the Battery at the southern tip of Manhattan were used to estimate downstream flows.

Historical and recent salinity, water temperature, and dissolved oxygen observations are derived from previous publications, as well as from the multi-utility Long River Surveys (Klauda et al. 1988) conducted between 1975 and 2000. These are followed by historical salinity, temperature, and dissolved oxygen trends at Poughkeepsie and in the vicinity of New York City. Factors influencing these parameters and a summary of flow and water quality changes over the last 50 years are also discussed.

Hudson River Basin

Basin Hydrology

The Hudson River basin drains an area of 34,540 km^2 (Abood et al. 1989). Most of this area lies in the east-central part of New York State, with small portions in Vermont, Massachusetts, Connecticut and New Jersey (Figure 1). Several locations of interest to this study are shown in Figure 2, which also illustrates the distance in river kilometers (river km) above the mouth of the River (the Battery, at river km 0). In addition, Figure 2 locates the multi-utility Long River study areas (Klauda et al. 1988), conducted from 1975 through 2000, and power plants.

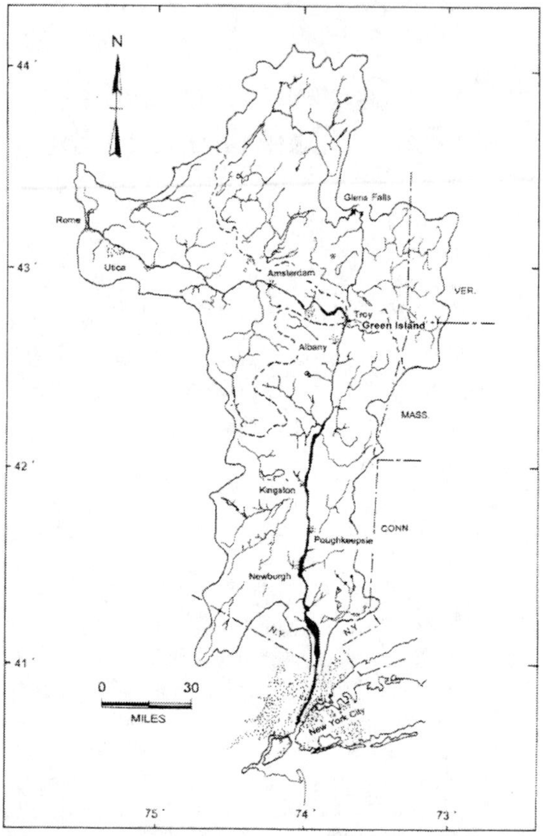

Figure 1. Map of the Hudson River drainage basin.

The Hudson River basin is bounded on the north by the St. Lawrence and Lake Champlain drainage basins; on the east by the Connecticut and Housatonic River basins, and the Connecticut coastal area; on the west by the Delaware, Susquehanna, Oswego and Black River basins; and on the south by the basins of small streams tributary to the Hudson River in New York Harbor. Its watershed extends 206 km east-to-west and 383 km north-to-south. The Hudson's source is Lake Tear-of-the-Clouds in the Adirondack Mountains of northern New York State and from there it flows generally south for 507 km to its mouth at the Battery where it discharges into New York Upper Bay.

Major tributaries entering the river's mainstem include the Mohawk River, Hoosic River, Kinderhook Creek, Indian River, Sacandaga River, Esopus Creek and Rondout Creek. For convenience, the Hudson River basin has been separated into three principal drainage areas: the Upper Hudson, the Mohawk River, and the Lower Hudson sub-basins. The division between the Upper and Lower Hudson basin is at the Federal Dam at Troy, some 248 km above the Battery. The Federal Dam represents the head of tidewater. Researchers have considered the Battery (as well as the Narrows 8 km south of the Battery) as its mouth.

The Upper Hudson River flows mainly south–southeast to the confluence with the Sacandaga River, where it turns to the east. At Hudson Falls, it turns again to the south. The river's total length to Green Island is about 240 km and it drains approximately

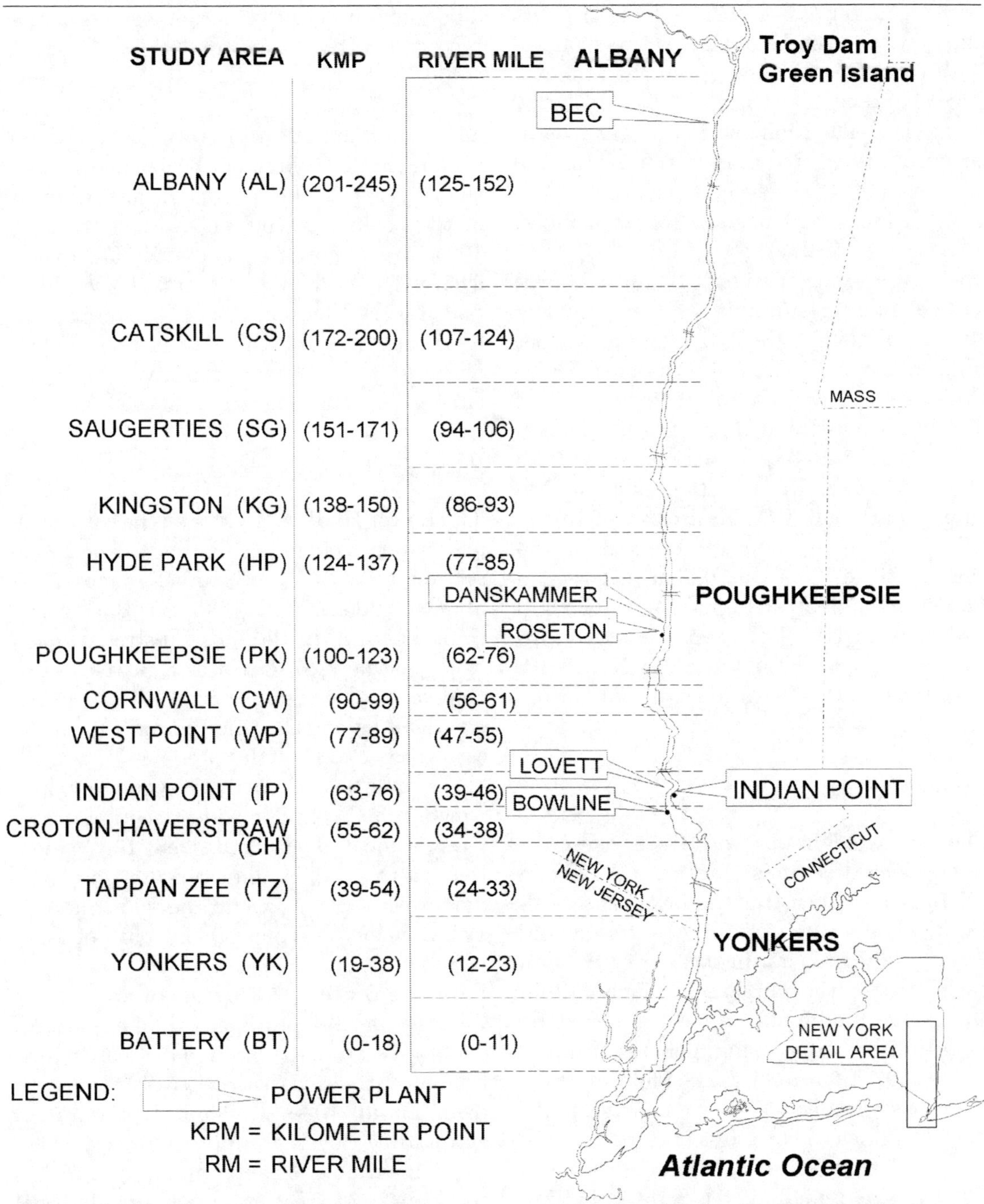

Figure 2. Hudson River multi-utility, Long River study areas, 1975-2000. Power generating stations are shown in the callout boxes.

11,984 square km. From its source to Troy Dam, the Hudson River drops about 552 m, resulting in an average bottom slope of about 4.4 m per km.

The Mohawk River has its source in the hills near the boundary between Lewis and Oneida Counties, New York. It flows in a southerly direction to Rome, and then follows a general east–southeast course to its junction with the Hudson River at Cohoes, New York. The total length of the Mohawk River is about 250 km and it drains 8,967 square km. The Mohawk River falls irregularly from its source at elevation about 550 m above mean sea level to elevation 4.4 m where it joins the Hudson River at Cohoes.

The Lower Hudson River, the primary water body of interest in this paper, commences at the junction of the Mohawk and Upper Hudson Rivers at Troy and discharges into Upper New York Bay. All of this section of the river is tidal and a substantial portion of it is subject to ocean-derived salt intrusion. The Lower Hudson River is about 248 km in length and it drains an area of 13,667 km^2. Its average slope, represented by the half tide level, is about 0.6 m in 241 km. The slope is greatest from Troy to Catskill and is smallest between Catskill and Tarrytown. From Tarrytown to New York City, the half tide elevation drops about 0.2 m.

Basin Climate

The climate of the Hudson River basin may be considered as moist continental (Abood 1974). The Upper Hudson basin has comparatively long, cold and snowy winters and short, mild summers. In the Lower Hudson basin, where the climate is much milder due to the modifying influence exerted by the Hudson Valley, summers are usually longer and winters milder. The Mohawk River basin has variable weather conditions with characteristics of both areas.

The average annual air temperature within the basin ranges from 10°C in the southern portion to 4.4°C in the Adirondack Mountains (Abood 1974). The corresponding average July temperature varies from about 24–18°C, and the average January temperature is about –1°C and –9°C, respectively. The maximum and minimum temperatures recorded in the basin are 41°C and –41°C, respectively.

The mean annual precipitation varies from 86 cm in the center of the basin to more than 127 cm in the Adirondacks, with most of the basin receiving an average of about 102 cm. Precipitation is distributed rather evenly throughout the year, with a slight rise during the summer. The average annual snowfall for the basin ranges from about 76 cm in New York City to over 330 cm in the Adirondack Mountains.

River Channel Geometry

The Lower Hudson River is a relatively deep and straight channel (Figure 3). The cross-sectional area of the channel ranges from about 23,336 m^2 in Haverstraw Bay to less than 4,645 m^2 in the most upstream reach of the estuary. Surface width in the Lower Hudson varies significantly and somewhat erratically along the longitudinal axis of the estuary because of the presence of several embayments, primarily Haverstraw and Newburgh Bays. The widest section of the river is located at Haverstraw Bay, where the width reaches almost 5 km, and the narrowest section, where the width is only 175 m, is located some 10 km south of Troy.

From the Battery to the head of Haverstraw Bay, the mean depth (mean tidal elevation), defined as the cross-sectional area divided by the surface width, generally decreases from about 10 m to 5 m. Upstream of Haverstraw Bay, it abruptly increases, reaching a maximum of approximately 27 m in the vicinity of West Point, where individual values approach some 60 m.

Tidal Characteristics

The Lower Hudson River is a tidal estuary between the Federal Dam at Troy and New

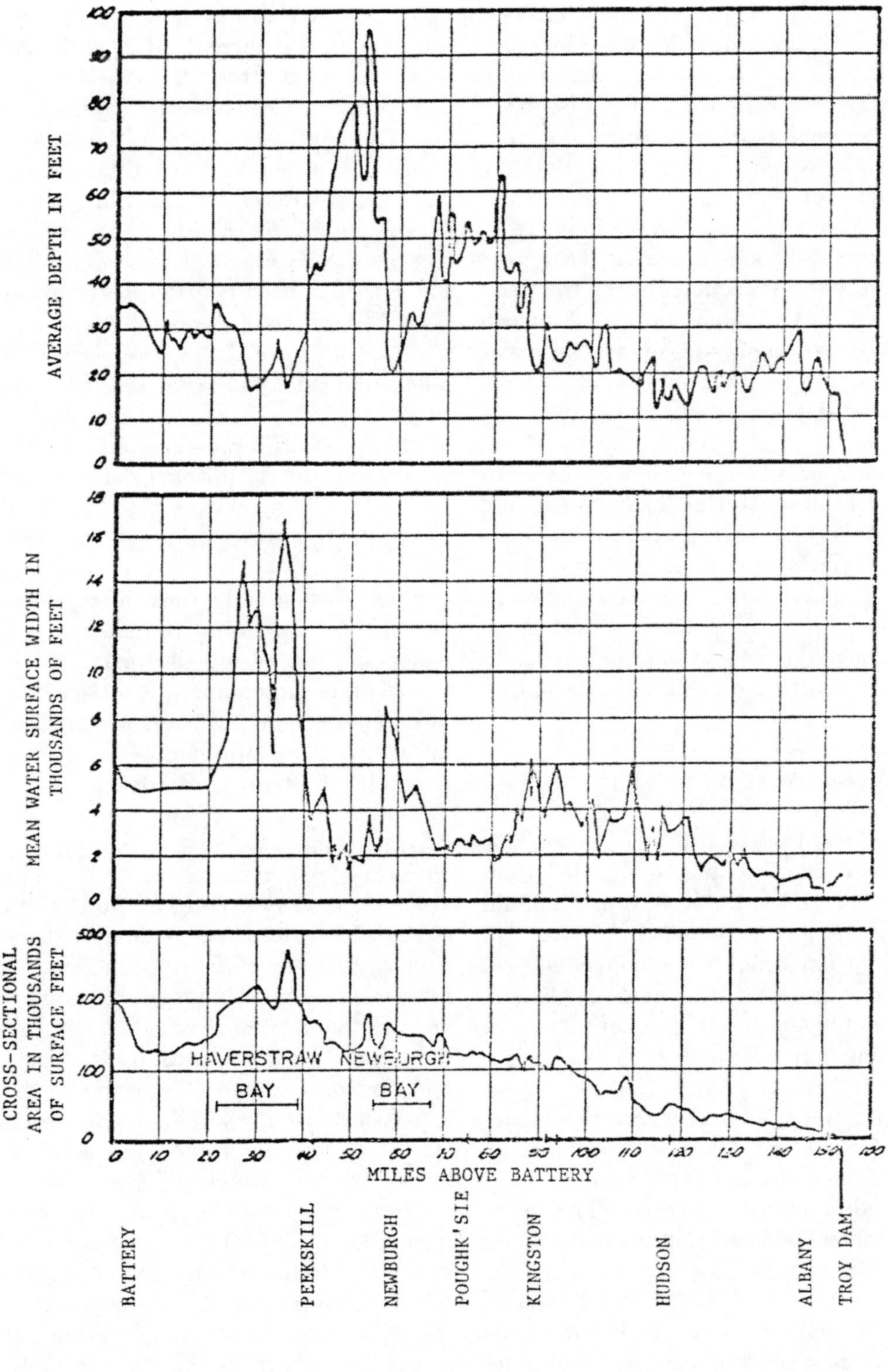

Figure 3. Hudson River Channel physical characteristics.

York Harbor. It is also classified (Abood 1974) as a dampened, reflected tidal wave regimen, with the dampening occurring as the energy of the tidal wave is dissipated by channel friction as the oceanic tidal wave progresses upstream. Reflection includes the propagation of secondary waves as a result of channel obstructions. Complete reflection occurs at the Federal Dam at Troy, and additional wave reflections result from significant changes in channel width. As width increases, wave amplitudes tend to decrease, whereas a decrease in channel width causes an increase in wave amplitudes.

Tidal behavior at any section is the composite effect of ocean tide, channel friction, and wave reflection. The primary ocean tides are also variable, with maximum amplitude occurring during spring tide and minimum amplitude during neap tide. Variations in freshwater discharge and barometric conditions also contribute to changes in amplitude.

The half tide level indicates the average slope in the river, and the total fall from Troy to the sea is about 0.6 m. As the tide moves upstream, its range diminishes from about 1.3 m at the Battery to a minimum of about 0.8 m near Storm King (river km 90). The tidal range then increases to a second maximum of 1.25 m near Catskill and then remains fairly constant (1.2 m) for the remaining distance to Troy.

The spring tide is characterized by a higher range of high and low water elevations, about 1.6 m at the Battery, 0.95 m at West Point, and about 1.5 m at Troy. The variation of mean sectional tidal velocity along the river was first observed in the 1929 USC&GS study (USC&GS 1934). Both ebb and flood strengths were averaged across the river cross section, and mean absolute velocity over a tidal cycle was obtained by averaging the section-averaged ebb and flood strengths. The Hudson River ebb strength (maximum ebb current) is generally greater than the flood strength in the entire estuary, primarily because the freshwater flow and the ebb flow proceed in the same direction. The mean sectional tidal velocity in the Hudson River ranges from just above 60 cm/s near the mouth, 30 cm/s from the Tappan Zee Bridge to Kingston, to just below 60 cm/s near Hudson, and zero at the Troy Federal Dam (Abood 1974).

The mean tidal flow decreases from a maximum of 12,035 m^3/s at the Battery to zero at the Federal Dam at Troy (river km 247). The values of the mean tidal flow are affected by the freshwater flow, particularly in the upstream reaches of the estuary.

Freshwater flow influences the tidal characteristics by increasing the surface ebb velocity and decreasing the surface flood velocity while prolonging the surface ebb and abbreviating the surface flood cycle. The influence of freshwater flow on these characteristics decreases with depth.

Basin Sediments

A major source of sediments in the Hudson River is its watershed, particularly from basin runoff erosion. The Hudson River's sediment load is approximately 800,000 tons/year. Secondary sources include wastewater discharges, bank erosion, dredged material and nonpoint sources. Suspended sediment concentrations in New York Harbor range from 3 to 276 mg/l and average about 30 mg/l. The unit weight of dry sediments in the lower reaches of the Hudson River ranges from about 400–800 kgm/m^3 with wet density of two to three times dry densities counterparts (Abood 1974). Abood (1979) estimated a theoretical sedimentation rate of about 3 cm/year to over 2 m/year in the most downstream reaches of the Hudson River and New York Harbor.

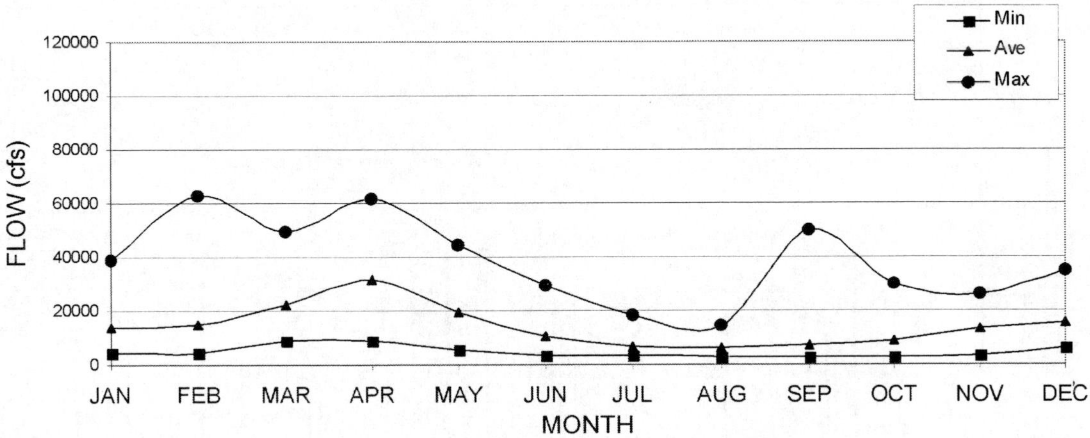

Figure 4. Hudson River historical monthly average flows (Green Island) 1946-2003.

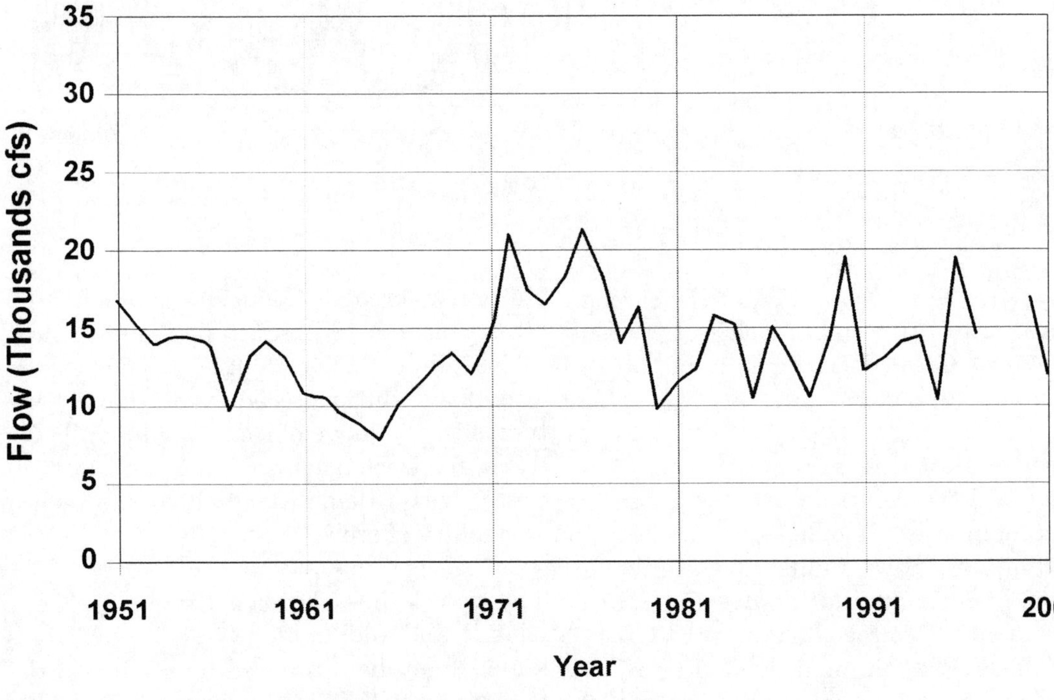

Figure 5. Hudson River annual flow—(Green Island) 1951–2001.

Freshwater Flow

The major portion (about 75% under normal summer, and 65% under mean annual conditions) of freshwater flow enters the estuary at its head at Troy. The remaining portion of freshwater flow consists of contributions by tributaries discharging largely into the upper reach of the estuary. Measurement of the freshwater flow in the estuary is not possible because of tidal oscillation. Lower Hudson flow histograms in the tidal portion of the river are usually constructed by measurements of Hudson River flows at Green Island (the most downstream USGS gauging station above tidewater [see

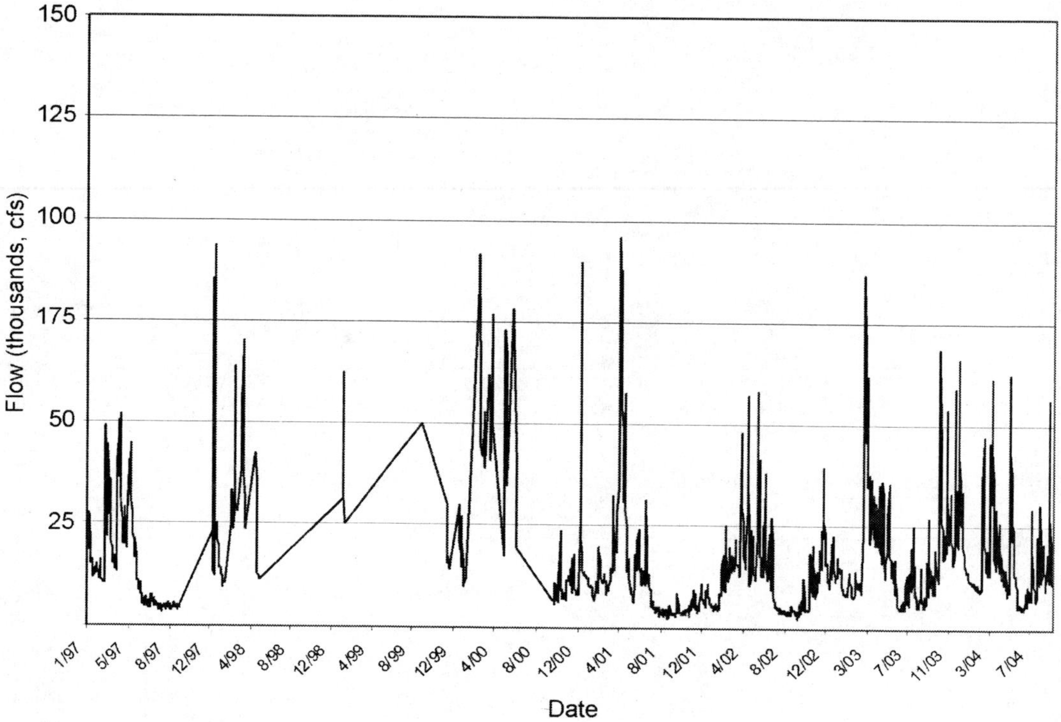

Figure 6. Hudson River daily flow—(Green Island).

Figure 1]), and by empirically developed flow and travel time relationships between the flows at Green Island over a period of 57 years (1946–2003).

Freshwater flow varies over the year, with maximum flows occurring primarily during the spring months (March, April and May), and minimum flows usually beginning in June and continuing until November. The overall monthly average flow range at Green Island from 1946 through 2003 (Figure 4) extends from a minimum of 83 m^3/s (2,912 cfs) during August to a maximum of 1,776 m^3/s (62,733 cfs) during February, with an annual average of 404 m^3/s (14,252 cfs).

During the 1960s, the entire northeastern region of the United States experienced a severe drought characterized by extremely low flows. The most severe drought in the Hudson River estuary was observed in 1964, when the freshwater flow in the lower Hudson averaged 3,500 cfs, or about 100 m^3/s , during the 6-month period from June to November. This drought receded in the late 1960s; during recent years flows have been slightly above normal (Figure 5). Daily flows at Green Island during the most recent 7 years demonstrate high short-term variability (Figure 6).

The travel time between the head (Green Island) and Indian Point (located at river km 69) may be estimated by the use of the following relationships (Abood 1977):

$$t_{LH} = 256.14 / Q_{LH}^{0.4} \qquad (1)$$

in which:

t_{LH} = travel time between Green Island and Indian Point in days;

Q_{LH} = estimated Lower Hudson freshwater flow in cubic feet per second.

Table 1. Relationship between Green Island freshwater flow and lower Hudson River flow at Poughkeepsie and the Battery.

		$(Q_{LH} = a + b\,Q_{GI})$				
	Poughkeepsie			Battery		
Month	a	b	R^2	a	b	R^2
Jan	–413	1.477	0.96	-353	1.648	0.94
Feb	–1179	1.600	0.92	1105	1.600	0.90
Mar	4718	1.319	0.92	5302	1.503	0.88
Apr	1115	1.311	0.93	1626	1.432	0.88
May	–154	1.356	0.98	85	1.483	0.96
Jun	711	1.231	0.91	1228	1.310	0.86
Jul	–12	1.228	0.94	52	1.317	0.95
Aug	–951	1.358	0.96	-1062	1.452	0.93
Sept	–192	1.237	0.87	-319	1.336	1.80
Oct	–1968	1.569	0.96	-2544	1.748	0.94
Nov	–964	1.516	0.95	-1678	1.741	0.92
Dec	–958	1.421	0.90	-1008	1.621	0.94

[a] Estimated by the USGS. Travel time not taken into account. Isolated points outside indicated range not used. Feb. values rounded.

This equation was derived from Manning's equation (negligible surface width with flow stages, $R_h \approx H$, $n \neq f(Q)$ assumed) and is based on the results of a detailed Hudson River flood routing study.

The contribution of the drainage area below Green Island (Q_{GI}) to the Lower Hudson freshwater flow (Q_{LH}) has been empirically determined (Abood et al. 1989) with the use of USGS estimates of ungauged and gauged tributaries from 1947 through 1965. Because of significant differences among calendar months and downstream location, 24 equations were developed (Table 1). Although the correlation coefficients in Table 1 are high (80–98%), flows generated using these relationships should represent only estimates of Lower Hudson River flows and should be considered accordingly. For convenience, these relationships may be approximated by:

$$Q_{LH} = a + bQ_{GI} \quad (2)$$

Equation (4) was used to convert Green Island flows to Lower Hudson River counterparts at Poughkeepsie and the Battery (Figure 7). The long term (1946 through 2003) mean annual flow in the Hudson River has increased by 4–5% since 1988; the most recent data set (prior to this paper) reported by Abood et al. (1989).

Salinity Intrusion

The lower reach of the Hudson River is subject to salinity intrusion. During drought periods the salinity front (100 mg/l) can extend some 130 km above the ocean entrance. Due to tidal motion in the ocean, turbulent eddies mix the lighter ocean-produced saltwater upwards. This action dilutes the landward flowing saltwater and increases the density of the seaward flowing freshwater. As a result, the seaward flow in the upper layer increases as more ocean-derived water intrudes upstream in the lower layer.

Within the salt-intruded reach, the Hudson

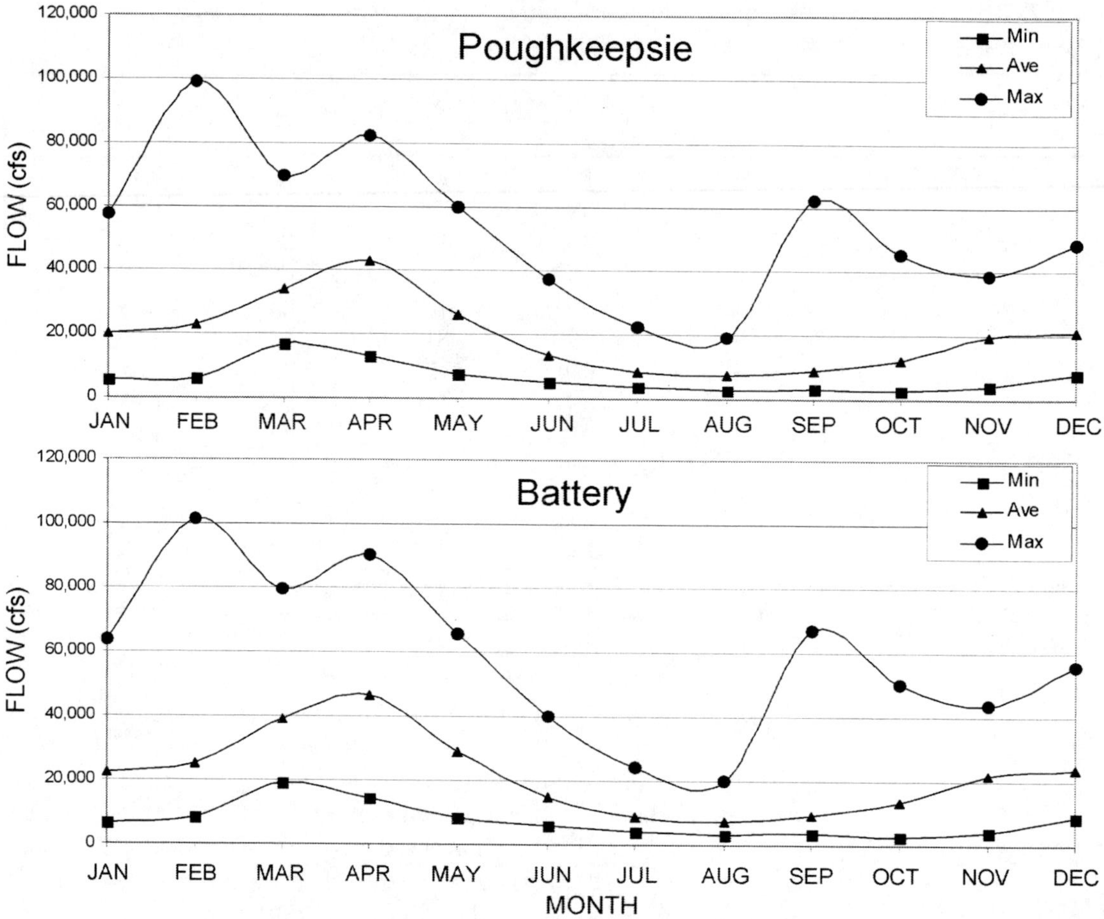

Figure 7. Hudson River historical monthly flows at Poughkeepsie and the Battery, 1946-2003.

is partially stratified. Generally, the Hudson is subject to a net upstream movement in its lower layer and a net downstream movement in its upper layer. This circulation is induced by density differences caused by salinity distribution. Upper layer flow values ranging from 10 to 40 times the corresponding freshwater flow have been reported (Abood 1977).

A set of generalized Hudson River mean salinity (averaged over tidal cycle and cross-sectional area) profiles, expressed as mean salinity versus Lower Hudson freshwater flow at fixed locations, is given in Figure 8 (Abood 1974).

The length of the salt-intruded reaches, identified as the 0.1 ppt isosal, may be estimated from the following relationship, developed by the Abood (1977):

$$L = 135\ Q_{LH}^{-.38}\ for\ Q_{LH} \leq 27\ kcfs \quad (3)$$

$$L = 1948\ Q_{LH}^{-1.19}\ for\ Q_{LH} \leq 27\ kcfs \quad (4)$$

where:

L = location of 0.1 ppt isosal in miles above the Battery;

Q_{LH} = Lower Hudson flow in 1,000 cfs

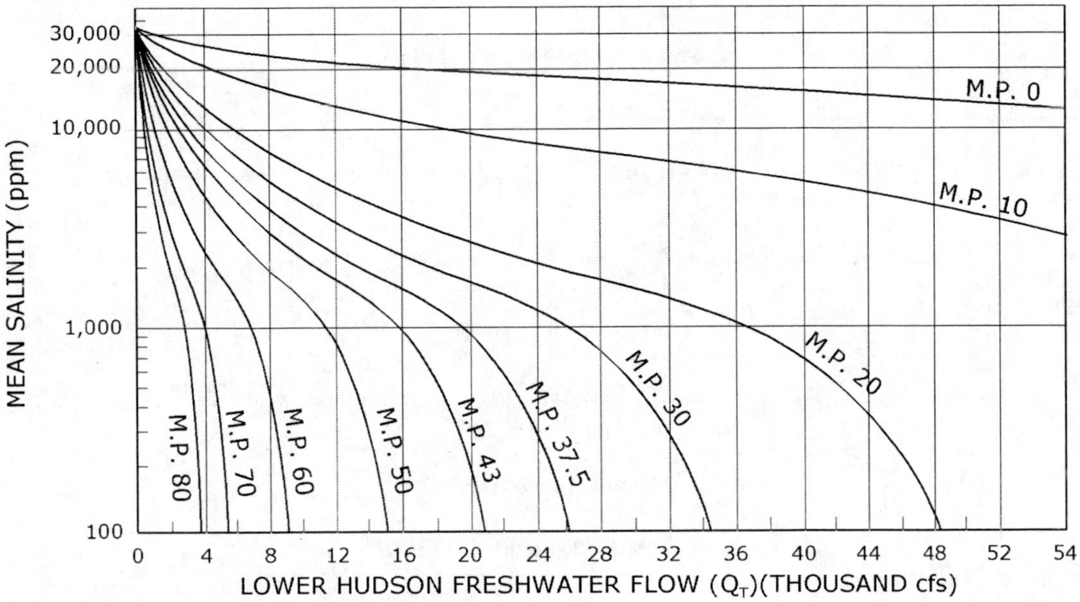

Figure 8. Salinity vs flow rate at mile points.

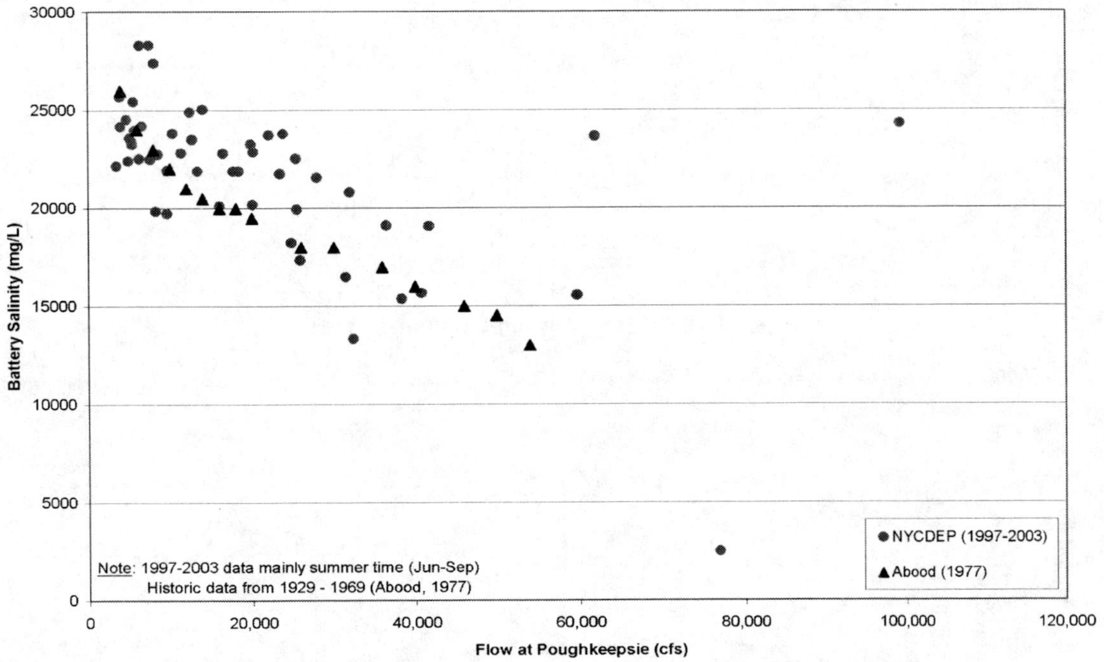

Figure 9. Hudson River salinity at the Battery vs Lower Hudson flow at Poughkeepsie.

These relationships were developed using 20 sets of salinity observations made over a 40-year period (1929 through 1969; Abood 1977). These data were synthesized to take into account the nature, characteristics, and extent of the individual surveys and to in-

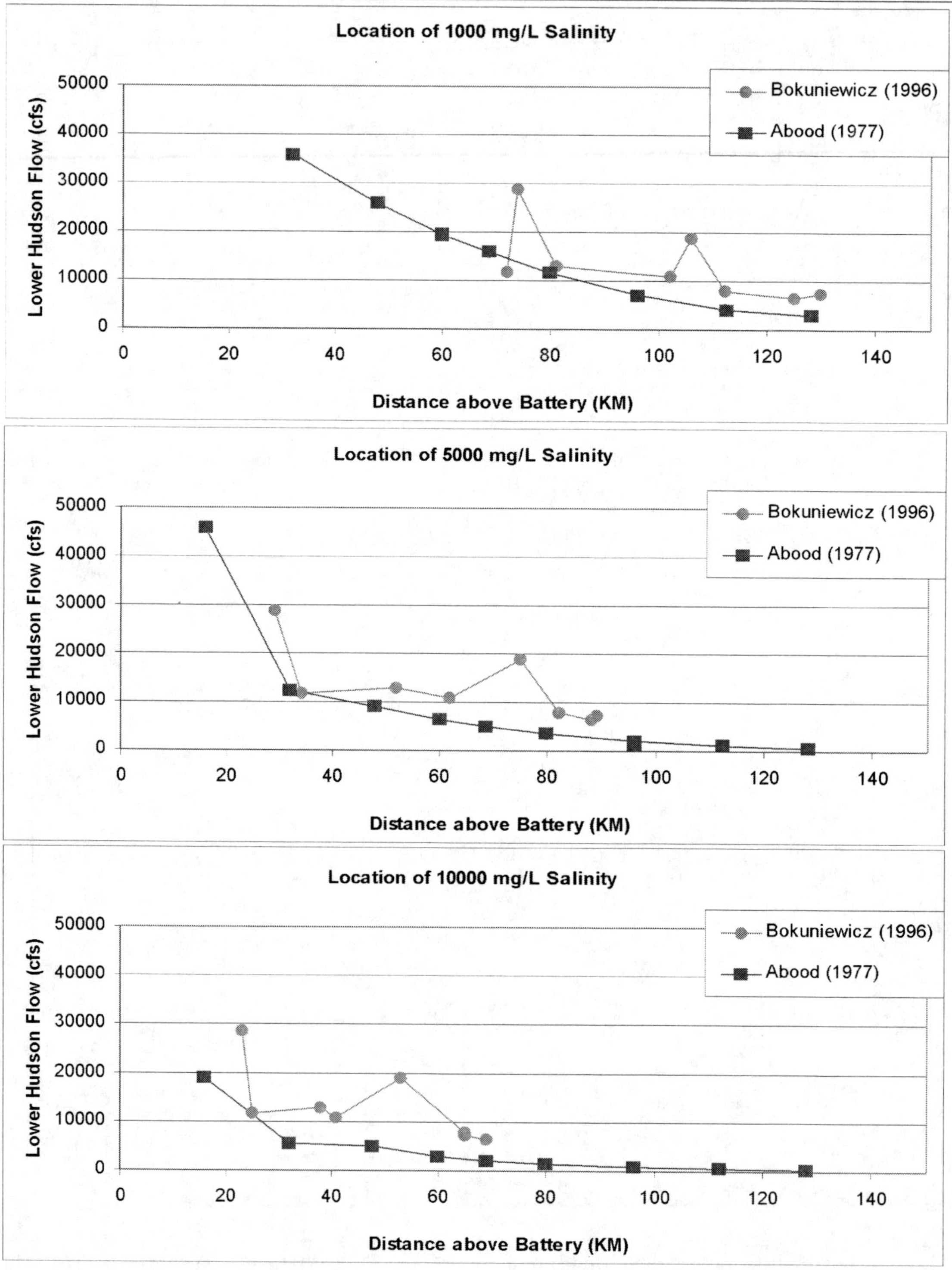

Figure 10. Salinity versus distance above the Battery.

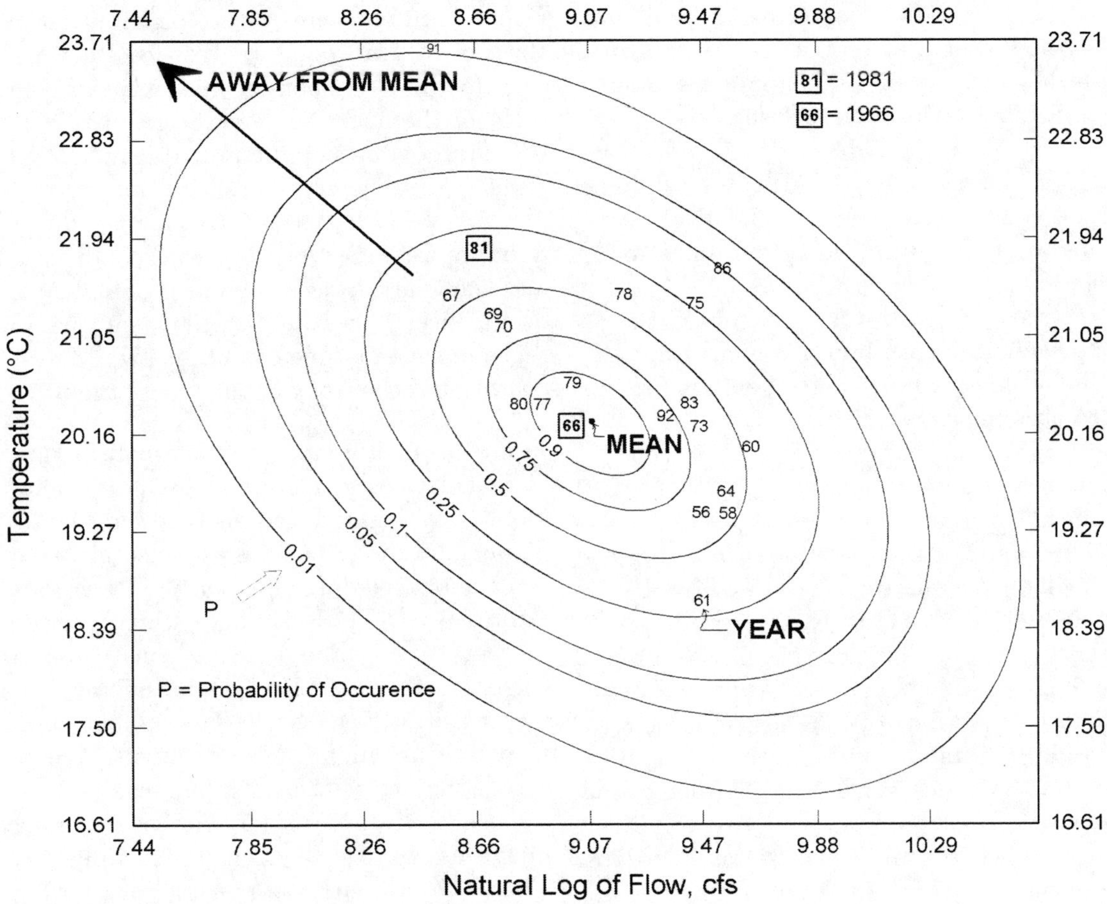

Figure 11. Natural log of flow versus temperature.

corporate differences in key factors influencing salinity intrusion in the Hudson River, including:

1. Sampling time within a tidal cycle. Abood (1977) has shown a sinusoidal relationship between salinity and flow with a maximum and a minimum occurring at high water slack water and low water slack, respectively.

2. Sampling point location with a cross-section. The Hudson is partially stratified (maximum salinity values near the bottom and minimum values near the surface). There is also a minor increase in salinity from the west shore towards the east.

3. Tidal range variation. Salinity is greater at neap tide than at spring tide.

4. Meteorological conditions. Key parameters include wind speed and direction, and, more importantly, an increase in mean sea level (either natural or due to hurricanes). An increase in sea level causes a rise in water elevation accompanied by an increase in the landward directed flow of seawater.

5. Freshwater flow conditions. As reflected in equations 1 and 2, and in Figure 8, the most important and controlling factor influencing salinity intrusion is the influence of freshwater in the Hudson River. This includes not only Upper Hudson and

Mohawk River flows at Green Island, but also the Lower Hudson tributaries and variable lag time from its head to its mouth (see Freshwater section, above).

6. Field instruments. The use of salinometers, conductivity meters, chemical titrations, or density can influence field readings.

A meaningful correlation among different salinity surveys must take these six factors into account (Abood 1977).

More recent observations made by NYCDEP from 1997 through 2003 and by Bokuniewicz (1996) from 1994 and 1995 indicate that although the overall trends produced by above-cited relationships (Figure 8) are still appropriate, recent observations (1997–2003) show an increase in salinity for a given Lower Hudson River flow of roughly 15%. Figures 9 and 10 compare several recent salinity/flow observations versus their earlier counterparts. These comparisons are crude in nature since the more recent salinity data were not subjected to detailed analysis incorporating the above-listed six key factors. The apparent increase in Hudson River salinity intrusion must, therefore, be viewed with caution. A more thorough investigation is needed to confirm this finding.

Water Temperature

Four categories of parameters influence the magnitude and distribution of water temperatures in the Hudson River (Abood et al. 1976). These include atmospheric conditions (radiation, evaporation, and conduction), hydrodynamic conditions (channel geometry, flow, and dispersion), boundary conditions (temperatures of the ocean and freshwater inputs), and anthropogenic inputs (primarily power plants using the River as a cooling water source). All of these parameters affect temperature distribution in the Lower Hudson; however, feed-forward and feedback dependencies exist among most of them. An example is given in Figure 11 showing a joint probability plot of June Poughkeepsie flows and water temperatures (1951–1992). The centroid in this figure represents average (mean) conditions, and the oval curves represent probability of occurrence. For example, 1966 appears to represent mean June conditions, while 1981 conditions occur only about 30% of the time.

Inputs from human activities include municipal discharges, industrial discharges, and, more significantly, those of power plants. Some man-made sources exercise either a heating or a cooling effect on the River. For example, a 1-MGD sewage effluent whose temperature is about 18°C, constitutes a cooling factor in the Hudson during the summer months, because River ambient temperatures are higher. The greatest percentage of the artificial heat input results from the use of River water for condenser cooling purposes for the electric power industry. There are presently six power plants with a total electric generation of approximately 6,000 Mwe using the Hudson River as a cooling water source; these plants use a cooling water flow of approximately 290 m^3/s, or 4,600,000 gallons/min, and constitute a heat input of approximately 8×10^{11} Btu/d. Data from the multi-utility Long River Surveys were used to construct a riverwide temperature profile indicative of average August conditions covering the period from 1975 through 2000 (Figure 12), as well as April, May, and June.

A long-term daily water temperature data set was used to construct a 50-year trend at Poughkeepsie. The mid-Hudson reach (Poughkeepsie) appears to have experienced an increasing trend of average temperatures from 1951 through 2002 (Figure 13). However, a significant variability existed during

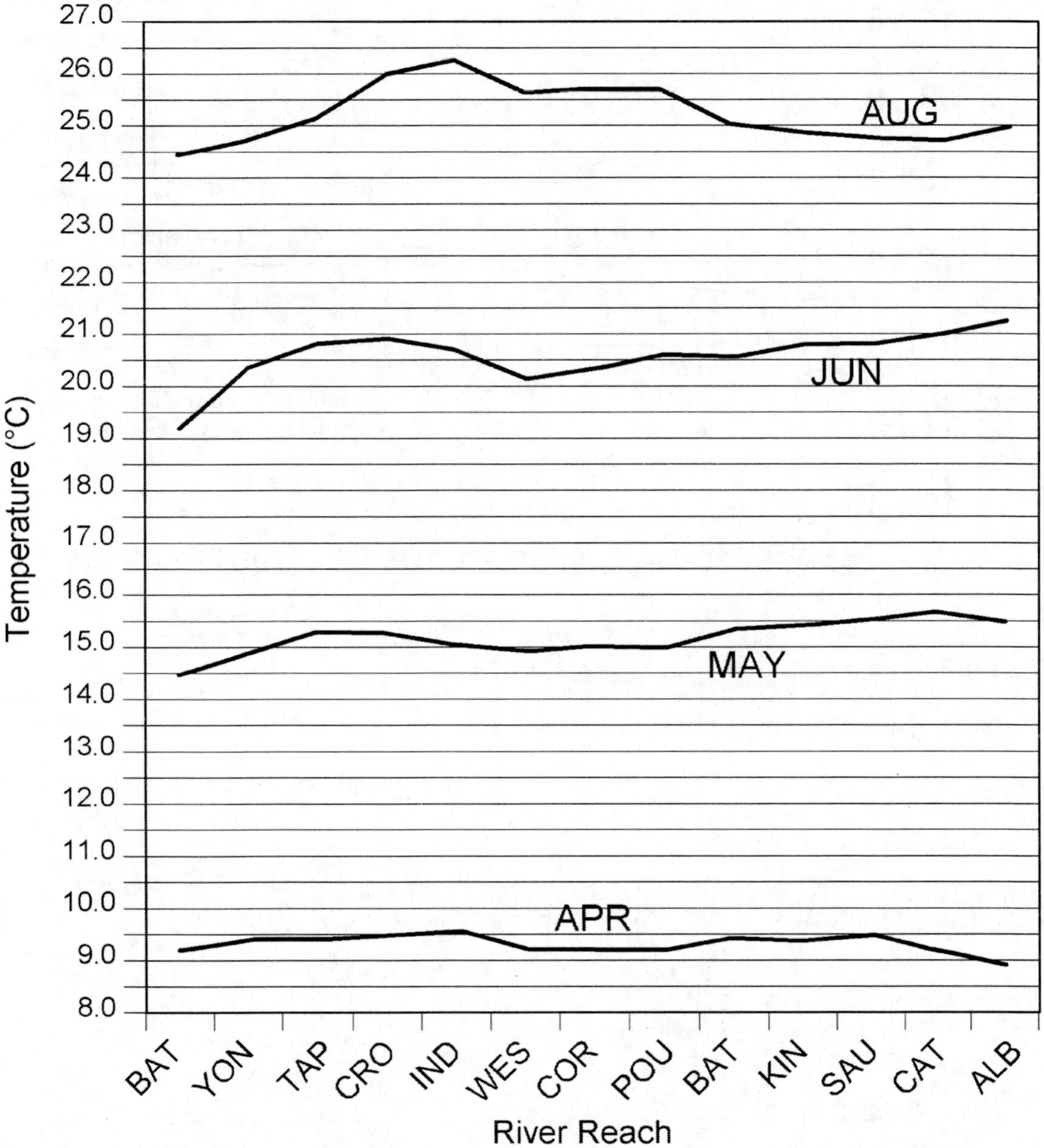

Figure 12. Temperature by region for April, May, June, and August (1975–2000).

this period. The trend (r^2 of only about 7%) may be expressed as follows:

$$Y=0.0106X-8.556 \tag{5}$$

in which:

Y=average daily temperature at Poughkeepsie in °C; X = calendar year.

The annual and 5-year running mean temperatures at Poughkeepsie from 1908 through 2002 (Figure 14) show a relatively steep decline over the 15-year period from about 1910 through 1935 of 2°C, followed by a more gradual warming trend of about

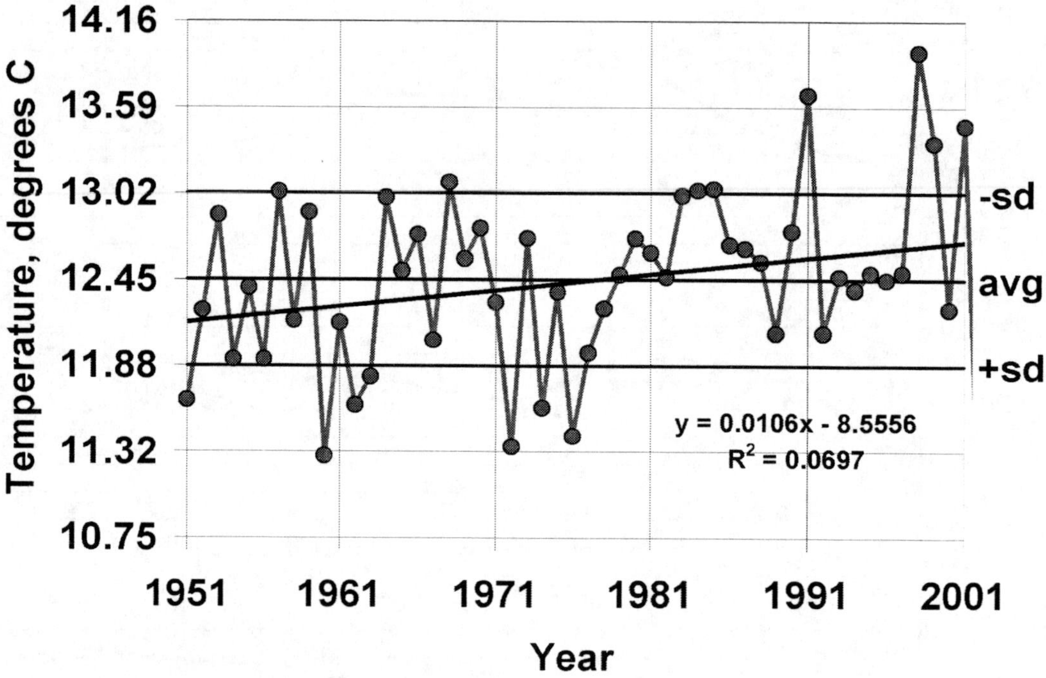

Figure 13. Average temperature (Poughkeepsie) 1951–2001.

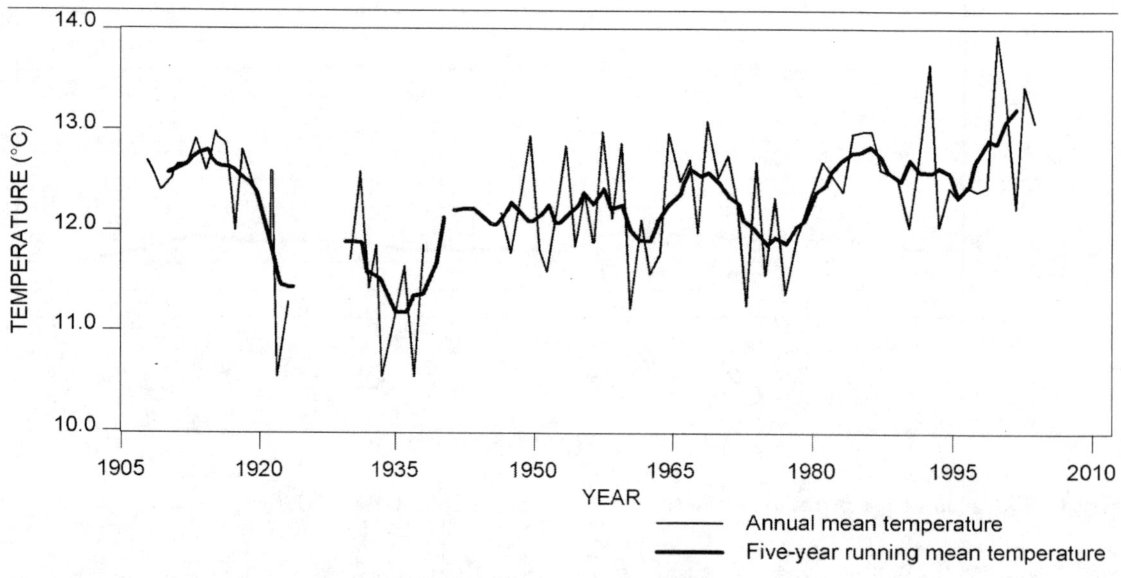

Figure 14. Average water temperature (Poughkeepsie) 1908–2002.

the same amount over the subsequent 67 years (1935–2002). Thus, in general, the warming trend evident in recent data and documented in equation 5 has apparently returned temperatures to levels previously observed in the early 1900s.

A long-term seasonal profile showing weekly average Poughkeepsie and a two-standard deviation band covering the period from 1951

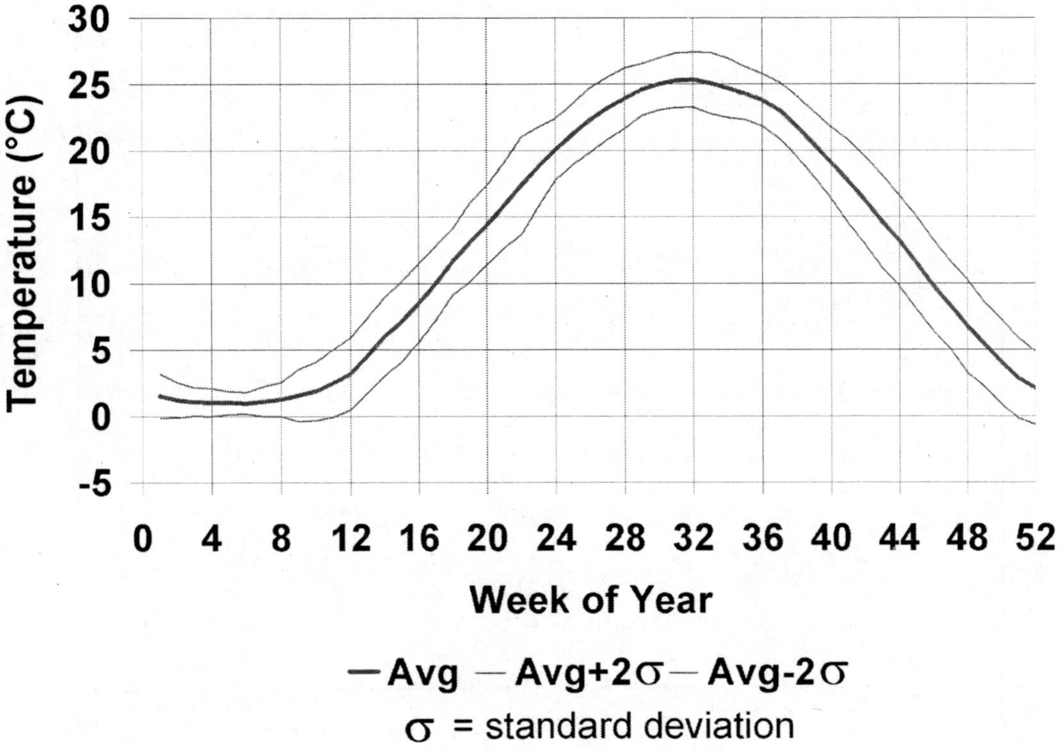

Figure 15. Weekly average water temperature (Poughkeepsie) 1951–2001.

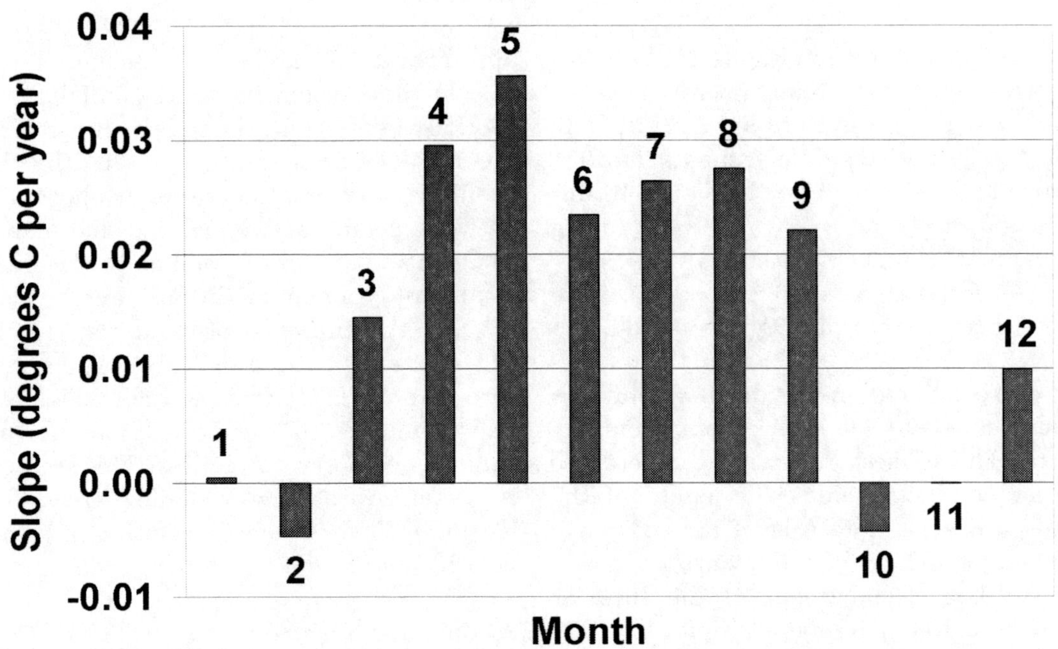

Figure 16. Change in monthly average temperatures (Poughkeepsie) 1951–2001.

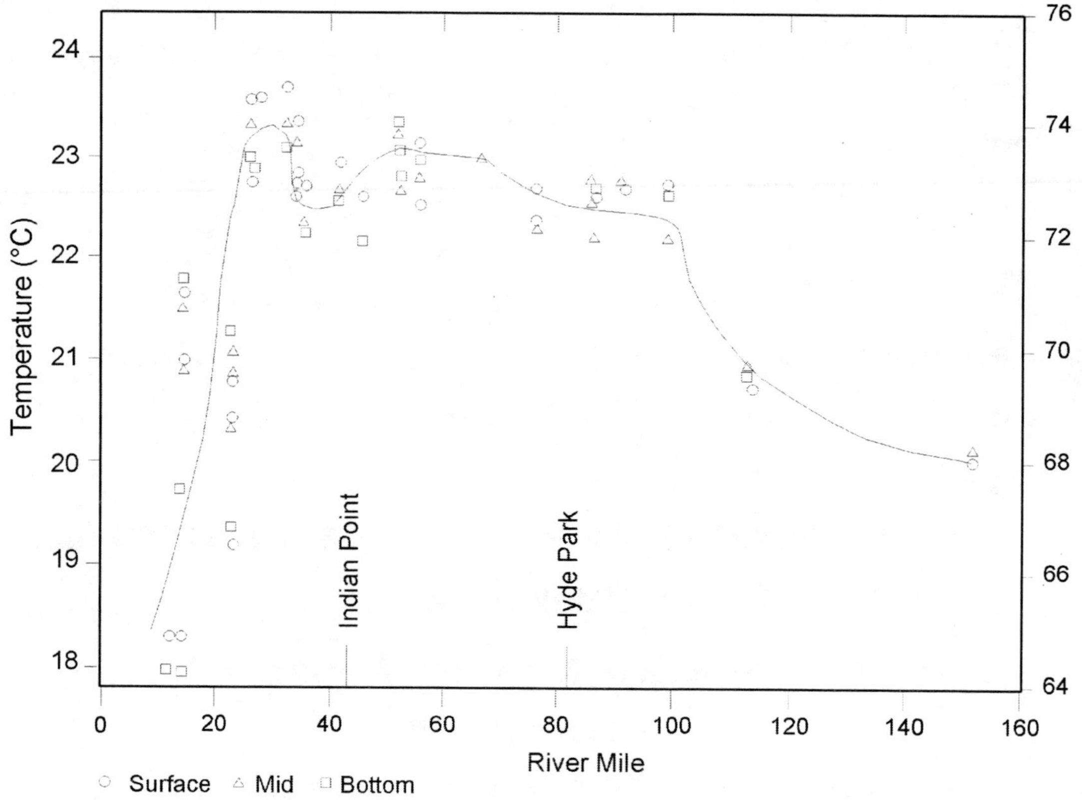

Figure 17. 1929 Temperature profile in the Hudson River.

through 2002 is given in Figure 15. Changes in monthly average temperatures and deviations by month observed at the Poughkeepsie Waterworks facility from 1951 through 2002 are depicted in Figure 16. Increases up to about 0.03°C per year during spring and summer months, as well as decreases of up to about 0.02°C per year were experienced during this 50-year period.

We also compared present day temperatures to earlier (premajor power plant) counterparts. The earliest riverwide temperature survey conducted along the length of the Hudson River in late August through early September 1929 as part of a current survey by the U.S. Department of the Interior (USC&GS 1934). Temperatures for the 1929 survey were about 18°C at the Battery, increased to about 23°C in the Nyack area (a significant increase of about 5.5°C, over an approximate 32 km river segment in two weeks), due primarily to the influence of cooler seawater on the Hudson River (Figure 17). The next trend illustrated by this profile is a gradual decrease of about 1°C between Nyack and Kingston, due mainly to climatic patterns as well as to the existing hydrodynamic conditions (lower velocity and the presence of shallows). In the last segment, a rapid decrease in temperatures (about 2°C) was recorded. This was due primarily to the hydrodynamic and channel geometry differences. This segment is deeper and has fewer shallow areas than the lower River reaches, resulting in higher velocity and less heat exchange.

A comparison between this 1929 trend (Figure 17) and a similar one depicting August temperatures from 1975 to 2000 (Figure 12) taken from the Long River Surveys, shows the influence of the above-described parameters. Both periods show a rapid increase from the Battery up to approximately the Indian Point vicinity, followed by a more gradual decrease near Hyde Park and a more pronounced decrease further upstream. These trends seem to suggest that boundary (ocean, source, and tributary) temperatures and hydrodynamic and atmospheric parameters seem to exert more significant influences on longitudinal trends during August than anthropogenic inputs.

Dissolved Oxygen

Two long-term sets of dissolved oxygen observations considered in this paper were collected in conjunction with biological studies of the fish populations of the Hudson River (see Figure 2) and by the New York City Department of Environmental Protection (DEP), covering the lower reaches of the Hudson River. Figure 18 depicts the average August dissolved oxygen observations (DO in mg/l and % saturation) taken during the Long River surveys from 1975 through 2000 between the Battery and Albany. There has been a substantial increase in DO since the early 1980s (Figure 19, derived from the DEP data set), resulting from the significant upgrades to the Yonkers and North River Sewage Treatment Plants in the lower reach of the Hudson River. Both data sets show remarkable water quality improvements when compared with earlier observations.

A closer look at summertime DO levels is given in Figure 20. This figure composes August measurements made in 1967 and 1978 against New York State standards and modeling counterparts from various governmental agencies, utilities and private industry were averaged by milepoint to provide the 1970–1972 values (LMS 1976). The 1978 results are averages by milepoint of data collected during the Hudson River Field Week II (HRFW II), a one-week riverwide sampling effort by industries, utilities, consultants, governmental agencies, colleges, and others, and coordinated by the Hudson River Research Council (HRRC 1980).

The 1967 and 1978 values are the monthly averages for August of the respective years. The 1970–1972 flow is the mean of the monthly average flows for August from the three years considered. The flows corresponding to these profiles range from 5,749–6,823 cfs, or 163–193 m^3/s; this tight range of flow makes this comparison reasonable. Freshwater flows affect the available dilution and the movement downstream of oxygen-consuming discharges, as well as temperature and salinity, which in turn affect the DO saturation level; that is, the maximum amount of DO that the water can retain. Other parameters affecting DO levels include waste loads, temperature, and overall climactic conditions.

Although detailed cause and effect comparison among these profiles and the Long River survey data (Figure 18) have not been made, a number of overall reach and Riverwide observations based on these profiles can be made below. The most obvious feature in Figure 20 is the sag in the region just downstream of Albany (river km 193–241 [river mile 120–150]). DO levels were below the state standard in 1967 and approached 0.0 mg/l in 1970–1972. In 1974, two waste treatment plants came on line in the Albany area, treating about 80% of the city's domestic wastes. Dissolved oxygen levels observed in 1978 were above 6.5 mg/l and were higher than the state standard (Figure 20).

In the New York City area (river km 0–16), the 1967, 1970–1972, and 1978 values (Figure 20) are almost identical and include the presence of another sag below the State standard. Construction delays of two major sew-

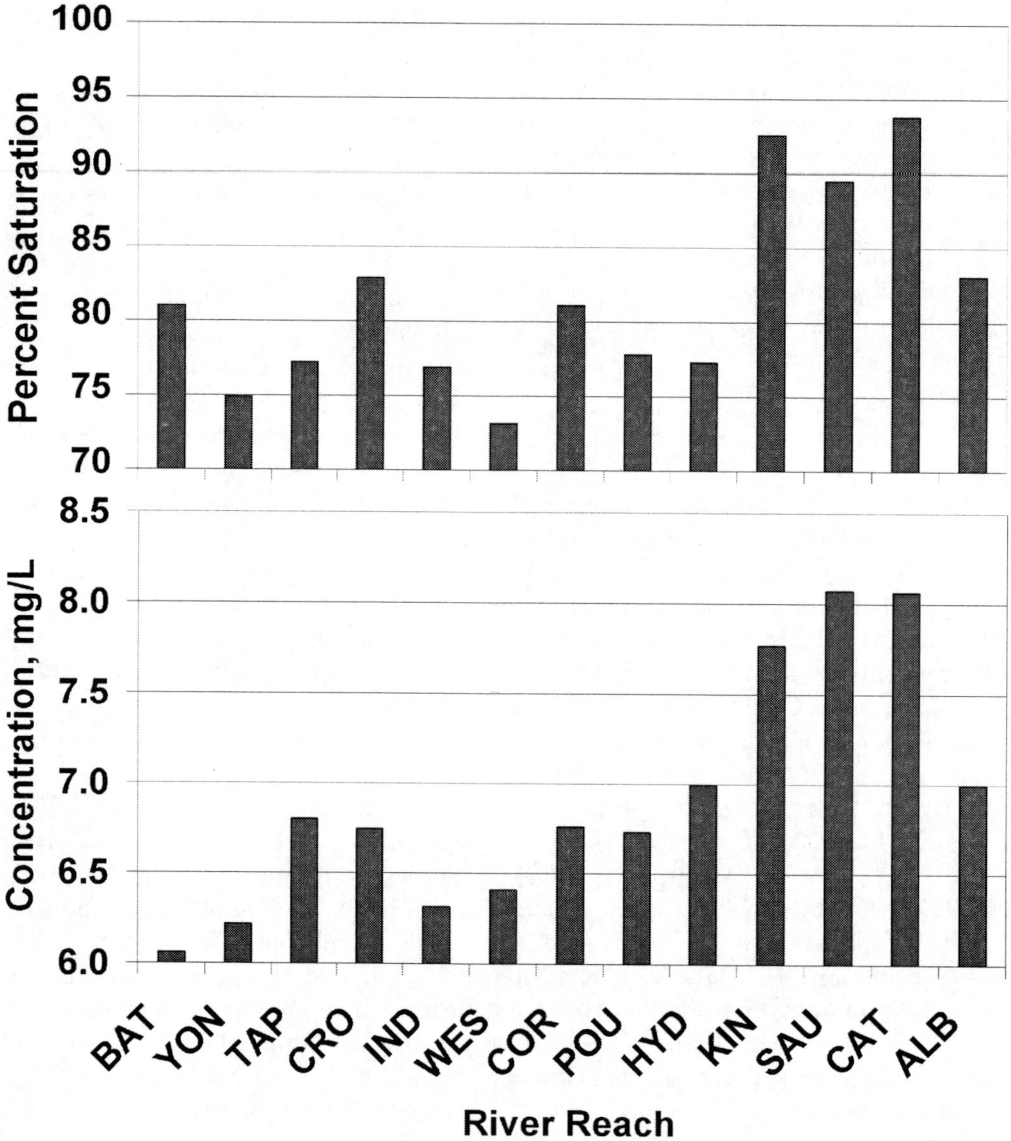

Figure 18. DO profile in August by Hudson River region (1975–2000).

age treatment plants—the North River Plant in Manhattan and the Passaic Valley STP in New Jersey—are partially responsible for the failure to achieve the projected 1977 best practicable technology (BPT) level. However, substantial increases in DO were observed in the vicinity of New York City since the 1980s and continue to rise after completion of the Yonkers and North River Water Pollution Control Plants (Figure 19).

In the mid-river region, the DO levels during the 1970s are irregular, but for all practical purposes they are the same. There appears to have been a slight increase over the 1967 level, but profiles for this region are well above the State standard.

The overall water quality levels, as measured by DO levels in the Hudson River have improved substantially. In the area north of New York City extending as far

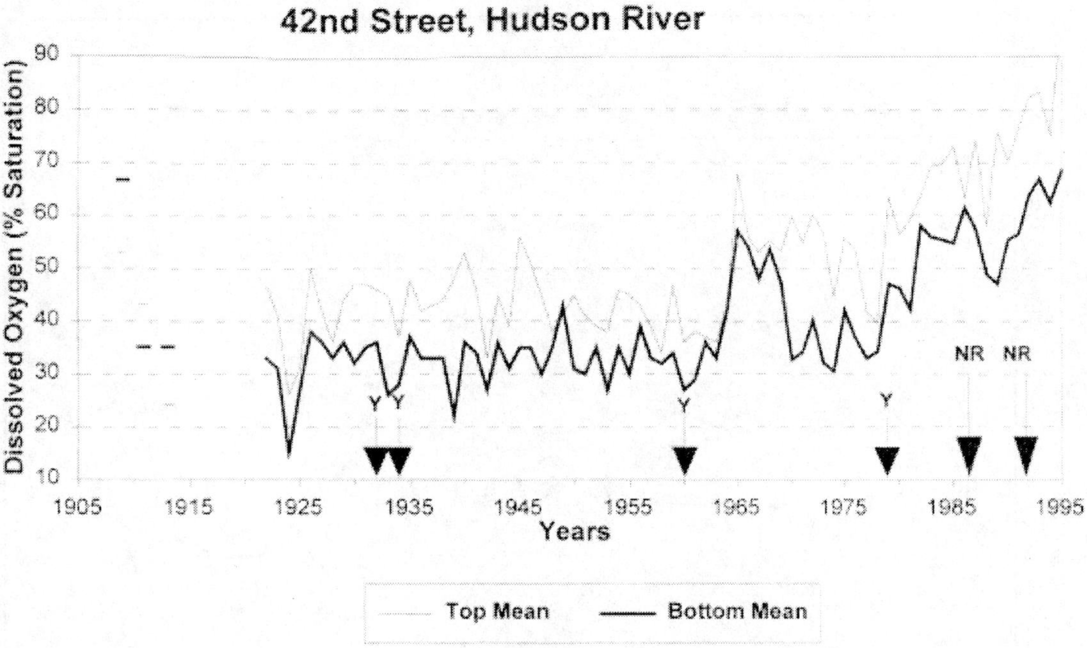

*Dates of construction and significant upgrades to the Yonkers (Y) and North River (NR) water pollution control plants are also depicted.

Figure 19. Dissolved oxygen percent saturation at 42nd Street, New York City.

north as the Federal Dam at Troy, wastewater treatment facilities, have resulted in water quality levels that are more capable of supporting aquatic life throughout the year. In the New York City area and around Albany, the low DO conditions that existed of the summer months during the early 1960s have improved significantly.

Conclusions

The water chemistry and physics that make up the biological environment of the Hudson River basin and estuary experience changes on many scales, both temporally and geographically. However, the system also has gone through notable environmental changes from a relatively pristine precolonial state, through the severe chemical and sewage contamination of its industrial era, to its partial recovery, due largely to the benefits of the Clean Water Act of 1972 (Waldman 1999; Limburg et al. 2006). This has resulted in a cleaner habitat for fishes, with reduced contaminant effects (Wirgin and Waldman, in press). The return to New York Harbor of marine borers, particularly teredos, in recent years (Abood et al. 1995) is another indication of improved water quality; these harmful invertebrates were largely absent in the harbor in the 1900s because of contamination. Although its changes through time in pollution levels may be the best known environmental factors that affected fishes, we document changes in temperature, freshwater runoff, and salinity that likely also may have been felt by fish populations. Among these is the apparent demise of the river's population of rainbow smelt *Osmerus mordax*, probably due to warming (Waldman 2006).

References

Abood, K. A. 1974. Circulation in the Hudson Estuary. Annals of the New York Academy of Sciences

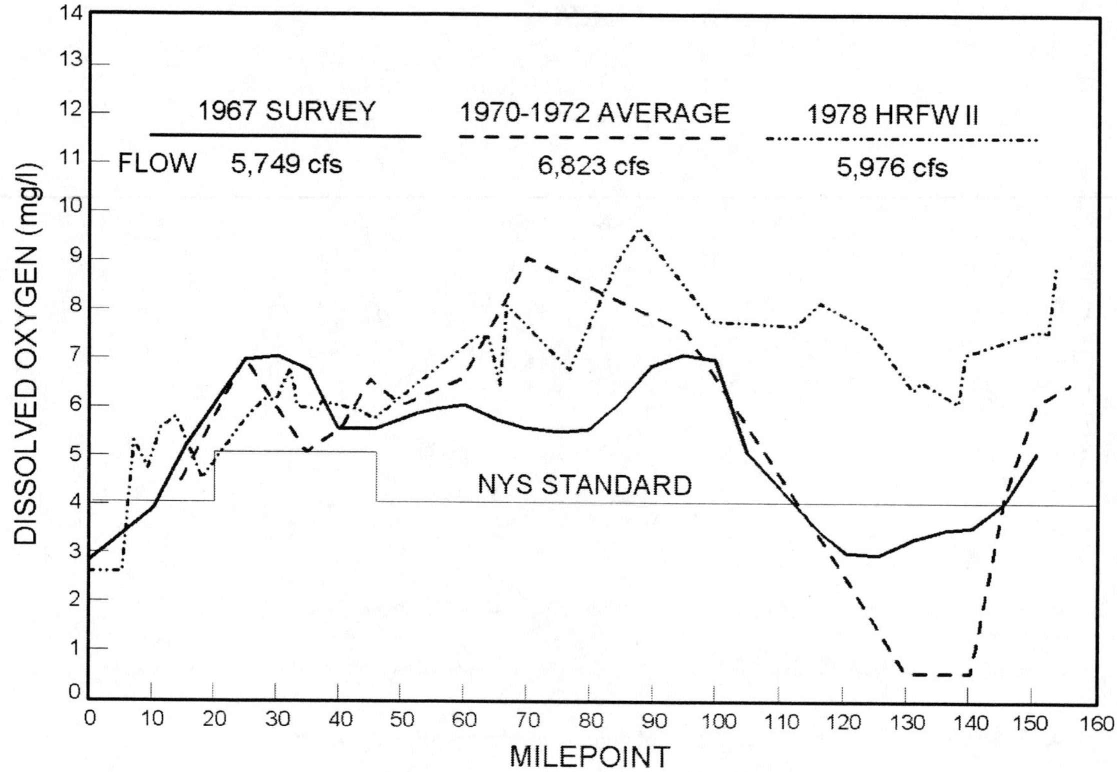

Figure 20. August dissolved oxygen profiles (1967–1978).

250:39–111.

Abood, K. A. 1977. Evaluation of circulation in partially stratified estuaries as typified by the Hudson River. Doctoral dissertation., Rutgers University, New Brunswick, New Jersey.

Abood, K. A., G. A. Apicella, and A. W. Wells. 1989. General evaluation of Hudson River freshwater flow trends. Pages 3–28 *in* C. L. Smith, editor. Estuarine research in the 1980s. State University of New York Press, Albany.

Abood, K. A., M. J. Ganas, and A. Matlin. 1995. The teredos are coming, the toredos are coming! Pages 677–690 *in* Proceedings of the Committee on Ports and Harbors of the Waterway, Port, Coastal, and Ocean Engineering Division/American Society of Civil Engineers, March 13–15, 1995, Tampa.

Abood, K. A., E. A. Maikish, and R. R. Kimmel. 1976. Investigation of ambient temperature distribution in the Hudson River. Hudson River Environmental Society Fourth Hudson River Symposium, March 28:1976, Hudson River Environmental Society, Poughkeepsie, New York.

Bokuniewicz, H. 1996. Building the turbidity maximum in the Hudson River Estuary. Special Data Report 115, Marine Science Research Center, State University of New York, Stony Brook.

Daniels, R. A., K. E. Limburg, R. E. Schmidt, D. L. Strayer, and R. C. Chambers. 2005. Changes in fish assemblages in the tidal Hudson River, New York. Pages 471–503 *in* J. N. Rinne, R. M. Hughes, and B. Calamusso, editors. Historical changes in large river fish assemblages of America. American Fisheries Society, Symposium 45, Bethesda. Maryland.

HRRC (Hudson River Research Council). 1980. Results of Hudson River field weeks: April 1977 and August 1978. Hudson River Research Council, Albany, New York.

Klauda, R. J., P. H. Muessig, and J. A. Matousek. 1988. Fisheries data sets compiled by utility-sponsored research in the Hudson River estuary. Pages 7–85 *in* C. L. Smith, editor. Fisheries research in the Hudson River. State University of New York Press, Albany.

Limburg, K. E., K. A. Hattala, A. W. Kahnle, and J. R.

Waldman. 2006. Fisheries of the Hudson River estuary. Pages 189–204 *in* J. S. Levinton and J. R. Waldman, editors. The Hudson River Estuary. Cambridge University Press, New York.

LMS (Lawler, Matusky & Skelly Engineers). 1976. Environmental impact assessment: water quality analyses: Hudson River. Prepared for the National Commission on Water Quality, NTIS, Springfield, Virginia.

USC&GS (United States Coast and Geodetic Survey). 1934. Tides and currents in the Hudson River. Special publication 180. U. S. Department of Commerce, Washington DC.

Waldman, J. 1999. Heartbeats in the muck: the history, sea life, and environment of New York Harbor. Lyons Press, New York.

Waldman, J. R. 2006. The diadromous fish fauna of the Hudson River: life histories, conservation concerns, and research avenues. Pages 171–188 *in* J. S. Levinton and J. R. Waldman, editors. The Hudson River Estuary. Cambridge University Press, New York.

Wirgin, I,. and J. R. Waldman. In press. Bioaccumulation and toxicities of aromatic hydrocarbon contaminants at different trophic levels of the Hudson River ecosystem. *in*: W. N. Rom, editor. Environmental and Occupational Medicine, 4th editor., Lippincott-Raven Publishers.

American Fisheries Society Symposium 51:63–74, 2006

Hydrologic Control of External Carbon Loads and Primary Production in the Tidal Freshwater Hudson

Nina F. Caraco and Jonathan J. Cole
Institute of Ecosystem Studies
Box AB, Millbrook, New York 12545 USA
caracon@ecostudies.org

Abstract.—Organic material can enter aquatic systems by de novo gross primary production (GPP) from inorganic carbon within the system (autochthonous sources) or by import from the surrounding watershed and upstream aquatic systems (allochthonous sources). We analyzed the spatio–temporal variation in these two organic carbon sources for the 150-km long tidal freshwater Hudson during 2 years (1999 and 2000). Gross primary production was calculated from dawn–dusk sampling at 2-km intervals over the length of the Estuary for dissolved oxygen, pH, and DIC. Allochthonous inputs (All) were also calculated for the same periods based on organic carbon measurements and hydrologic flows. The relative importance of these two organic carbon sources (GPP/All) varied by over 30-fold across space and time. This variation was related to hydrologic inputs; on average 1999 (the second driest summer in the past 15 years) had 10-fold greater GPP/All than did 2000 (the wettest summer in 15 years). Associated with the high GPP/All in 2000 there was higher carbon dioxide supersaturation and net heterotrophy as compared to 1999. We suggest that hydrologically driven variation in GPP/All may also impact the food web and the dominant organic carbon source for fish and invertebrate consumers in the river. Natural abundance ^{14}C measurements could potentially trace these two food sources as they vary in space and time.

Introduction

At the base of aquatic food webs are two sources of organic matter, that produced within the aquatic system (autochthonous inputs) and that imported from the surrounding terrestrial watershed and upstream aquatic systems (allochthonous inputs). The relative input of these organic sources structures the aquatic food web, determines the balance between autotrophic and heterotrophic processes, and impacts dissolved oxygen and carbon dioxide concentrations of aquatic systems (Vannote et al. 1980; Howarth et al. 1996; Caraco et al. 2000; Caraco and Cole 2004). For lotic systems, an influential conceptual model for viewing the relative importance of allochthonous and autochthonous inputs is the river continuum (Vannote et al. 1980). Small streams in close proximity to surrounding wooded watersheds are generally dominated by allochthonous inputs while in larger rivers autochthonous sources can gain greater importance

(McTammany et al. 2003). Within lotic systems, however, the relative importance of autochthonous sources may vary greatly not only between systems (Webster and Meyer 1997; Caraco and Cole 2004) but also within single systems across time and space (Peterson et al. 1986; McTammany et al. 2003). Across time hydrologic inputs vary substantially and this variation in hydrologic input can influence both allochthonous inputs from the surrounding watershed (Findlay et al. 1991a; Jassby et al. 1993) and primary production within the system (Cole et al. 1992; Howarth et al. 2000; Cole and Caraco 2006). Additionally for large rivers there can be substantial spatial variability in the importance of allochthonous versus autochthonous inputs at a given time (Findlay et al. 1991a; Caraco et al. 1997). This variation needs to be considered when assessing the importance of allochthonous versus autochthonous inputs at the scale of the whole system.

Here we use data from two consecutive growing seasons to examine the spatio–temporal variation in autochthonous and allochthonous carbon inputs to the tidal freshwater Hudson River. The selected years (1999 and 2000) have average growing season hydrologic inputs that differ by threefold but do not differ in other factors known to impact allochthonous inputs or primary production (GPP) in the river.

Study Site

The tidal freshwater Hudson (TFW Hudson) extends for nearly 150-km from just north of Albany, New York to Newburgh, New York. Water enters this stretch mainly from the Mohawk and Upper Hudson rivers with smaller tributary inputs accounting for about 25% of the input from the combined Mohawk and Upper Hudson inputs (Abood et al. 1992; Findlay et al. 1998). Considering this stretch of river as a whole, on an annual basis allochthonous carbon inputs are substantially larger than average annual autochthonous inputs (Howarth et al. 1996). Far less is known, however, about how the relative importance of these inputs as they vary spatially and temporally within the TFW Hudson.

Allochthonous inputs to the TFW Hudson from both dissolved organic carbon (DOC) and particulate organic carbon (POC) are input primarily from the nontidal upper Hudson and Mohawk Rivers (Findlay et al. 1998). Inputs from sewage, adjoining marshes, and smaller tributaries account for less than 25% of inputs from these combined tributaries. Over the past century the allochthonous loads may have been influenced by change in land-use, population growth, and sewage treatment (Howarth et al. 1991; Swaney et al. 1996). These changes are not occurring extremely rapidly, however, and between 1999 and 2000 there were likely no substantial changes in land use, sewage treatment, or population density in the Hudson watershed.

Autochthonous carbon inputs in the TFW Hudson come from the primary production of both phytoplankton and macrophytes (Cole and Caraco 2006). Phytoplankton production has been dramatically impacted by the zebra mussel invasion, which began in the early 1990s (Caraco et al. 1997). During this time zebra mussel grazing has varied by nearly threefold from 2.6 to 7.1 m d^{-1} and biomass of phytoplankton in the river varies with this grazing pressure (Caraco et al. 2006). For 1999 and 2000 the grazing pressure by the zebra mussels *Dreissena polymorpha* was almost identical at 2.8 and 3.1 m d^{-1}, respectively (Caraco et al. 2006).

Macrophyte production comes from both submerged vegetation (dominantly *Vallisneria americana*, Findlay et al. 1997, unpublished data; Caraco and Cole 2002) and floating leaved vegetation (*Trapa natans* Findlay et al. 1997; Caraco and Cole 2002). Over the past 100 years, the total macrophyte production may have declined due to dredging and loss of shallows in some regions and at the same time some areas occupied by submerged vegetation have been taken over by the invasive *Trapa natans* (Kiviat 1993). During the past decade there has been no evidence of large changes in abundance of either of these macrophytes, although it is likely that variation in average light extinction in the river could impact production of submerged vegetation (Caraco and Cole 2002).

Methods

Sampling and Analysis

During May–August 1999 and 2000, we sampled the entire TFW Hudson River during monthly transects (Figure 1, Figure 2). Transects were run at ca. 60-km h^{-1} with the entire transect taking 2.5 h from the start at Albany to the end at Newburgh. Samples were taken from self-filling on board 10 L containers that filled at 10 s intervals. Automatic recording instruments (YSI-Endeco UPG-6000 Sondes, Cole and Caraco in press) were used to measure dissolved oxygen (DO), temperature, pH, and turbidity at 1-min intervals (ca. every km). At 2–3 km intervals, samples were taken for chl-*a* and at 4–6 km intervals samples were additionally taken for DOC and DIC (Raymond et al. 1997; Findlay et al. 1998). All samples were from ca. 0.1 m depth (depth of self filling tube). On each sampling date we completed two transects starting at Albany and ending at Newburgh. Transects started at 0530 hours (dawn) and 1730 hours (dusk), allowing for calculation of 12-h (ca. dawn–dusk) changes in DO and pH based on over 100 locations in the river.

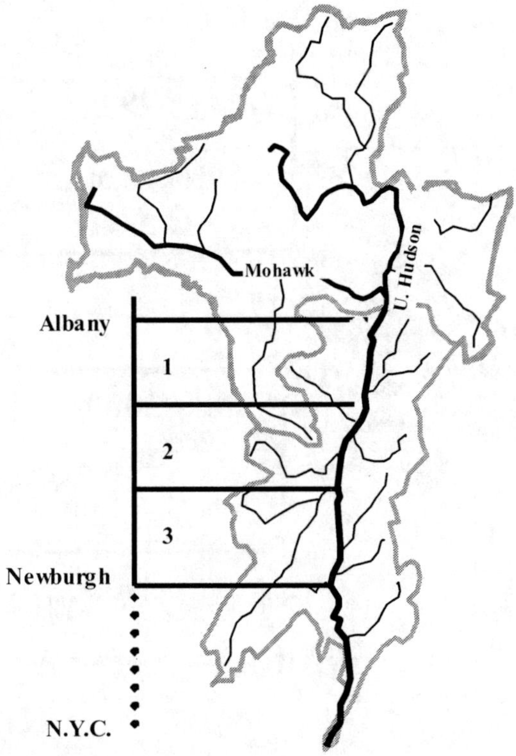

Figure 1. The Hudson River and its watershed. Between Albany, NY and Newbrugh, NY the river is both tidal and freshwater; tidal freshwaters cover a distance of nearly 150 km. The spatial variation of carbon inputs for this stretch of the river was budgeted in three approximately 50-km sections labeled 1-3.

Estimates of Allochthonous and Autochthonous Inputs

We divided the 150-km TFW Hudson into three 50-km reaches for constructing budgets of allochthonous and autochthonous inputs for the eight time periods (2 years ´ 4 months). Thus, 24 budgets exist but our comparison is largely between years and sections (i.e., six budgets that include seasonal averages). Each of the 24 monthly GPP estimates is, therefore, based on over 30 dawn–dusk pH and DO measurements and each seasonal estimate of GPP is based on over 120 measurements.

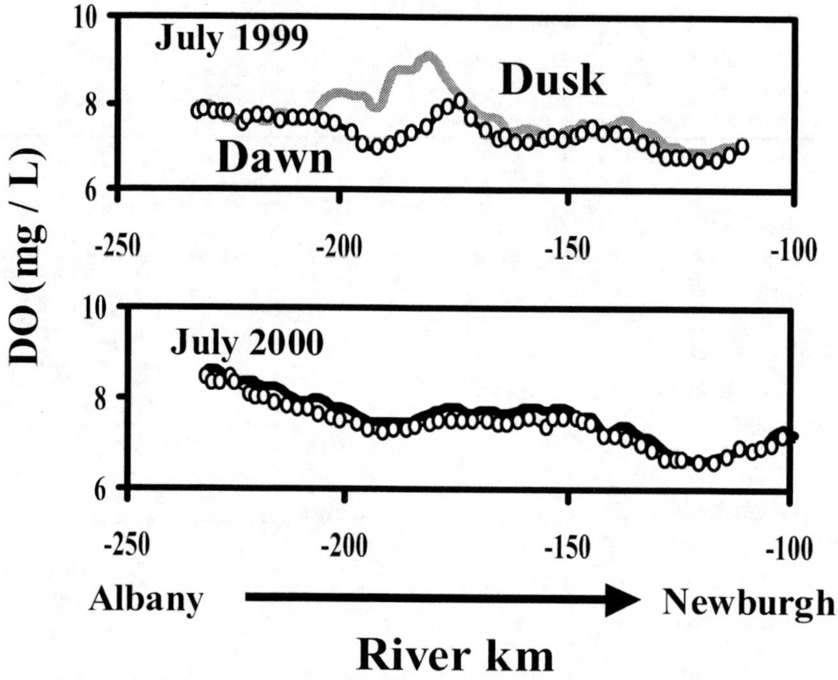

Figure 2. Example dawn-dusk longitudinal transects for dissolved oxygen (DO) in the tidal Hudson River for July in both 1999 and 2000. Note the far lower dawn-to-dusk change in 2000 as compared to 1999. This difference is captured by the extensive high-speed spatial sampling.

Hydrologic inputs to each study reach are monthly averages for Green Island corrected for downstream inputs along the TFW Hudson (Abood et al. 1992; Findlay et al. 1998). Residence time of each box is then the volume of each section divided by the hydrologic input and ignores any tidal exchange between the three sections of the river. Monthly allochthonous inputs of DOC to each study reach are calculated as the average DOC concentration at the north end of the study reach (based on a 5 station average) times the average monthly hydrologic flow. This assumes that DOC inputs and hydrologic inputs are proportional down river (see Findlay et al. 1998). Inputs of POC are estimated from turbidity based on 173 data points in post zebra mussel years. Such that:

$$\text{POC (mg L}^{-1}) = 0.1 \text{ x Turbidity (NTU)}^{0.7}$$

This relationship explained 56% of the variation in POC and is consistent with decreasing POC content of suspended sediments with increasing suspended sediment concentration (Meybeck 1982). For all three stretches of the TFW Hudson DOC and POC concentrations (in mg L^{-1}) were converted to daily areal loads (g $C{\cdot}m^{-2}{\cdot}d^{-1}$) by dividing by the surface area of each section of the river.

Total system GPP was estimated from the average of dawn–dusk changes in DO and CO_2. CO_2 was calculated from pH, tempera-

Table 1. Average values for a dry (1999) and wet (2000) year for the three budgetary sections of the tidal Hudson River (Figure 1). Averages are based on 4 monthly transects and multiple stations within each budgetary section. RT is residence time. Net Het. is net heterotrophy and is in g C·m^{-2}(of river)·d^{-1}.

	1999			2000		
	1	2	3	1	2	3
RT (days)	11.2	28.7	43.7	3.6	9.3	14.1
DOC (mg/L)	3.6	3.5	3.2	3.8	3.8	3.6
POC (mg/L)	0.5	0.7	0.7	1.5	1.7	3.6
Chl-*a* (μg/L)	3.1	15.1	8.2	1.6	2.6	4.1
DIC (μM)	1237	1217	1205	1305	1393	1339
CO_2	49.4	37.4	48.1	57.6	73.3	73.7
ΔCO_2	−12.5	−12.2	−7.5	−6.1	−0.8	−5.9
ΔDO(μM)	11.5	14.5	8.6	3.9	−0.5	1.0
Net Het.	0.47	0.36	0.46	0.55	0.7	0.71

ture, and DIC (Raymond et al. 1997; Cole and Caraco 2006). For calculations with oxygen we assumed a photosynthetic quotient of 1 by atoms. Volumetric changes in DO and CO_2 were calculated as the product of surface water changes and depth of the river section. To calculate GPP from daytime net system production we multiplied daytime areal changes by two—which assumes nighttime respiration is equal to daytime respiration and dawn-to-dusk DO and CO_2 are constant.

Results

The three ca. 50-km sections of the river (Figure 1) have average depths of 6.0, 6.2 and 11.3 m, and surface areas of 20, 57, and 55 km^2, for sections 1, 2 and 3, respectively. Average flows increased down river and varied across river sections between 142 and 185 m^{3}s^{-1} in 1999 and between 431 and 561 m^{3}s^{-1} in 2000. Despite the higher flows in the downriver sections, residence times were shorter in upriver sections due to lower volumes. The average residence time of the sections increased downriver and varied between 11 and 43 d in 1999 and 3.5 and 14 d in 2000 (Table 1).

Chlorophyll-a concentration (chl-*a*) averaged 5.6 μg/L across years and sections of the river but there was both strong spatial and year-to-year variation in this average (Table 1). For 1999 the chl-*a* varied between 3 and 15 μg/L across sections while in 2000 it varied between 1.5 and 4 μg/L. DIC averaged 1282 μM and had comparatively little variation in space or time (Table 1). Dissolved organic carbon concentrations averaged 3.6 mg/L and despite relatively small variation (3.2–3.8 mg/L) showed a relatively clear pattern of higher concentrations at the top of the river (section 1) and higher concentrations in 2000 as compared to 1999. Calculated POC concentrations averaged 1.1 mg/L and varied between 0.5 and 1.6 mg L^{-1}. Concentrations were higher in 2000 than 1999 but little spatial pattern was evident.

Allochthonous inputs of TOC varied between 0.6 and 12.5 g C·m^{-2}(of river area)·d^{-1} and both spatial and temporal variation were large. TOC inputs were on average highest at the upper most stretch of the Hudson and decreased down river in both years by over twofold (Figure 3). Average allochthonous loads were approximately fourfold lower in 1999 than in 2000 for all three river sections. Considering both spa-

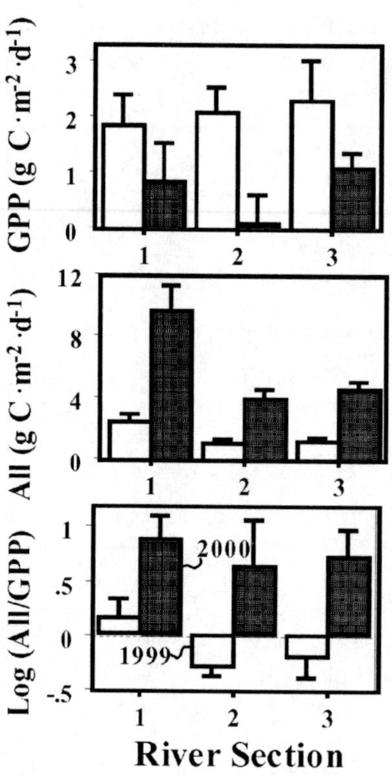

Figure 3. Carbon inputs from internal primary production (GPP, top panel) and external allochthonous carbon loads (All, mid-panel) for three sections of the tidal Hudson (Figure 1) and for two years that varied in hydrologic loads (1999 and 2000). Also shown is the log ratio of these two inputs (bottom panel).

tial and temporal samples, allochthonous inputs were strongly negatively related to residence time of the river section (Figure 4) as it varied with river section, year and month. The relationship to residence time is in large part driven by freshwater flow, which is used to calculate both loads and residence time, but is also in part driven by somewhat higher concentrations of DOC at higher flows and shorter residence times. The average dawn-to-dusk DO increase was 6.3 μM while the average calculated CO_2 drop was 7.5 μM. The variation in both diurnal DO and CO_2 changes differed strongly between years but there was little consistent difference across space. Overall there was a strong correlation between dawn-dusk DO and CO_2 changes that was not significantly different from a 1:1 (by moles) relationship (Figure 5). Thus, GPP calculated from either DO or CO_2 changes would be nearly identical (our calculation of GPP is the average between these two nearly identical estimates).

GPP averaged 1.4 g C·m^{-2}·d^{-1} but varied between 0.1 and 2.3 g C·m^{-2}·d^{-1} across years and sections of the river (Figure 3). There was strong year-to-year variation in GPP with 1999 having on average over threefold greater GPP than 2000. Spatial variation was, however, not consistent between years. In 1999 there was on average a slight increase in GPP from up to downriver sections. In 2000 the middle section of the river had the lowest average calculated GPP. Across years and times, GPP was positively correlated with chl-*a* in the water column ($p < 0.05$, linear regression) but chl-*a* explained only 31% of the variance in GPP.

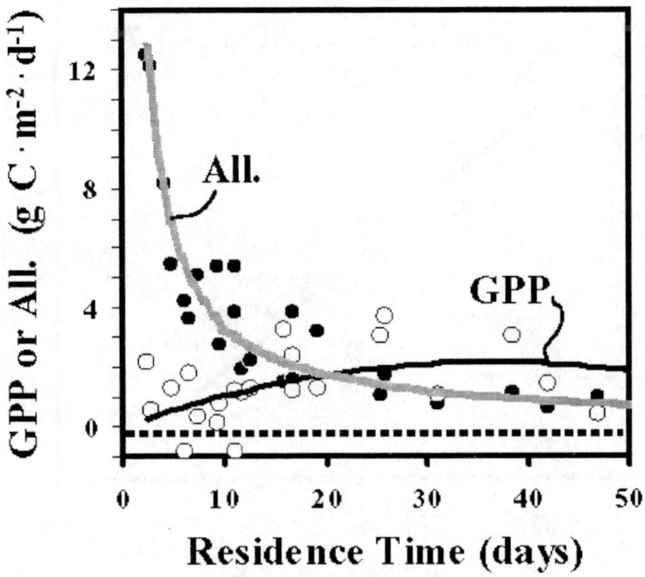

Figure 4. Gross primary production (GPP–open circles) and allochthonous carbon inputs (All–filled circles) vs. residence time of budgetary sections of the tidal Hudson River (Figure 1). For both GPP and All there are 24 data points representing the 3 river sections x 4 months x 2 years. The line fit to the All values is a negative exponential that explains 82% of the variance. GPP is fit with a second order polynomial that explains 33% of the variance.

Overall GPP was also positively correlated with residence time of river section as it varied with time. This relationship, however, was not as strong as the negative relationship between residence time and allochthonous inputs (Figure 4). Of all the GPP measurements, only 2 out of 24 had apparent negative values (Figure 4). These negative values occurred in May and August of 2000 in the middle section of the river where overall GPP was very low (Figure 3) and could be due to very small analytical errors in DO and pH measurements since GPP cannot actually be less than 0.

The relative importance of allochthonous inputs to GPP (All/GPP) varied greatly across space and time (Figure 3). For 1999 only the north section had All/GPP ratios greater than 1 suggesting that in most of the river GPP exceeds inputs of terrestrial material during the summer growing season. On the other hand for 2000 All/GPP was greater than 3 in all sections of the river suggesting strong dominance by allochthonous inputs. For 2000 as for 1999 the north stretch of the Hudson (section 1) had the greatest relative importance of allochthonous inputs. Considering the entire river there was also a switch from autochthonous to allochthonous dominance between years. The area-weighted All/GPP ratio was 0.5 in 1999 and 5 in 2000.

Discussion

For the Hudson and many other rivers the majority of microbial and consumer production occurs during the warm summer months (Maranger et al. 2005). During this time temperatures are warmer and microbes, invertebrates and fish increase metabolism substantially. Further, most resident and anadromous

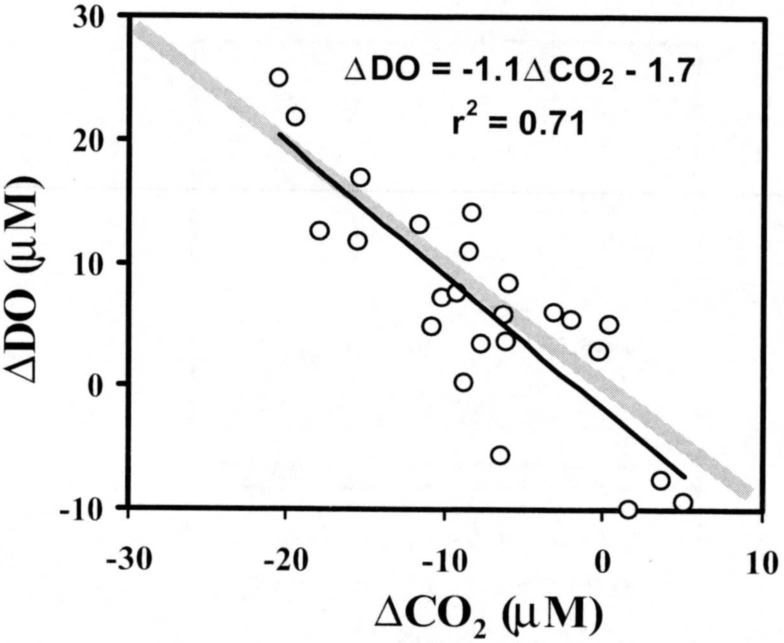

Figure 5. Relationship between dawn-dusk changes in CO_2 (DCO_2) vs changes in DO (DDO). Each point is the average for one of 3 budgetary sections of the Hudson (Figure 1) during a monthly transect run and is based on over 30 points. The black line is the regression line and the grey line is the 1:1 line.

species of fish reproduce during late spring to early summer and food supplied during May–August is, therefore, a critical driver of growth and survival of young fish (Limburg et al. 1999; Jordan et al. 2003). During this time period the relative inputs of allochthonous to autochthonous inputs can vary substantially between years and across locations.

The spatial variation in the relative importance of allochthonous versus autochthonous inputs is driven by higher allochthonous loads in the north stretch of the river where DOC concentration is somewhat higher and water exchanges more rapidly. The year-to-year variation is driven by variation in both allochthonous loads and GPP which are in turn influenced by between year differences in hydrology. Hydrologic loads in 1999 and 2000 differed by over threefold with 1999 being a dryer than average year and 2000 being wetter than average (Limburg et al. 1986; Abood et al. 1992). In fact, 2000 was the wettest summer (May–August) and 1999 was the second driest summer over a 15-year record (1987–2002). The low hydrologic input in 1999 not only resulted in comparatively low loadings of allochthonous inputs but also was associated with substantially greater phytoplankton biomass and primary production. The greater phytoplankton production in 1999 than 2000 was likely a result of higher than average water clarity in dry as compared to wet years; other factors known to influence primary production in the Hudson remained similar between years. Zebra mussel grazing, which can strongly influence phytoplankton production was almost identical in 1999 and 2000, nutrient concentrations were actually somewhat greater in 2000 as compared to 1999 (Caraco et al. 2006), and there was not a substantial loss in shallow habitat for aquatic vegetation.

The dominance of autochthonous inputs in the Hudson during some times may seem at odds with past data emphasizing the dominance of allochthonous inputs to this system (Findlay et al. 1991b; Howarth et al. 1996) particularly since the invasion of the zebra mussel when phytoplankton GPP declined (Caraco et al. 1997; Cole and Caraco 2006). It should be emphasized that the system-level GPP given here includes not only phytoplankton but also periphyton and at least submerged macrophytes (Gladden et al. 1988). Additionally, the net primary production (NPP) given in some calculations may be substantially lower than the GPP as respiration by primary producers themselves can consume a substantial proportion of GPP. In the Hudson for post zebra mussel years, estimates of the ratio of phytoplankton GPP:NPP vary between 7 and 2 across different studies (Caraco et al. 1997; Cole and Caraco 2006). Finally, most prior comparisons of allochthonous inputs to GPP have been based primarily on annual estimates rather than for the growing season (as we have done here) when most metabolism and secondary production occur most rapidly in the river.

Because the importance of allochthonous and autochthonous loads vary substantially between years, it suggests that net heterotrophy and food sources to consumers in the river could also vary substantially between years. Net heterotrophy can be estimated from CO_2 oversaturation and gas exchange (Raymond et al. 1997). For the Hudson this net heterotrophy varied between 0.36 and 0.47 g $C{\cdot}m^{-2}{\cdot}d^{-1}$ in 1999 and 0.55 and 0.71 g $C{\cdot}m^{-2}{\cdot}d^{-1}$ (Table 1). Thus, net heterotrophy does indeed appear to be impacted by the higher allochthonous loads and/or lower autochthonous inputs in 2000 as compared to 1999. However, the less than two-fold difference in net heterotrophy is substantially smaller than the approximately fourfold difference in allochthonous loads and 10-fold difference in All/GPP between years. The comparatively small impact on system heterotrophy is likely explained to a large degree by a lower metabolism of allochthonous inputs under high flow–low residence time conditions (Caraco and Cole 2004). However, lower temperatures in 2000 as compared to 1999 (20°C and 23°C averages, respectively) and potentially poorer quality of allochthonous loads could play a role in the proportionally lower metabolism of allochthonous inputs in 2000 as compared to 1999.

In addition to altering the degree of net heterotrophy in the system, the variation in relative importance of allochthonous to autochthonous loads can cause a change in food web structure and dominance of food sources supporting higher food levels. That is, in wet years it might be expected that allochthonous inputs that are consumed directly or processed through the microbial food web could be more important. In some systems the importance of allochthonous versus autochthonous food sources can be directly assessed by stable isotopes (e.g., ^{13}C and ^{15}N, Hamilton et al. 1982; Hesslein et al. 1991); in the Hudson these isotopes do not distinguish well allochthonous versus autochthonous food sources (Caraco et al. 1998). Phytoplankton, floating leaved vegetation, terrestrial vegetation, and allochthonous carbon inputs all have $\delta^{13}C$ values that vary between –25 and –29‰ (Caraco et al. 1998). Only submersed vegetation has $\delta^{13}C$ values that are distinguishable from any of these food sources (Figure 6). Further, $\delta^{15}N$ values in the Hudson do not seem to add unambiguous information on the organic source supporting the aquatic food chain (Caraco et al. 1998).

A possible other source of information is the use of natural level $\Delta^{14}C$ measurements (Schell 1999; Cole and Caraco 2001). Preliminary ^{14}C measurements of DOC and

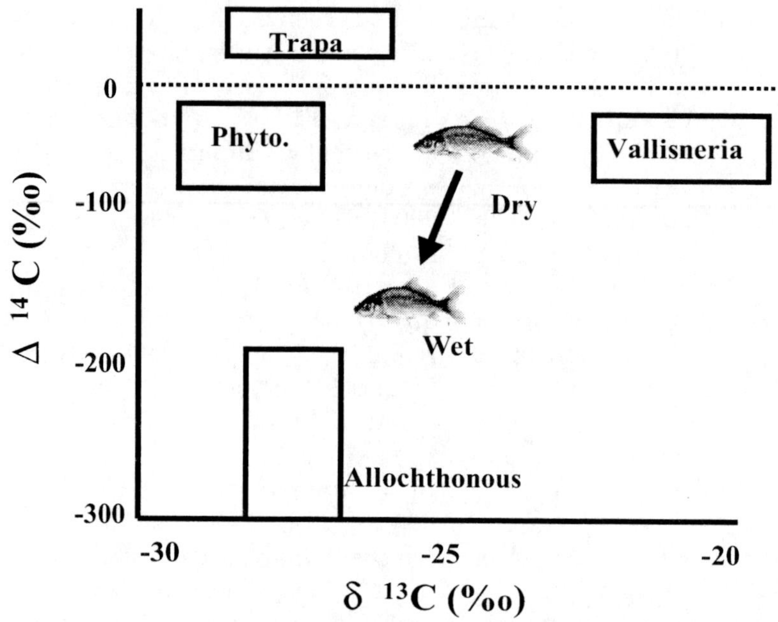

Figure 6. $\delta^{13}C$ and $\Delta^{14}C$ content of autotrophic primary production by phytoplankton and the two dominant macrophytes and allochthonous inputs. $\delta^{13}C$ values are from Caraco et al. (1998). $\Delta^{14}C$ values of allochthonous material are based on measurements of $\Delta^{14}C$-DOC and $\Delta^{14}C$ POC at river km 245 (Cole and Caraco 2001) while phytoplankton and macrophytes are based on $\Delta^{14}C$ values of inorganic carbon sources for production. For phytoplankton and submersed *Vallisneria* this is the $\Delta^{14}C$ of DIC in the river (Cole and Caraco 2001), for floating leaved *Trapa* it is the $\Delta^{14}C$ of the atmosphere which due to bomb inputs has a positive $\Delta^{14}C$ signal. The fish in this depiction is hypothesized to change its isotopic signal as food sources vary with hydrology.

POC suggest that organic matter entering the northernmost end of the TFW Hudson are substantially depleted in ^{14}C as compared to present day ^{14}C values or even as compared to prebomb modern ^{14}C values ($\Delta^{14}C = 0‰$). $\Delta^{14}C$ POC values have been measured as low as –500 ppt and $\Delta^{14}C$-TOC values can be between –200 and –300‰ (Cole and Caraco 2001; Raymond et al. 2004). As material moves downstream $\Delta^{14}C$ values of organic matter increase due to a combination of decomposition of ^{14}C-depleted organic material and addition of material with less depleted ^{14}C values (e.g., GPP, Cole and Caraco 2001). A likely explanation for the ^{14}C-depleted values at the northern end of the tidal Hudson is that they represent old organic material with ages up to 5000 years old (Cole and Caraco 2001). The old age of the organic matter is interesting in its implications for food quality of allochthonous material entering the Hudson (Cole and Caraco 2001). The difference in age and ^{14}C content of allochthonous versus autochthonous carbon sources also provide a potentially useful tool for tracing the importance of these two food sources to the Hudson. It seems likely that the relative importance will vary not only with feeding strategy of consumers but could also vary strongly with location and between years in the Hudson.

The TFW Hudson is an important location for the production of anadromous fish including herring, shad, and striped bass (Limburg et al. 1986). These fish rely on organic carbon in this section of the Hudson from May to August when they mature from eggs to young-of-year that can potentially leave the river to enter coastal waters. Despite much research on carbon metabolism and fish biology in the TFW Hudson it is still not known what the carbon sources are that ultimately fuel the growth of fish in the river. It is known, however, that the proximate food sources of secondary consumers vary greatly in space and time in the river (Jordan et al. 2003) and this could be driven by changes at the base of the food chain. ^{14}C as well as other tracers could help answer this question and help to better understand the spatio-temporal variation in fish diet, growth and survival (Jordan et al. 2003; Strayer et al. 2004). Our results suggest that the importance of allochthonous and autochthonous carbon could have strong spatio–temporal variation that must be considered when tracking the carbon sources that fuel fish production in the river.

In the future global warming will likely decrease summertime runoff and increase residence time of the river (Howarth et al. 2000). This change in itself would be expected to decrease the importance of allochthonous sources relative to autochthonous sources of carbon in the river. However, changes in grazing pressure, land-use influenced changes in water-turbidity, and organic carbon could all influence this potential hydrology driven change. If the relative importance of allochthony to GPP does change in the future it could result in significant food web changes. Examining the food web structure as it presently varies with hydrology may help to understand these potential climate driven changes.

Acknowledgments

We thank G. Lampman for his sampling and analytical assistance. Funding was provided by the National Science Foundation, New York Sea Grant, and the Hudson River Foundation. The work is a contribution to the Institute of Ecosystem Studies.

References

Abood, K. A., G. A. Apicella, and A. W. Wells. 1992. General evaluation of Hudson River freshwater flow trends. Pages 3–28 *in* C. L. Smith, editor. Estuarine research in the 1980s. State University of New York Press, Albany.

Caraco, N. F., and J. J. Cole. 2002. Contrasting impacts of a native and alien macrophyte on dissolved oxygen in a large river. Ecological Applications 12:1496–1509.

Caraco, N. F., and J. J. Cole. 2004. When terrestrial carbon is sent down the river: importance of allochthonous carbon inputs to the metabolism of lakes and rivers. Pages 301–316 *in* G. A. Polis and M. E. Power, editors. Food webs at the landscape level. University of Chicago Press, Chicago.

Caraco, N. F., J. J. Cole, S. E. G. Findlay, D. T. Fischer, G. G. Lampman, M. L. Pace , and D. L. Strayer. 2000. Dissolved oxygen declines in the Hudson River associated with the invasion of the zebra mussel (*Dreissena polymorpha*). Environmental Science and Technology 34:1204–1210.

Caraco, N. F., J. J. Cole, P. A. Raymond, D. L. Strayer, M. L. Pace, S. E. G. Findlay, and D. T. Fischer. 1997. Zebra mussel invasion in a large, turbid river: phytoplankton response to increased grazing. Ecology 78:588–602.

Caraco, N. F., G. Lampman, J. J. Cole, K. E. Limburg, M. L. Pace, and D. Fischer. 1998. Microbial assimilation of DIN in a nitrogen rich estuary: implications for food quality and isotope studies. Marine Ecology Progress Series 18:59–71.

Caraco, N. F., J. J. Cole, and D. L. Strayer. 2006. Top down control from the bottom: regulation of eutrophication in a large river by benthic grazing. Limnology and Oceanography 51:664–670.

Cole, J. J., N. F. Caraco, and B. L. Peierls. 1992. Can phytoplankton maintain a positive carbon balance in a turbid, freshwater, tidal estuary? Limnology and Oceanography 37:1608–1617.

Cole, J. J., and N. F. Caraco. 2001. Carbon in catchments: connecting terrestrial carbon losses with aquatic metabolism. Marine Freshwater Research 52:101–110.

Cole, J. J., and N. F. Caraco. 2006. Primary production and its regulation in the tidal-freshwater Hudson River. Pages 107–120 *in* J. S. Levinton and J. R. Waldman, editors. The Hudson River Estuary.

Cambridge Press, New York.

Findlay, S., M. Pace, and D. Lints. 1991a. Variability and transport of suspended sediment, particulate and dissolved organic carbon in the tidal freshwater Hudson River. Biogeochemistry 12:149–169.

Findlay, S., M. L. Pace, D. Lints, J. J. Cole, and N. F. Caraco. 1991b. Weak coupling of bacterial and algal production in a heterotrophic ecosystem: the Hudson River estuary. Limnology and Oceanography 36:268–278.

Findlay, S., R. L. Sinsabaugh, D. T. Fischer, and P. Franchini. 1998. Sources of dissolved organic carbon supporting planktonic bacterial production in the tidal freshwater Hudson River. Ecosystems 1:227–239.

Gladden, J. B., F. R. Cantelmo, J. M. Croom, and R. Shapot. 1988. Evaluation of the Hudson River Ecosystem in relation to the dynamics of fish populations. Pages 37–52 *in* L. W. Barnthouse, R. J. Klauda, D. S. Vaughan, and R. L. Kendall, editors. Science, law, and Hudson River Power plants: a case study in environmental impact assessment. American Fisheries Society, Monograph 4, Bethesda, Maryland.

Hamilton, S. K., W. M. Lewis, and S. J. Sippel. 1982. Energy sources for aquatic animals in the Orinoco River floodplain: evidence from stable isotopes. Oecologia 89:324–330.

Hesslein, R. H., M. J. Capel, D. E. Fox, and K. A. Hallard. 1991. Stable isotopes of sulfur, carbon, and nitrogen as indicators of trophic level and fish migration in the lower Mackenzie River basin, Canada. Canadian Journal of Fisheries and Aquatic Sciences 48:2258–2265.

Howarth, R. W., J. R. Fruci, and D. Sherman. 1991. Inputs of sediment and carbon to an estuarine ecosystem: influence of land use. Ecological Applications 1:27–39.

Howarth, R. W., R. Schneider, and D. Swaney. 1996. Metabolism and organic carbon fluxes in the tidal freshwater Hudson River. Estuaries 19:848–865.

Howarth, R. W., D. P. Swaney, T. J. Butler, and R. Marino. 2000. Climatic control on eutrophication of the Hudson River Estuary. Ecosystems 3:210–215.

Jassby, A. D., J. E. Cloern, and T. M. Powell. 1993. Organic carbon sources and sinks in San Francisco Bay-variability induced by river flow. Marine Ecology Progress Series 95:39–54.

Jordan, R. C., D. V. Howe, T. P. Hurst, and F. Juanes. 2003. Feeding habits of age-0 striped bass, *Morone saxatilis*, in the mid-Hudson River estuary: temporal, spatial and otogenetic variation. Estuaries 26:1486–1493.

Kiviat, E. 1993. Under the spreading water-chestnut. Hudsonia 9:1–7.

Limburg, K. E., M. A. Moran, and W. H. McDowell. 1986. The Hudson River ecosystem. Springer Verlag, New York.

Limburg, K. E., M. L. Pace, and K. K. Arend. 1999. Growth, mortality, and recruitment of larval *Morone* spp. in relations to food availability and temperature in the Hudson River. Fisheries Bulletin 97:80–91.

Maranger, R. J., M. L. Pace, P. A. del Giorgio, N. F. Caraco, and J. J. Cole. 2005. A spatial analysis of bacterial production, respiration, and carbon consumption in a river-estuarine ecosystem. Ecosystems 8:318–330.

McTammany, M. E., J. R. Webster, E. F. Benfield, and M. A. Neatrour. 2003. Longitudinal patterns of metabolism in a southern Appalachian river. Journal of the North American Benthological Society 22:359–370.

Meybeck, M. 1982. Carbon, nitrogen, and phosphorus transport by world rivers. American Journal of Science 282:401–450.

Peterson, B. J., J. E. Hobbie, and T. L. Corliss. 1986. Carbon flow in a tundra stream ecosystem. Canadian Journal of Fisheries and Aquatic Sciences 43:1259–1270.

Raymond, P. A., J. E. Bauer, N. F. Caraco, J. J. Cole, B. Longworth, and S. T. Petsch. 2004. Controls on the variability of organic matter and dissolved inorganic carbon age in Northeast U.S. Rivers. Marine Chemistry 92:353–366.

Raymond, P. A., N. F. Caraco, and J. J. Cole. 1997. CO_2 concentration and atmospheric flux in the Hudson River. Estuaries 20:381–390.

Schell, D. M. 1999. Carbon-13 and carbon-14 abundances in Alaskan aquatic organisms: delayed production from peat in Arctic food webs. Science 219:1068–1071.

Strayer, D. L., K. Hattala, and A. Kahnle. 2004. Effects of an invasive bivalve (*Dreissena polymorpha*) on fish populations in the Hudson River estuary. Canadian Journal of Fisheries and Aquatic Sciences 61:924–941.

Swaney, D. P., D. Sherman, and R. W. Howarth. 1996. Modeling water, sediment and organic carbon discharges in the Hudson-Mohawk Basin: coupling to terrestrial sources. Estuaries 19:633–647.

Vannote, R. L., G. W. Minshall, K. W. Cummins, J. R. Sedell, and C. E. Cushing. 1980. The river continuum concept. Canadian Journal of Fisheries and Aquatic Sciences 37:130–137.

Webster, J. R., and J. L. Meyer. 1997. Organic matter budgets for streams: a synthesis. Journal of the North American Benthological Society 16:141–161.

American Fisheries Society Symposium 51:75–112, 2006

Some Historical Changes in the Patterns of Population and Land Use in the Hudson River Watershed

Dennis P. Swaney
Cornell University, Department of Ecology and Evolutionary Biology
Ithaca, New York 14853 USA
dps1@cornell.edu

Karin E. Limburg and Karen Stainbrook
SUNY College of Environmental Science and Forestry
1 Forestry Way, Syracuse, New York 13210 USA
klimburg@esf.edu
k_m_stainbrook@yahoo.com

Abstract.—Using a combination of data sources and historic or contemporary accounts, we describe and document changes in the Hudson River watershed's population size, agricultural and forested land uses, and the construction of dams, largely since the time of European colonization. Population within the watershed has grown from 230,000 at the time of the first census in 1790 to around 5 million today (not including parts of those boroughs of New York City outside the watershed, such as Queens). The watershed was almost entirely forested in 1609, with minor amounts of Indian agriculture. By 1880, approximately 68% of the watershed was farmland, but as soil productivity declined and industry created other jobs, much cleared land gradually reverted to secondary forest. Most land not in agriculture was forested and exploited first for lumber and tanbark and, later on, pulpwood for paper. The tanning industry existed from the 1700s, but reached its height in the mid-1800s, collapsing from a combination of resource (hemlock) exhaustion and market forces. Finally, available records list nearly 800 dams, ranging from 0.6 m to 213 m (Ashokan Reservoir) in height and with maximum storage of 1.07 km^3 (Sacandaga Reservoir), that were constructed from the early 18th century until 1993. The environmental legacies of these changes include effects on hydrology, soils, vegetation, biogeochemical cycling, sediment loading, and ecological relationships

Introduction

Anyone who has done environmental research in northeastern U.S. watersheds or for that matter, has ever walked through the countryside, eventually ponders the local history of the place and the effect of earlier inhabitants on the local environment. It is commonplace to discover the ruins of homesteads and stone fences while walking in local forests and to observe the remnants of millworks and their impoundments in the waters of local streams. Walking along the brackish river's edge, one finds oyster shells from beds dating back before European settlement in reaches where they are currently absent.

What are the connections between the landscape, its flows of water, sediment and nutrients, and the creatures that live in its forests and waters? Such questions prodded us to learn a bit more about changes in the land use and land cover of the Hudson-Mohawk watershed, and to begin to assemble some data sets to characterize these changes more fully. We present some of this work below, with the caveat that we are not historians and so our efforts are subject to criticism by those more expert in reconstructing the historical record.

Our primary aim was to document changes in regional land use noted in census data at the county level and to express these changes in terms of the corresponding change in the watershed and subwatershed (hydrologic unit) land uses. Thus, any discussion of changes in land cover associated with aboriginal populations in the pre-European settlement period is largely beyond the scope of this paper, and we refer the interested reader to some previous excellent work done in the northeastern United States (Cronon 1983; Foster 1999). Some work on the historical reconstruction of pollutant loads to the Hudson-Raritan has already been done (cf. Ayres and Rod 1986; Rod 1986; Rod et al. 1989). We focus primarily on alterations caused by agriculture, forest use, and dam construction, leaving such topics as industrialization and urbanization for others.

Methods

County level census statistics were obtained from the Bureau of the Census for both population (U.S. Census Bureau, various years) and agricultural land use (U.S. Department of Commerce, various years) in New York and adjoining states with areas in the Hudson-Mohawk watershed (i.e., New Jersey, Connecticut, Massachusetts and Vermont). U.S. Geological Survey (USGS) 8-digit Hydrologic Catalogue Units (HUC) coverages of subwatersheds of the Hudson-Mohawk drainage were overlaid on county boundaries obtained from the Cornell University Geospatial Information Repository (CUGIR) using ARCView 3.2 (ESRI 1999) to obtain the proportion of areas of each county falling within each HUC and for the entire Hudson-Mohawk watershed (Table A.1). These proportions were then used as multipliers on county population and land use areas to determine area-weighted estimates for each HUC. A potential difficulty with this approach is historical change in county boundaries. While the county boundaries of the region have been largely stable for the last 100 years or so, the period from the first national census (1790) to 1910 saw numerous changes in county boundaries, including the formation of new counties at the expense of old ones. We used the Historical U.S. County Boundary (HUSCO) data set to evaluate areas of counties for the years of the decennial census in this period (Earle et al. 1999). The period between the earliest settlements (1600s) to the first national census saw even more radical changes in political boundaries, but change in this period is largely beyond the scope of quantitative analysis of this work. For census years in which county boundaries are very different from current boundaries, area-weightings had to be calculated separately, as shown in Table A.2 of the Appendix.

To assess changes in population and population density, we used county level data from the decennial U.S. census records and the New York State Data Center (New York State Data Center 2000; U.S. Census Bureau, various years), from which total population counts of each county were obtained. Similarly, the Censuses of Agriculture were used to determine changes in acreage of land in farms over time. Figures on county and subwatershed population and land in farms are given in Tables A.3–A.6 of the Appendix.

Forested land area by county is more difficult to obtain and was determined from several different sources, depending upon the period of interest. These included Hough (1878) for the year 1875, Recknagel (1923) for the year 1920, Stout (1958) for the year 1955, USDA Forest Resource Bulletins NE-20, NE-34, NE-44, NE-46, and NE-48 for the period 1968–1972, and Forest Resource Inventory Reports NE-112, NE-132, NE-145, NE-147, and NE-148 for the late 1980s through mid-1990s. These latter inventories are updated approximately every 10 years, rotating through the states within a jurisdictional region of the U.S. Forest Service. Forest land areas in counties and subwatersheds are shown in Tables A.7-A.8 of the Appendix.

The U.S. Army Corps of Engineers maintains the National Inventory of Dams (NID) for all dams greater than 2 m in height with a storage greater than 61,700 m^3, or greater than 8 m in height with a storage greater than 18,500 m^3, or of any size likely to pose a significant threat downstream (Graf 1999). The U.S. Environmental Protection Agency incorporates this information into its Better Assessment Science Integrating Point and Nonpoint Sources (BASINS) software for watershed analysis (USEPA 2001). For these analyses, we extracted the dam data for the Hudson-Mohawk basin. Dam data organized by subwatershed and size are shown in Table A.9. The NID includes the dates of dam construction.

It is worth noting that the watershed is not a natural unit for the study of populations, and in the case of the Hudson, the estimate of "watershed population" is heavily weighted by the counties nearest New York City. As a result, for these counties, even small errors resulting from the assumption of uniform spatial distribution of population, or from the proportion of county areas falling within the watershed boundaries, may have significant effects on the total population estimates. However, our aim is not to further the analysis of New York demographics, but rather to assess the spatiotemporal variation of major environmental drivers, and as of this writing, we believe the use of county level data for this purpose is the state of the art.

Results and Discussion

Pre-European Settlement and Colonial Periods

"There was not an unbroken forest here when the first settlers came; as the fires of the Indians, in their pursuit of game, had destroyed the timber on the dry lands, except a few specimens of oak, white wood and wild cherry, some of which attained great size. On the plains were scattered small oaks which had sprung up after the fires, and by the creeks and in wet lands there were large buttonwood and black ash trees, while all the streams were overhung with a mass of alders and willows. The mountains, it has been said, were covered with a less dense growth of wood than at present. It is evident that in the valleys, the white wood or tulip tree, and the wild cherry have given place to other trees,

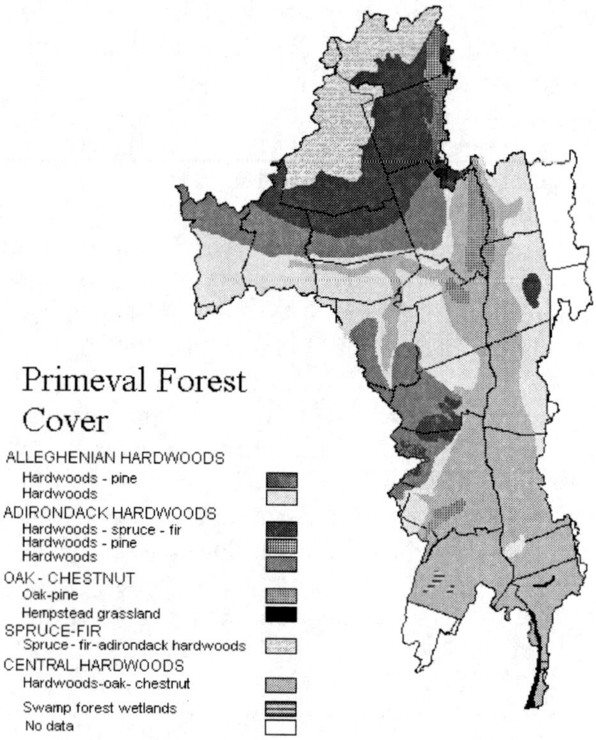

Figure 1. Distribution of primeval forest cover in the region (redrawn from Hamilton et al. 1980). Alleghenian hardwoods include primarily beech, sugar maple, hemlock, white pine, and basswood; Adirondack hardwoods are primarily beech, sugar maple, yellow birch, hemlock, and white pine.

as the elm; and that on the mountains, the chestnut has greatly increased. The mountains, being burned over also by the Indians, were so bare, that the wild deer were plainly seen from the valleys below."
- Early History of Amenia. Newton Reed, 1875.

Aboriginal peoples have been present in the area for at least the last 7,000 years, at first as hunters and gatherers, and later creating semi-permanent settlements and practicing agriculture (Hamilton et al. 1980; Cronon 1983). As the above quote suggests, lands of the watershed were largely forested (Figure 1) with the exception of areas cleared by the inhabitants, often by burning, in the process of hunting and planting crops (Day 1953).

The earliest European settlement in the Hudson region was New Netherland, established by the Dutch West India Company, a follow-up to the Dutch East India Company that had sponsored Henry Hudson's 1609 searches for a northwest passage to Asia (Shorto 2004). New Amsterdam, its main population center, was established first as a fort at the southern tip of the island of the Manhattes (Lenape) Indians in 1623; Fort Orange (Albany) was established that same year. Peter Minuit, Director of the New Netherland colony from 1626–1633, purchased Manhattan Island to gain greater control over it. Early settlement patterns followed the main waterways (Figure 2), as these were the main means of transporting goods through the region. Settlement up the Hudson Valley was slow and interspersed

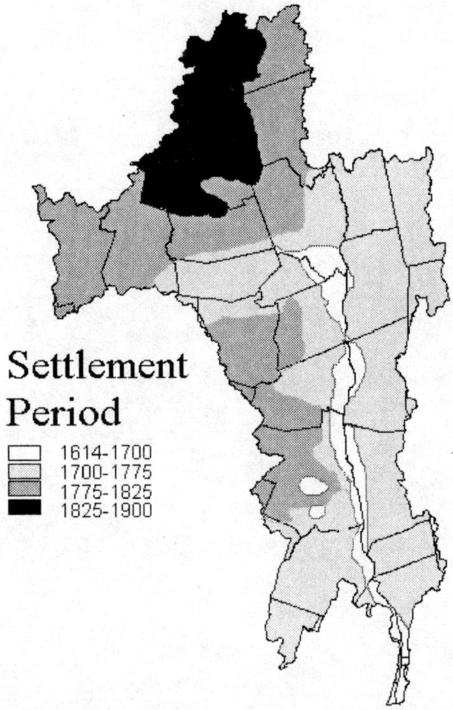

Figure 2. Patterns of settlement in the watershed (redrawn from Fox, 1900).

with conflicts with both English and Indians. Eventually, the Dutch yielded their possessions to the English in 1664, and New Amsterdam was renamed for the Duke of York.

Just as with the English settlers in New England, the Dutch were impressed by the abundance of wildlife, the plenitude of forests, and the beauty of the landscape. Montanus (1671) described in some (often fabulous, Figure 3) detail wolves, lions (panthers), black bear, deer, elk, "muskcat" (probably muskrat), beaver, rattlesnakes, and many kinds of waterfowl and birds of prey. Among the birds of the forest, he noted that "pigeons fly in such flocks that the Indians designedly remove to their breeding places, where the young birds, pushed by hundreds from their nests, serve for food during a long month for the whole family." Turkey was in great abundance, and the streams were filled with many fish species. Boyle (1969) lists many similar accounts.

From the start, tree clearing was necessary to "improve the land" and establish agriculture. To this end, the West India Company constructed three sawmills in its new territory (Anonymous 1647). With machinery imported from Holland, these were constructed to operate with either wind or water power (Fox 1900). One was located on Governor's Island and another on Sawmill Creek, draining into the East River (Fox 1900). Other sawmills were constructed for purposes of clearing in the 1600s in Albany, Westchester (on the Sawmill River), Kinderhook, Troy, Castleton, and Catskill (Fox 1900). Trees were girdled and cut with axes, and wood was used for buildings, fence rails, and posts. A great deal of wood in the early settlements was simply burned to clear the land.

Figure 3. Wild beasts of New Netherland, depicted by Montanus (1671). Note the mixture of real and fabulous Hudson Valley species.

Population

Population growth was slow during the period of Dutch settlement (1600s) and English colonization up through most of the 18th century, as large land grants and patents had been given to a few families such as the van Cortlandts, the Livingstons, the Beekmans, and the van Rensselaers, who maintained near-feudal conditions in the early part of that century (O'Callaghan 1850; Ellis et al. 1973). After the American Revolution, watershed population grew steadily. Population growth was exponential from 1790, when the first national census data available indicate about 230,000 people in the watershed, until around 1920, when the population reached about 3.4 million (Figure 4a). The engine of population change for the watershed has always been New York City. While the total watershed population is dominated by that of the metropolitan area (here, seen as the lower Hudson HUC, Figure 4b), population farther north in the watershed has always followed the transportation corridors.

Construction began on the Erie Canal in 1817; it was opened in 1825, and development of feeder canals and enlargements continued through the 19th century, facilitating trade and settlement in rural areas. By the 1840s, railroads had begun to change population distributions in the counties along the main trunklines paralleling the Hudson and expanded the effective metropolitan area of New York City. Most of the New York State population today falls in

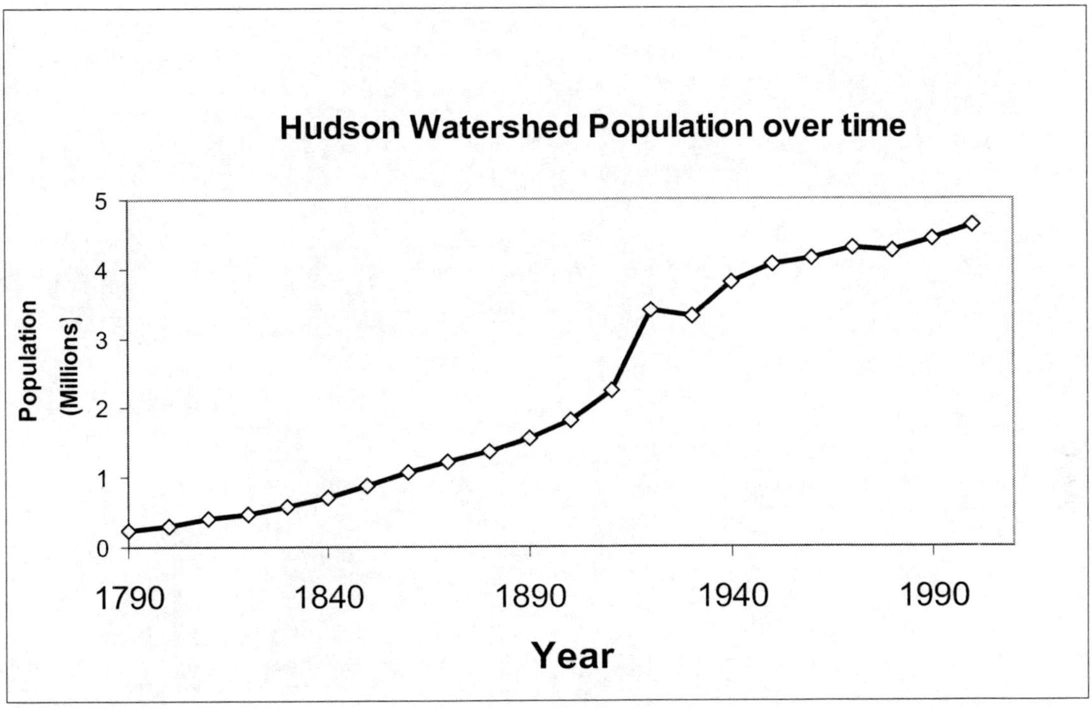

Figure 4a. Population change in the Hudson watershed, 1790-2000.

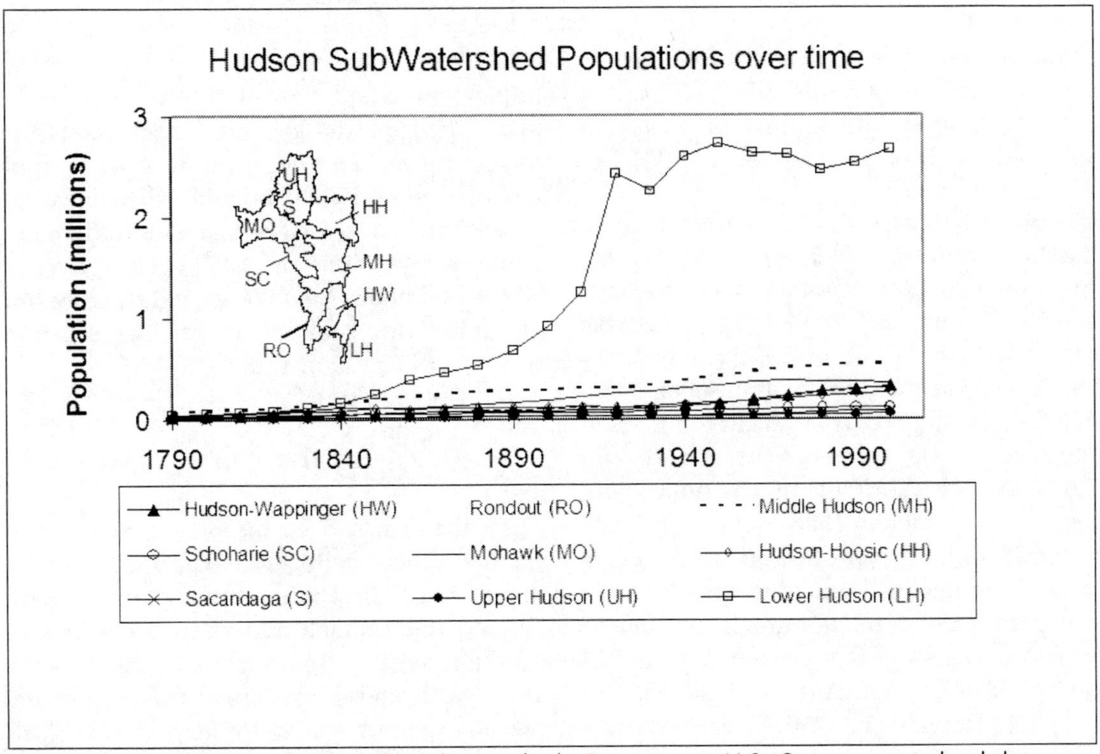

Figure 4b. Breakdown of change by subwatershed. Data source: U.S. Census, county level data.

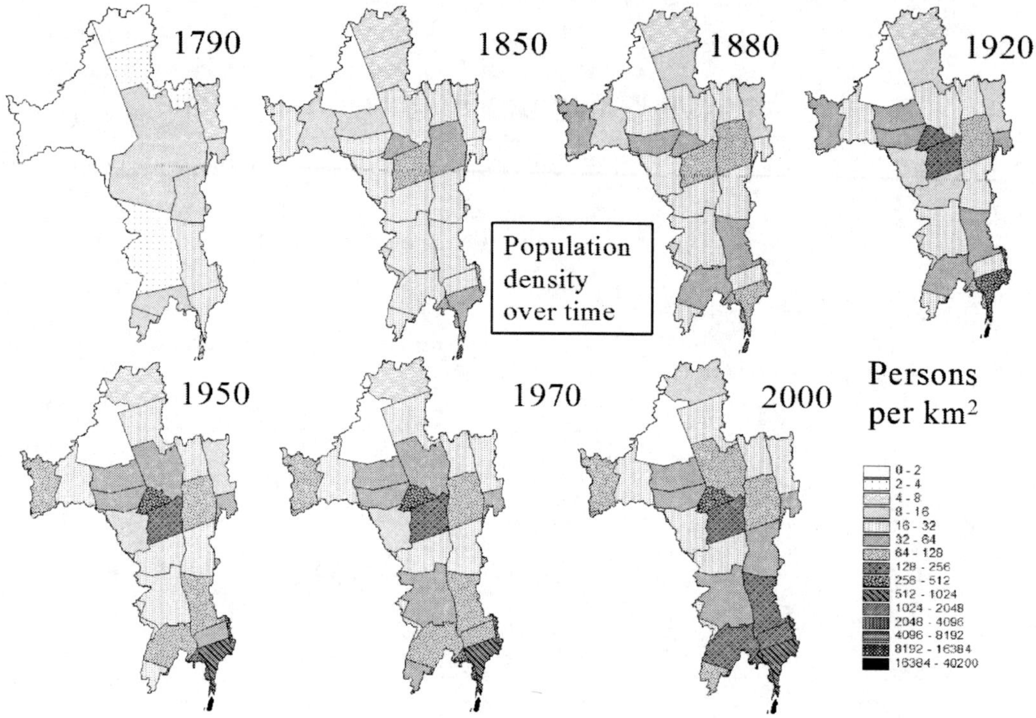

Figure 5. Changes in spatial distribution of population density over time at the county level.

a band along the Hudson/Mohawk rivers, and extending westward along the transportation routes, which lie along the old canal beds to Buffalo (Ellis et al. 1967).

Between 1920 and 1930, a period which saw the beginning of the Great Depression, the watershed population appears to have declined (Figure 4a, b). While it is worth noting that in the fall of 1918 the country experienced the great Spanish influenza pandemic, in which 675,000 individuals lost their lives in the United States, and over 50,000 in New York State (Brainerd and Siegler 2002; Eichel 1923), the pandemic was not a major driver of population trends. Rather, the decline coincides with barriers to the influx of foreign immigrants raised by the Immigration Restriction Act of 1921 and the Immigration Act of 1924 (Ellis et al. 1967). Population shifts were also due to changes in the economy: as regional agriculture declined, rural population in the upper watershed moved to cities and suburbs as people looked for better economic opportunities. In the lower Hudson, population has remained steady with decadal-scale variation over the last hundred years, while from around the 1930 census, overall watershed population has grown steadily but at a lower rate than in the previous century, to around 5 million today.

Agriculture

Agriculture has been important in the watershed since well before European settlement. Aboriginal inhabitants planted corn, beans, and squash and tapped maple forests for sugar (Hedrick 1933; Ellis et al. 1967). The Dutch settlers of the precolonial period planted wheat for export, and this remained a major cash crop in the region

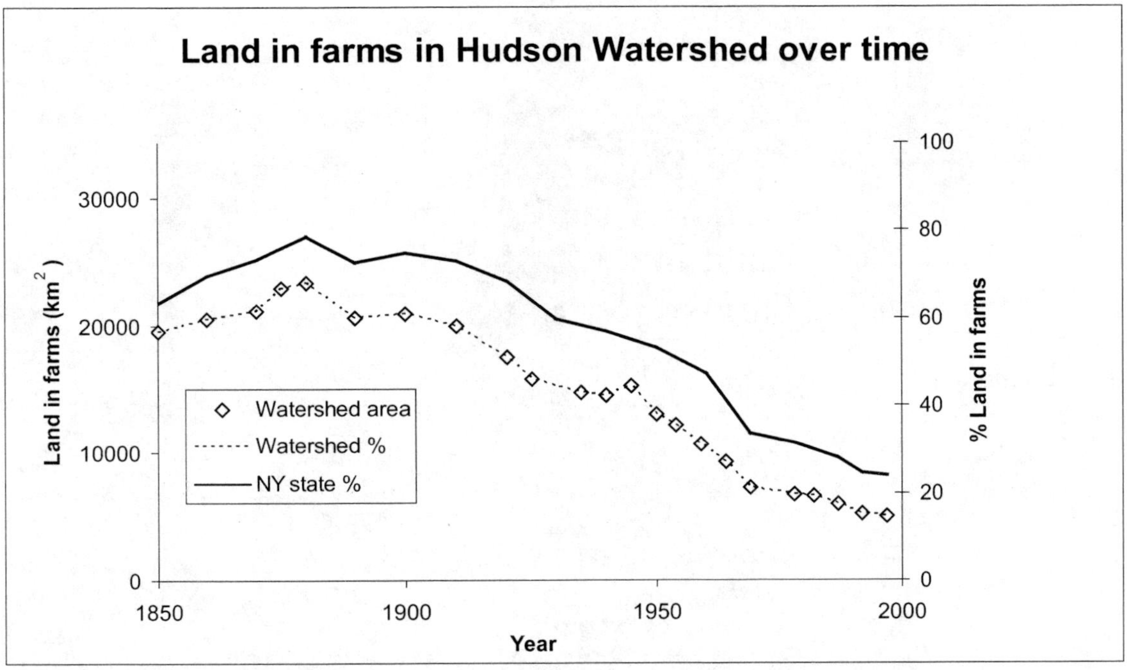

Figure 6. Agricultural land use in the watershed peaked in the 1870s-1880s. Data source: U.S. Census of Agriculture, county level data.

through the first decades of the 1800s when it became unprofitable because of the effects of the Hessian fly, wheat rust, and other insect and disease pests (Hedrick 1933). By the 1850s, dairy farming had surpassed wheat in economic importance in the state. Corn for livestock feed and other small grains replaced wheat as it became less profitable (Ellis et al. 1967). Land in farms grew steadily in the early part of the 19th century and apparently peaked in the watershed around 1880 at about 68% of the total area (Figure 6), about 11% less than the corresponding peak of the state overall. During the peak years of agriculture in the watershed, land in farms exceeded 90% in several counties, though it is important to note that farmland included "unimproved land" (woodlots, etc), and trees harvested for lumber and other uses were a significant source of farm income, especially early in the 19th century. The trend thereafter until the mid-20th century is negative—a steady decline until the late 1960s and a less steep decline from then until the present day value of around 15% of the watershed area. A slight upward trend in the late 1940s reflects agricultural demands of the war years. This pattern has reflected that of New York State to a remarkable degree, though the proportion of watershed land in farms has been somewhat lower than the corresponding figure for the state as a whole for at least the last 150 years (Figure 6).

Agricultural land uses have never been uniformly distributed over the watershed (Figure 7). Orchards have thrived along the Hudson from the days of earliest settlement. The New York dairy industry began in Dutchess, Herkimer, Oneida, and Orange counties in the early decades of the 1800s and spread to other counties of the watershed within the next 30 years (Hedrick 1933). Beginning in the 19th century, some areas in the watershed became known for specialty crops (flax in the Hoosic Valley, hops in Otsego County). Rye was grown in the

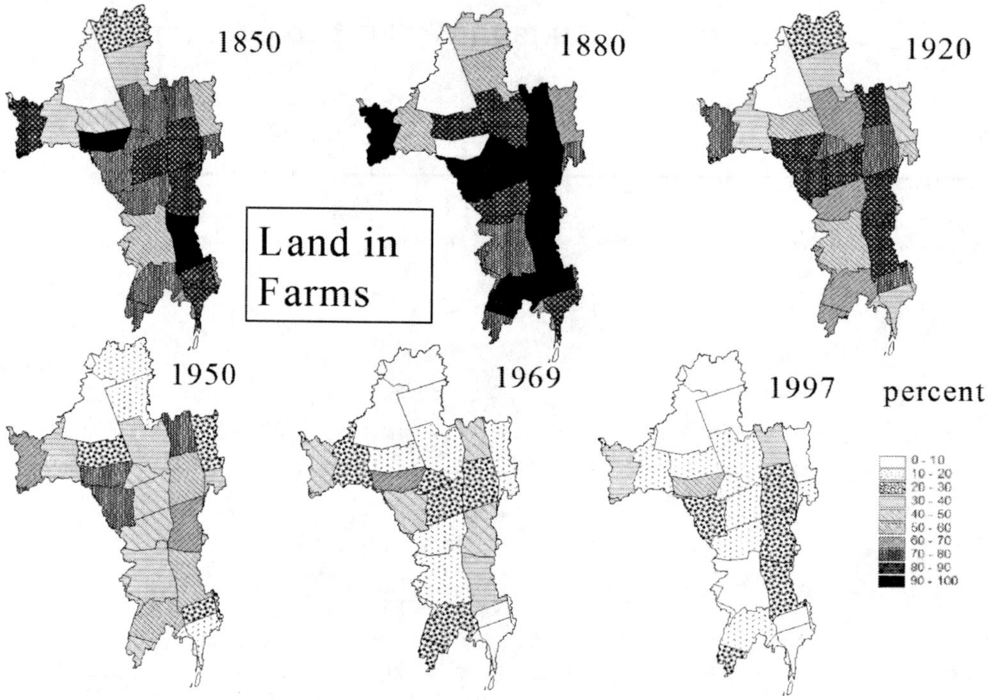

Figure 7. Spatial distribution of agriculture as percent of land use in the watershed, 1875-1997. Data source: U.S. Census of Agriculture, county level data.

Hudson Valley to meet the demand of distilleries (Ellis et al. 1967). At the peak of agricultural intensity, over 90% of the land area of counties along the Hudson River corridor was allocated to farms (Figure 7). The northern counties of the watershed remained relatively unforested at that time, as did the southernmost tip of the watershed in New Jersey, though for very different reasons. Logging operations and associated fires deforested large areas of the Adirondacks, Catskills, and Hudson River Highlands (Fox 1900; Moon 1909; see next section). The southernmost tip of the watershed was experiencing explosive population growth as the New York City metropolitan area expanded outward in response to migration from other regions of the United States as well as immigration from abroad (Ellis et al. 1967).

As in most other parts of the United States, regional urban and suburban populations have increased at the expense of rural populations throughout the 20th century (Ellis et al. 1967). Today, while the counties of the region all have substantially less land in farms, the distribution reflects that of the height of agricultural activity in the region: counties with the greatest proportion of agricultural land tend to follow the main river corridors. Land in the north is largely reserved as parkland, and the watersheds of the New York City water supply system in the Catskills and the counties immediately north of New York City are controlled to some extent by regulations of the New York City Department of Environmental Protection. The suburbs of the New York City metropolitan area have grown to cover most of northern New Jersey and the southern counties of the watershed.

Finally, while a detailed discussion of the

effects of agricultural technology is beyond the scope of this study, a major aspect of 20th century agriculture was the enormous growth in pesticide and fertilizer use, especially following World War II. In the Hudson-Raritan basin, organochlorine pesticide loading grew from around 1,400 kg in 1945, peaking in the early 1970s (12,400 kg in 1971), and then declining in the 1980s (Ayres and Rod 1986). The total nitrogen load from fertilizer in the counties of the Hudson/ Mohawk watershed also grew rapidly from the 1940s (5,000 metric tons N) until the mid 1980s (28,000 metric tons N), but has declined, with fluctuations, since then (Alexander and Smith 1990; Battaglin and Goolsby 1994). It should also be clear that conventional agriculture is not the only source of fertilizers and pesticides; beginning in the 1950s, commercial and residential lawns have rivaled row crop agriculture in the intensity of pesticide and fertilizer use, though quantitative load estimates are more difficult to assess.

Forests

The history of New York, including the Hudson River watershed, is intimately tied to its forests. After the last glaciation, succession resulted in an oak and chestnut dominated forest in the Hudson Valley, northern hardwoods (composed primarily of beech, maple, yellow birch, hemlock, and white pine) in the Catskills and lower Adirondack regions, and a spruce, fir, and paper birch complex dominated the higher elevations of the Adirondacks and Catskills (Hamilton et al. 1980). Indians of the region manipulated the forests to enhance productivity of resources, girdling trees, and burning areas to open up the woods for wildlife and creating habitat for fruit-bearing plants (Day 1953). When early European settlers arrived, they encountered large areas in the forests of southern New England and the lower Hudson that were "remarkably open, almost parklike at times" (Cronon 1983; p. 25). The northern forests were denser.

As suggested above, the early settlement and colonial days were largely periods of land clearing in which Dutch, and later English, settlers cleared land for farms and established sawmills along local streams. Revenues from wood exceeded those from crops. Early lumbering, beyond the needs of the colonials, sought out first the ancient white pines that were used by both the Dutch and English royal navies as ships' masts and timbers. Fox (1900) noted that the tallest pines were about 255 ft (89 m) tall and over 2 m wide and that Pine Street in New York City was named for the "many magnificent pines" on the farm of January Jansen Damen. In addition to white pine, oak and hickory were exported, first to Amsterdam and later to England, which had suffered wood shortages since the 1500s (Cronon 1983).

After the American Revolution, and consequent routing of the Iroquois, the colonists expanded northward and westward. As demand for wood increased, sawmills were erected wherever sufficient supplies of wood and water occurred (Hamilton et al. 1980). Logs cut close enough to the sawmills were dragged in by draft animals, but as trees became depleted, the mills could be moved to other sites with more trees. In this manner, forests were selectively cut for decades, with mills successively moved to more remote places.

Tanning was a parallel, major forest-based industry that arose early on in the Hudson Valley and had a significant impact on forested ecosystems. Colonial tanning operations began in New England, using oak, chestnut, and hemlock bark as the source of tannins. As the supplies dwindled there, the industry pushed westward into the

Hudson Valley, eventually heading up into Michigan, Wisconsin, and ultimately the West Coast. By 1810, there were 867 tanneries in New York State (McMartin 1992), many of them in the Hudson Valley and most of them small scale. However, the large financial interests of "The Swamp," a group of New York City leather businessmen named for that section of the city where tanning was practiced, selected and backed the development of large-scale operations, first in the Catskills and later in the Adirondacks. Colonel William Edwards, a tanner and grandson of the famed Boston minister Jonathan Edwards, not only received financial backing from The Swamp but also lobbied to have the laws changed to allow tannery corporations to operate in Green and Delaware counties (but not Ulster and Sullivan, where anti-tanning sentiments were higher; Evers 1972). His New York Tannery opened in 1816 in Edwardsville (now Hunter). An 1820 painting depicts the tannery by the Schohariekill, showing the once-forested land laid to waste all around (Evers 1972). Zadock Pratt, for whom Prattsville is named, developed the prototype tannery that became common in the Adirondacks (McMartin 1992).

Hemlock came to be the preferred resource, as it grew in great stands on north-facing slopes in the Catskills and on lower slopes in the Adirondacks. Tanning used only the hemlock bark and not the rest of the tree, which was of little value for timber. Thus, it was a wasteful industry, with trees felled and only the bark peeled for use. The stems were left to decay, and were described as looking like the bleached bones of giants. Tanning also required a plentiful supply of water to run the operations and into which to sluice the wastes, which were considerable. The putrid smells and severe organic pollution of tanneries ruined many rivers and streams while they operated. As the Catskills came to be depleted of hemlock, operations moved into the Adirondacks. A general trend from small-scale to increasingly larger operations took place from the 1830s until the end of the tanning era in New York in the 1890s (McMartin 1992). Finally, as the hemlocks became increasingly scarcer and newer technology came online, tanneries moved closer to the sources of leather and the industry collapsed in New York.

While the tanning industry declined, logging for timber and pulp continued to grow throughout the 19th century. Rafts were used to float sawed timber down to markets to the north, west, and southeast (Fox 1900). Because of the tides, rafts could not be floated on the Hudson below Albany, and so the wood was loaded onto ships that sailed down to New York City. Log driving was commenced in 1813 to send timber down to the mills in Glens Falls. This innovation was quickly repeated and became the main means of getting logs to the mills. Throughout the early to mid-1800s, rivers became declared public transportation highways through acts of the legislature. Lake levels throughout the Adirondacks were raised with "spill-dams" that were removed in spring to augment the freshets and move the logs downstream. The towns of Glens Falls, Sandy Hill, and Fort Edward became the center for the emerging wood products industry. In 1849, the Hudson River Boom Association set up "booms" or collecting and sorting points (Figure 8). Logging companies would place unique brands on their logs to identify them. Many accounts exist documenting the bravery of the log drivers, who would risk their lives breaking up log-jams.

The advent of rail provided a new means of moving timber out of the Northern Forest. The first railroad into the Adirondacks was opened in 1868 (Donaldson 1921). Although railroads were built at first to move in equip-

Figure 8. Photograph of the Glens Fall Boom holding logs from the Northern Forest. Source: Fox (1900).

ment, they were soon put to use in transporting out wood (Hamilton et al. 1980). Rail carried much of the hardwood that was too heavy to float down rivers and thus facilitated the exploitation of relatively unused species (Fox 1900). Steam engines, with hot coals and sparks flying, were also responsible for many of the forest fires that plagued the region.

In 1867, the sulfite process for reducing the fiber from wood pulp (produced by grinding wood) was introduced, and wood fiber became a new source of material in papermaking, which had previously been made with rag and cotton fibers (Recknagel 1923). As this industry began to take off, smaller trees could be exploited. Lumbering became increasingly less selective. Timber production peaked in the 1870s, and pulpwood production peaked in 1908 (Canham 1981). Although quantitative data were unavailable prior to 1870, the time trend (Figure 9) shows that 20th century wood production was never more than half of the 19th century highs, again suggesting the severe overexploitation of the forests in the 1800s.

The history of forest use in New York up to the late 1800s was a picture of exploitation and general lack of concern for the welfare of the forested ecosystems. Forest destruction began to raise fears of large-scale environmental degradation. Early environmentalists noted the increased sediment loads in the Hudson (Hamilton et al. 1980) and

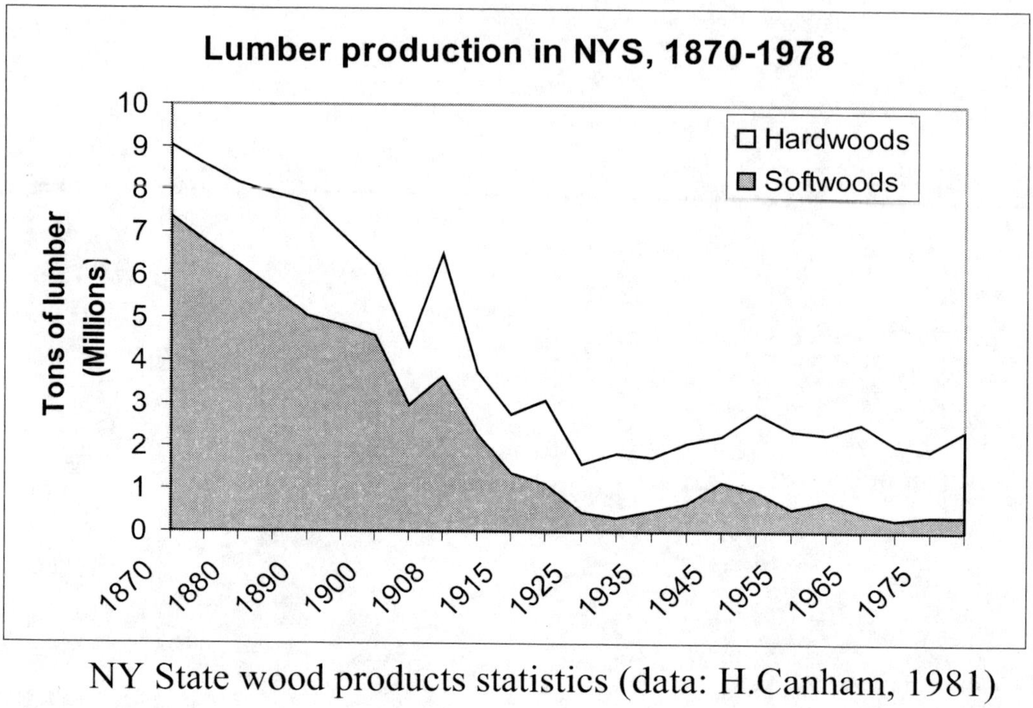

Figure 9. Temporal trends in production of lumber and pulpwood in the Northern Forest. Data source: Canham (1981).

streams, warmed from lack of shade, were losing their trout and other coldwater species. At the same time, forest scientists in Europe were becoming aware of the connections between deforestation in the Alps and the lowering water level of Lake Geneva (Hough 1878), and the nascent field of scientific forestry began prescribing new practices to reduce the environmental impacts of the industry. The depredations of logging in New York inspired a movement to preserve the forest, lest it be completely ruined. Spurred on by the eloquent lobbying of Verplanck Colvin, the state surveyor who mapped the Adirondacks, in 1885 the legislature enacted a law to establish permanent forest preserves and develop a system of forest fire protection. Although this "forever wild" law prevented the sale of lands, it did not prevent tree harvesting. The preserve began with the acquisition of tax-delinquent lands, and from the 1890s onward, also began a program of land purchasing and reforestation. This activity was administered by the newly consolidated Forest, Fish and Game Commission, which worked to train a new generation of "scientific" foresters. The work of the Commission also involved fire prevention and fire fighting. Droughts, in combination with the practice of leaving intact treetops after cutting the logs, promoted fire hazards; sparks from trains, lightning, from careless campers caused massive fires in the early 1900s (Figure 10). The Commission's reports motivated legislation for preventing fires, with stiffer fines, modification to the rail rights-of-way, and fire towers as some of the innovations (Whipple 1909). Another preserve in the Hudson River Highlands was created in 1909, in response to extensive clear-cutting for fueling the brickyards, as well as susceptibility to fires created by poor stand conditions (Moon 1909).

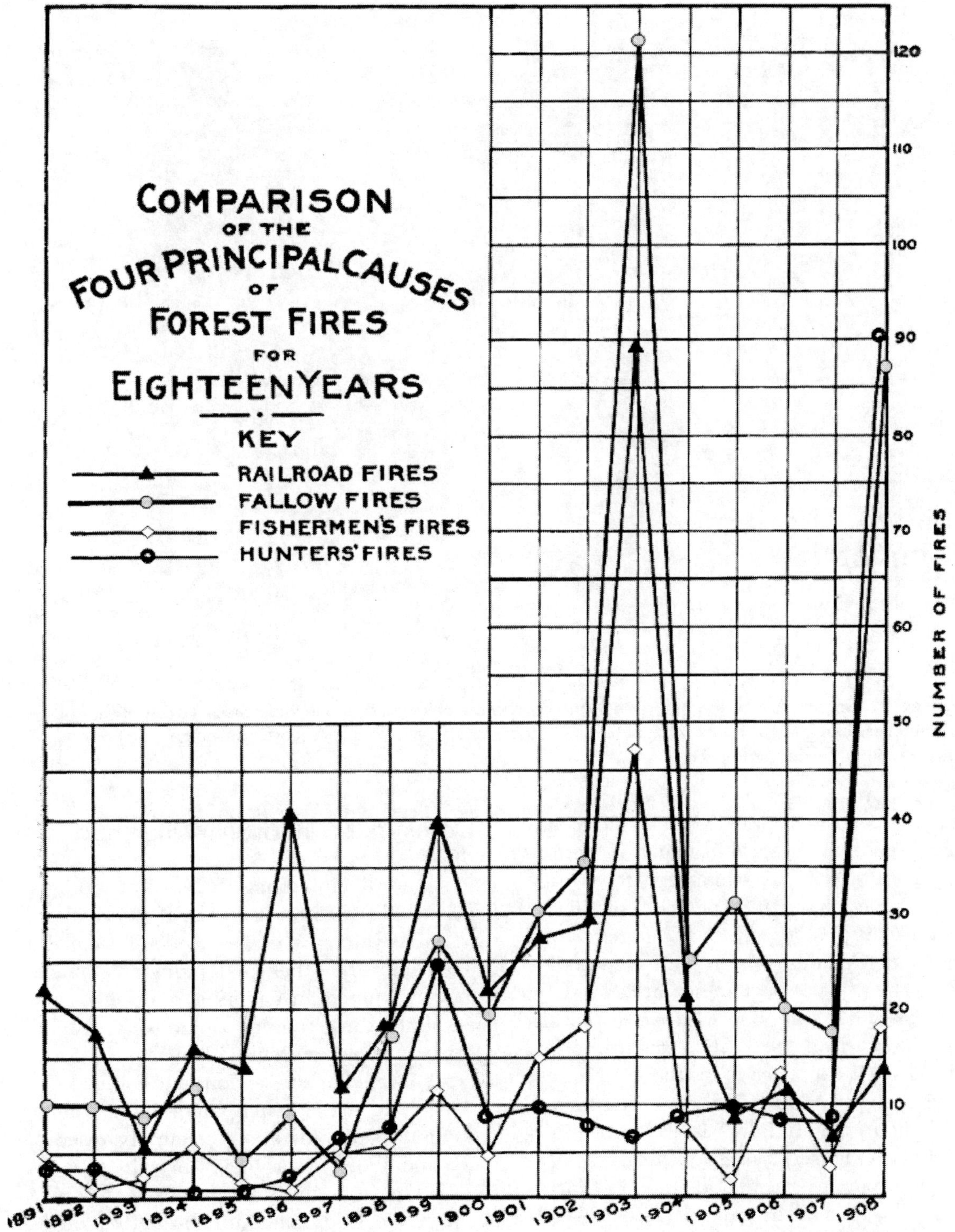

Figure 10. Causes of forest fires, from the 1908 Report of the Forest, Fish and Game Commission. Severe droughts were associated with the high-fire years (symbols have been added to the original figure).

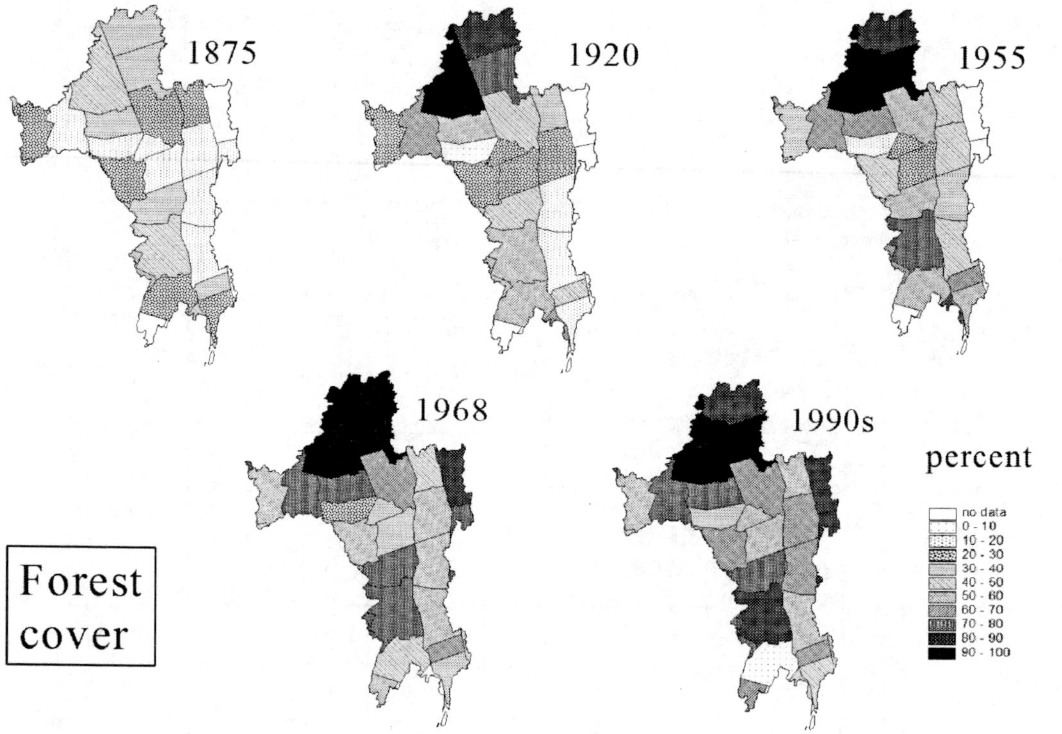

Figure 11. Spatial distribution of forested land as percent of land use in the watershed, 1875-1990. Data sources: Hough (1878), Recknagel (1923), Stout (1958), and U.S.D.A. Forest Resource Bulletins and Forest Resource Inventory Reports.

Better management practices in the 20th century promoted regrowth of many of the forests. Forests were also managed for watershed protection, with recognition of this as early as the first report of the Forest, Fish and Game Commission. Programs of woodlot management were begun and carried out by farmers with help from state foresters (Hamilton et al. 1980). Forest inventories, carried out every decade by the U.S. Forest Service, show a pattern of recovering forests throughout the Hudson basin counties into the 1990s (Figure 11). This pattern was also driven in part by the abandonment of farmland, as agrarianism gave way to a fossil-fuel supported economy, with resources increasingly drawn in from outside.

Dams in the Hudson Watershed

Dams were likely among the earliest modifications of streams within the Hudson watershed, as waterpower was needed for driving sawmills and other small industry. Dams were constructed for a number of other purposes, including tanning operations, raising water levels to ensure adequate flow for log drives (see Forests section) and for feeder canals along the Erie and Champlain canal systems, flood control, water supply reservoirs, and hydropower. Inevitably, these systems often serve multiple purposes and, in age (e.g., the Eddyville Dam on the Rondout Creek), sometimes outlive their main purpose.

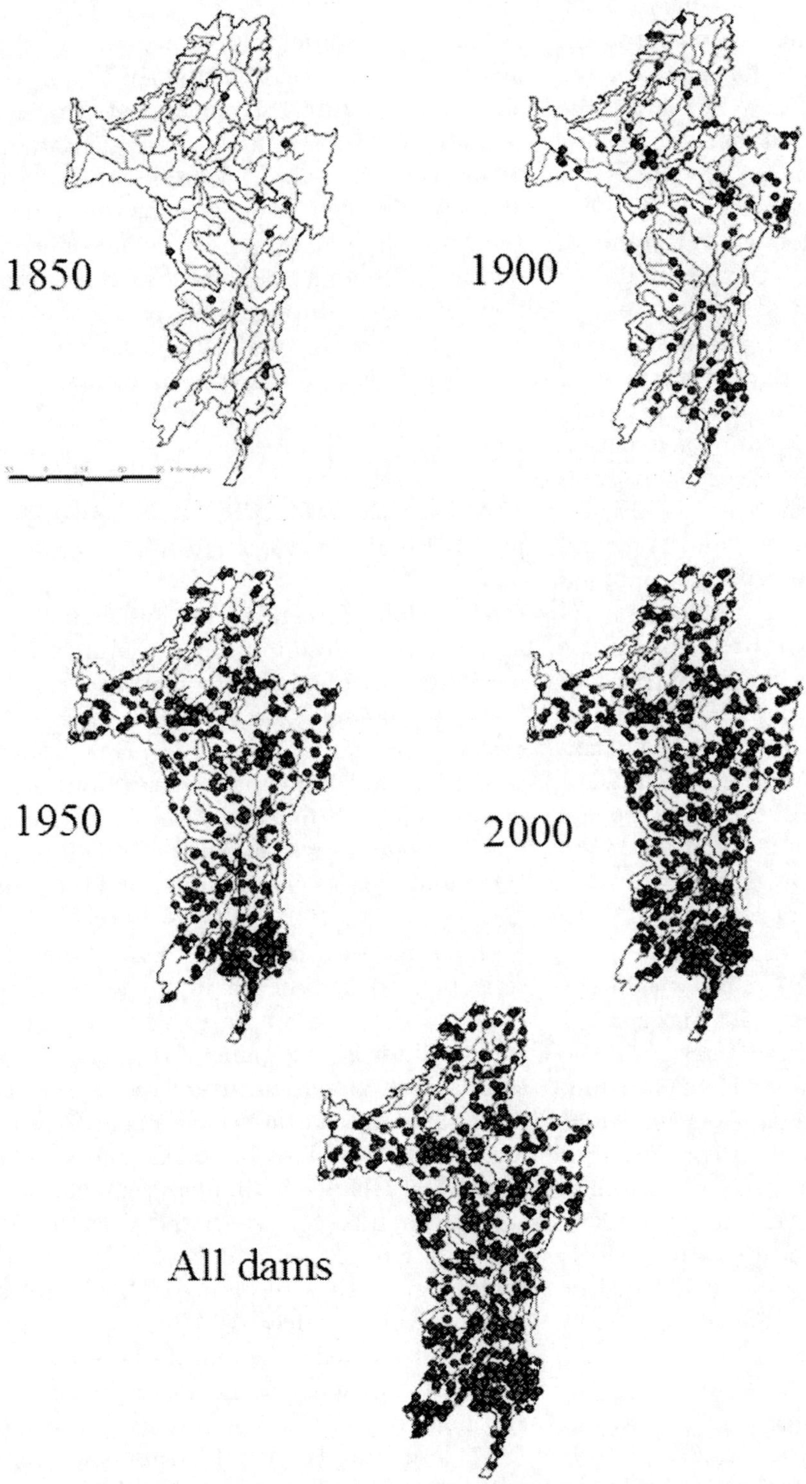

Figure 12. Accumulation of dams over time in the Hudson watershed. Top four maps show the cumulative change in numbers of dams, by dates of completion, every 50 years from 1850 onwards. Bottom map shows all dams, inclusive of those lacking dates. Data source: BASINS (USEPA 2001).

The BASINS data set includes the dates of completion of most dams in its database, but 128 dams (16% of the database for the Hudson drainage) lack dates. Some of these may well be among the earlier dams. Furthermore, the database is not an exhaustive list of all dams in the watershed (Graf 1999). Nevertheless, the spatiotemporal trend of dam construction (using records with dates) is dramatic (Figure 12) and appears to capture the general pattern of the entire data set. The period of 1900–1930 was an age of rapid dam construction in the Hudson and presaged the era of great dam construction in the West. Many of the largest dams were built in this period in the Hudson (Figure 13A), including the Kensico and Ashokan Reservoir dams in 1916 with a combined storage capacity of 0.8 km^3 (629,060 acre-feet) and the Sacandaga Reservoir, completed in 1930 both for flood control and hydroelectric generation, with a displacement of 1.07 km^3 (866,000 acre-feet). Examining the rate of change of cumulative storage capacity (Figure 13B), it appears that most of the damming potential of the basin was built out before 1950, with relatively minor additions since then.

Today there are some 797 dams within the Hudson watershed proper, with another 357 extant in the surrounding basins of the Hackensack-Palisades, Sandy Hook, Raritan, and Bronx. Most of the dams range in height between 2 and 10 m (Table A9), the tallest being the Ashokan Dam at Olive Branch (213 ft, 64.9 m). The "damming intensity" (number of dams per subcatchment area) is greatest in the Lower Hudson (0.069 km^{-2} or 0.18 mi^{-2}) and lowest in the Sacandaga (0.011 km^{-2} or 0.03 mi^{-2}).

We note that the database likely does not include very small dams (ca. 0.5 m tall) and impoundments and that these could also be abundant and significant in the overall hydrology of the watershed. Even without including these effects, when the cumulative impoundment is expressed in units of average annual flow of the Hudson at the Troy dam (Figure 13B), it appears that impoundments represent the equivalent about 4 months of flow. If we take this as an average of the effect on streams throughout the watershed, dams currently have the capacity to impound one-third of the annual flow of water through the catchment and consequently retard and dampen the response of streamflow to precipitation events.

Hydrological, Biogeochemical, and Ecological Implications of Changes in the Hudson/Mohawk Watershed

Clearing the forests of the watershed for agriculture undoubtedly had the most dramatic impact on the watershed as a whole. Some simple modeling results have suggested that the fully forested primeval watershed would have seen erosive soil losses around one-eighth of current rates and, during the peak days of agricultural activity of the 1800s, more than double modern rates (Swaney et al. 1996). The combined effect of clearing, burning, and then tilling the soils poured sediments into the streams, removed the litter and organic matter, and eventually wore out some of the soils. Some simulations of agricultural landuse in the northeast United States have suggested that long-term depletions of soil N and C, due to the disturbance of biomass removals, are greater than losses associated with timber harvesting (Ollinger et al. 2002). Early leaders of agriculture recognized that careless agricultural practices depleted the soils of nutrients and complained of the effects of poor management on the condition of soils and livestock of the region (e.g., Ellis et al. 1967; quoting G. W. Featherstonhaugh 1819), and the 19th century migration of farmers from the region to the western frontier was largely a matter of literally moving to

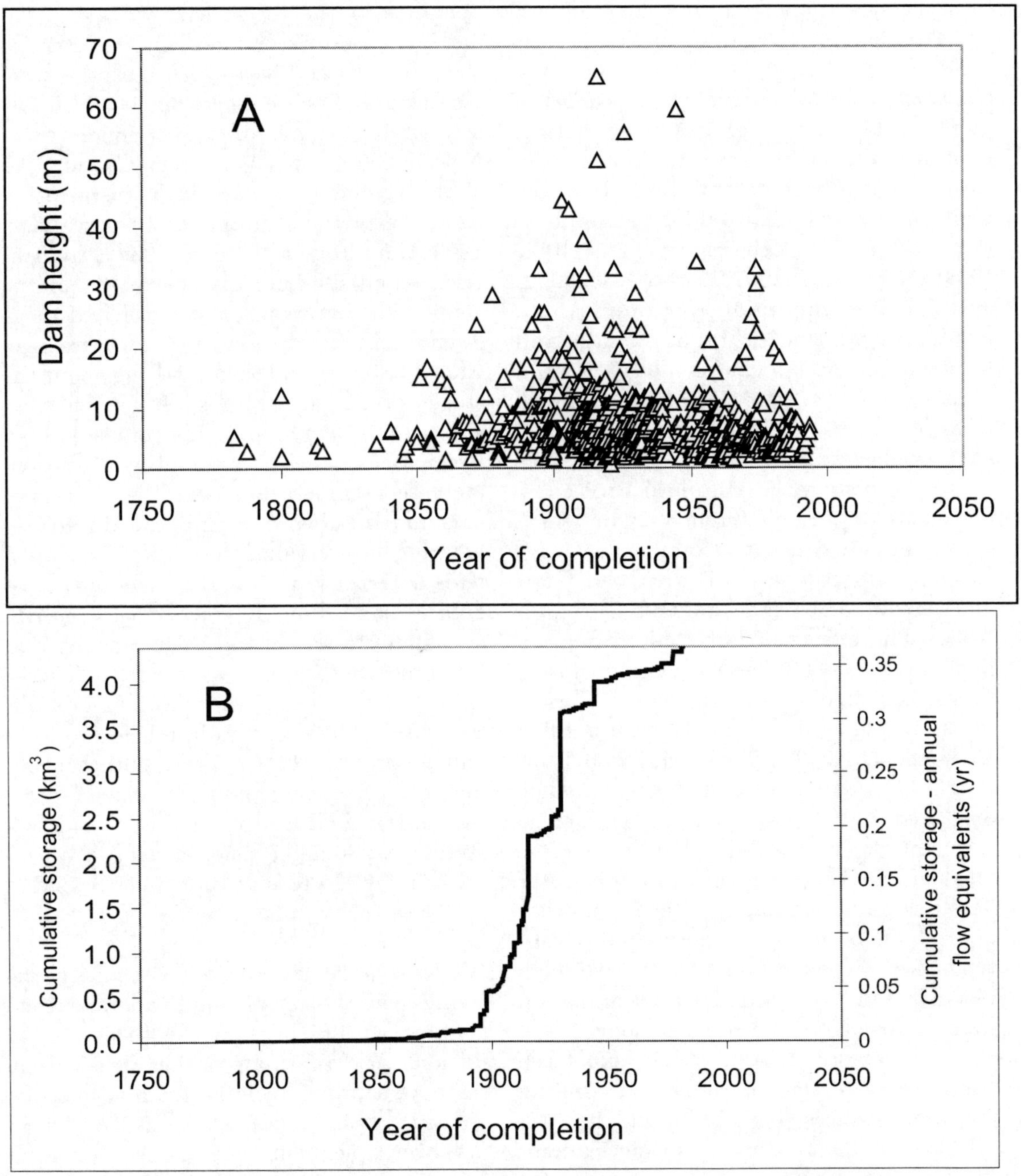

Figure 13. Time series of (B) the heights and (B) the cumulative storage capacity of dams in the Hudson-Raritan drainage. Data source: BASINS (USEPA 2001).

greener pastures. While tile drainage of agricultural fields began in the United States in Seneca County, New York, to the west of the Hudson watershed, as early as the 1830s, farmers in several counties within the watershed had adopted this technology by the 1860s (Weaver 1964). Today, the benefits of tile drainage to crop yield in wet, heavy soils are recognized, as are the consequent increases in water discharge and agricultural chemical transport from fields (McIsaac and Hu, 2004).

Changes in the forests had their own set of impacts. The early high-grading practices for white pine at first, and then other species, ecologically represented selective species removal, which would have implications both for biogeochemistry (e.g., Hobbie 1992; Chapin et al. 1997; Lovett et al. 2002) and forest community dynamics (e.g., Canham and Pacala 1995; Pacala and Deutschman 1995; Chapin et al. 1997). Similarly, the practice of cutting hemlock trees, removing the bark, and leaving the stems to decay might have had effects similar to those shown at Hubbard Brook Experimental Forest watershed 2, in which stream nitrate concentration increased significantly in response to trees being cut and left in place (Likens and Bormann 1995). The effects of forest harvest practices, fires, and land use change on forest biogeochemistry are active areas of ecological research (e.g., Likens and Bormann 1995; Aber and Driscoll 1997; Waring and Running 1998; Goodale et al. 2000; Goodale and Aber 2001; Latty et al. 2004). Generally, biogeochemical functions become "leakier" in the years immediately after harvesting and especially after fires; other effects can be detected more than a century after the fact (Goodale et al. 2000; Goodale and Aber 2001; Latty et al. 2004). Many other factors, such as geology, stand age, and forestry practices may affect the particular direction of change, depending on circumstances (e.g., Yanai et al. 2000; Lovett et al. 2002). Changes in species composition could also have implications for the quality of allochthonous materials delivered to lake, stream, and ultimately estuarine food webs (Pace et al. 2004).

Removing trees also affects hydrology, by altering evapotranspiration among other things. Although transpiration is lowered, openings in the forest can warm the soil and increase evaporation. The foresters of the late 19th and early 20th centuries were well aware of the connections between forest cutting and hydrology (Hough 1878; Newell 1900), and Verplanck Colvin used these arguments as part of the justification to establish the Adirondack Preserve (Fox 1900; Hamilton et al. 1980). The early fisheries scientists were also cognizant of the connection between forest cover and moderation of streamflow and temperature. Charles Stevenson (1899), writing about the status of the anadromous American shad *Alosa sapidissima*, drew the connection between deforestation, agriculture, siltation, increased temperatures, and the decline of this marine-dwelling animal. He wrote: "During heavy rains the plowed soil upon the hillsides is easily washed into gullies...[and] quickly conveyed to the rivers, filling them beyond their capacity and bringing into them masses of earth and other debris, thus covering the spawning grounds (p. 113)." The great Adirondack droughts and forest fires of the early 20th century may even have affected the water temperature of the tidal Hudson River as far downstream as Poughkeepsie, causing a warming of a few degrees at that time (Abood et al. 2006, this volume).

While we have not considered explicitly the changes in the urban landscape and road network within the watershed, it is easy to demonstrate that population density can serve as a proxy variable for road density, as they vary in space (e.g., county to county) and over time. Consequently, it is apparent that the changes in population density within the watershed indicate corresponding changes in road and building density and, therefore, impervious surface density. The resulting implications for runoff and water quality are obvious: faster responses to precipitation and snowmelt events, higher road salt loads in winter and spring, elevated water temperature from warm pavement in the summer, and increased dissolved and particulate substances related to automobile traffic. On the other hand, the high organic nutrient loads that were ubiquitous in the roads and streets of the "horse and buggy days" have essentially vanished.

Modern transportation networks affect more than just the hydrology and chemistry of the local environment, serving as both barriers to wildlife movement and conduits for invasive species, increasing wildlife mortality, and so forth. Forman and Alexander (1998) suggest that a road density of 0.6 km/km^2 is the maximum for a "naturally functioning landscape." According to New York Department of Transportation (DOT) figures, of all the counties falling all or partly within the watershed, only two met this criterion in 2001: Essex and Hamilton, in the Adirondacks (New York State Department of Transportation 2002).

The environmental impacts of dams receive much attention today (McCully 1996). The alteration of hydrological regimes is an obvious effect, and large dams worldwide have been documented to slow the release of waters draining off the continents, with concomitant increased evaporation (Vörösmarty et al. 1997). Even small dams, when taken in the aggregate, may have measurable hydrologic effects. Smith et al. (2002) note that small impoundments in the contiguous United States, most of them in the East, account for 20% of the standing water area and affect hydrology, sedimentation, and biogeochemistry. Dams cause increased sedimentation upstream in reservoirs and increased erosion downstream. Dean and Gorham (1998) estimated that reservoirs worldwide have sedimentation rates 4.7-fold greater than lakes. It is also recognized that many biogeochemical processes (e.g., denitrification and settling of particulates) are sensitive to the residence time of waters flowing through the landscape; longer residence times result in more settling and nutrient "processing." A model-based analysis of the effect of reservoirs on denitrification by Seitzinger et al. (2002) found very little impact at the catchment scale, in part because dams are often sited upstream of nitrogen sources in watersheds. On the other hand, Humborg et al. (1997) found dams not only to affect river biogeochemistry (e.g., to alter dissolved nutrient ratios), but to propagate effects on ecosystem structure and productivity in recipient marine waters.

Finally, dams can have devastating ecological effects, which have been best documented on fisheries. Dams have contributed to the downfall of Pacific salmon through the loss of spawning habitat (Nehlsen et al. 1991), and Limburg et al. (2003) estimated that over 4,000 km of American shad spawning habitat have been lost due to damming North American East Coast rivers. A number of studies have documented the potentially important links of the transport of marine-derived nutrients to freshwater ecosystems (e.g., Richey et al. 1975; Durbin et al. 1979; Bilby et al. 1996; Garman and Macko 1998; MacAvoy et al. 2000), providing both direct and indirect subsidies to stream and riparian zone ecosystems; these links are generally truncated by dams. Pringle (1997) and Freeman et al. (2003) document the disruption by dams of many other ecological interactions, and even short-term evolution has been documented by the splitting of populations by dams (Morita and Suzuki 1999). At this point, we can only guess at the environmental and ecological changes effected by dams in the Hudson-Mohawk watershed, but undoubtedly, given the degree and magnitude of damming, the effects have been considerable.

Table 1 summarizes the environmental impact of various human activities in the wa-

Table 1. Changes in watershed characteristics and associated effects.

Pre-European settlement	
Aboriginal agriculture	Localized fires and associated change in biomass, habitat, and nutrient remobilization
Pre-Colonial and Colonial settlement	
Clearing of land for agriculture	Removal of forest cover; changes in habitat; increase in sediment loads, streamflow
19th Century	
Tanning	Preferential clearing of hemlock forests, increased organic loads to watershed streams
Logging	Partial to whole clearing of forests; corresponding increases in soil temperature, sediment loads; alterations of water balance, habitat; modification of streamflow by damming and logging operations
Agriculture (clearing, drainage, animal agriculture, tillage, crop rotations, other management practices)	Clearing of forest land; increase in sediment loads, soil nutrient loss, streamflow, organic nutrient loads (manure); increase in N fixing crops (alfalfa)
Canal and dam development	Introduction of waterborne invasive species; wetland drainage, alteration of flow directions of watercourses; increased nutrient loads to waterways; changes in water residence time and flow regime by damming
Railroad development	Increased access to forests and risk of fire
Road development	Increases in impervious surface and consequent runoff
Rise of urbanization and industrialization	Increased pollution from sewage and factory waste

tershed as they have changed over time from the pre-European settlement period through modern times. Some, such as canals, road development, urbanization, and specific agricultural practices, have not been addressed in detail in this paper, but we feel it important to acknowledge their influence.

Table 1 continued.

20th century	
Further dam development for water supply (in New York and other municipalities) and power	Elimination of extreme flows and other changes in flow regime and sediment transport; increase in overall water residence time and evaporation
Development of highways and other road construction	Increases in impervious surface and consequent runoff, and related effects; increased wildlife mortality and other effects on plant and animal populations
Agricultural decline	Old-field succession, forest regrowth, expansion of suburban development
Changing agricultural practices (rise of fertilizer and agrochemicals, tillage, etc)	Increased inorganic nutrient (fertilizer) and changes in organic nutrient (manure) loads
Growth of urban and suburban development ("sprawl")	Increased impervious surface, runoff, pollutants; sediment loads from construction; stream channelization

Conclusion

Although our analysis can only be considered as incomplete, we have outlined some of the changes human activities have wrought on the Hudson River watershed at different periods of its recent history. Without doubt, it would require at least one book to discuss all of them in adequate detail. Nevertheless, we hope that our work will remind readers of the historic legacies shaping the landscape and suggest to them that current activities in the watershed will also have implications for the environment of the region in centuries to come.

Acknowledgments

We thank Christy Goodale, Charlie Canham, Stephen V. Smith, and two anonymous reviewers for comments on the manuscript and Stephen V. Smith for advice on GIS analysis. Partial funding came from National Science Foundation Grant No. DEB-0238121 to KL.

References

Aber, J. D., and C. T. Driscoll. 1997. Effects of land use, climate variation and N deposition on N cycling and C storage in northern hardwood forests. Global Biogeochemical Cycles 11:639–648.

Abood, K. A., T. L. Englert, S. G. Metzger, C.V. Beckers, Jr., T. J. Groninger, and S. Mallavaram. 2006. Current and evolving physical and chemical conditions in the Hudson River estuary. Pages 39–62 *in* J. R. Waldman, K. E. Limburg, and D. L. Strayer, editors. Hudson River fishes and their environment. American Fisheries Society, Symposium 51, Bethesda, Maryland.

Alexander, R. B., and R. A. Smith. 1990. County-level estimates of nitrogen and phosphorus fertilizer use in the United States, 1945 to 1985. USGS Open-File Report 90–130.

Anonymous. 1647. Journal of New Netherland. Translated and published in O'Callaghan, E. B. 1851. The documentary history of the state of New York, volume IV. Charles Van Benthuysen, Public Printer, Albany, New York.

Ayres, R. U., and S. R. Rod. 1986. Patterns of pollution in the Hudson-Raritan Basin. Environment 28:14–43.

Battaglin, W. A., and D. A. Goolsby. 1994. U.S. Geological Survey Water-Resources Investigations Report 94–4176.

Bilby, R. E., B. R. Fransen, and P. A. Bisson. 1996. Incorporation of nitrogen and carbon from spawning coho salmon into the trophic system of small streams: evidence from stable isotopes. Canadian Journal of Fisheries and Aquatic Sciences 53:164–173.

Boyle, R. H. 1969. The Hudson River: a natural and unnatural history. W. W. Norton and Co., Inc., New York.

Brainerd, E., and M. Siegler. 2002. The economic effects of the 1918 influenza epidemic. Centre for Economic Policy Research, CEPR discussion paper 3791. London. Available: http://ideas.repec.org/p/cpr/ceprdp/3791.html (December 2004).

Canham, C. D., and S. W. Pacala. 1995. Linking tree population dynamics and forest ecosystem processes. Pages 84–93 *in* C. G. Jones and J. H. Lawton, editors. Linking species and ecosystems. Chapman and Hall, Inc., New York.

Canham, H. O. 1981. Forest wood products. NYS Forest Resource Assessment Report No. 7. New York State Department of Environmental Conservation, Division of Lands and Forests, Forest Resources Planning Program, Albany.

Chapin, F. S. III, B. S. Walker, R. J. Hobbs, D. U. Hooper, J. H. Lawton, O. E. Sala, and D. Tilman. 1997. Biotic control over the functioning of ecosystems. Science 277:500–504.

Cronon, W. 1983. Changes in the land: Indians, colonists, and the ecology of New England. Hill and Wang, New York.

Day, G. M. 1953. The Indian as an ecological factor in the Northeastern forest. Ecology 34:329–346.

Dean, W. E., and E. Gorham. 1998. Magnitude and significance of carbon burial in lakes, reservoirs, and peatlands. Geology 26:535–538.

Donaldson, A. L. 1921. A history of the Adirondacks. The Century Co., New York.

Durbin, A. G., S. W. Nixon, and C. A. Oviatt. 1979. Effects of the spawning migration of the alewife, *Alosa pseudoharengus*, on freshwater ecosystems. Ecology 60:8–17.

Earle, C., C. Cao, J. Heppen, and S. Otterstrom. 1999. The Historical United States County Boundary Files 1790–1999 on CD-ROM. Geoscience Publications. Louisiana State University, Baton Rouge. Cornell University Press, Ithaca, New York.

Eichel, O. R. 1923. A special report on the mortality from influenza in New York State during the epidemic of 1918–19. New York State Department of Health, Albany.

Ellis, D. M., J. A. Frost, H. C. Syrett and H. J. Carman. 1967. A history of New York State. Cornell University Press, Ithaca, New York.

Environmetal Systems Research Institute. 1999. ARCVIEW GIS 3.2. Environmental Systems Research Institute, Redlands, California.

Evers, A. 1972. The Catskills: from wilderness to Woodstock. Doubleday & Co., Inc., Garden City, New York.

Forman, R. T. T., and L. E. Alexander. 1998. Roads and their major ecological effects. Annual Review of Ecology and Systematics 29:207–231.

Foster, D. R. 1999. Thoreau's country. Journey through a transformed landscape. Harvard University Press, Cambridge, Massachusettes.

Fox, W. F. 1900. History of the lumber industry in the state of New York. Pages 237–283 *in* Sixth Annual Report of the Forest, Fish and Game Commission. James B. Lyon, State Printer, Albany, New York.

Freeman, M. C., C. M. Pringle, E. A. Greathouse, and B. J. Freeman. 2003. Ecosystem-level consequences of migratory faunal depletion caused by dams. Pages 255–266 *in* K. E. Limburg and J. R. Waldman, editors. Biodiversity, status, and conservation of the world's shads. American Fisheries Society, Symposium 35, Bethesda, Maryland.

Garman, G. C., and S. A. Macko. 1998. Contribution of marine-derived organic matter to an Atlantic coast, freshwater, tidal stream by anadromous clupeid fishes. Journal of the North American Benthological Society 17:277–285.

Goodale, C. L., and J. D. Aber. 2001. The long-term effects of land-use history on nitrogen cycling in northern hardwood forests. Ecological Applications 11:253–267.

Goodale, C. L., J. D. Aber, and W. H. McDowell. 2000. The long-term effects of disturbance on organic and inorganic nitrogen export in the White Mountains, New Hampshire. Ecosystems 3:433–450.

Graf, W. L. 1999. Dam nation: a geographic census of American dams and their large-scale hydrologic impacts. Water Resources Research 35:1305–1311.

Hamilton, L., B. Askew, and A. Odell. 1980. Forest history. New York State Forest Resources Assessment: Report No. 1. Prepared for the New York State Department of Environmental Conservation, Division of Lands and Forests, Forest Resources Planning Program, Albany.

Hedrick, U. P. 1933. A history of agriculture in the state of New York. Hill and Wang, New York.

Hobbie, S. E. 1992. Effects of plants on nutrient cycling. Trends in Ecology and Evolution 7:336–339.

Hough, F. B. 1878. First report upon forestry for 1877. U.S. Government Printing Office, Washington D.C.

Humborg, C., V. Ittekkot, A. Cociasu, and B. Bodungen. 1997. Effect of Danube River dam on Black Sea biogeochemistry and ecosystem structure. Nature (London) 386:385–388.

Latty, E. F., C. D. Canham, and P. L. Marks. 2004. The effects of land-use history on soil properties and nutrient dynamics in northern hardwood forests of the Adirondack Mountains. Ecosystems 7:193–207.

Likens, G. E., and F. H. Bormann. 1995. Biogeochemistry of a forested ecosystem. 2nd edition. Springer-Verlag, New York.

Limburg, K. E., K. A. Hattala, and A. W. Kahnle. 2003. American shad in its native range. Pages 125–140 *in* K. E. Limburg and J. R. Waldman, editors. Biodiversity, status, and conservation of the world's shads. American Fisheries Society, Symposium 35, Bethesda, Maryland.

Lovett, G. M., K. C. Weathers, and M. A. Arthur. 2002. Control of nitrogen loss from forested watersheds by soil carbon:nitrogen ratio and tree species composition. Ecosystems 5:712–718.

MacAvoy, S. E., S. A. Macko, S. P. McIninch, and G. C. Garman. 2000. Marine nutrient contributions to freshwater apex predators. Oecologia 122:568–573.

McCully, P. 1996. Silenced rivers: the ecology and politics of large dams. Zed Books, London.

McIsaac, G. F., and X. Hu. 2004. Net N input and riverine N export from Illinois agricultural watersheds with and without extensive tile drainage. Biogeochemistry 70:253–273.

McMartin, B. 1992. Hides, hemlocks and Adirondack history. North Country Books, Utica, New York.

Montanus, A. 1671. Description of New Netherland. Translated from "De Nieuwe en Onbekende Weereld: of Beschryving van America en 't Zuidland" and published in O'Callaghan, E. B. 1851. The documentary history of the state of New York, volume IV. Charles Van Benthuysen, Public Printer, Albany, New York.

Moon, F. F. 1909. Report on the highlands of the Hudson Forest reservation. Pages 217–233 *in* Annual reports of the Forest, Fish and Game Commissioner for 1907–1908-1909. J. B. Lyon, State Printer, Albany, New York.

Morita, K., and T. Suzuki. 1999. Shifts of food habit and jaw position of white-spotted charr after damming. Journal of Fish Biology 55:1156–1162.

Nehlsen, W., J. E. Williams, and J. A. Lichatowich. 1991. Pacific salmon at the crossroads: stocks at risk from California, Oregon, Idaho, and Washington. Fisheries 16(2):4–21.

New York State Data Center. 2000. Population of New York State by County 1790 to 1990. New York State Department of Economic Development, Albany.

New York State Department of Transportation. 2002. 2001 Highway mileage summary. Available: http://www.dot.state.ny.us/tech_serv/high/highwaym.html (December 2004).

Newell, F. H. 1900. A discussion of conservative lumbering and the water-supply. Pages 163–167 in Sixth Annual Report of the Forest, Fish and Game Commission. James B. Lyon, State Printer, Albany.

NYCDOH (New York City Department of Health). 1918. Annual Report of the New York City Dept of Health, New York.

O'Callaghan, E. B., editor. 1850. The documentary history of the state of New York. Volume III. Weed, Parsons & Co., Albany.

Ollinger, S. V., J. D. Aber, P. B. Reich, and R. J. Freuder. 2002. Interactive effects of nitrogen deposition, tropospheric ozone, elevated CO_2 and land use history on the carbon dynamics of northern hardwood forests. Global Change Biology 8:545–562.

Pacala, S. W., and D. H. Deutschman. 1995. Details that matter: the spatial distribution of individual trees maintains forest ecosystem function. Oikos 74:357–365.

Pace, M. L., J. J. Cole, S. R. Carpenter, J. F. Kitchell, J. R. Hodgson, M. C. Van de Bogert, D. L. Bade, E. S. Kritzberg, and D. Bastviken. 2004. Whole-lake carbon-13 additions reveal terrestrial support of aquatic food webs. Nature (London) 427:240–243.

Pringle, C. M. 1997. Exploring how disturbance is transmitted upstream: going against the flow. Journal of the North American Benthological Society 16:425–438.

Recknagel, A. B. 1923. The forests of New York State. The Macmillan Company, New York.

Reed, N. 1875. Early History of Amenia. 1st ed, DeLacey and Wiley, Amenia, New York. 2nd ed (1964), Harlem Valley Times, Amenia, New York. 108 p.

Richey, J. E., M. A. Perkins, and C. R. Goldman. 1975. Effects of kokanee salmon (*Oncorhynchus nerka*) decomposition on the ecology of a subalpine stream. Journal of the Fisheries Research Board of Canada 32:817–820.

Rod, S. R. 1986. Historical reconstruction of pollutant loadings in river basins. Pages165–170 *in* R. Hanham, W. G. Vogt, and M. H. Mickle, editors. Modeling and simulation, v 17 part 1. Proceedings of the 17th Annual Pittsburgh Conference.

Rod, S. R., R. U. Ayres, and M. Small. 1989. Reconstruction of historical loadings of heavy metals and chlorinated hydrocarbon pesticides in the Hudson-Raritan basin, 1880–1980. Final Report to the Hudson River Foundation.

Shorto, R. 2004. The island at the center of the world: the epic story of Dutch Manhattan and the forgotten colony that shaped America. Doubleday, New York.

Smith, S. V., W. H. Renwick, J. D. Bartley, and R. W. Buddemeier. 2002. Distribution and significance of small, artificial water bodies across the United States landscape. The Science of the Total Environment 299:21–36.

Stevenson, C. H. 1899. The shad fisheries of the Atlantic coast of the United States. Pages 101–269 *in* U.S. Commission of Fish and Fisheries, Part XXIV. Report of the Commissioner for the year ending June 30, 1898. Government Printing Office, Washington, D.C.

Stout, L.J. 1958. Atlas of Forestry in New York. State Univ. College of Forestry at Syracuse University

Swaney. D. P., D. Sherman, and R. W. Howarth. 1996. Modeling water, sediment and organic carbon discharges in the Hudson-Mohawk Basin: coupling to terrestrial sources. Estuaries 19:633–647.

U. S. Census Bureau. 2000 Census Data. Available at: http://www.census.gov (December 2004).

U. S. Census of Agriculture: 1925. United States Department of Commerce, United States Government Printing Office, Washington DC.

U. S. Census of Agriculture: 1935. United States Department of Commerce, United States Government Printing Office, Washington DC

U. S. Census of Agriculture: 1945. 1946. volume 1 Counties and State Economic Areas. Part 2. Middle Atlantic States. United States Department of Commerce, United States Government Printing Office, Washington DC.

U. S. Census of Agriculture:1954. 1956. volume 1 Counties and State Economic Areas. Part 2. Middle Atlantic States. United States Department of Commerce, United States Government Printing Office, Washington DC.

U. S. Census of Agriculture:1964. 1966. volume 1 Statistics for the States and Counties. Part 7. New York. United States Department of Commerce, United States Government Printing Office, Washington DC.

U. S. Census of Agriculture:1974. 1977. volume 1 State Reports. Part 32. New York. United States Department of Commerce, United States Government Printing Office, Washington DC.

U. S. Census of Agriculture:1987. 1989. volume 1 Geographic Area Series. Part 32. New York State and County Data. United States Department of Commerce, United States Government Printing Office, Washington DC.

United States Department of Agriculture, Forest Service. 1970–1977. Forest Resource Bulletins NE-20 (NY-1968), NE-34 (NJ-1972), NE-44 (CT-1972), NE-46 (VT-1973), NE-48 (MA-1972).

United States Department of Agriculture, Forest Service. 1990–1999. Forest Resource Inventory Reports NE-112 (NJ-1987), NE-132 (NY-1993), NE-145 (VT-1997), NE-147(CT-1998), NE-148 (MA-1998).

USEPA. 2001. Better Assessment Science Integrating Point and Nonpoint Sources. BASINS Version 3.0. Users' Manual. Office of Water. United States Environmental Protection Agency Report EPA 823B01001. Washington.

Vörösmarty, C. J., K. P. Sharma, B. M. Fekete, A. H. Copeland, J. Holden, J. Marble, and J. A. Lough. 1997. The storage and aging of continental runoff in large reservoir systems of the world. Ambio 26:210–219.

Waring, R. H., and S. W. Running. 1998. Forest Ecosystems: Analysis at Multiple Scales. 2nd edition. Academic Press, San Diego.

Weaver, M. 1964. History of Tile Drainage (in America prior to 1900). M. M. Weaver, Waterloo, NY.

Whipple, J. S. 1909. Report of the Superintendent of Forests for the year 1909. State of New York Forest, Fish and Game Commission, Annual Reports 1907–1908-1909. J. B. Lyon Company, Albany.

Yanai, R. D., M. A. Arthur, T. G. Siccama, and C. A. Federer. 2000. Challenges of measuring forest floor organic matter dynamics: repeated measures from a chronosequence. Forest Ecology and Management 138:273–283.

Table A.1 Modern (1990) values of county areas and the proportions falling within the Hudson-Mohawk watershed and its hydrological catalogue units (HUCs).

County	State	Area (km^2)	Upper Hudson	Sacandaga	Hudson-Hoosic	Mohawk	Schoharie	Middle Hudson	Rondout	Hudson-Wappinger	Lower Hudson	total
Fairfield	CT	1621									0.064	0.064
Berkshire	MA	2411			0.158			0.040				0.198
Bergen	NJ	606									0.041	0.041
Hudson	NJ	122									0.018	0.018
Passaic	NJ	479							0.036			0.036
Sussex	NJ	1349							0.360			0.360
Albany	NY	1357			0.000	0.063	0.141	0.796				1.000
Bronx	NY	109									0.175	0.175
Columbia	NY	1647						0.956		0.000		0.956
Delaware	NY	3745					0.010	0.000				0.010
Dutchess	NY	2077						0.103		0.662	0.044	0.809
Essex	NY	4654	0.414									0.414
Fulton	NY	1285		0.291		0.709						1.000
Greene	NY	1678					0.363	0.596				0.959
Hamilton	NY	4457	0.214	0.273		0.145						0.632
Herkimer	NY	3657				0.462						0.462
Lewis	NY	3305				0.015						0.015
Madison	NY	1699				0.030						0.030
Montgomery	NY	1049				0.834	0.165					1.000
New York	NY	73									0.988	0.988

Table A.1 continued

County	State	Area (km²)	Upper Hudson	Sacandaga	Hudson-Hoosic	Mohawk	Schoharie	Middle Hudson	Rondout	Hudson-Wappinger	Lower Hudson	total
									0.988	0.988		
Oneida	NY	3142				0.468						0.468
Orange	NY	2113							0.456	0.312	0.018	0.474
Otsego	NY	2598				0.010	0.006					0.015
Putnam	NY	601								0.181	0.818	1.000
Rensselaer	NY	1694			0.406	0.000		0.594				1.000
Rockland	NY	451								0.001	0.365	0.366
Saratoga	NY	2103	0.014	0.272	0.580	0.134						1.000
Schenectady	NY	534				0.619	0.130	0.252				1.000
Schoharie	NY	1611				0.040	0.805	0.067				0.912
Sullivan	NY	2512							0.090			0.090
Ulster	NY	2919						0.317	0.459	0.084		0.860
Warren	NY	2253	0.543	0.189	0.055							0.787
Washington	NY	2165			0.512							0.512
Westchester	NY	1121									0.662	0.662
Bennington	VT	1751			0.708							0.708
Rutland	VT	2414			0.002							0.002
Windham	VT	2044			0.001							0.001

Table A.2 Estimated proportions of county areas falling within the Hudson-Mohawk watershed over time (to 3 decimal places).

County	State	Year							
		1790	1820	1850	1880	1920	1950	1970	2000
Fairfield	CT	0.065	0.058	0.058	0.058	0.058	0.064	0.064	0.064
Berkshire	MA	0.197	0.178	0.179	0.179	0.179	0.198	0.198	0.198
Bergen	NJ	0.038	0.034	0.036	0.037	0.037	0.041	0.041	0.041
Hudson	NJ	0.000	0.000	0.015	0.016	0.016	0.018	0.018	0.018
Passaic	NJ	0.000	0.000	0.030	0.032	0.032	0.036	0.036	0.036
Sussex	NJ	0.219	0.213	0.324	0.326	0.326	0.360	0.360	0.360
Albany	NY	0.999	0.903	0.903	0.903	0.903	1.000	1.000	1.000
Bronx	NY	0.000				0.161	0.175	0.175	0.175
Columbia	NY	0.956	0.864	0.864	0.864	0.864	0.956	0.956	0.956
Delaware	NY	0.000	0.009	0.009	0.009	0.009	0.010	0.010	0.010
Dutchess	NY	0.852	0.732	0.731	0.731	0.731	0.809	0.809	0.809
Essex[1]	NY	0.201	0.381	0.382	0.380	0.380	0.414	0.414	0.414
Fulton	NY	0.000	0.000	0.903	0.903	0.903	1.000	1.000	1.000
Greene	NY	0.000	0.867	0.866	0.866	0.866	0.959	0.959	0.959
Hamilton	NY	0.000	0.567	0.567	0.566	0.566	0.632	0.632	0.632
Herkimer	NY	0.000	0.418	0.418	0.418	0.418	0.462	0.462	0.462
Kings	NY	0.000	0.000	0.000	0.000	0.000	0.000	0.000	0.000
Lewis	NY	0.000	0.013	0.013	0.013	0.013	0.015	0.015	0.015
Madison	NY	0.000	0.026	0.026	0.027	0.027	0.030	0.030	0.030
Montgomery	NY	0.167	0.903	0.903	0.903	0.903	1.000	1.000	1.000
New York	NY	0.470	0.358	0.358	0.367	0.892	0.988	0.988	0.988
Oneida	NY	0.000	0.423	0.423	0.422	0.422	0.468	0.468	0.468
Orange	NY	0.648	0.710	0.710	0.710	0.428	0.474	0.474	0.474
Otsego	NY	0.000	0.013	0.013	0.014	0.014	0.015	0.015	0.015
Putnam	NY	0.000	0.903	0.903	0.903	0.903	1.000	1.000	1.000
Queens	NY	0.000	0.000	0.000	0.000	0.000	0.000	0.000	0.000
Rensselaer	NY	0.000	0.903	0.903	0.903	0.903	1.000	1.000	1.000
Rockland	NY	0.000	0.333	0.333	0.331	0.331	0.366	0.366	0.366

[1] 1790 values for Clinton County, from which Essex County was later formed.

Table A.2 continued

County	State	Year							
		1790	1820	1850	1880	1920	1950	1970	2000
Saratoga	**NY**	0.000	0.903	0.903	0.903	0.903	1.000	1.000	1.000
Schenectady	**NY**	0.000	0.903	0.903	0.904	0.904	1.000	1.000	1.000
Schoharie	**NY**	0.000	0.821	0.823	0.824	0.824	0.912	0.912	0.912
Sullivan	**NY**	0.000	0.082	0.082	0.082	0.082	0.090	0.090	0.090
Ulster	**NY**	0.454	0.777	0.777	0.777	0.777	0.860	0.860	0.860
Warren	**NY**	0.000	0.710	0.710	0.711	0.711	0.787	0.787	0.787
Washington	**NY**	0.617	0.464	0.464	0.463	0.463	0.512	0.512	0.512
Westchester	**NY**	0.661	0.597	0.597	0.598	0.598	0.662	0.662	0.662
Bennington	**VT**	0.707	0.641	0.639	0.639	0.639	0.708	0.708	0.708
Rutland	**VT**	0.002	0.003	0.001	0.001	0.001	0.002	0.002	0.002
Windham	**VT**	0.001	0.001	0.001	0.001	0.001	0.001	0.001	0.001

Table A.3 Population of counties of Hudson/Mohawk watershed.

County	State	Year							
		1790	1820	1850	1880	1920	1950	1970	2000
Fairfield	CT	36290	42739	59775	112042	320936	504342	792814	882567
Berkshire	MA	30263	35720	49591	69032	113033	132966	149402	134953
Bergen	NJ	12601	18178	14725	36786	210703	539139	897148	884118
Hudson	NJ	-	-	21822	187944	629154	647437	607839	608975
Passaic	NJ	-	-	22569	68860	259174	337093	460782	489049
Sussex	NJ	19500	32752	22989	23539	24905	34423	77528	144166
Albany	NY	75980	38116	93279	154890	186106	239386	286742	294565
Bronx	NY	-	-	-	-	732016	1451277	1471701	1332650
Columbia	NY	27496	38330	43073	47928	38930	43182	51519	63094
Delaware	NY	-	26587	39834	42721	42774	44420	44718	48055
Dutchess	NY	45276	46615	58992	79184	91747	136781	222295	280150
Essex[2]	NY	1615[1]	12811	31148	34515	31871	35086	34631	38851
Fulton	NY	-	-	20171	30985	44927	51021	52637	55073
Greene	NY	-	22996	33126	32695	25796	28745	33136	48195
Hamilton	NY	-	1251	2188	3923	3970	4105	4714	5379
Herkimer	NY	-	31017	38244	42669	64962	61407	67633	64427
Kings	NY	4549	11187	138882	599495	2018356	2738175	2602012	2465326
Lewis	NY	-	9227	24564	31416	23704	22521	23644	26944
Madison	NY	-	32208	43072	44112	39535	46214	62864	69441
Montgomery	NY	28852	37569	31992	38315	57928	59594	55883	49708
New York	NY	33111	123706	515547	1206299	2284103	1960101	1539233	1537195

[1] 1790 values for Clinton County, from which Essex County was later formed.

Table A.3 continued

County	State	Year							
		1790	1820	1850	1880	1920	1950	1970	2000
Oneida	**NY**	-	50997	99566	115475	182833	222855	273037	235469
Orange	**NY**	18477	41213	57145	88220	119844	152255	221657	341367
Otsego	**NY**	-	44856	48638	51397	46200	50763	56181	61676
Putnam	**NY**	-	11268	14138	15181	10802	20307	56696	95745
Queens	**NY**	16013	21519	36833	90574	469042	1550849	1987174	2229379
Rensselaer	**NY**	-	40153	73363	115328	113129	132607	152510	152538
Rockland	**NY**	-	8837	16962	27690	45548	89276	229903	286753
Saratoga	**NY**	-	36052	45646	55156	60029	74869	121764	200635
Schenectady	**NY**	-	13081	20054	23538	109363	142497	161078	146555
Schoharie	**NY**	-	23154	33548	32910	21303	22703	24750	31582
Sullivan	**NY**	-	8900	25088	32491	33163	40731	52580	73966
Ulster	**NY**	29370	30934	59384	85838	74979	92621	141241	177749
Warren	**NY**	-	9453	17199	25179	31673	39205	49402	63303
Washington	**NY**	14077	38831	44750	47871	44888	47144	52725	61042
Westchester	**NY**	23978	32638	58263	108988	344436	625816	894406	923459
Bennington	**VT**	12206	16125	49591	21950	21577	24115	29282	36994
Rutland	**VT**	15590	29983	33059	41829	46213	45905	52637	63400
Windham	**VT**	17572	28457	29062	26763	26373	28749	33476	44216

Table A.4 Estimated population of watersheds (HUCs) of Hudson/Mohawk watershed.

Subwatershed	Time							
	1790	1820	1850	1880	1920	1950	1970	2000
Upper Hudson	5150	10197	21281	26891	29140	37708	43833	54356
Sacandaga	7783	16428	19965	26974	32952	43732	59062	84051
Hudson-Hoosic	37912	67504	111102	118562	125303	161731	206697	260679
Mohawk	9441	76661	120588	147749	258877	335476	381366	361641
Schoharie	15490	33810	54873	63452	69693	91451	103184	112403
Middle Hudson	63095	113066	192406	274192	302450	414075	503923	543985
Rondout	16710	37214	58515	84538	99732	140184	215218	313554
Hudson-Wappinger	30956	43664	58177	81241	96137	149582	238645	324474
Lower Hudson	46202	80492	243235	543088	2426964	2727641	2611766	2667309
Whole Watershed	232740	479035	880142	1366688	3441248	4101579	4363694	4722454

Table A.5 Land in farms (km^2) in counties of Hudson/Mohawk watershed.

County	State	Year						
		1850	1870	1880	1920	1950	1969	1997
Fairfield	CT	1072	993	1266	913	391	92	48
Berkshire	MA	1811	1561	1944	1442	818	430	254
Bergen	NJ	440	426	380	150	54	26	11
Hudson	NJ	69	11	12	3	4	0	0
Passaic	NJ	390	387	352	130	38	6	9
Sussex	NJ	989	937	1027	868	663	398	295
Albany	NY	1204	1265	1239	1126	651	349	230
Bronx	NY	0	0	0	5	1	0	0
Columbia	NY	1455	1510	1653	1378	1052	706	465
Delaware	NY	2610	3073	2982	3015	2550	1472	743
Dutchess	NY	1923	1862	1986	1767	1229	656	432
Essex	NY	1229	1848	1692	1257	796	295	195
Fulton	NY	666	1066	1148	794	373	197	139
Greene	NY	1272	1372	1439	1144	695	284	197
Hamilton	NY	152	397	402	147	30	0	3
Herkimer	NY	1377	1540	2173	1353	1142	801	574
Kings	NY	84	49	42	4	1	0	0
Lewis	NY	943	1757	1874	1769	1298	887	727
Madison	NY	1574	1576	1595	1467	1285	955	752
Montgomery	NY	968	980	1160	919	828	653	546
New York	NY	11	5	10	1	0	0	0

Table A.5 continued.

County	State	Year						
		1850	1870	1880	1920	1950	1969	1997
Oneida	**NY**	2696	2645	2871	2475	1963	1294	875
Orange	**NY**	1715	1692	2023	1418	1108	636	384
Otsego	**NY**	2218	2454	2594	2320	1938	1318	838
Putnam	**NY**	489	515	563	457	174	56	14
Queens	**NY**	687	624	673	57	2	0	
Rensselaer	**NY**	1415	1452	1718	1245	921	493	401
Rockland	**NY**	313	227	321	162	70	16	2
Saratoga	**NY**	1671	1661	1867	1363	811	401	295
Schenectady	**NY**	428	501	508	415	245	140	74
Schoharie	**NY**	1259	1511	1602	1396	1170	699	448
Sullivan	**NY**	956	1599	1909	1679	777	375	235
Ulster	**NY**	1785	1860	2188	1654	921	371	279
Warren	**NY**	898	1405	1361	867	298	43	37
Washington	**NY**	1627	1884	1987	1757	1558	1082	789
Westchester	**NY**	1020	983	1035	421	196	61	30
Bennington	**VT**	906	771	1103	839	517	251	131
Rutland	**VT**	1800	1786	1923	1766	1361	796	509
Windham	**VT**	1697	1757	1768	1342	892	334	189

Table A.6 Estimated land in farms (km^2) in subwatersheds (HUCs) of Hudson/Mohawk watershed.

Watershed	Year						
	1850	1870	1880	1920	1950	1969	1997
Upper Hudson	957	1486	1408	1256	633	309	123
Sacandaga	776	1027	1093	888	504	308	143
Hudson-Hoosic	3032	3065	3583	3288	2452	1784	1011
Mohawk	3520	3930	4513	3751	3083	2484	1696
Schoharie	1715	1954	2066	1922	1647	1215	682
Middle Hudson	4498	4703	5113	4764	3584	2533	1351
Rondout	1862	1918	2244	2053	1425	1081	496
Hudson-Wappinger	1848	1816	2016	1837	1296	1072	510
Lower Hudson	1264	1233	1343	1113	260	238	79
Whole Watershed	19472	21133	23379	20872	14507	10667	5946

Table A.7 Forested lands (km²) in counties of Hudson/Mohawk watershed. – indicates no data. 1968 and 1993 include estimates based on reports from succeeding years

County	State	Year				
		1875	1920	1955	1968	1993
Fairfield	CT	-	-	-	756	600
Berkshire	MA	-	-	-	-	-
Bergen	NJ	-	-	-	-	-
Hudson	NJ	-	-	-	-	-
Passaic	NJ	-	-	-	-	-
Sussex	NJ	-	-	-	789	850
Albany	NY	245	324	384	539	691
Bronx	NY	0	0	0	0	0
Columbia	NY	251	324	638	842	1016
Delaware	NY	1572	1578	1974	2492	2708
Dutchess	NY	399	405	955	1071	1121
Essex	NY	1651	3845	4115	4322	4184
Fulton	NY	539	708	888	994	987
Greene	NY	581	688	1000	1182	1329
Hamilton	NY	2233	4249	4422	4374	4367
Herkimer	NY	537	2509	2410	2776	2776
Kings	NY	8	0	0	0	0
Lewis	NY	1603	1862	2261	2518	2497
Madison	NY	366	304	464	776	806
Montgomery	NY	146	101	194	264	394
New York	NY	0	0	0	0	0
Oneida	NY	823	951	1151	1581	1843
Orange	NY	611	1214	1123	1026	1159
Otsego	NY	620	648	905	1355	1493
Putnam	NY	210	283	418	398	417
Queens	NY	220	0	0	0	0
Rensselaer	NY	336	445	746	996	1034
Rockland	NY	241	324	333	0	210
Saratoga	NY	600	911	1172	1350	1423
Schenectady	NY	93	142	122	190	283
Schoharie	NY	396	405	732	871	1081
Sullivan	NY	1348	1518	1697	1813	1968
Ulster	NY	1409	1740	2334	2167	2359
Warren	NY	878	1680	2074	2120	2159
Washington	NY	513	708	884	1062	1232
Westchester	NY	261	142	598	406	487
Bennington	VT	-	-	-	1522	1529
Rutland	VT	-	-	-	1837	1990
Windham	VT	-	-	-	2022	1765

Table A.8 Estimated forested land (km²) in subwatersheds of Hudson/Mohawk watershed.

Watershed	Year				
	1875	1920	1955	1960s-70s	1980s-90s
Upper Hudson	1477	3078	3790	3892	3856
Sacandaga	996	1756	2175	2249	2273
Hudson-Hoosic	718	1051	1549	2931	3082
Mohawk	1513	2805	3428	4024	4337
Schoharie	559	610	1078	1305	1585
Middle Hudson	1373	1716	2874	3436	3937
Rondout	947	1347	1738	1912	2096
Hudson-Wappinger	552	763	1254	1283	1377
Lower Hudson	417	436	921	708	848
Whole Watershed	8551	13562	18807	21740	23390

Table A.9 Numbers of dams in each hydrologic subunit of the Hudson River basin.

Hydrologic	Area	Number of dams						
Upper Hudson	4222	1	2	34	16	2	0	55
Sacandaga	2720	0	2	15	8	3	3	31
Hudson-Hoosic	4869	0	3	35	25	12	7	82
Mohawk	6605	0	6	37	31	16	17	107
Schoharie	2401	0	4	22	17	1	6	50
Middle Hudson	6190	0	2	51	50	13	7	123
Rondout	3082	0	3	61	41	6	2	113
Hudson-Wappingers	2403	1	4	45	40	11	3	104
Lower Hudson	1864	0	5	62	37	14	11	129
Bronx[2]	(?)	0	0	0	1	1	1	3

[1] Includes only those dams falling within the Hudson part of this hydrologic subunit.

American Fisheries Society Symposium 51:113–128, 2006

Hudson River Paleoecology from Marshes: Environmental Change and Its Implications for Fisheries

Dorothy M. Peteet
NASA/Goddard Institute for Space Studies
2880 Broadway, New York, New York 10025 USA
Lamont-Doherty Earth Observatory
61 Route 9W, Palisades, New York 10964 USA
peteet@ldeo.columbia.edu

Dee Cabaniss Pederson
Lamont-Doherty Earth Observatory
61 Route 9W, Palisades, New York 10964 USA

Dorothy Kurdyla and Tom Guilderson
Lawrence-Livermore National Laboratory, Center for Accelerated Mass Spectometry
7000 East Avenue, L-403, Livermore, California 94550 USA

Abstract.—Hudson riverine and coastal marshes provide a paleoecological archive consisting of information on climate and land use at both the local and watershed scales. The timing of formation of these marshes is documented using accelerator mass spectrometry (AMS) ^{14}C dating of identified plant macrofossils in basal marsh organic sediments. While the Staten Island marsh is oldest and dates to 11,000 years before present (BP), Piermont, Iona, and Croton marshes date to the mid-Holocene, and the Jamaica Bay marshes formed most recently. Pollen and spores, charcoal, and plant macrofossils in the marsh sediments document marked climatic shifts as well as anthropogenic impact in the region. Assessment of the inorganic and organic content of the sediments in the marshes reveals a pattern of decreasing inorganic supply with the arrival of the Europeans, possibly due to the construction of numerous Hudson River tributary dams. Piermont Marsh, because of its sensitive location in the Hudson River, records droughts and wet intervals through species which have specific salinity affinities. Throughout the marsh records, the ragweed *Ambrosia* pollen rise marks the anthropogenic impact at the landscape scale. The recorded changes in hydrology and salinity that occurred throughout the centuries and millennia would have had profound effects throughout the food web and estuarine ecosystem. Fish populations would have been affected by changes in the food supply due to shifts in runoff affecting turbidity and light penetration in the river. Local vegetation changes within marshes may also have affected juvenile fish populations.

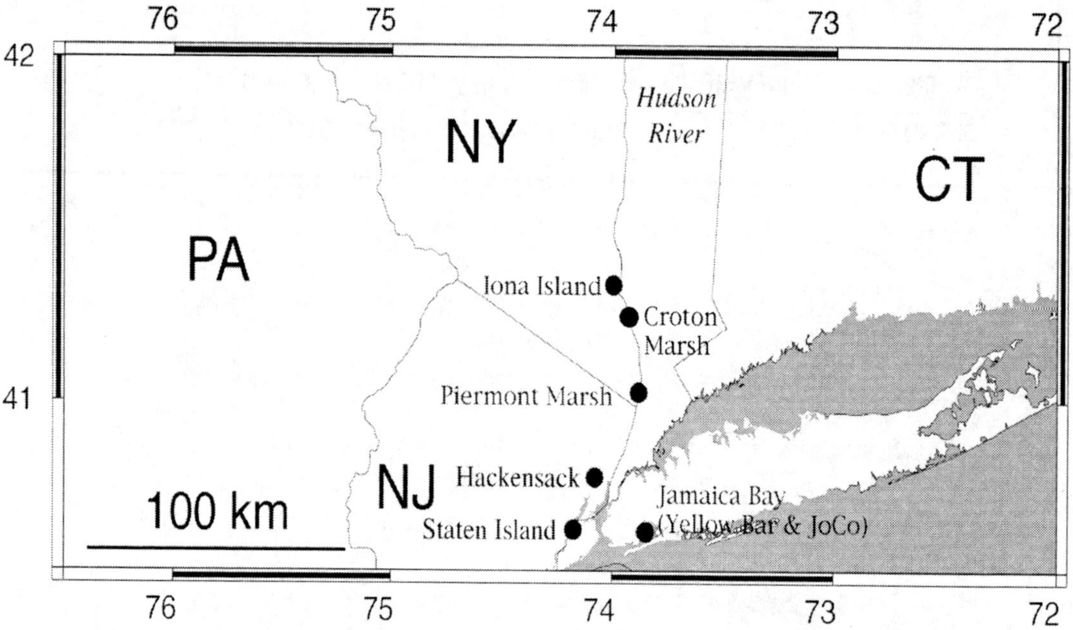

Figure 1. Suite of Hudson River tidal marsh sites from which we have collected cores.

Introduction

Tidal river marshes, among the most productive ecosystems in the world, are ecologically important for providing a nursery habitat for aquatic biota, protecting the shoreline, and mitigating floods. However, they are also unique archives of past environmental conditions in the estuary. Unlocking the history of regional vegetation and local marsh species reveals a wealth of information concerning past climate and biodiversity for the Hudson landscape. Despite decades of paleoenvironmental research on Atlantic seaboard tidal marshes, including the Chesapeake (Brush et al. 1982; Brush 1984, 1986, 1989; Willard and Korejwo 1999; Pasternack et al. 2001; Willard et al. 2003) and along the Connecticut coast (Thomas and Varekamp 1991; Varekamp and Thomas 1998), very little is known about the history of Hudson River estuarine marshes, with previous paleoecological studies limited to exploratory stratigraphic study (Newman et al. 1987; Wong and Peteet 1999). However, extensive wetland losses have occurred throughout the estuary, and, in Jamaica Bay, approximately 25% of the original marsh remains after human development in this century alone (Englebright 1975). Estimates of the percent marsh loss in three remaining Jamaica Bay salt marshes from 1959–1998 is about 12% (Hartig et al. 2002). These losses in biota including invertebrate, fish, and bird habitat signal a decline in the health of the estuary.

The adjacent Long Island pollen, spore, and macroscopic marsh investigations include records back to about 500 years ago (Heusser et al. 1975; Clark and Patterson 1985; Clark 1986), where human influence is recorded in the uppermost sediments. The nearby Hackensack marshes (Figure 1) also record substantial shifts in native vegetation in the pollen and macrofossil records that have been interpreted from several sites. Heusser (1949, 1963), and more recently, Carmichael (1980), documented the 2,500-year history of vegetational change in the Hackensack tidal marsh (Meadowlands), New Jersey. The peat stratigraphy there traces the development of the marsh from a freshwater alder swamp to a brackish marsh with frequent tidal inundation. The Hackensack Meadowlands macrofossil record shows a large impact of Europeans on the region as common reed *Phragmites* and narrow-leaf cattail *Typha* seeds replace members of the Cyperaceae family (*Scirpus, Cyperus, Eleocharis*) in upper sediments (Carmichael 1980). The increased biodiversity exemplified by the Meadowlands prior to major impact of invasive species illustrates the major shift in plant communities providing habitat for the tidal food web.

Our goals are to describe the age of formation of the lower Hudson tidal marshes, to review significant recent findings that have been documented in them, and to evaluate the importance of drought for the future of biota in the estuary. Particular emphasis is given to the role of anthropogenic influence in the estuary, and its ongoing effects on vegetation and the marsh environment. Specific questions we address are

1)What are the ages of the lower Hudson tidal marshes and how deep is the peat sequestered in them?

2) What is the history of inorganic and organic content in the Hudson marshes over time?

3) What do the marshes record about vegetational and climate change in New York and New Jersey?

4) What is the relationship of Hudson Marsh stratigraphy to previous regional paleoclimate records?

5) How are paleoecological shifts as indicated by Hudson marshes important for fisheries, especially in light of future droughts?

Hudson Marsh Importance for Paleoenvironmental Studies

Unlike river bottom sediments, which are subject to frequent erosion and redeposition, most marshes are dynamic in that their growth can keep pace with sea level and provide a fairly constant and uninterrupted record of sediment accumulation. In contrast to most coastal marshes of the northeastern United States, which are characterized by a thin blanket of sediment ranging from 2 to 3 m in depth (Niering et al. 1977; Varekamp and Thomas 1998), sedimentary profiles in marshes of the Hudson River can be very deep, ranging from 10 to 12 m over the last 4000–6000 years (Newman et al. 1987; Wong and Peteet 1999; Peteet, unpublished data). These high-depositional estuarine wetlands provide the ideal environmental setting for identifying abrupt short-term shifts in climate because of the opportunity to sample at high temporal resolution. Brackish marshes are also extremely sensitive to shifts in salinity.

Riverine marshes are largely controlled by the hydrologic regime, which can be highly variable, depending on the balance between water inputs (e.g., precipitation, snowmelt) and outputs (evaporation, evapotranspiration) (Mitsch and Gosselink 1993). The Hudson estuary has an extremely low stability ratio (0.4) which is defined as the ratio of the volume of an estuary (in km^3) to the mean flow of freshwater in m^3/s into the system (Simpson et al. 1973). This means that it has a very rapid response to storms and hurricanes, with the salt front moving long distances very rapidly (Cooper et al. 1988). Salinity is the dominant factor in determining the distribution of most plant species along the estuarine gradient (Odum 1988). Riverine marshes such as Piermont, Croton, and Iona Marsh in the lower Hudson (Figure 1) thus are probably the most sensitive marshes on the East Coast in terms of major shifts in salinity from sustained droughts. Species composition, plant cover, and allocation of aboveground and belowground biomass in marshes can shift markedly and rapidly in response to hydrologic changes (Dunton et al. 2001). A suite of marshes along a salinity gradient thus provides a powerful resource to link stratigraphic changes throughout the lower estuary, also enabling a better understanding of the processes that link the sites in the adjacent uplands in the watershed to deposition in the marshes.

Hudson Marsh Sites and Regional Marsh Records

We now are working on a suite of cores from tidal marshes in the Hudson estuary (Figure 1) ranging from the Hudson mouth (Staten Island, Jamaica Bay) to about 70 km upriver (Iona Island). The sites described here, from south to north, include the Jamaica Bay JoCo and Yellow Bar marshes, Arthur Kill Marsh in Staten Island, Piermont Marsh, Croton Point Marsh, and Iona Island Marsh. These are all tidally influenced but arrayed at increasing distances from the New York Bight along a wide salinity gradient. Plants characteristic of the marshes range from typical maritime salt meadow cordgrass *Spartina alterniflora*, salt meadow hay *S. patens*, and inland saltgrass *Distichlis spicata* in the coastal marshes (Jamaica Bay, Staten Island) to the more brackish marshes that have higher species diversity (Winogrond 1997) with a large component of narrow-leaf cattail and bulrush *Scirpus* species (Piermont, Iona) as well as the ubiquitous common reed. Recent research on the modern plant communities and their dramatic changes due to human impact in the last few decades is best documented in Piermont, one of the four National Estuarine Research Reserve (NERR) sites in the Hudson River. From 1965–1991, common reed colonies increased their coverage of the marsh from 35% to 66%, destroying much of the biodiversity which characterized this brackish setting (Lehr 1967; Winogrond 1997). The pattern of common reed invasion in recent decades (Carmichael 1980; Clark and Patterson 1985; Wong and Peteet 1999; Pederson et al. 2005) suggests that it is a result of human impact, and its recent dominance has been attributed to introduced varieties that vary genotypically from the native types (Saltonstall 2002).

Methods

Tidal marsh sites were cored manually using primarily the modified Livingstone piston corer (Wright et al. 1984), which is 5 cm in diameter. To retrieve the top meter of the sediment, which is sometimes problematic with the Livingstone, a clear plastic pipe was hammered into the peat and dug out with a shovel and the Hiller peat corer was used to eliminate compaction.

Chronological Control

We use ^{137}Cs for dating of the "bomb spike" in 1963, and Accelerator Mass Spectrometry (AMS) radiocarbon dating of in situ plant macrofossils, which is the best possible method of dating this type of stratigraphy. Because the seeds take in carbon from the atmosphere, there is no correction for marine or terrestrial old carbon which results in a large uncertainty in estuarine cores (Colman et al. 2002; Rubenstone and Peteet, Lamont Doherty Earth Observatory, unpublished). Radiocarbon dating of basal organic material just atop mineral matter indicates the time of formation of the marsh in the estuary.

Sediment Stratigraphy, Loss-on-Ignition (LOI)

All sediment cores are described in terms of the sediment type (peat, silt, clay) and weighed samples are dried overnight and then combusted at 450°C to determine percent loss-on-ignition (LOI) (modified from Dean 1974). The LOI values are a proxy for organic content and, hence, the carbon content of the cores. Carbon accumulation rates can then be calculated using dry bulk density of the sediments and precise sedimentation rates calculated from AMS ^{14}C dates in the cores.

Pollen and Spore Analysis, Microscopic Charcoal

Samples were processed for pollen and spores using the procedures of Heusser and Stock (1984), which include soaking the samples in KOH, HF, screening with 7- and 150-mm screens, and the use of glacial acetic acid, acetolysis mixture, and silicone oil mounts (Faegri and Iversen 1975). Exotic *Lycopodium* tablets were added to the samples to determine pollen concentration. A minimum of 300 pollen grains were counted and all palynomorphs tallied; where local pollen was abundant, we counted at least 300 upland types. Pollen identification was aided by a modern reference collection and reference books (Faegri and Iversen 1975; Moore and Webb 1978; Lewis et al. 1983). Both percentage and influx diagrams were constructed using TiliaGraph (Grimm 1992). Microscopic charcoal pieces greater than 50 µm x 10 µm (500 µm^2), used to infer fire (Clark and Patterson 1985) were counted along with pollen grains and *Lycopodium*. Charcoal: pollen ratios (C:P) were calculated for the Piermont core to clearly demonstrate the role of this disturbance at the site:

$$C{:}P = \frac{\text{charcoal area } (\mu m^2)}{\#\text{ pollen grains}}$$

Plant Macrofossils, Foraminifera, and Charcoal

Macrofossils were analyzed according to the procedures of Watts and Winter (1966). This includes soaking samples in 10% KOH overnight to disaggregate the sediment, then washing them through 500-and 150-µm brass screens. Plant macrofossils are a crucial component of the vegetation reconstructions because they can be identified to the species level and, therefore, more definitively portray the ecological conditions. They unequivocally establish the presence of a species at the site, and selected macrofossils are used for AMS ^{14}C dating to give precise timing of changes in vegetation and climate through time. Tidal marsh peat is composed of peat rhizomes as well as reproductive plant parts. Keys such as Niering et al. (1977) were used to identify peat types in the samples. Foraminifera were screened and picked in water, then identified to assess salinity changes for selected samples. Charcoal pieces larger than 500 µm were counted as part of the macrofossil analysis.

Results and Discussion

What is the age of the lower Hudson tidal marshes and how deep is the peat sequestered in them?

From 1998–2003, cores were extracted from marsh sites throughout the lower Hudson River, ranging from Jamaica Bay at the mouth of the estuary (Figure 1) to Iona Island, about 70 km north of the Battery. Table 1 documents peat depth and basal core age from the suite of Hudson marshes where sediment cores have been extracted, peat depths recorded, and basal peat ages determined through AMS ^{14}C dating on identified plant macrofossils contained in the sediment cores. At all sites except Piermont, inorganic sediments were identified as basal

Table 1. Location, peat depth, and basal ^{14}C age of Hudson marshes.

Tidal Marsh	Latitude/ Longitude	Peat Depth	Basal ^{14}C Age
JoCo	40° 37'N	2 m	>460 <2000
Jamaica Bay	73° 47'W		
Yellow Bar	40° 37'N	0.8 m	>450
Jamaica Bay	73° 50'W		
Arthur Kill	40° 36'N	8.0 m	11,100
Staten Island	74° 13'W		
Hackensack	40° 48'N	3.7 m	2,610
New Jersey	74° 04'W		
Piermont N	41° 00'N	13.7 m	5,700
Piermont	73° 55'W		
Croton Marsh	41° 14'N	10 m	4,630
	73° 50"W		
Iona Marsh	41° 18'N	10 m	5,500
	73° 58'W		

mineral matter on which the marsh generation took place. The large variability in peat depth and age suggests that the origin of marsh formation is quite diverse. Basal dates from two Jamaica Bay sites indicate that the Jamaica Bay marshes were formed within the last millennium, suggesting that they likely owe their origin to recent protection of the area from open ocean wave action.

In contrast, Staten Island's Arthur Kill marsh at the same latitude is older than the estuary itself, as its freshwater sediments reveal a fossil assemblage including violet *Viola*, pondweed *Potamogeton*, and alder *Alnus* seeds and willow *Salix* buds, indicating an age of 11,100 years. The freshwater origin of the Arthur Kill marsh in Staten Island is significant because it predates sea level rise to the site. Additional implications for this study include interpretations of the timing of isostatic uplift in coastal New York or changes in river channel geomorphology.

Of much younger origin than the Arthur Kill marsh are the upriver brackish sites such as Piermont, Croton, and Iona Island marshes. These wetlands all formed in the mid-Holocene, probably due to the influx of the marine water into the Hudson River canyon as eustatic sea level rose and isostastic rebound favored marshes keeping pace with sea level. The basal material from Piermont (13.77 m) lacks any identifiable macrofossils, but the oldest age obtained from seeds is 5,700 years (at 13.70 m depth). We have not yet reached the base of the marsh at this site. Basal material from the Croton Point marsh (10 m core) includes fibrous woody material, Foraminifera, and *Zanichellia* seeds and other brackish vegetational components. The basal sediment from the Iona Island (10 m core) has vegetation such as *Scirpus* and *Cyperus* seeds, implying a brackish environment. Reasons for the remarkably high sedimentation rate at Piermont marsh includes possible basal clay compaction below the marsh itself, high plant production due to river and marine nutrients at the site, and tectonic activity. The generation of these marsh sites would have substantially increased the nursery habitat for those fish that use these habitats for food and protection of young. This study provides the groundwork for more

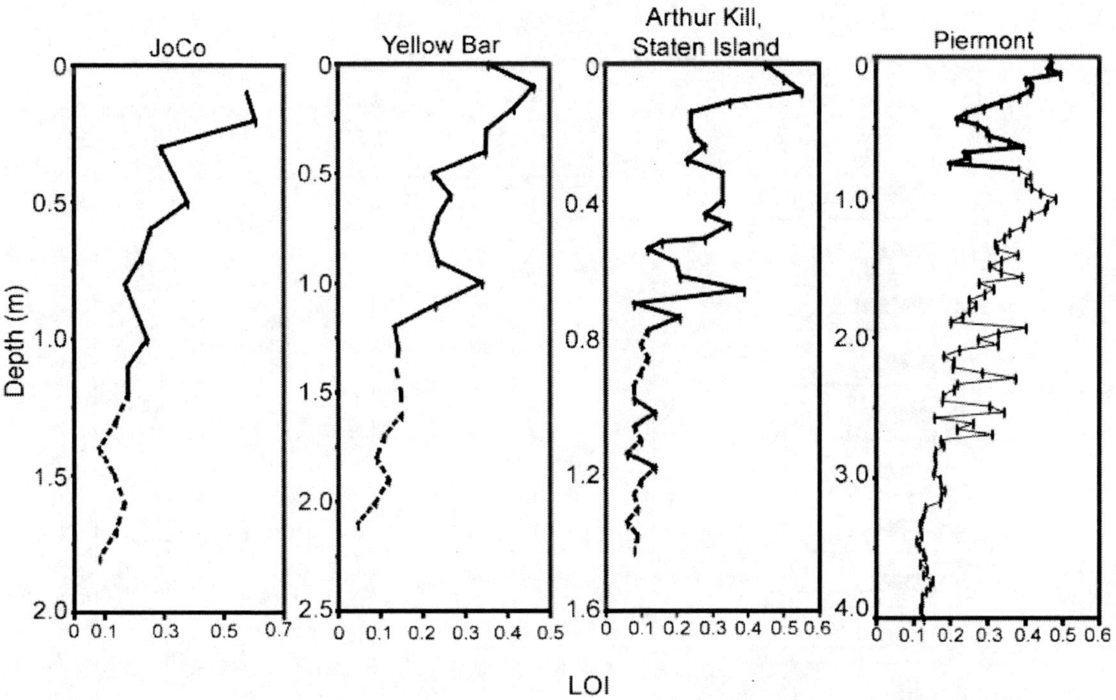

Figure 2. Loss-on-ignition (LOI) with depth for four tidal marsh sites in the Hudson Estuary. Solid line indicates anthropogenic impact. Between 0.8–1.2 m depth in each core, pollen analysis and ^{14}C dating indicates European arrival, and the organic percent increases with fluctuations up to the present

high-resolution studies of these marshes to understand the fluctuations in salinity caused by climate (i.e., precipitation and evaporation) or possibly tectonic change or relative sea level rise.

What is the history of inorganic and organic content in the marshes?

Marsh LOI profiles (Figure 2) indicate a significant increase in percent organic matter from the arrival of Europeans to the uppermost modern sediment. However, subsequent research on masses in all of the marshes (Figures 3 and 4) indicates that the trend of increasing LOI with European arrival results from a dramatic decline in inorganic content instead of an increase in organic content. The Yellow Bar, Jamaica Bay core (Figure 3) shows a sharp decline in inorganic content at 110–115 cm, the same interval as the European impact as shown from the ragweed rise (Peteet unpublished). The last 1500 years in Piermont Marsh (Figure 4) demonstrates a fairly constant organic content while the inorganic content is much greater during the Medieval Warming in the lower part of the diagram (230–160 cm), probably due to increased upland runoff, and in contrast increases only slightly in the settlement era between 76 and 36 cm (Pederson et al. 2005).

One hypothesis for the decline in inorganic flux to the sites with anthropogenic influence is the construction of numerous dams on the Hudson tributaries by the settlers and continuing to the present, with almost 800 remaining to this day (Swaney et al. 2006, this volume). While the dams would have blocked the heavier inorganic sands and silts, the lighter organic material would continue downriver with the flow. Other possibilities include changes in source contri-

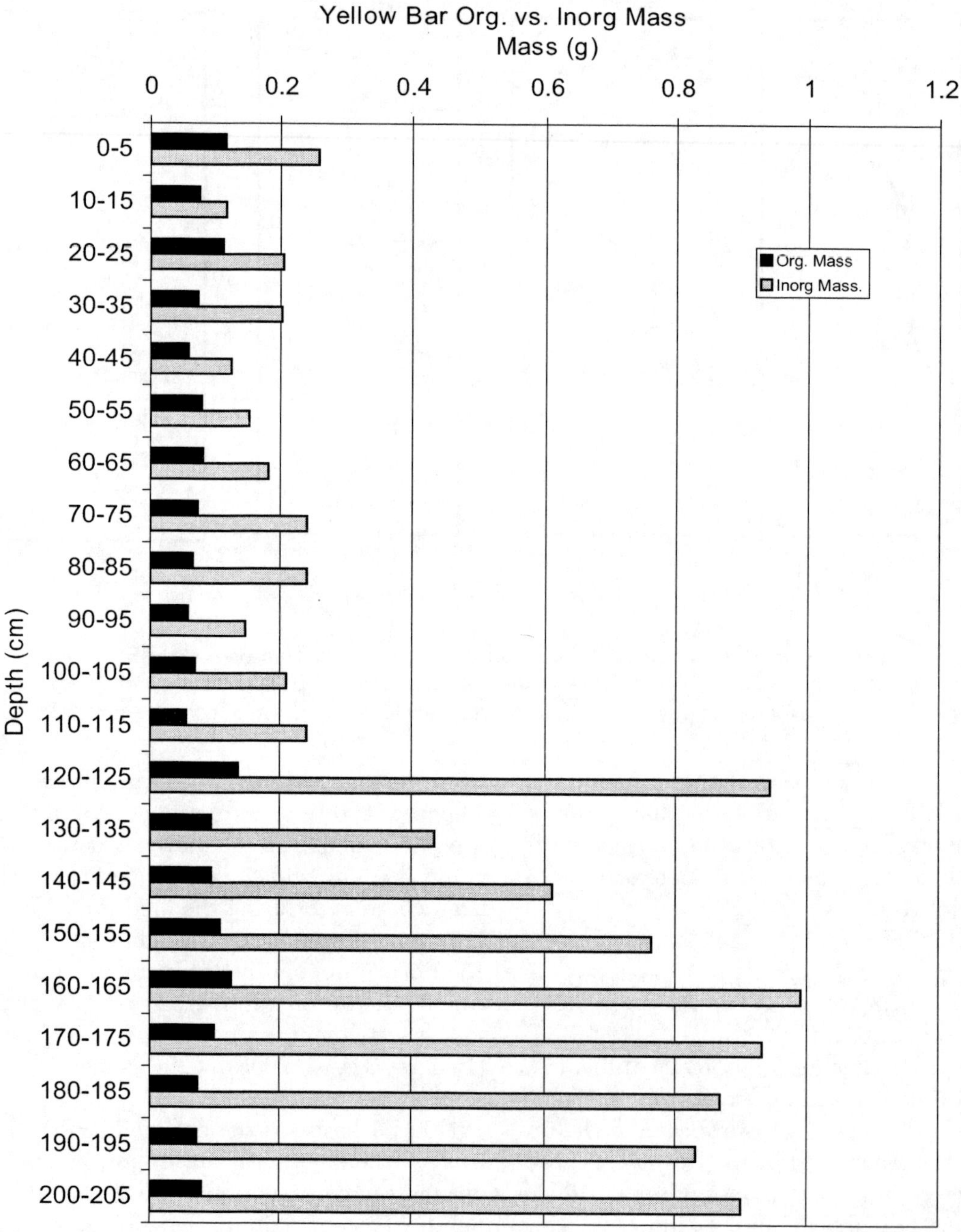

Figure 3. Sediment component masses based on dry sample mass in Yellow Bar, Jamaica Bay core. Inorganic content drops sharply at 110 cm with European impact.

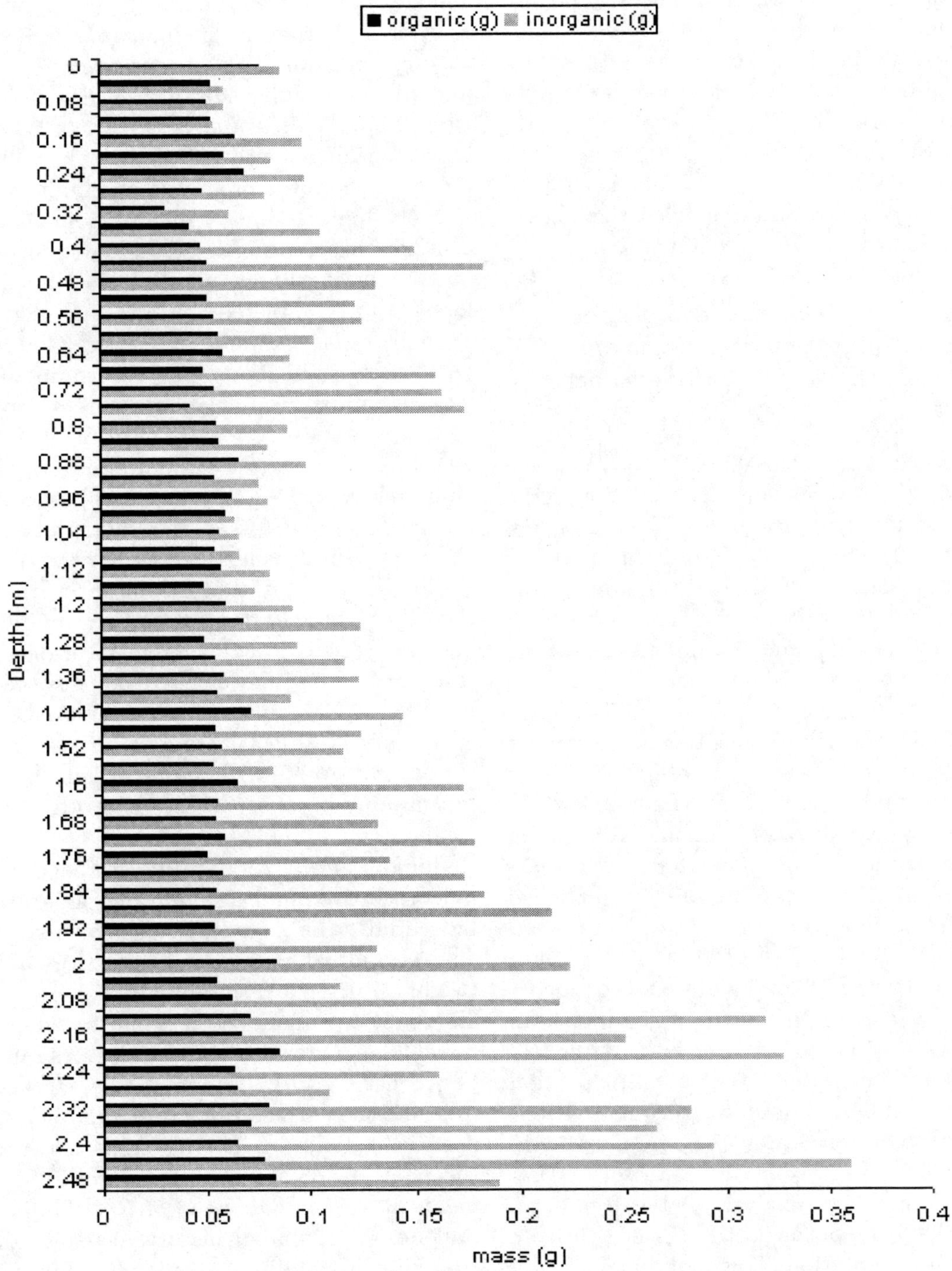

Figure 4. Sediment component masses based on dry sample mass in upper portion of northern Piermont Marsh core. While the Medieval Warm interval from 230 to 160 cm has relatively high inorganic mass, the upper settlement era of high ragweed (*Ambrosia*) from 76 to 36 cm also has a slight increase in inorganic mass (adapted from Pederson et al. 2005).

butions from the watershed with land use change, although land use changes with settlement in the Chesapeake estuary resulted in major increases in inorganic sediment as well as eutrophication which increased sedimentation rates (Cooper and Brush 1991). A third possibility is that local vegetational shifts changed the matrix for trapping inorganic sediments.

What do the marshes record about regional vegetational and climate change?—Example from Piermont Marsh

Piermont Marsh (40 °00 N, 73 °55 W), protected by the Hudson River National Estuarine Research Reserve (HRNERR), extends over 110 ha, is about 40 km north of the southern tip of Manhattan, and is bordered on the west by Tallman State Park (Figure 1). The mean tidal range of the marsh is 0.98 m and the mean salinity is 6 practical salinity units (psu) (Winogrond 1997). Piermont Marsh is the northernmost marsh on the Hudson River with native salt-tolerant marsh grasses (Lehr 1967). We have two existing cores in the high marsh communities (*Spartina patens, Distichlis spicata*) from the site. In 1998 we retrieved a 11.2 m core from the site with an AMS date at that depth of 4100 ^{14}C yr BP. Subsequently, we retrieved additional sediments in the same hole down to 13.77 m with an AMS date on seeds of 5700 + 45 ^{14}C yr BP. A second core site in the southern portion of the marsh was sampled in 2001. Preliminary exploratory study of Piermont Marsh, Piermont, NY (Wong and Peteet 1999) resulted in coarse analysis of the top 11 m (4100 ^{14}C yrs) of the northern Piermont core at a 1-m resolution. This preliminary study reported the occurrence of a wide diversity of marsh macrofossils which record variations in salinity (freshwater aquatics such as *Chara* and *Nitella* versus halophytes such as *Salicornia* in the Chenopodiaceae). Large changes in charcoal abundance are present, as well as variations in high marsh Foraminifera such as *Trochammina macrescens* and higher salinity low marsh *Elphidium*. Preliminary pollen analysis, also at the 1-m resolution, showed major shifts in pollen percentages downcore, with intriguing, extremely high *Tsuga* (hemlock) pollen percentages near the base of the core. Subsequent ongoing higher resolution of macrofossils (not completed) on a 2.2-m section of the core (11.2–9 m depth, 4200–3400 ^{14}C years BP) indicates rapid environmental changes, illustrated at about 940 cm depth by a dramatic drop in LOI from 80% to 10%, the disappearance of Foraminifera and charcoal particles, and the dominance of freshwater algae fossils. We interpret this change as indicating increased river flow and higher precipitation alternating with intervals of low river flow (more Foraminifera, higher charcoal, drought). We have also identified *Salicornia* and charred *Scirpus* seeds in these drought intervals. The presence of *Salicornia* is especially important because it does not grow in Piermont Marsh today, presumably because the site is too fresh.

Pederson et al. (2005) investigated the paleorecord from Piermont marsh at the bidecadal scale over the last 1500 years. The simplified pollen and spore diagram (Figure 5) from stratigraphy in Pederson et al. (2005) documents both climatic and land use change. Our downcore LOI values range between 20% and 60% (Figure 5), suggesting that the surface was always marsh and not subject to channel erosion, as our measurements indicate modern channel sediments have an LOI value of 5%. The last millennium of marsh history also includes a striking drought from about 800–1300 years BP (Zone P-1, Figure 5), as demonstrated by high values in more drought-tolerant pine *Pinus* and hickory *Carya* concurrent with low values in oak *Quercus* and other more drought-sensitive species such as hemlock

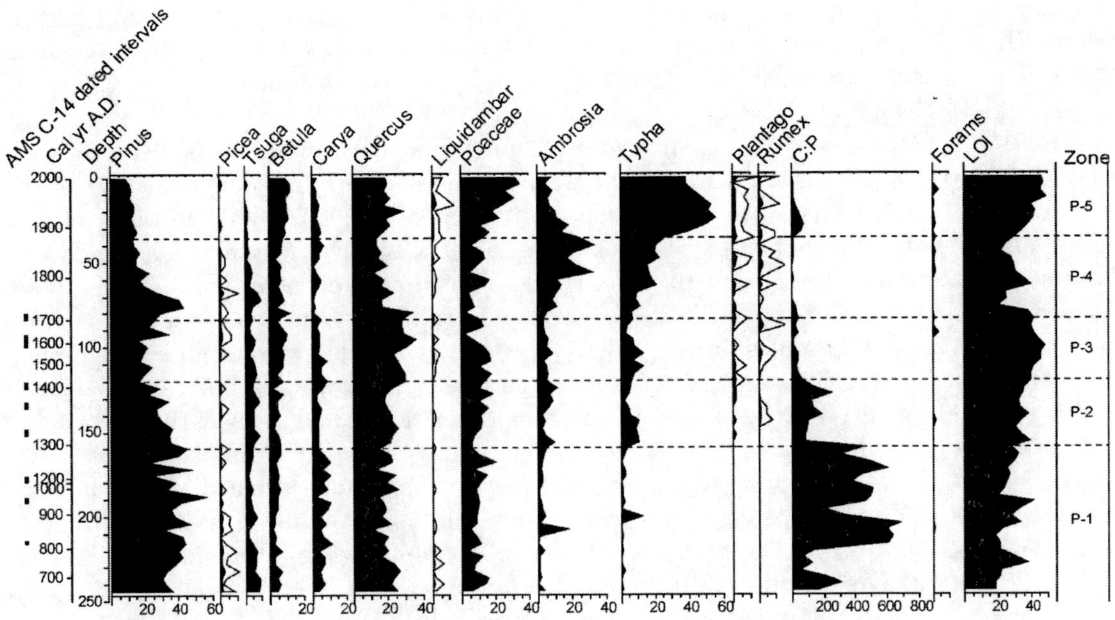

Figure 5. Pollen and spore percentage diagram of selected taxa along with Foraminifera and LOI from upper portion of northern Piermont core (adapted from Pederson et al. 2005).

Tsuga and birch *Betula*, large increases in charcoal, low pollen accumulation, and the absence of Foraminifera (Pederson et al. 2005).

What is the relationship of Hudson marsh stratigraphy to previous regional paleoclimate records?

The timing of this dry period in Piermont corresponds to the classic Medieval Warm Period (MWP) as defined by Lamb (1982) to be between approximately A.D. 800–1200. We have correlated this drought in our region with salinity records from Chesapeake Bay (Cronin et al. 2000; Cronin et al. 2003), pollen records from lakes and ponds in Massachusetts (Fuller et al. 1998), and tree ring records of east coast drought (Stahle et al. 1988). Cronin et al. (2000) noted that large fluctuations in salinity in the Chesapeake estuary included 14 wet–dry cycles in the last 500 years, including 16th and early 17th century megadroughts that exceeded 20th century droughts in their severity. Cronin et al. (2000) further stated that these salinity oscillations equate to changes in regional precipitation and streamflow of 40–50% if relationships among rainfall, streamflow, and salinity were similar to those of the last 200 years. The consequences of severe and prolonged droughts for the future of the lower Hudson and the character of the estuarine habitat would be extreme. Determining the presence of such sustained droughts is particularly critical to evaluate their potential future effects on the Hudson estuary, both in the open river and in the wetland communities.

How are paleoecological shifts indicated by Hudson marshes important for fisheries, especially in light of future droughts?

Freshwater flow may be the single largest factor influencing trophic functioning in the Hudson estuary (Gladden et al. 1988). The food web is supported by allochthonous carbon input (69% of total) as well as phytoplankton and submersed macrophytes (Howarth et al. 1996; Caraco et al. 2000). Our results demonstrate that major changes in the hydrological cycle have taken place over millennia, with long-term megadroughts that must have affected the entire estuary as well as the local marsh biota. As Howarth et al. (1991) noted, carbon and sediment inputs to the estuary have major influences on ecosystem metabolism and light penetration. Larvae and juveniles of many fish in the Hudson feed heavily on plankton and drifting benthic invertebrates in the water column (i.e., American shad *Alosa sapidissima*, blueback herring *Alosa aestivalis*, alewife *Alosa pseudoharengus*, striped bass *Morone saxatilis*; Strayer and Smith 2001). Strayer et al. (2004) have shown that freshwater flow is significantly correlated with distributions of many fish species in the Hudson. In years of higher freshwater flow, in general, fish populations are centered further downriver and reveal slow apparent growth. The increased flow in wet years presumably is responsible for poor development of phytoplankton because of high advective losses and light limitation (Caraco et al. 1997; Pace and Lonsdale 2005). Drought years with increases in salinity would favor movement of open water fish upriver (Strayer et al. 2004).

However, large watershed changes such as drought and fire contribute sediment in runoff that affects river turbidity, resulting in deposition of high numbers of inorganic particles in the marshes, as demonstrated by the Piermont record (Pederson et al. 2005). Thus, while Howarth et al. (1991) proposed that sediment fluxes were much lower prior to European settlement, our data indicate that inorganic flux prior to European settlement was much higher, especially during the Medieval drought. Depending on the grain size of the inorganic material in suspension, light may have been more limited than today. A focused research question is to determine the phytoplankton contribution to the food web in drought years with high inorganic runoff.

The composition of the river benthic communities is clearly a strong function of salinity (Strayer 2005). While marine animals such as the bivalves *Mya* and *Macoma* dominate in the river near its mouth, above Newburgh the benthos is dominated by freshwater species of oligochaetes, insects, and bivalves. However, the brackish zone is characterized by a blurring of the fauna along the salinity gradient (Simpson et al. 1984; Strayer 2005), suggesting that wet and dry years shift the biota. Sustained droughts would have favored movement of the benthic communities upriver as well as the open water fish.

Tidal marsh resident fish such as various killifishes (*Fundulus* spp.) and caridean shrimps (*Palaemonetes* spp.) would also be affected by drought. The spotfin killifish *Fundulus luciae*, although reported from salt and brackish marshes of the eastern United States, was found only recently in Piermont Marsh in the upper intertidal regions (Yozzo and Ottman 2003). At Piermont, increased rainfall throughout the season was thought to increase the retention area for killifish in small ponds in the upper marsh area (Yozzo and Ottman 2003). Sustained drought would thus presumably limit the distribution of these fish in the tidal marshes.

Drought would also affect the local food web in the marshes because of the shifts in plant composition, as well as aboveground and belowground biomass allocation of the vegetation in the marshes. Land use change since Europeans settled the region reveal the magnitude of the shifts that can take place over decades and centuries. For example, mixed communities of grasses and sedges in the Hackensack and Piermont Marshes were replaced by the dominant dense common reed stands today (Carmichael 1980; Winogrond 1997; Pederson et al. 2005). This replacement increased stem density, plant height, and detrital accumulation, which combine to reduce light at the marsh surface (Meyer et al. 2001). The expansion of common reed at the expense of salt meadow cordgrass has also been linked to the reduced frequency and duration of flooding in southern Connecticut salt marshes (Osgood et al. 2003) through increases in elevation and decrease in marsh microtopography (Windham and Lathrop 1999), which affected invertebrates and juvenile fish. In Piermont Marsh, nekton use of three common reed-dominated sites (erosional creekbank, depositional creekbank, interior) was comparable to that measured in wetlands elsewhere on the Hudson River, where common reed was not dominant (Hanson et al. 2002). A more recent study of Piermont and Iona marshes indicated reduced larval survival of mummichog *Fundulus heteroclitus* within common reed stands (Osgood et al. 2005). These authors suggest that this reduction in larval density could decrease secondary production within the system. Thus, the energy flow from the marshes would affect the open waters of the estuary.

In summary, our results suggest that the Hudson marshes have undergone large shifts in vegetation over centuries and millennia due to megadroughts and wetter time intervals which would have affected Hudson fisheries. While these long-term changes are natural, they are important for understanding future risks of climate change to the hydrology of the estuary and to the large urban populations in the lower Hudson Valley. The more recent anthropogenic impacts, such as the invasion of common reed, are important for diminishing biodiversity in the wetlands as well as reducing larval fishes and thus the food web in the larger estuary.

Acknowledgments

Support for this project included funding from a NASA Director's Discretionary Fund grant, the Hudson River Foundation, and the LDEO Investment Fund. We thank many people for field help in extracting these cores, including Jennifer Epstein, Katie Fitzgibbons, Louisa Liberman, Beate Liepert, Ross McPhee, Mike Merley, Neil Pederson, James Rubenstone, and Jennifer Wong. Sample material for this project was stored in the Lamont Doherty Earth Observatory cold room. Support for the collection and curating facilities of the repository is provided by the National Science Foundation (GrantCE00–02380) and the Office of Naval Research (Grant N00014–O2–1-0073). This is LDEO contribution #6911.

References

Brush, G. S. 1984. Patterns of recent sediment accumulation in Chesapeake Bay (Virginia-Maryland, USA) tributaries. Chemical Geology 44:227–242.

Brush, G. S. 1986. Geology and paleoecology of Chesapeake Bay: a long-term monitoring tool for management. Journal of the Washington Academy of Sciences 76:146–160.

Brush, G. S. 1989. Rates and patterns of estuarine sediment accumulation. Limnology and Oceanography 34:1235–1246.

Brush, G. S., E. A. Martin, R. S. DeFries, and C. A. Rice. 1982. Comparisons of ^{210}Pb and pollen methods for determining rates of estuarine sediment accumulation. Quaternary Research 18:196–217.

Caraco, N. F., J. J. Cole, S. E. G. Findlay, D. T. Fischer, G. G. Lampman, M. L. Pace, and D. L. Strayer. 2000. Dissolved oxygen declines associated with the invasion of the zebra mussel (*Dreissena polymorpha*). Environmental Science and Technology 34:1204–1210.

Caraco, N. F., J. J. Cole, P. A. Raymond, D. L. Strayer, M. L. Pace, S. E. G. Findlay, and D. T. Fischer. 1997. Zebra mussel invasion in a large, turbid river: phytoplankton response to increased grazing. Ecology 78:588–602.

Carmichael (Peteet), D. 1980. A record of environmental change during recent millennia in the Hackensack tidal marsh, New Jersey. Bulletin of the Torrey Botanical Club 107:514–524.

Clark, J. S. 1986. Late-Holocene vegetation and coastal processes at a Long Island tidal marsh. Journal of Ecology 74:561–578.

Clark, J. S., and W. A. Patterson III. 1985. The development of a tidal marsh: upland and oceanic influences. Ecological Monographs 55:189–217.

Colman, S. M., P. C. Baucom, J. F. Bratton, T. M. Cronin, J. P. McGeehin, D. Willard, A. Zimmerman, and P. R. Vogt. 2002. Radiocarbon dating, chronologic framework, and changes in accumulation rates of Holocene estuarine sediments from Chesapeake Bay. Quaternary Research 57:58–70.

Cooper, J. C., F. R. Cantelmo, and C. E. Newton. 1988. Overview of the Hudson estuary. Pages 11–24 *in* L. W. Barnthouse, R. J. Klauda, D. S. Vaughn, and R. L. Kendall, editors. Science, law, and Hudson River power: a case study in environmental impact assessment. American Fisheries Society, Monograph 4, Bethesda, Maryland.

Cooper, S. R., and G. S. Brush. 1991. Long-term history of Chesapeake Bay anoxia. Science 254:992–996.

Cronin, T. M., G. S. Dwyer, T. Kamiya, S. Schwede, and D. A. Willard. 2003. Medieval Warm Period, Little Ice Age and 20th century temperature variability from Chesapeake Bay. Global and Planetary Change 36:17–29.

Cronin, T., D. Willard, A. Karlsen, S. Ishman, S. Verardo, J. McGeehin, R. Kerhin, C. Holmes, S. Colman, and A. Zimmerman. 2000. Climatic variability in the eastern United States over the last millennium from Chesapeake Bay sediments. Geology 28:3–6.

Dean, W. E. 1974. Determination of carbonate and organic matter in calcareous sediments and sedimentary rocks by loss on ignition: comparison with other methods. Journal of Sedimentary Petrology 44:242–248.

Dunton, K. H., B. Hardegree, and T. Whitledge. 2001. Response of estuarine marsh vegetation to interannual variations in precipitation. Estuaries 24:851–861.

Englebright, S. 1975. Jamaica Bay: a case study of geo-environmental stresses: a guidebook to field excursions. New York State Geological Association, Hofstra University, Hempstead.

Faegri, K., and J. Iversen. 1975. Textbook of pollen analysis. Hafner Publishing Company, New York.

Fuller, J. L., D. R. Foster, J. S. McLachlan, and N. Drake. 1998. Impact of human activity on regional forest composition and dynamics in Central New England. Ecosystems 1:76–95.

Gladden, J. B., F. R. Cantelmo, J. M. Croom, and R. Shapot. 1988. Evaluation of the Hudson River ecosystem in relation to the dynamics of fish populations. Pages 37–52 *in* L. W. Barnthouse, R. J. Klauda, D. S. Vaughn, and R. L. Kendall, editors. Science, law, and Hudson River power: a case study in environmental impact assessment. American Fisheries Society Monograph 4, Bethesda, Maryland.

Grimm, E. C. 1992. TILIA and Tilia-Graph software, version 2. Illinois State University, Normal.

Hanson, S. R., D. T. Osgood, and D. J. Yozzo. 2002. Nekton use of a *Phragmites australis* marsh on the Hudson River, New York. Wetlands 22:326–337.

Hartig, E. K., V. Gornitz, A. Kolker, F. Mushacke, and D. Fallon. 2002. Anthropogenic and climate-change impacts on salt marshes of Jamaica Bay, New York City. Wetlands 22:71–89.

Heusser, C. J. 1949. History of an estuarine bog at Secaucus, New Jersey. Bulletin of the Torrey Botanical Club 76:385–406.

Heusser, C. J. 1963. Pollen diagrams from three former cedar bogs in the Hackensack tidal marsh, northeastern New Jersey. Bulletin of the Torrey Botanical Club 90:16–28.

Heusser, L. E., C. J. Heusser, and D. Weiss. 1975. Man's influence on the development of the estuarine marsh, Flax Pond, Long Island, New York. Bulletin of the Torrey Botanical Club 102:61–66.

Heusser, L. E. and C. E. Stock. 1984. Preparation techniques for concentrating pollen from marine sediments and other sediments with low pollen density. Palynology 8:225–227.

Howarth, R. W., J. R. Fruci, and D. Sherman. 1991. Inputs of sediment and carbon to an estuarine ecosystem: influence of land use. Ecological Applications 1:27–39.

Howarth, R. W., R. Schneider, and D. Swaney. 1996. Metabolism and organic carbon fluxes in the tidal freshwater Hudson River. Estuaries 19:848–865.

Lamb, H. H. 1982. Climate, history and the modern world. Methuen, London.

Lehr, J. H. 1967. The marshes at Piermont, New York: a field report. Sarracenia 11:31–34.

Lewis, W. H., P. Vinay, and V. E. Zenger. 1983. Airborne and allergenic pollen of North America. The Johns Hopkins University Press, Baltimore, Maryland.

Meyer, D. L., J. M. Johnson, and J. W. Gill. 2001. Comparison of nekton use of *Phragmites australis* and *S. alterniflora* marshes in the Chesapeake Bay, USA. Marine Ecology Progress Series 209:71–84.

Mitsch, W. J. and J. G. Gosselink. 1993. Wetlands, 2nd edition. Van Nostrand Reinhold Company, New York.

Moore, P. D., and J. A. Webb. 1978. An illustrated guide to pollen analysis. Wiley, New York.

Newman, W. S., L. J. Cinquemani, J. A. Sperling, L. F. Marcus, and R. Pardi. 1987. Holocene neotectonics and the Ramapo Fault sea-level anomaly: a study of varying marine transgression rates in the Lower Hudson Estuary, New York and New Jersey. In D. Nummendal, O. Pilkey, and J. Howard, editors. Sea level fluctuation and coastal evolution. Society of Economic Paleontologists and Minerologists, Special Publication No. 41, Tulsa, Oklahoma.

Niering, W. A., R. S. Warren, and C. G. Weymouth. 1977. Our dynamic tidal marshes: vegetation changes as revealed by peat analysis. Connecticut Arboretum Bulletin 22.

Odum, H. T. 1988. Self-organization, transformity, and information. Science 242:1132–1139.

Osgood, D. T., D. J. Yozzo, R. M. Chambers, D. Jacobson, T. Hoffman, and J. Wnek. 2003. Tidal hydrology and habitat utilization by resident nekton in *Phragmites* and non-*Phragmites* marshes. Estuaries 26:522–533.

Osgood, D. T., D. J. Yozzo, R. M. Chambers, S. Pianka, C. LePage, and J. Lewis a. 2006. Patterns of habitat utilization by resident nekton in *Phragmites* and *Typha* marshes of the Hudson River Estuary. Pages 151–174 in J. R Waldman, K. E. Limburg, and D. L. Strayer, editors. Hudson River fishes and their environment. American Fisheries Society, Symposium 51, Bethesda, Maryland.

Pace, M. L., and D. J. Lonsdale. 2005. Ecology of the Hudson zooplankton community. Pages 217–229 *in* J. S. Levinton and J. R.. Waldman, editors. The Hudson River estuary. Cambridge University. Press, New York.

Pasternack, G. B., G. S. Brush, and W. B. Hilgartner. 2001. Impact of historic land-use change on sediment delivery to a Chesapeake Bay subestuarine delta. Earth Surface Processes and Landforms 26:1–19, John Wiley and Sons, Ltd., London.

Pederson, D., D. Peteet, D. Kurdyla, and T. Guilderson. 2005. Medieval Warming, Little Ice Age, and European impact on the environment during the last millennium in the lower Hudson Valley, New York, USA. Quaternary Research 63:238–249.

Saltonstall, K. 2002. Cryptic invasion by a non-native genotype of the common reed, *Phragmites australis* into North America Proceedings of the National Academy of Sciences 99:2445–2449.

Simpson, H. J., R. Bopp, and D. Thurber. 1973. Salt movement patterns in the Hudson. In Proceedings of the 3rd Symposium of Hudson River Ecology. Hudson River Environmental Society, New Paltz, New York.

Simpson, K. W., R. W. Bode, J. P. Fagnani, and D. Denicola. 1984. The freshwater macrobenthos of the main channel, Hudson River, part B. Biology, taxonomy, and distribution of resident species. Final report to the Hudson River Foundation, Albany, New York.

Stahle, D. W., M. K. Cleveland, and J. G. Hehr. 1988. North Carolina climate change reconstructed from tree rings: A.D. 372 to 1985. Science 240:1517–1519.

Strayer, D. L. 2006. The benthic animal communities of the riverine Hudson Estuary. Pages 266–278 *in* J.S. Levinton and J. R. Waldman, editors. The Hudson estuary. Cambridge University Press, New York.

Strayer, D. L., K. A. Hattala, and A. W. Kahnle. 2004. Effects of an invasive bivalve (*Dreissena polymorpha*) on fish in the Hudson River estuary. Canadian Journal of Fisheries and Aquatic Sciences 61:924–941.

Strayer, D. L., and L. C. Smith. 2001. The zoobenthos of the freshwater tidal Hudson River and its response to the zebra mussel (*Dreissena polymorpha*) invasion. Archiv für Hydrobiologie Supplementband 139:1–52..

Swaney, D. P., K. Limburg, and K. Stainbrook. 2006. Some historical changes in the pattern of population and land use in the Hudson River watershed. Pages 75–112 *in* J. R. Waldman, K. E. Limburg, and D. L. Strayer, editors. Hudson River fishes and their environment. American Fisheries Society, Symposium 51, Bethesda, Maryland.

Thomas, E., and J. C. Varekamp. 1991. Paleo-environmental analysis of marsh sequences (Clinton, Connecticut): Evidence for a punctuated rise in sea level during the latest Holocene. Journal of Coastal Research (Special Issue) 11:125–158.

Varekamp, J. C., and E. Thomas. 1998. Climate change and the rise and fall of sea level over the millennium. EOS 79(6):69,74–75.

Watts, W. A., and T. C. Winter. 1966. Plant macrofossils from Kirchner Marsh, Minnesota: a paleoecological study. Geological Society of America Bulletin 77:1339–1360.

Willard, D. A., and D. A. Korejwo. 1999. Holocene palynology from Marion-Dufresne cores MD99–2209 and 2207 from Chesapeake Bay: Impacts of climate and historic land-use change. Pages 78–86 *in* T. M. Cronin, editors. Initial report on IMAGES V cruise of the Marion-Dufresne to the Chesapeake Bay June 20–22, 1999. Open file report 00–306. United States Geological Survey, Reston, Virginia.

Willard, D. A., T. M. Cronin, and S. Verardo. 2003. Late-Holocene climate and ecosystem history from Chesapeake Bay sediment cores, USA. The Holocene 13:201–214, Hodder Arnold, London.

Windham, L., and R. Lathrop. 1999. Effects of *Phragmites australis* (common reed) invasions on above-ground biomass and soil properties in brackish tidal marshes of the Mullica River, New Jersey. Estuaries 22:927–935.

Winogrond, H. G. 1997. Invasion of *Phragmites australis* in the tidal marshes of the Hudson River, Master's thesis. Bard College, Annandale-on-Hudson, New York.

Wong J. K., and D. Peteet. 1999. Environmental history of Piermont Marsh, Hudson, River, NY. Pages 1–30, section III *in* W. C. Nieder and J. R. Waldman, editors. Final Reports of the Tibor T. Polgar Fellowship Program, 1998. Hudson River Foundation, Albany, New York.

Wright, H. E., Jr., D. H. Mann, and P. H. Glaser. 1984. Piston corers for peat and lake sediments. Ecology 65:657–659.

Yozzo, D. J., and F. Ottman. 2003. New distribution records for the spotfin killifish, *Fundulus luciae* (Baird), in the lower Hudson River estuary and adjacent waters Northeastern Naturalist 10:399–408.

American Fisheries Society Symposium 51:129–150, 2006

Biodiversity and Zoogeography of the Fishes of the Hudson River Watershed and Estuary

JOHN R. WALDMAN

Queens College, Biology Department
65-30 Kissena Boulevard, Flushing, New York 11367 USA
john.waldman@qc.cuny.edu

THOMAS R. LAKE

New York State Hudson River Estuary Program
21 South Putt Corners Road, New Paltz, New York 12561 USA

ROBERT E. SCHMIDT

Simon's Rock College of Bard
84 Alford Road, Great Barrington, Massachusetts 01230 USA

Abstract.—The Hudson River Estuary (defined here as the Hudson River drainage and New York Harbor) is home to a large and diverse ichthyofauna. Estimates of species richness reflect both their geographic boundaries and time periods. The most complete estimate is for the Hudson River drainage north of the southern tip of Manhattan, where, as of 2005, 212 fish species have been recorded. This includes 11 new forms not reported in the most recently published tally (1990). We categorize the fishes of the Hudson River drainage as derived from 12 zoogeographic or anthropogenic sources (including species for which we make no judgment [n = 26]), the largest contributions from which include temperate marine strays (n = 65), introduced species (n = 28), and freshwater species that survived Pleistocene glaciations in Atlantic coastal refugia (n = 21). Additional species appear to have invaded from the Mississippi refugia, some naturally (n = 11) and some later, via canals (n = 11). Only ten diadromous fishes occur in the estuary, but many of these are, or have been, commercially and recreationally important (e.g., Atlantic sturgeon *Acipenser oxyrinchus*, American shad *Alosa sapidissima*, and striped bass *Morone saxatilis*). Extremely high seasonal temperature changes in the main-channel Hudson River foster a seasonally dynamic ichthyofauna with relatively few species occurring year round. However, the small number of resident estuarine fishes (n = 8) often occur in high abundances. Species richness peaks between June and September and reaches a minimum in winter. Long-term data indicate that although species richness has increased with the additions of new species, diversity is decreasing because of the decrease in population size of certain species, especially native cyprinids. The Hudson estuary hosts a population of one federally endangered species, shortnose sturgeon *Acipenser brevirostrum*, which is flourishing. Only one species, the anadromous rainbow smelt *Osmerus mordax* appears to have become extirpated in the Hudson Estuary.

Introduction

The Hudson River watershed and the Hudson River Estuary have a diverse ichthyofauna derived from many zoogeographic sources. These fishes have been unusually well studied because of their occurrence in a region of high human population density and because of the management issues that arise from such proximity. Records also extend relatively well back in time because of the early European settlement of the region (Daniels et al. 2005). However, much of the data on its fishes is piecemeal; in this paper we attempt to synthesize information and characterize this assemblage across the entire system.

The major features of the Hudson River system include the tributaries that feed the mainstem Hudson River (i.e., its upland watershed), the mainstem Hudson River, and its estuary, which includes New York Harbor. The Hudson River begins in New York's Adirondack Mountains and drains a total of about 35,000 km^2 from portions of five states. It is tidal to the federal dam at Troy, New York, 243 km inland from the southern tip of Manhattan at the Battery (Cooper et al. 1988). Its estuary may be defined as the tidal Hudson River downriver from the federal dam and including tidal waters of its tributaries, in addition to New York Harbor. New York Harbor has been inconsistently defined (Waldman 1999). However, we consider New York Harbor to include those waters landward of the Sandy Hook–Breezy Point transect to the Battery, the East River, Arthur Kill, Newark Bay, lower Hackensack and Passaic rivers, Raritan Bay and the lower Raritan River, and any smaller brackish or marine water bodies within this complex.

This complex geography and associated nomenclature have affected accounts of the ichthyofaunal diversity of the Hudson River system. In their review of fishes reported from the salt and brackish waters of New York, Briggs and Waldman (2002) listed 335 species, many of which also have been found in the Hudson Estuary. Beebe and Savidge (1988) tallied 140 species of fishes from the tidal river between the federal dam and the George Washington Bridge, 19 km upriver of the Battery. Smith (1985) reported approximately 150 species occurring between the southern tip of Manhattan Island and the federal dam; of these, about 80 are freshwater or diadromous forms and 50 occur regularly in inshore areas. Smith and Lake (1990) developed an annotated list of fishes documented from the entire Hudson River drainage south to the Battery. They noted 201 species, including many that are found far from saline waters. One of us (TRL) maintains a current list of all fish species found in the Hudson River watershed north of the Battery. As of July 2005, this list stood at 212 species. Our objectives are to provide a synthesis of diverse information of the biodiversity and zoogeography of the fishes of the Hudson River estuary and an update on the new species additions. However, because of better documentation, we place greater focus on the ichthyofauna of the Hudson River watershed and estuarine waters north of the Battery than on its estuarine waters outside of the mainstem Hudson.

Historical Studies

Mitchill (1815) prepared a detailed systematic account of fishes found around Long Island and New York City, followed by a species list for the region (Mitchill 1818). DeKay (1842) published a list of New York's fishes, including many collected from the lower Hudson River, as part of the state's first natural history survey. Smith (1897) prepared a descriptive list of fresh and brackish water fishes from the New York City area. Historically, the most comprehensive work on New York's fishes, both freshwater and marine, was Bean's (1903) monograph, which includes taxonomic and life history information.

Prior to the compilation by Briggs and Waldman (2002), the most complete list of New York's marine fishes was by Nichols and Breder (1927). The fishes of New York Harbor were described in a short paper by Breder (1938). The first broad surveys of Hudson River fishes were conducted by the New York State Conservation Department in the 1930s; Greeley (1937) presented results for the lower river and Greeley (1939) for Long Island and Staten Island drainages to New York Harbor. Other useful papers about the Hudson's fishes include Breder and Nigrelli (1934); Alevras (1973); Friedman and Hamilton (1980); Young et al. (1982); and Waldman (1985).

Major Components of the Ichthyofauna of the Hudson River Drainage

Classification of species to zoogeographic origin allows for better understanding of the development of a faunal community. We classify the ichthyofauna of the Hudson drainage according to its probable zoogeographic origin (Table 1), using the scheme of Smith and Lake (1990), but including new species records and new zoogeographic information.

(1) **Introduced** (28 species). Species released into the system from elsewhere. Examples of purposeful introductions that have resulted in reproducing populations include brown trout *Salmo trutta* and common carp *Cyprinus carpio*; purposeful introductions that failed include Atlantic salmon *Salmo salar*. Individual specimens may also be illegally released, such as the aquarium hobby species pirapitinga *Piaractus brachypomus* and alligator gar *Atractosteus spatula* (Bragin et al. 2005).

(2) **Freshwater species from an Atlantic coastal refugium** (21 species). Species whose ranges indicate natural postPleistocene colonization from an Atlantic coastal refugium (Schmidt 1986). Chain pickerel *Esox niger* and tessellated darter *Etheostoma olmstedi* are of this category.

(3) **Freshwater species from the Mississippi refugium** (11 species). Species whose ranges indicate natural postPleistocene colonization from the Mississippi refugium (Schmidt 1986). An example is stonecat *Noturus flavus*.

(4) **Freshwater species from the Mississippi refugium that are likely canal immigrants** (11 species). Species whose ranges indicate recent colonization from the Mississippi refugium through humanmade canals. Among these are freshwater drum *Aplodinotus grunniens* and central mudminnow *Umbra limi*.

(5) **Periglacial freshwater species** (14 species). Coldwater species that colonized via movement along the meltwater bodies behind receding Pleistocene glaciers. Examples include brook trout *Salvelinus fontinalis* and brook stickleback *Culaea inconstans*.

(6) **Freshwater species of uncertain origin** (11 species). Zoogeographic information not sufficient to make a confident determination. This group includes white sucker *Catostomus commersonii* and brown bullhead *Ameiurus nebulosus*.

(7) **Diadromous species** (10 species). Diadromous fishes are either anadromous or catadromous. Anadromous fishes are those that spawn in freshwater but migrate to the sea (e.g., American shad, striped bass, and Atlantic sturgeon). Catadromous fishes

Table 1. Modified Smith & Lake (1990) list for the Hudson River watershed (north of southern tip of Manhattan), updated to 212 species (new ones underlined). (1) Introduced; (2) Freshwater species from an Atlantic coastal refugium; (3) Freshwater species from the Mississippi refugium; (4) Freshwater species from the Mississippi refugium that are likely canal immigrants; (5) Periglacial freshwater species; (6) Freshwater species of uncertain origin; (7) Diadromous species; (8) Estuarine species; (9) Permanent or seasonal resident marine species; (10) Temperate marine strays; (11) Tropical marine strays; (12) Boreal marine strays.

Common name	Latin name	Probable origin
Lampreys - Petromyzontidae		
silver lamprey	*Ichthyomyzon unicuspis*	[4]
American brook lamprey	*Lampetra appendix*	[2]
sea lamprey	*Petromyzon marinus*	[7]
Requiem sharks - Carcharhinidae		
shark	*Carcharhinus* sp.	[10]
Houndsharks - Triakidae		
smooth dogfish	*Mustelus canis*	[10]
Dogfish sharks - Squalidae		
spiny dogfish	*Squalus acanthias*	[10]
Skates - Rajidae		
little skate	*Raja erinacea*	[10]
barndoor skate	*Raja laevis*	[10]
Sturgeons - Acipenseridae		
shortnose sturgeon	*Acipenser brevirostrum*	[9]
Atlantic sturgeon	*Acipenser oxyrinchus*	[7]
Gars – Lepisostidae		
longnose gar	*Lepisosteus osseus*	[1]
Bowfins - Amiidae		
bowfin	*Amia calva*	[1]
Tenpounders- Elopidae		
ladyfish	*Elops saurus*	[11]
Bonefishes - Albulidae		
<u>bonefish</u>	*Albula vulpes*	[11]
Conger eels - Congridae		
conger eel	*Conger oceanicus*	[10]
Freshwater eels - Anguillidae		
American eel	*Anguilla rostrata*	[7]
Snake eels - Ophichthidae		
speckled worm eel	*Myrophis punctatus*	[10]
Herrings - Clupeidae		
blueback herring	*Alosa aestivalis*	[7]
hickory shad	*Alosa mediocris*	[7]
alewife	*Alosa pseudoharengus*	[7]
American shad	*Alosa sapidissima*	[7]
Atlantic menhaden	*Brevoortia tyrannus*	[9]
Atlantic herring	*Clupea harengus harengus*	[10]
gizzard shad	*Dorosoma cepedianum*	[4]
round herring	*Etrumeus teres*	[10]
Anchovies - Engraulidae		
striped anchovy	*Anchoa hepsetus*	[10]

Table 1. continued.

Common name	Latin name	Probable origin
bay anchovy	*Anchoa mitchilli*	[8]
Mudminnows - Umbridae		
central mudminnow	*Umbra limi*	[4]
eastern mudminnow	*Umbra pygmaea*	[2]
Pikes - Esocidae		
redfin pickerel	*Esox americanus americanus*	[2]
northern pike	*Esox lucius*	[1]
chain pickerel	*Esox niger*	[2]
Bullhead Catfishes - Ictaluridae		
white catfish	*Ameiurus catus*	[2]
yellow bullhead	*Ameiurus natalis*	[6]
brown bullhead	*Ameiurus nebulosus*	[6]
channel catfish	*Ictalurus punctatus*	[1]
stonecat	*Noturus flavus*	[3]
tadpole madtom	*Noturus gyrinus*	[2]
margined madtom	*Noturus insignis*	[1]
brindled madtom	*Noturus miurus*	[4]
Suckers - Catostomidae		
longnose sucker	*Catostomus catostomus*	[6]
white sucker	*Catostomus commersonii*	[6]
creek chubsucker	*Erimyzon oblongus*	[2]
northern hog sucker	*Hypentelium nigricans*	[6]
shorthead redhorse	*Moxostoma macrolepidotum*	[3]
Characins - Characidae		
<u>pirapatinga</u>	*Piaractus brachypomus*	[1]
Minnows - Cyprinidae		
central stoneroller	*Campostoma anomalum*	[4]
goldfish	*Carassius auratus*	[1]
redside dace	*Clinostomus elongatus*	[3]
lake chub	*Couesius plumbeus*	[5]
grass carp	*Ctenopharyngodon idella*	[1]
satinfin shiner	*Cyprinella analostana*	[2]
spotfin shiner	*Cyprinella spiloptera*	[3]
common carp	*Cyprinus carpio*	[1]
cutlip minnow	*Exoglossum maxillingua*	[2]
brassy minnow	*Hybognathus hankinsoni*	[5]
eastern silvery minnow	*Hybognathus regius*	[2]
common shiner	*Luxilus cornutus*	[6]
pearl dace	*Margariscus margarita*	[5]
hornyhead chub	*Nocomis biguttatus*	[4]
golden shiner	*Notemigonus crysoleucas*	[6]
comely shiner	*Notropis amoenus*	[2]
emerald shiner	*Notropis atherinoides*	[4]

Table 1. continued.

Common name	Latin name	Probable origin
bridle shiner	*Notropis bifrenatus*	[2]
blackchin shiner	*Notropis heterodon*	[3]
blacknose shiner	*Notropis heterolepis*	[3]
spottail shiner	*Notropis hudsonius*	[6]
rosyface shiner	*Notropis rubellus*	[3]
sand shiner	*Notropis stramineus*	[1]
northern redbelly dace	*Phoxinus eos*	[5]
finescale dace	*Phoxinus neogaeus*	[5]
bluntnose minnow	*Pimephales notatus*	[3]
fathead minnow	*Pimephales promelas*	[3]
eastern blacknose dace	*Rhinichthys atratulus*	[2]
longnose dace	*Rhinichthys cataractae*	[6]
bitterling	*Rhodeus sericeus*	[1]
rudd	*Scardinius erythrophthalmus*	[1]
creek chub	*Semotilus atromaculatus*	[6]
fallfish	*Semotilus corporalis*	[2]
Trouts - Salmonidae		
cisco	*Coregonus artedi*	[5]
lake whitefish	*Coregonus clupeaformis*	[5]
round whitefish	*Prosopium cylindraceum*	[5]
rainbow trout	*Oncorhynchus mykiss*	[1]
sockeye salmon	*Oncorhynchus nerka*	[1]
Chinook salmon	*Oncorhynchus tshawytscha*	[1]
Atlantic salmon	*Salmo salar*	[1]
brown trout	*Salmo trutta*	[1]
brook trout	*Salvelinus fontinalis*	[5]
lake trout	*Salvelinus namaycush*	[5]
Smelts - Osmeridae		
rainbow smelt	*Osmerus mordax*	[7]
Lizardfishes - Synodontidae		
inshore lizardfish	*Synodus foetens*	[10]
Toadfishes - Batrachoididae		
oyster toadfish	*Opsanus tau*	[10]
Trout-perches - Percopsidae		
trout-perch	*Percopsis omiscomaycus*	[5]
Goosefishes - Lophiidae		
goosefish	*Lophius americanus*	[10]
Codfishes - Gadidae		
Atlantic cod	*Gadus morhua*	[10]
Atlantic tomcod	*Microgadus tomcod*	[7]
pollock	*Pollachius virens*	[10]

Table 1. continued.

Common name	Latin name	Probable origin
Phycid Hakes - Phycidae		
fourbeard rockling	*Enchelyopus cimbrius*	[10]
red hake	*Urophycis chuss*	[10]
spotted hake	*Urophycis regia*	[10]
white hake	*Urophycis tenuis*	[10]
Merlucciid Hakes - Merlucciidae		
silver hake	*Merluccius bilinearis*	[10]
Cusk-eels - Ophidiidae		
striped cusk-eel	*Ophiodon marginatum*	[10]
Needlefishes - Belonidae		
Atlantic needlefish	*Strongylura marina*	[9]
houndfish	*Tylosurus crocodilus*	[11]
Pupfishes - Cyprinodontidae		
sheepshead minnow	*Cyprinodon variegatus*	[1]
Topminnows - Fundulidae		
banded killifish	*Fundulus diaphanus*	[2]
mummichog	*Fundulus heteroclitus*	[8]
<u>spotfin killifish</u>	*Fundulus luciae*	[8]
striped killifish	*Fundulus majalis*	[10]
Livebearers - Poeciliidae		
<u>western mosquitofish</u>	*Gambusia affinis*	[1]
Silversides - Atherinidae		
brook silverside	*Labidesthes sicculus*	[4]
rough silverside	*Membras martinica*	[8]
inland silverside	*Menidia beryllina*	[8]
Atlantic silverside	*Menidia menidia*	[10]
Cornetfishes - Fistularidae		
bluespotted cornetfish	*Fistularia tabacaria*	[10]
Sticklebacks - Gasterosteidae		
fourspine stickleback	*Apeltes quadracus*	[8]
brook stickleback	*Culaea inconstans*	[5]
threespine stickleback	*Gasterosteus aculeatus*	[9]
ninespine stickleback	*Pungitius pungitius*	[5]
Pipefishes - Syngnathidae		
lined seahorse	*Hippocampus erectus*	[9]
northern pipefish	*Syngnathus fuscus*	[9]
Sea Basses - Serranidae		
black sea bass	*Centropristis striata*	[10]
gag	*Mycteroperca microlepis*	[11]
Temperate Basses - Moronidae		
white perch	*Morone americana*	[8]

Table 1. continued.

Common name	Latin name	Probable origin
white bass	*Morone chrysops*	[4]
striped bass	*Morone saxatilis*	[7]
Sunfishes - Centrarchidae		
rock bass	*Ambloplites rupestris*	[1]
bluespotted sunfish	*Enneacanthus gloriosus*	[2]
banded sunfish	*Enneacanthus obesus*	[2]
redbreast sunfish	*Lepomis auritus*	[2]
green sunfish	*Lepomis cyanellus*	[1]
pumpkinseed	*Lepomis gibbosus*	[6]
warmouth	*Lepomis gulosus*	[1]
bluegill	*Lepomis macrochirus*	[1]
smallmouth bass	*Micropterus dolomieu*	[1]
largemouth bass	*Micropterus salmoides*	[1]
white crappie	*Pomoxis annularis*	[1]
black crappie	*Pomoxis nigromaculatus*	[1]
Perches - Percidae		
greenside darter	*Etheostoma blennioides pholidotum*	[3]
fantail darter	*Etheostoma flabellare*	[3]
tessellated darter	*Etheostoma olmstedi*	[2]
yellow perch	*Perca flavescens*	[2]
logperch	*Percina caprodes*	[4]
shield darter	*Percina peltata*	[2]
walleye	*Sander vitreus*	[1]
Bigeyes - Priacanthidae		
short bigeye	*Pristigenys alta*	[10]
Bluefishes - Pomatomidae		
bluefish	*Pomatomus saltatrix*	[9]
Cobias - Rachycentridae		
cobia	*Rachycentron canadum*	[10]
Remoras - Echeneidae		
sharksucker	*Echeneis naucrates*	[10]
Jacks - Carangidae		
crevalle jack	*Caranx hippos*	[9]
Atlantic moonfish	*Selene setapinnis*	[10]
lookdown	*Selene vomer*	[10]
permit	*Trachinotus falcatus*	[10]
Snappers - Lutjanidae		
schoolmaster	*Lutjanus apodus*	[11]
gray snapper	*Lutjanus griseus*	[11]
Mojarras - Gerreidae		
spotfin mojarra	*Eucinostomus argenteus*	[10]

Table 1. continued.

Common name	Latin name	Probable origin
Grunts - Haemulidae		
pigfish	*Orthopristis chrysoptera*	[10]
Porgies - Sparidae		
sheepshead	*Archosargus probatocephalus*	[10]
pinfish	*Lagodon rhomboides*	[10]
scup	*Stenotomus chrysops*	[10]
Drums - Sciaenidae		
freshwater drum	*Aplodinotus grunniens*	[4]
silver perch	*Bairdiella chrysoura*	[10]
weakfish	*Cynoscion regalis*	[9]
spot	*Leiostomus xanthurus*	[9]
northern kingfish	*Menticirrhus saxatilis*	[10]
Atlantic croaker	*Micropogonias undulatus*	[10]
Butterflyfishes - Chaetodontidae		
foureye butterflyfish	*Chaetodon capistratus*	[11]
spotfin butterflyfish	*Chaetodon ocellatus*	[11]
Wrasses - Labridae		
tautog	*Tautoga onitis*	[10]
cunner	*Tautogolabrus adspersus*	[10]
Mullets - Mugilidae		
striped mullet	*Mugil cephalus*	[9]
white mullet	*Mugil curema*	[9]
Stargazers - Uranoscopidae		
northern stargazer	*Atroscopus guttatus*	[10]
Combtooth Blennies - Blenniidae		
feather blenny	*Hypsoblennius hentz*	[11]
freckled blenny	*Hypsoblennius ionthas*	[11]
Gunnels - Pholidae		
rock gunnel	*Pholis gunnellus*	[10]
Sand Lances - Ammodytidae		
American sand lance	*Ammodytes americanus*	[9]
Sleepers - Eleotridae		
fat sleeper	*Dormitator maculatus*	[11]
Gobies - Gobiidae		
highfin goby	*Gobionellus oceanicus*	[11]
naked goby	*Gobiosoma bosc*	[9]
seaboard goby	*Gobiosoma ginsburgi*	[10]
Cutlassfishes - Trichiuridae		
Atlantic cutlassfish	*Trichiurus lepturus*	[10]
Mackerels - Scombridae		
Spanish mackerel	*Scomberomorus maculatus*	[10]

Table 1. continued.

Common name	Latin name	Probable origin
Atlantic mackerel	*Scomber scombrus*	[10]
Barracudas - Sphyraenidae		
northern sennet	*Sphyraena borealis*	[10]
guaguanche	*Sphyraena gauchancho*	[10]
Butterfishes - Stromateidae		
butterfish	*Peprilus triacanthus*	[10]
Sculpins - Cottidae		
slimy sculpin	*Cottus cognatus*	[5]
sea raven	*Hemitripterus americanus*	[10]
grubby	*Myoxocephalus aenaeus*	[9]
longhorn sculpin	*Myoxocephalus octodecemspinosus*	[10]
Searobins - Triglidae		
northern searobin	*Prionotus carolinus*	[10]
striped searobin	*Prionotus evolans*	[10]
Snailfishes and Lumpfishes - Cyclopteridae		
seasnail	*Liparis atlanticus*	[10]
lumpfish	*Cyclopterus lumpus*	[12]
Flying Gurnards - Dactylopteridae		
flying gurnard	*Dactylopterus volitans*	[11]
Lefteye Flounders - Bothidae		
Gulf Stream flounder	*Citharichthys arctifrons*	[10]
smallmouth flounder	*Etropus microstomus*	[10]
summer flounder	*Paralichthys dentatus*	[9]
fourspot flounder	*Paralichthys oblongus*	[10]
windowpane	*Scopthalmus aquosus*	[10]
Righteye Flounders - Pleuronectidae		
winter flounder	*Pseudopleuronectes americanus*	[9]
yellowtail flounder	*Limanda ferruginea*	[10]
American Soles - Achiridae		
hogchoker	*Trinectes maculatus*	[8]
Tonguefishes - Cynoglossidae		
blackcheek tonguefish	*Symphurus plagiusa*	[10]
northern tonguefish	*Symphurus pusillus*	[10]
Filefishes - Monacanthidae		
orange filefish	*Aluterus schoepfi*	[10]
planehead filefish	*Stephanolepis hispidus*	[11]
Boxfishes - Ostraciidae		
scrawled cowfish	*Acanthostracion quadricornis*	[10]
Puffers - Tetraodontidae		
smooth puffer	*Lagocephalus laevigatus*	[11]
northern puffer	*Sphoeroides maculatus*	[10]
Porcupinefishes - Diodontidae		
striped burrfish	*Chilomycterus schoepfi*	[10]

spawn at sea and reside in freshwater; the American eel *Anguilla rostrata* is the only catadromous species in the Hudson estuary.

(8) **Estuarine species** (8 species). Estuarine species have broad salinity and temperature tolerances. Because of these constraints, there are not many of these species but they may be highly abundant. Examples include mummichog *Fundulus heteroclitus*, white perch *Morone americana*, and hogchoker *Trinectes maculatus.*

(9) **Permanent or seasonal resident marine species** (17 species). Those that remain within local or immediately-offshore waters all year, such as lined seahorse *Hippocampus erectus* and tautog *Tautoga onitis.*

(10) **Temperate marine strays** (65 species). Typically pelagic or deepwater species not usually seen in estuaries. Examples include little tunny *Euthynnus alletteratus* and Atlantic cod *Gadus morhua.*

(11) **Tropical marine strays** (15 species). Tropical or semitropical vagrants that almost always occur in the Hudson region as juveniles. These include bonefish *Albula vulpes*, Atlantic flyingfish *Cheilopogon melanurus*, scamp *Mycteroperca phenax*, African pompano *Alectis ciliaris*, and gray angelfish *Pomacanthus arcuatus* among many others.

(12) **Boreal marine strays** (1 species). Coldwater species from northern waters. Only lumpfish *Cyclopterus lumpus* fits this category.

Community Composition

The ichthyofauna of the Hudson estuary has received little analysis of community composition; however, what studies have occurred showed considerable variability in space and time. Regional structure is evident, as shown by Woodhead et al. (1992) who used multivariate techniques to analyze catches from a trawl survey of the lower Hudson River estuary between 1984 and 1986. The 72 fish species recorded clustered as four primary groups. These groups showed some seasonal variation along the salinity gradient that ranged from the southern limit of lower New York Bay to the Hudson River near Peekskill, New York. Seasonal variation in community structure also was found among juvenile fishes in the lower Hudson River (off Manhattan) by Able et al. (1998). They noted an early summer group dominated by Atlantic tomcod *Microgadus tomcod* and winter flounder *Pseudopleuronectes americanus* and a late summer group dominated by striped bass and black sea bass *Centropristis striata.*

On a coarse level, Hudson River fishes also form communities according to primary habitats within the drainage. Daniels et al. (2005) observed that for 210 species found in the drainage (including the Mohawk River), 129 occur in the main channel of the tidal portion of the river, with the remaining 81 confined to tributaries of the lower Hudson River or the upper Hudson or Mohawk systems. Of the species present in the main channel of the tidal portion, 49 are primarily marine visitors and 80 are either resident or diadromous forms.

Gladden et al. (1988) described seasonal occurrences of resident and migratory fishes in the tidal Hudson River, 1974–1977. Among their findings was higher annual variability among migrant than resident species. They also found that maximum species richness occurred between June and September and lowest species richness in December (but later winter months were not sampled). However, only 31 species were captured almost every month throughout the 3-year period. Given that 100 fishes were collected over this span and that at least 140 were known from that river reach at that time (Beebe and Savidge 1988), Gladden et al. (1988) suggested that many species only use the estuary for short periods. Of the 31 species consistently observed, 22 were considered residents and 9 were considered migrants. Gladden et al. (1988) proposed that temperature and freshwater inflow were the major abiotic factors in determining the abundances of migratory and many resident species.

Hurst et al. (2004) did later find a statistical association between annual variation in river flow from June through August and late-summer and autumn community composition of nearshore fishes in the mesohaline region of the river. In 21 years (1979–1999) of beach seine sampling they encountered 60 species, the most abundant of which were silversides *Menidia* spp., striped bass, white perch, American shad, and blueback herring *Alosa aestivalis*. They also observed unexplained long-term declines in catch rates of freshwater and estuarine species and a dramatic increase in the abundance of silversides.

Community composition also has varied temporally across decades. Daniels (1995) compared late 20th century nearshore fish abundance data from the freshwater reach of the Hudson with information from the 1936 survey by Greeley (1937). Daniels found that, although species richness remained relatively constant, species composition and relative abundance has changed considerably. The two dominant resident species in 1936, spottail shiner *Notropis hudsonius* and white perch, made up 34% of the individuals in the assemblage, whereas their relative abundance in a 1990 survey had nearly doubled (64%). The relative abundance of an additional five persistent species declined between the two sampling dates, but only slightly. However, in 1936, 36 nondominant species made up 26% of the catch, whereas in 1990 the remaining species accounted for only 7%. Daniels concluded that although richness had increased with the addition of new species, diversity had decreased because of the decrease in population size of certain species, including native cyprinids, such as bridle shiner *Notropis bifrenatus*, common shiner *Luxilus cornutus*, spotfin shiner *Cyprinella spiloptera*, and creek chub *Semotilus atromaculatus*.

Influence of Geography, Oceanography, and Climate on Marine Species

The substantial diversity of fishes that occur in the Hudson River Estuary is due, in part, to its wide variety of habitats. New York Harbor is a mixing zone of waters from rivers, tidal straits, and the New York Bight. Long Island Sound mixes with ocean waters through its eastern outlet and with estuarine waters of New York Harbor via the East River at its western terminus.

Within the Hudson River, detectable salinity (i.e., the "salt front"; as defined as a salinity of 0.1 psu; salinities often are higher in bottom waters of the Hudson River) has ranged from the Bronx to the Kingston area, about 160 km upriver from the mouth of the river (Limburg et al. 1986).

New York lies mid-latitudinally along the North American coast and within the northern portion of the Virginian zoogeographic province, which is bounded on the south by Cape Hatteras and on the north by Cape Cod. A warm oceanic surface current, the Gulf Stream, flows northeasterly along the eastern margin of the New York Bight. The glacial channel of the Hudson River—the Hudson Canyon—cuts diagonally through the continental shelf at the edge of the New York Bight, bringing deep-sea fishes near the New York coast.

Temperatures of New York marine waters undergo pronounced seasonal changes that are among the greatest worldwide. During winter, coastal surface waters descend to levels not far above freezing (~1°C) and during summer they may reach 26°C; inshore bottom temperatures range between less than 1°C to about 21°C (Parr 1933). In the Hudson River, summer temperatures may be as high as 30°C in shallow areas (Limburg et al. 1986).

Interaction of the Biological and Physical Realms

The combination of geographic, oceanographic, and climatological influences fosters a largely mobile and migratory, rather than resident, ichthyofauna. Many species move north and south annually in conjunction with rising and falling water temperatures. These include anadromous fishes that spawn in the Hudson River or other rivers (e.g., American shad, alewife *Alosa pseudoharengus*, striped bass, and Atlantic sturgeon). Euryhaline fishes such as white perch and shortnose sturgeon *Acipenser brevirostrum* make seasonal movements across a salinity spectrum. Some species, such as bluefish *Pomatomus saltatrix* and Atlantic mackerel *Scombrus scombrus* may cross zoogeographic provinces. Others, such as summer flounder *Paralichthys dentatus*, black sea bass, and scup *Stenotomus chrysops* make less pronounced movements seasonally, retreating somewhat southward and outward to continental shelf waters in winter, but remaining within the Virginian Province.

Oceanic pelagic forms such as little tunny and Atlantic bonito *Sarda sarda* tend to occur in New York waters during the peak temperature period of late summer and early autumn. Ocean currents help disperse young-of-the-year of bluefish, which use coastal bays and rivers as nursery grounds. Warm-core rings from the Gulf Stream are believed to be largely responsible for the appearance each summer of juveniles of a great variety of fishes considered to be tropical or semitropical. The presence and proportions of each species vary each year, but regularly appearing "exotics" include bluespotted cornetfish *Fistularia tabacaria*, snowy grouper *Epinephelus niveatus*, lookdown *Selene vomer*, permit *Trachinotus falcatus*, and gray snapper *Lutjanus griseus*. Anecdotal observations indicate that the less-mobile tropical juveniles neither migrate south nor survive the autumn decline in water temperatures (McBride 1996). The only tropical or semitropical fishes occasionally seen in New York waters as adults, but not juveniles, are tarpon *Megalops atlanticus* and cubera snapper *Lutjanus cyanopterus* (Briggs and Waldman 2002); however, neither has yet been found in the Hudson estuary.

Zoogeography of Freshwater Fishes

The present day Hudson River watershed was covered by glaciers several times during the Pleistocene. The Wisconsinan epoch began approximately 50,000–70,000 years before present (BP) (Strahler 1966) and persisted until about 14,700–12,000 BP (Borns 1973; Strahler 1966). This most recent glacial advance eliminated fishes from the present day drainage. Thus, the focus of the zoogeography of native freshwater fishes in the Hudson watershed is on the processes of postglacial dispersal.

We have divided the fishes into primary division species (Myers 1938) that have little salinity tolerance and those which can disperse through saltwater avenues. We also divided dispersal into postglacial (within several thousand years of the Wisconsinan recession, summarized from Schmidt 1986) and recent (within the past several hundred years).

Postglacial Dispersal of Primary Division Fishes

Sea level at the peak of the Wisconsinan epoch was considerably lower than today (about –150 m; Emery and Garrison 1967). Sea level rose slowly as glaciers melted and receded, attaining –25 m by about 7,500 BP, and then rising more slowly (Rampino and Sanders 1980) to its current level about 3500 BP (Strahler 1966). Exposed areas on the Atlantic Coastal Plain and shallows south of the glacial margin were probably inhabited by fishes and these species had the opportunity to move north through swampy habitat and anastomosing river channels prior to inundation with salt water.

Several fish species in the Hudson watershed have clear zoogeographic affinities with populations on the Atlantic Coastal Plain. These species have a broad distribution (some are found in Florida) and are not taxonomically distinct from the southern coastal populations. Examples are bridle shiner *Notropis bifrenatus*, creek chubsucker *Erimyzon oblongus*, bluespotted sunfish *Enneacanthus gloriosus*, mud sunfish *Acantharcus pomotis*, tessellated darter *Etheostoma olmstedi*, and eastern mudminnow *Umbra pygmaea*. There also are extralimital species, such as swamp darter *Etheostoma fusiforme*, pirate perch *Aphredoderus sayanus*, blackbanded sunfish *Enneacanthus chaetodon,* and banded sunfish *E. obesus* in southern New Jersey and Long Island that did not quite manage to reach or persist in the Hudson watershed. These species appear to have moved north through swampy coastal plain habitat similar to that which still occurs in much of coastal Virginia and North Carolina. Such habitat must have been amenable to latitudinal dispersal given the small sizes and small home ranges of these species. There also are two species (cutlip minnow *Exoglossum maxillingua* and rosyside dace *Clinostomus funduloides*) that do not inhabit the coastal plain but which appear to have paralleled this latitudinal dispersal pattern, most likely through stream captures or other upland aquatic corridors.

A second set of freshwater species also have affinities with Atlantic coastal populations but are taxonomically distinct from those populations. This is the "northeastern fauna" that Schmidt (1986) proposed were isolated in refugia during the Wisconsinan advance, such as the exposed Georges Bank, and has subsequently moved into the Hudson and other northeastern rivers. Examples of these fishes include the freshwater fallfish *Semotilus corporalis* and yellow perch *Perca flavescens* (Todd and Hatcher 1993) and the estuarine Atlantic tomcod *Microgadus tomcod* and mummichog *Fundulus heteroclitus* (Brown and Chapman 1991). Also, Waldman et al. (2002) found genetic evidence for a northeastern refugium, most likely Georges Bank, for shortnose sturgeon.

The Mississippi Valley was a third refugium for Hudson River fishes (Underhill 1986). As glaciers receded, the outlet for the Laurentian Lakes migrated northeastward. At one time the Great Lakes drained through the Mohawk River Valley (12,000–13,000 BP; Miller 1914; Flint 1971), providing a direct route for fishes into the Hudson River drainage. Later, fishes could have moved less directly through the Champlain Valley until eustatic tilting closed that migration route around 11,800 BP (Flint 1971). Native Hudson watershed fishes with apparent Mississippian affinities include shorthead redhorse *Moxostoma macrolepidotum* and greenside darter *Etheostoma blennioides*.

Some fishes were able to persist in periglacial lakes during the glacial maxima. These environments were probably cold and harsh and we perceive that the boreal cold-water fauna were the most likely to be candidates for having inhabited these lakes. The fishes in these lakes could have included lake trout *Salvelinus namaycush*, ciscoes and whitefishes (*Prosopium* and *Coregonus*), burbot *Lota lota*, and two sticklebacks (ninespine stickleback *Pungitius pungitius* and brook stickleback *Culaea inconstans*). Pielou (1991) discussed the periglacial lake environment and suggested that the fauna may have been quite diverse. Dispersal of fishes from these lakes followed the glacial recession northward. Fishes moved through the Hudson Valley but only some (a cisco, whitefishes, and *Culaea*) persist today. Lake trout populations in periglacial lakes were isolated from each other for a considerable time and differentiated (Wilson and Hebert 1996; Wilson and Mandrak 2004). Native lake trout in the Hudson Valley would have originated in an Atlantic periglacial refuge.

There are several other, widely distributed native freshwater fishes that occur in the Hudson for which refugia have not been identified and which, for the most part, definitive taxonomic work has not occurred. These include white sucker *Catostomus commersonii*, common shiner *Luxilus cornutus*, eastern blacknose dace *Rhinichthys atratulus*, longnose dace *R. cataractae*, and creek chub *Semotilus atromaculatus*.

The Role of Canals

Canals are usually constructed to facilitate the passage of goods by boat. Canals can also make possible the passage of aquatic animals from one watershed to another, but they do not necessarily cause this to happen. Daniels (2001) questioned the role of early canals in dispersal of fishes and other aquatic organisms once thought to be explained by canals acting as dispersal corridors. Subsequent to Daniels (2001), molecular evidence revealed that sea lamprey *Petromyzon marinus* (Waldman et al. 2004) and alewife (Kuhn and Kornfield 2004), both hypothesized to have invaded Lake Ontario from the Hudson River through the Erie Canal system, are likely native to Lake Ontario.

McCauley (1829) observed that "dace" were newcomers to the Mohawk River, having appeared after the construction of the canal at Rome, New York, which connects the Mohawk with Wood Creek, tributary to Otsego Lake and, ultimately, Lake Ontario. McCauley probably was referring to hornyhead chub *Nocomis biguttatus*, which Smith (1985) also thought was a canal migrant. In the same paper McCauley discussed severe flooding of the Mohawk and Wood Creek, which also could have facilitated faunal dispersal. Thus, situations such as these may or may not reflect colonization via canals.

In contrast, starting in the 1890s (Daniels 2001), the canals began to be modified to pass larger vessels and emphasis was placed on locks that bypassed barriers and emptied water and boats into the main rivers rather than small parallel canals. Today, some of these canals are well suited to allow faunal passage and fishes and other aquatic organisms have taken advantage of the new opportunities to disperse. For instance, many blueback herring annually ascend the locks in Waterford and move up the Mohawk River to the vicinity of Rome (Schmidt et al. 2003). However, these locks only operate from May through October and are dry the remainder of the time, thus limiting potential dispersal through them.

Fishes that have used these modernized canals to disperse from Lake Champlain into the Hudson include central mudminnow *Umbra limi* (Schmidt and Daniels 2006), silver lamprey *Ichthyomyzon unicuspis*, logperch *Percina caprodes*, and freshwater drum *Aplodinotus grunniens*, although the latter two also could have arrived via the Mohawk Barge Canal. It also is likely that the brindled madtom *Noturus miurus* and white bass *Morone chrysops* invaded the Hudson drainage through the Barge Canal.

Human Introductions of Exotic Species

Most of the nonnative fishes in the Hudson River have appeared via the direct release of individuals by humans. Of the 80 species known from the freshwater sections of the main channel of the Hudson River, 30 are nonnative (Daniels et al. 2005). Some of these introductions occurred before anyone surveyed the fish fauna of the Hudson. For instance, DeKay (1842) discussed carp and rock bass *Ambloplites rupestris* from the Hudson River and Gill (1855) listed largemouth bass *Huro nigricans* = *Micropterus salmoides* in the Fulton Fish Market and smallmouth bass *Centrachus fasciatus* = *Micropterus dolomieu* from the Hudson River.

Later introductions to the Hudson watershed have been overseen by New York State agencies to enhance sport fishing (brown trout *Salmo trutta*, walleye *Sander vitreus*, and tiger muskellunge [northern pike *Esox lucius* × muskellunge *E. masquinongy*]), mosquito control in some locations (western mosquitofish *Gambusia affinis*), or as part of pond management activities (fathead minnow *Pimephales promelas*, green sunfish *Lepomis cyanellus*, and grass carp *Ctenopharyngodon idella*) (W. Keller, personal communication, NYS Department of Environmental Conservation). Introductions also have occurred without state agency consent, such as channel catfish *Ictalurus punctatus*, which is now well established in the Hudson. The repeated occurrence of the tropical pirapitinga *Piractus brachypomus* in the Hudson drainage is due to ongoing aquarium releases, as is the alligator gar reported from the Hackensack River (Bragin et al. 2005). Other species have been released for unknown reasons (e.g., warmouth *Lepomis gulosus*).

Recent Species Additions

Given the continuous nature and large extent of marine waters, species not seen before within a given area may suddenly be observed either because of a combination of chance and their actual rarity, because of amenable changes in oceanographic conditions, or because of an expansion in their range coincident with an increase in their abundance. One now prominent species in the brackish waters of the Hudson River estuary is the gizzard shad *Dorosoma cepedianum*. About a century ago, Bean (1903) stated that it was found in brackish waters from New York southward. Greeley (1937) did not mention this euryhaline, but largely estuarine species, among the common fishes found in a major survey of the Hudson River conducted in 1936, although Breder (1938) did report specimens from New York Harbor. George (1983) reviewed published reports and concluded that the first record of gizzard shad in the Hudson River did not occur until 1973, when 674 specimens were impinged on the water-intake screens of four power plants. Since then, gizzard shad have remained numerous in the Hudson River estuary (and have expanded their range to other coastal rivers as far north as the Kennebec River; Daniels et al. 2005) but it is not known whether this population is derived from marine sources or from inland waters via the Erie

Barge Canal, or both (Beebe and Savidge 1988).

The 11 documented additions for the Hudson River watershed north of the Battery since Smith and Lake (1990) are mostly temperate or tropical marine strays, including scrawled cowfish *Acanthostracion quadricornis* (Schmidt and Lake 2001), highfin goby *Gobionellus oceanicus*, feather blenny *Hypsoblennius hentz*, northern tonguefish *Symphurus pusillus*, schoolmaster *Lutjanus apodus*, sheepshead *Archosargus probatocephalus*, and bonefish *Albula vulpes*; but they also include a boreal marine stray, lumpfish *Cyclopterus lumpus*, two freshwater introductions, western mosquitofish and the aforementioned pirapitanga, and an estuarine species, spotfin killifish *Fundulus luciae*. However, spotfin killifish may have been present in the Hudson drainage for some time but went unnoticed because it occurs in rarely sampled high intertidal marsh habitats and because of its resemblance to mummichog (Yozzo and Ottman 2003).

Recent Species Losses

The regional diversity of marine fishes may decrease through regional losses of species, but to date, marine fishes have almost never become extinct in historical times (Casey and Myers 1998). These regional losses typically are felt as large decreases in numbers over extended periods, not as complete absences. Two species that were common in New York Harbor and nearby coastal waters in the 19th century, black drum *Pogonias cromis* and sheepshead (Bean 1903; Zeisel 1988) are almost never encountered there today. Both were at the northern end of their ranges and are closely associated with oyster reefs. It is likely that their present-day scarcity in New York waters is due to the loss of oyster reefs and to contractions of coastal populations. Sharks also were common in inshore waters of the New York Harbor and New York Bight in the 1800s. They were even seen and caught off docks in Manhattan, probably drawn there by dumped refuse (Zeisel 1990).

A few species experience very wide fluctuations in abundance that lead to near-absences in New York waters during some periods and high abundances in others. Spot *Leiostomus xanthurus* may appear in very large numbers for only a year or two and then not be seen for almost a decade; they are often called "Lafayettes" because one of their appearances coincided with the visit of the Marquis de Lafayette to New York City in 1824 (Waldman 1999). Bluefish and weakfish *Cynoscion regalis* have also varied in abundance from near absences to high abundances on a scale of decades (Bigelow and Schroeder 1953). At times, two typically southern fishes, Atlantic croaker *Micropogonias undulatus* and Spanish mackerel *Scomberomerus maculatus* were abundant enough to be important in commercial catches in Raritan Bay, the Atlantic croaker in the 1910s, 1920s, and the late 1930s and early 1940s and the Spanish mackerel in the 1880s and 1890s (McKenzie 1990).

Populations of anadromous species fishes may become extirpated. This appears to have occurred in the Hudson for rainbow smelt *Osmerus mordax*, a coldwater species at the southern edge of its range in the Hudson River estuary. Rainbow smelt once ran up both Hudson tributaries and the mainstem but declined in the late 20th century (Rose 1993), appearing to have disappeared from the system in the late 1990s (Daniels et al. 2005; Waldman 2006), probably due to warming temperatures. Also, whether Atlantic salmon were native to the Hudson River has long been debated (e.g., Webster 1982). Occasional specimens were seen in historical times and a late 19th century stocking program provided returns but no apparent reproduction (Waldman 1999). However, there is a fossil evidence for their presence (either anadromous or landlocked) in the more southern Delaware River watershed (Peteet et al. 1993).

Diversity, Nonnative Species, and Comparisons with Other Drainages

Carlson and Daniels (2004) inventoried the freshwater and diadromous fishes of New York State, noting 176 taxa (including subspecies) of 171 species. They treated the Hudson drainage as three watersheds: Upper and Lower Hudson, and the Mohawk. Of the 114 taxa identified in the Hudson drainage, 69 were native throughout, which meant that 39% were exotic in at least one of the three watersheds. They believed that among all New York drainages, the Hudson's ichthyofauna has been most affected by introductions of exotic species. For example, they estimated that 33% of the taxa in the Upper Hudson have been gained since 1940. In contrast, apparent total losses of native taxa in the entire drainage have been minor; only the brassy minnow *Hybognathus hankinsoni* is absent from recent surveys.

Carlson and Daniels (2004) also examined the freshwater fish fauna of the Hackensack and Passaic Rivers in New York. This highly urbanized watershed, which drains to Newark Bay, had 51 species reported, of which 33 are native. However, it also has had five native species extirpated since 1950, the highest percentage in all the watersheds of New York State.

Carlson and Daniels (2004) found that the proportion of freshwater and diadromous taxa gained since 1940 for the Hudson drainage was 18%, versus 11% for the Saint Lawrence drainage and 33% for the Susquehanna. They also estimated that 42% of fishes now found in the Lower Hudson watershed are exotic. Schmidt (1986) compared the native freshwater ichthyofauna of the Hudson River watershed to more northern Atlantic coastal rivers. Comparisons with nearby systems showed faunal resemblance indices of 69.4% with the Housatonic River and 75.2% with the Connecticut River.

Conservation

The biodiversity of New York State's marine ichthyofauna is most susceptible to stresses that influence abundance and species composition (i.e., species-wide or local extinctions are not likely), but substantial changes in the numbers of fish and their relative proportions can occur. The combined effect of commercial and recreational fishing may be the greatest stress on New York's marine fishes. In the Hudson Estuary, commercial fishing has declined but recreational fishing, particularly for striped bass, has increased greatly despite health advisories on consumption (McKenzie 1990; Limburg et al. 2006). In New York, only shortnose sturgeon are on the U.S. federal endangered species list and this species is now flourishing in the Hudson River (Waldman 2006). However, its congener, Atlantic sturgeon was heavily overfished in the late 1980s and early 1990s (Waldman et al. 1996). Atlantic sturgeon are now protected by a long-term fishing moratorium in U.S. waters.

Habitat alteration and degradation can affect the quality of spawning, nursery, feeding, and wintering grounds, and the Hudson estuary has been subjected to myriad habitat changes (Daniels et al. 2005). However, the importance of habitat has been recognized by the National Marine Fisheries Service as being integral to conserving marine fishes and habitat concerns are now being included in fishery management plans. In the New York region many efforts have been undertaken to restore intertidal marsh habitats (Waldman 1999) and New York State maintains an artificial reef construction program to provide new fish habitat.

Although the lethality and sublethal effects of pollution are difficult to quantify, contamination with chemical pollutants has resulted in elevated levels of cancer in Atlantic tomcod of the Hudson River (Dey et al. 1993) and in abnormalities of other fishes in New York waters (Sindermann et al. 1982). One threat that deserves vigilance against is the introduction of exotic species of fish and other taxa that can alter the ecology (e.g., Asian shore crab *Hemigrapsus sanguineus*; Ahl and Moss 1999) or parasitize fishes such as an Asian nematode worm found in American eel *Anguilla rostrata* in the Hudson River (Barse and Secor 1999). Also, the colonization of the freshwater portion of the Hudson River by the Eurasian zebra mussel *Dreissena polymorpha* has been shown to strongly modify the ecology of the system, which has instigated both increases and declines in abundance and growth of fishes and in their centers of distribution in the river (Strayer et al. 2004).

A more nebulous concern is the effects of global warming on fish species composition. Long-term data show increasing water temperatures in the Hudson (Ashizawa and Cole 1994; Abood et al. this volume). Climate change may be responsible both for colonization of the Hudson River by gizzard shad and the apparent loss of its rainbow smelt population.

References

Able, K. W., J. P. Manderson, and A. L. Studholme. 1998. The distribution of shallow water juvenile fishes in an urban estuary: the effects of manmade structures in the lower Hudson River. Estuaries 21(4B):731–744.

Abood, K. A., T. L. Englert, S. G. Metzger, C. V. Beckers, Jr., T. J. Groninger, and S. Mallavaram. 2006. Current and evolving physical and chemical conditions in the Hudson River estuary. Pages 39–61 *in* J. R. Waldman, K. E. Limburg, and D. L. Strayer. Hudson River fishes and their environment.

Ahl, R. S., and S. P. Moss. 1999. Status of the nonindigenous crab, *Hemigrapsus sanguineus*, at Greenwich Point, Connecticut. Northeastern Naturalist 6:221–224.

Alevras, R. A. 1973. Occurrence of a lookdown in the Hudson River. New York Fish and Game Journal 20(1):76.

Ashizawa, D., and J. J. Cole. 1994. Long-term temperature trends of the Hudson River: a study of the historical data. Estuaries 17(1B):166–171.

Barse, A. M., and D. H. Secor. 1999. An exotic nematode parasite of the American eel. Fisheries 24(2):6–10.

Bean, T. H. 1903. Catalogue of the fishes of New York. New York State Museum Bulletin 60, Zoology 9.

Beebe, C. A., and I. R. Savidge. 1988. Historical perspective on fish species composition and distribution in the Hudson River estuary. Pages 25–36 *in* L. W. Barnthouse, R. J. Klauda, D. S. Vaughan, and R. L. Kendall, editors. Science, law, and Hudson River power: a case study in environmental impact assessment. American Fisheries Society Monograph 4, Bethesda, Maryland.

Bigelow, H. B., and W. C. Schroeder. 1953. Fishes of the Gulf of Maine. Fishery Bulletin 74, Bulletin of U.S. Fish and Wildlife Service 53.

Borns, H. W., Jr. 1973. Late Wisconsin fluctuations of the Laurentide ice sheet in southern and eastern New England. Pages 37–44 *in* R. F. Black, R. P. Goldwait, and H. B. Williams, editors. The Wisconsin Stage. Geological Society of America Memoirs 136.

Bragin, A. B., J. Misuik, C. A. Woolcott, K. R. Barrett, and R. Jusino-Atrensino. 2005. A fishery resource inventory of the lower Hackensack River within the Hackensack Meadowlands District. Draft Report, New Jersey Meadowlands Commission, Meadowlands Environmental Research Institute.

Breder, C. M., Jr. 1938. The species of fish in New York Harbor. Bulletin of the New York Zoological Society 41(1):23–29.

Breder, C. M., Jr., and R. M. Nigrelli. 1934. Fish notes for 1933 and 1934 from Sandy Hook Bay and other New York localities. Copeia 1934:193–194.

Briggs, P. T., and J. R. Waldman. 2002. Annotated list

of fishes reported from the marine waters of New York. Northeastern Naturalist 9:47–80.

Brown, B. L., and R. W. Chapman. 1991. Gene flow and mitochondrial DNA variation in the killifish, *Fundulus heteroclitus*. Evolution 45:1147–1161.

Carlson, D. M., and R. A. Daniels. 2004. Status of fishes in New York: increases, declines and homogenization of watersheds. American Midland Naturalist 152:104–139.

Casey, J. M., and R. A. Myers. 1998. Near extinction of a large widely distributed fish. Science 281:690–692.

Cooper, J. C., F. R. Cantelmo, and C. E. Newton. 1988. Overview of the Hudson River estuary. Pages 11–24 *in* L. W. Barnthouse, R. J. Klauda, D. S. Vaughan, and R. L. Kendall, editors. Science, law, and Hudson River power: a case study in environmental impact assessment. American Fisheries Society Monograph 4, Bethesda, Maryland.

Daniels, R. A. 1995. Nearshore fish assemblage of the tidal Hudson River. Pages 260–263 *in* E. T. LaRoe, G. S. Farris, C. E. Puckett, P. D. Doran, and M. J. Mac, editors. Our living resources: a report to the nation on the distribution, abundance, and health of U.S. plants, animals, and ecosystems. U.S. Department of the Interior, National Biological Service, Washington, D.C.

Daniels, R. A. 2001. Untested assumptions: the role of canals in the dispersal of sea lamprey, alewife, and other fishes in the eastern United States. Environmental Biology of Fishes 60:309–329.

Daniels, R. A., K. E. Limburg, R. E. Schmidt, D. L. Strayer, and R. C. Chambers. 2005. Changes in fish assemblages in the tidal Hudson River, New York. Pages 471–503 *in* Rinne, J. N., R. M. Hughes, and B. Calamusso, editors. Historical changes in large river fish assemblages of America. American Fisheries Society, Symposium 45, Bethesda, Maryland.

DeKay, J. E. 1842. Zoology of New York. Part IV. Fishes. White and Visscher, Albany, New York.

Dey, W. P., T. H. Peck, C. E. Smith, and G. L. Kreamer. 1993. Epizoology of hepatic neoplasia in Atlantic tomcod (*Microgadus tomcod*) from the Hudson River estuary. Canadian Journal of Fisheries and Aquatic Sciences 50:1897–1907.

Emery, K. O., and L. E. Garrison. 1967. Sea levels 7,000–20,000 years ago. Science 157:684–687.

Flint, R. F. 1971. Glacial and quaternary geology. Wiley, New York.

Friedmann, B. R., and C. T. Hamilton. The fish life of Upper New York Bay. Underwater Naturalist 12(2):18–21.

George, C. J. 1983. Occurrence of the gizzard shad in the lower Mohawk Valley. New York Fish and Game Journal 30(1):113–114.

Gill, T. 1855. On the fishes of New York. Report of the Smithsonian Institution (1856):253–269, Washington D.C.

Gladden, J. B., F. R. Cantelmo, J. M. Croom, and R. Shapot. 1988. Evaluation of the Hudson River ecosystem in relation to the dynamics of fish populations. Pages 37–52 *in* in L. W. Barnthouse, R. J. Klauda, D. S. Vaughan, and R. L. Kendall, editors. Science, law, and Hudson River power: a case study in environmental impact assessment. American Fisheries Society Monograph 4, Bethesda, Maryland.

Greeley, J. R. 1937. Fishes of the area with an annotated list. Pages 45–103 *in* A biological survey of the lower Hudson watershed. Supplement to Twenty-sixth Annual Report, 1936. New York State Conservation Department, Albany.

Greeley, J. R. 1939. The fresh-water fishes of Long Island and Staten Island with annotated list. Pages 29–44 *in*: A biological survey of the fresh waters of Long Island. Supplement to Twenty-eighth Annual Report, 1938. New York State Conservation Department.

Hurst, T. P., K. A. McKown, and D. O. Conover. 2004. Interannual and long-term variation in the nearshore fish community of the mesohaline Hudson River estuary. Estuaries 27:659–669.

Kuhn, K., and I. Kornfield. 2004. Genetic differentiation of the alewife, *Alosa pseudoharengus*, in the Hudson River. Section V *in* W. C. Nieder and J. R. Waldman, editors. Final Reports of the Tibor T. Polgar Fellowship Program, 2003. Hudson River Foundation, New York.

Limburg, K. E., K. A. Hattala, A. W. Kahnle, and J. R. Waldman. 2006. Fisheries of the Hudson River estuary. Pages 189–204 *in* J. S. Levinton and J. R. Waldman, editors. The Hudson River Estuary. Cambridge University Press, New York.

Limburg, K. E., M. A. Moran, and W. H. McDowell. 1986. The Hudson River ecosystem. Springer-Verlag, New York.

McBride, R. 1996. On the rarity of banded butterflyfish in the mid-Atlantic. Underwater Naturalist 23(3):18–19.

McCauley, J. 1829. The natural, statistical, and civil history of the State of New York. Gould and Banks and William Gould and Co., Albany, New York.

McKenzie, Jr., C. L. 1990. History of the fisheries of

Raritan Bay, New York and New Jersey. Marine Fisheries Review 52(4):1–45.

Miller, W. S. 1914. The geological history of New York State. Kennikat Press, Port Washington, New York.

Mitchill, S. L. 1815. The fishes of New York, described and arranged. Transactions of the Literary and Philosophical Society of New York 1:355–492.

Mitchill, S. L. 1818. Memoir on ichthyology. The fishes of New York, described and arranged. American Monthly Magazine Critical Review 2:241–248.

Myers, G. S. 1938. Fresh-water fishes and West Indian zoogeography. Smithsonian Reports (1937):339–364.

Nichols, J. T., and C. M. Breder, Jr. 1927. The marine fishes of New York and southern New England. Zoologica IX:1–192.

Parr, A. E. 1933. A geographic-ecological analysis of the seasonal changes in temperature conditions in shallow water along the coast of the United States. Bulletin of the Bingham Oceanographic Collection of Yale University 4:1–90.

Peteet, D. M., R. A. Daniels, L. E. Heusser, J. S. Vogel, J. R. Southern, and D. E. Nelson. 1993. Late-glacial pollen, macrofossils and fish remains in Northeastern U.S.A.—the Younger Dryas oscillation. Quaternary Science Reviews 12:597–612.

Pielou, E. C. 1991. After the Ice Age: the return of life to glaciated North America. University of Chicago Press, Chicago, Illinois.

Rampino, M. R., and J. E. Sanders. 1980. Holocene transgression in south-central Long Island, New York. Journal of Sedimentary Petrology 50:1063–1080.

Rose, F. P. 1993. Have all the smelt gone somewhere? Assessing changes in population size of the rainbow smelt (*Osmerus mordax*) in the Hudson River estuary. Master's Thesis, Bard College, Annandale-on-Hudson, New York.

Schmidt, R. E. 1986. Zoogeography of the northern Appalachians. Pages 137–159 *in* C. H. Hocutt and E. O. Wiley, editors. The Zoogeography of North American Freshwater Fishes. Wiley, New York, New York.

Schmidt, R. E., and R. A. Daniels. 2006. Hybridization in Umbridae in the Hudson River, New York, with designation of neotypes for *Umbra limi and. Umbra pygmaea*. Zootaxa 1113:1–20.

Schmidt, R. E., and T. R. Lake. 2001. A benthic juvenile scrawled cowfish (*Acanthostracion quadricornis*) from the lower Hudson River Estuary. Northeastern Naturalist 8:343–346.

Schmidt, R. E., B. M. Jessop, and J. E. Hightower. 2003. Status of river herring stocks in large rivers. Pages 171–182 *in* K. E. Limburg and J. R. Waldman, editors. Biodiversity, status, and conservation of the world's shads. American Fisheries Society, Symposium 35, Bethesda, Maryland.

Sindermann, C. J., S. C. Esser, E. Gould, B. B. McCain, J. L. McHugh, R. P. Morgan II, R. A. Murchelano, M. J. Sherwood, and P. R. Spitzer. 1982. Effects of pollutants on fishes. Pages 23–38 *in* G. F. Mayer, editor. Ecological stress and the New York Bight: science and management. Estuarine Research Federation, Columbia, South Carolina.

Smith, C. L. 1985. The inland fishes of New York State. New York State Department of Environmental Conservation, Albany.

Smith, C. L., and T. R. Lake. 1990. Documentation of the Hudson River fish fauna. American Museum Novitates 2981:1–17.

Smith, E. 1897. The fishes of the fresh and brackish waters in the vicinity of New York City. Proceedings of the Linnaean Society of New York 9:1–51.

Strahler, A. N. 1966. A geologist's view of Cape Cod. The Natural History Press, Garden City, New York.

Strayer, D. L., K. A. Hattala, and A. W. Kahnle. 2004. Effects of an invasive bivalve (*Dreissena polymorpha*) on fish in the Hudson River estuary. Canadian Journal of Fisheries and Aquatic Sciences 61:924–941.

Todd, T. N., and C. O. Hatcher. 1993. Genetic variability and glacial origins of yellow perch (*Perca flavescens*) in North America. Canadian Journal of Fisheries and Aquatic Sciences 50:1828–1834.

Underhill, J. C. 1986. The fish fauna of the Laurentian Great Lakes, the St. Lawrence lowlands, Newfoundland and Labrador. Pages 105–136 *in* C. H. Hocutt and E. O. Wiley, editors. The Zoogeography of North American Freshwater Fishes. Wiley, New York.

Waldman, J. R. 1985. Ichthyological notes from the Westway Study. Underwater Naturalist 15(3):19–22.

Waldman, J. 1999. Heartbeats in the muck: the history, sea life, and environment of New York Harbor. Lyons Press, New York.

Waldman, J. R. 2006. The diadromous fish fauna of the Hudson River: life histories, conservation concerns, and research avenues. Pages 171–188 *in* J. S. Levinton and J. R. Waldman, editors. The Hudson River Estuary. Cambridge University Press, New York.

Waldman, J. R., C. Grunwald, N. K. Roy, and I. I.

Wirgin. 2004. Mitochondrial DNA analysis indicates sea lampreys are indigenous to Lake Ontario. Transactions of the American Fisheries Society 133:950–960.

Waldman, J. R., C. Grunwald, J. Stabile, and I. I. Wirgin. 2002. Impacts of life history and biogeography on the genetic stock structure of Atlantic sturgeon *Acipenser oxyrinchus oxyrinchus*, Gulf sturgeon *A. o. desotoi*, and shortnose sturgeon. *A. brevirostrum.* Journal of Applied Ichthyology 18:509–518.

Waldman, J. R., J. T. Hart, and I. I. Wirgin. 1996. Stock composition of the New York Bight Atlantic sturgeon fishery based on analysis of mitochondrial DNA. Transactions of the American Fisheries Society 125:364–371.

Webster, D. A. 1982. Early history of the Atlantic salmon in New York. New York Fish and Game Journal 29(1):26–44.

Wilson, C. C., and P. D. N. Hebert. 1996. Phylogeographic origins of lake trout (*Salvelinus namaycush*) in eastern North America. Canadian Journal of Fisheries and Aquatic Sciences 53:2764–2775.

Wilson, C. C., and N. E. Mandrak. 2004. History and evolution of lake trout in Shield lakes: past and future challenges. Pages 21–35 *in* J. Gunn, R. J. Steadman, and R. Ryder, editors. Boreal Shield watersheds: lake trout ecosystems in a changing environment. Lewis Publishers, Boca Raton, Florida.

Woodhead, P. M. J., F. J. Rohlf, and M. A. O'Hare. 1992. The structure of the fish community and distributions of major species in the lower Hudson estuary and New York harbor. Final Report to the Hudson River Foundation, New York.

Young, B. H., I. H. Morrow, and S. R. Wanner. 1982. First record of the bluespotted cornetfish in the Hudson River. New York Fish and Game Journal 29(1):106.

Yozzo, D. J., and F. Ottman. 2003. New distribution records for the spotfin killifish, *Fundulus luciae* (Baird), in the lower Hudson River estuary and adjacent waters. Northeastern Naturalist 10:399–408.

Zeisel, W. 1988. A history of recreational fishing on the Hudson River from colonial times to 1920. Final Report to the Hudson River Foundation, New York

Zeisel, W. N., Jr. 1990. Shark!!! and other sport fish once abundant in New York Harbor. Seaport (autumn):36–39, South Street Seaport Museum, New York.

American Fisheries Society Symposium 51:151–173, 2006

Patterns of Habitat Utilization by Resident Nekton in *Phragmites* and *Typha* Marshes of the Hudson River Estuary

DAVID T. OSGOOD

Albright College, Department of Biology, Post Office Box 15234
Reading, Pennsylvania 19612-5234 USA
dosgood@alb.edu

DAVID J. YOZZO

Barry A. Vittor and Associates, Inc., 1973 Ulster Avenue
Lake Katrine, New York 12449 USA

RANDOLPH M. CHAMBERS

College of William and Mary, Department of Biology
Post Office Box 8795, Williamsburg, Virginia 23187 USA

SETH PIANKA AND CHRISTINA LEPAGE

University of New Haven, Department of Biology and Environmenal Science
300 Orange Avenue, West Haven, Connecticut 06516 USA

JASON LEWIS

U.S Geological Survey, 810 Bear Tavern Road, Suite 206
West Trenton, New Jersey 08628 USA

Abstract.—We compared nekton density and benthic prey availability over a range of flooding conditions within common reed *Phragmites australis* and narrowleaf cattail *Typha angustifolia* at a mesohaline and an oligohaline marsh on the Hudson River Estuary. Nekton were sampled using lift nets at high and low elevations. Marsh surface nekton sampling occurred on 16 high tide events from May–October 1999 and 2000. Hydrology (depth, duration, and frequency) was simultaneously measured. Benthic macroinvertebrate density and richness were measured from sediment cores in June, August, and October, coincident with lift net collections. A total of 690 individuals representing nine species, mostly mummichog *Fundulus heteroclitus* and daggerblade grass shrimp *Palaemonetes pugio* were captured within both vegetation types. Mean nekton density (individuals 6 m^{-2} $\pm$ SE) on the marsh surface was not significantly different among reed at the mesohaline marsh (4.1 $\pm$ 1.3) and reed (3.4 $\pm$ 0.6) and cattail (4.1 $\pm$ 0.9) at the oligohaline marsh. Nekton density did not vary predictably across the measured range of flooding depth and duration. Nekton density was also not predictably related to benthic macroinvertebrate density or taxonomic richness. Significantly more juvenile mummichog were collected within cattail relative to reed, suggesting that spawning success of this species may be lower in reed. Adult resident fish and macrocrustacean species are distributed similarly within and between reed and cattail when they experience comparable flooding regimes. The present study represents one of the first direct measures of nekton density within reed and corroborates previous patterns for nekton abundance between reed and nonreed vegetation stands across a wide salinity regime.

Introduction

Common reed *Phragmites australis* is an invasive clonal wetland plant that has become increasingly prevalent within tidal marshes of the northeastern United States over the last century (Orson et al. 1987; Chambers et al. 1999; Chambers et al. 2003). The invasion has been attributed to introduction of a nonnative genetic strain of the plant (Saltonstall 2002). Invasion by reed is particularly relevant to the Hudson River estuary, which has experienced invasion by many species with little understanding of the ecological consequences (Mills et al. 1996).

Common reed grows aggressively and often displaces a variety of salt and brackish marsh plant species including salt meadow hay *Spartina patens*, salt meadow cordgrass *Spartina alterniflora*, and narrow-leaf cattail *Typha angustifolia* (Chambers et al. 1999). Changes in hydrogeomorphology (intertidal elevation, creek density) following reed invasion have led to speculation that access by marsh resident nekton (e.g., mummichog *Fundulus heteroclitus* and grass shrimp *Palaemonetes pugio*) within reed may be diminished (Weinstein and Balletto 1999).

Prior studies in northeastern and mid-Atlantic salt marshes have found similarities in species composition and abundance of marsh-resident and estuarine-dependent nekton occupying reed and nonreed marshes (Fell et al. 1998; Rilling et al. 1998; Meyer et al. 2001; Warren et al. 2001; Raposa and Roman 2001; Hanson et al. 2002; Fell et al. 2003; Osgood et al. 2003). In contrast, the specific abundance of larval and juvenile resident nekton within reed is often lower than nonreed marshes (Able and Hagan 2000; Meyer et al. 2001; Able et al. 2003; Fell et al. 2003; Osgood et al. 2003), despite successful spawning within reed stands (Able and Hagan 2003). A reduction in surface microtopography, irregular tidal flooding, and a relative lack of benthic prey are potential biotic and hydrogeomorphic factors that may impair spawning success by intertidal marsh fundulids occupying reed stands (Able and Hagan 2000; Able and Hagan 2003; Raichel et al. 2003; Osgood et al. 2003). Previous studies, due to methodological constraints, have not paired individual measurements of nekton density with other data (Fell et al. 1998; Rilling et al. 1998; Meyer et al. 2001; Warren et al. 2001; Raposa and Roman 2001; Fell et al. 2003). Therefore, patterns of nekton utilization have not been adequately linked to location-specific variables (e.g., flooding depth, duration, frequency or benthic prey availability) within reed-impacted marshes.

Reed marsh surfaces, when accessible, do serve as foraging sites for nekton (Fell et al. 1998; Meyer et al. 2001; Fell et al. 2003). The composition of mummichog diets differed between reed and nonreed marsh surfaces in Connecticut, but the amount of consumed biomass was similar (Fell et al. 1998; Rilling et al. 1998). Raichel et al. (2003) documented a greater frequency of large mummichog prey (amphipods, gastropods) in reed relative to salt marsh cordgrass marshes in New Jersey. Angradi et al. (2001) demonstrated that benthic macroinvertebrate community composition differed between reed and salt marsh cordgrass stands in a New Jersey brackish marsh. Further, salt marsh cordgrass stands had higher densities of benthic macroinvertebrates relative to adjacent reed stands and differences in density and taxonomic composition of the benthic community increased with distance from the marsh–water boundary (Angradi et al. 2001). Despite these differences, it is unclear if the distribution of benthic macroinvertebrates in reed marshes influences the distribution of marsh surface nekton.

Expansion rates of reed in tidal marshes vary across the estuarine salinity gradient, with higher rates of expansion in brackish marshes (5–20‰) relative to tidal freshwater or high salinity marshes (Chambers et al. 1999; Warren et al. 2001). This can result in some marshes with contiguous stands of reed and others with sparse, smaller stands (Winogrond and Kiviat 1997; Warren et al. 2001; Lathrop et al. 2003). Over time, reed increases elevation (decreasing flooding) and decreases marsh surface microtopography relative to nonreed stands (Windham and Lathrop 1999) most likely by accumulating organic biomass (Rooth and Stevenson 2000), but not necessarily inorganic sediments (Leonard et al. 2002) at rates faster than other vegetation types. Older, well-developed stands of reed also have a reduced number of low order tidal creeks as these features are filled during expansion (Weinstein and Balletto 1999; Lathrop et al. 2003). Variable stand age and configuration among marshes throughout the estuary mean that impaired habitat function associated with hydrogeomorphic changes may also vary along the salinity gradient.

The present study was conducted within reed and cattail stands in mesohaline and oligohaline Hudson River marshes. Most prior studies compare nekton abundance within reed stands to nonreed stands. In contrast, our objective was to directly quantify nekton density within the different vegetation types. In this manner we were also able to pair nekton density with corresponding measures of tidal flooding and macroinvertebrate prey availability. This is one of only two other nekton studies (Hanson et al. 2002; Yozzo et al. 1999) within reed stands on the Hudson River estuary.

Methods

Study Site

This study was conducted at the Piermont Marsh and Iona Island Marsh components of the Hudson River National Estuarine Research Reserve (Figures 1 and 2). Piermont Marsh is a 400 ha tidal mesohaline marsh (5–10‰) located along the west shoreline of the Hudson River, 37 km north of the New York Battery. Piermont Marsh was invaded by common reed between 1790 and 1858 (Winogrond and Kiviat 1997). Reed stands increased to cover almost 65% of the total marsh area by 1991, converting many mixed cattail species and reed communities to monospecific stands of reed (Winogrond and Kiviat 1997). Iona Island Marsh, located on the western shore of the Hudson River, 26 km north of Piermont Marsh, encompasses a 225 ha area of intertidal marsh and upland habitat and is located at the confluence of the oligohaline and tidal freshwater sections of the estuary (0–5‰). Vegetation cover includes a mix of ~ 70% reed and ~ 30% cattail. Broadleaf emergent plants are also present, but comprise a minor component of the vegetation community. Reed has been expanding in Iona Island Marsh since the mid-1960s (Winogrond and Kiviat 1997).

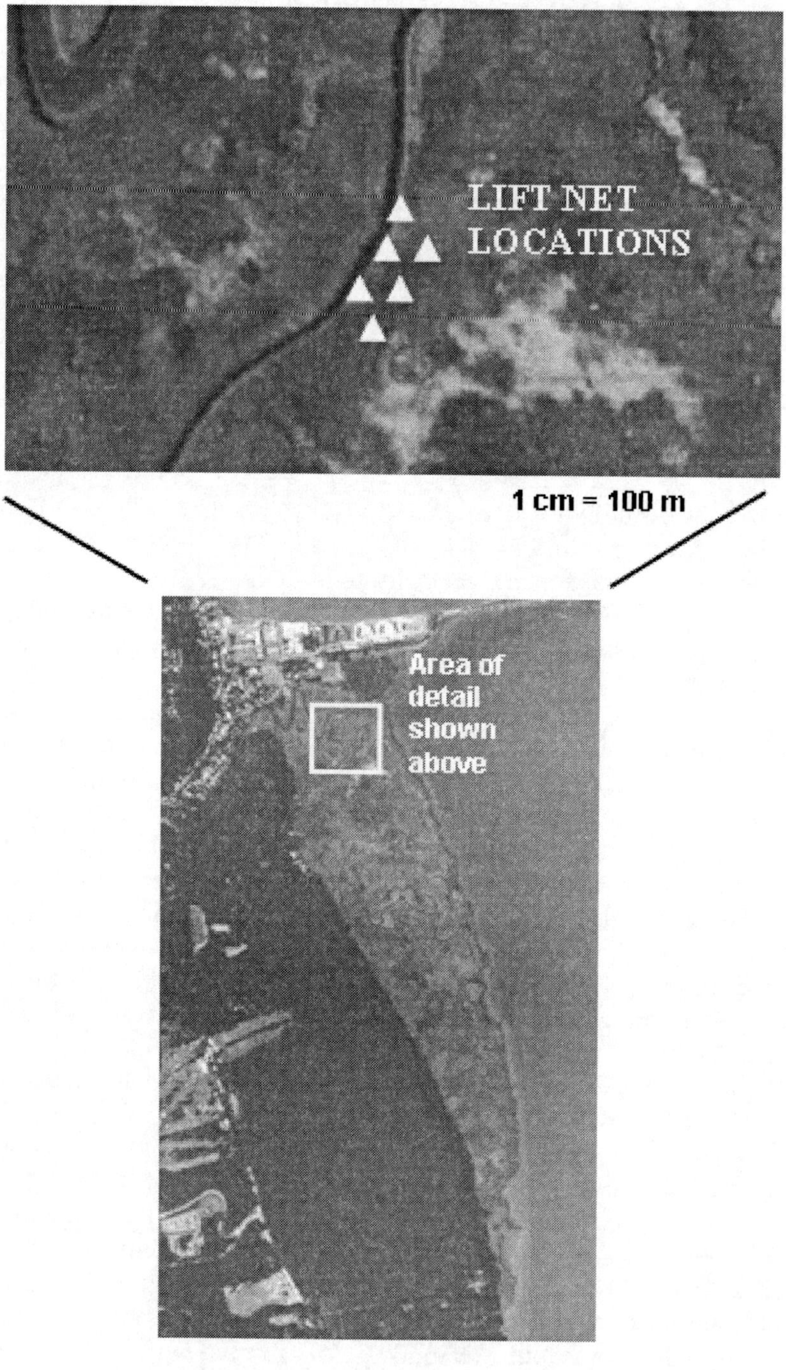

Figure 1. Piermont Marsh site map depicting lift net sampling and well locations within common reed stands (triangles).

Study design

Six sampling locations at Piermont Marsh, each consisting of one lift net and one well (see below for details on field instrumentation), were established within reed stands at random distances along one side of a 100 m long section of tidal creek (Figure 1). Three sampling locations each were positioned at high and low elevations as predetermined through a standard topographic survey (using surveying transit and leveling rod). Average elevation difference between high and low elevations was ~ 4 cm. Most locations within Piermont Marsh flood irregularly (personal observation) so this elevation difference was typical of the most frequently flooded section of the marsh. All sampling locations were within 20 m of the creek. Nonreed vegetation at Piermont Marsh is restricted to the interior sections of the marsh and does not flood except during extreme spring tides. Six sampling locations each were established within monospecific reed and cattail stands at Iona Island Marsh (Figure 2). Within both vegetation types, three sampling locations were situated at high and low elevations as predetermined through a topographic survey. Average difference between high and low elevations was 11.5 cm. Sampling locations were situated within 20 m of a tidal creek.

Hydrologic measurements

Marsh surface flooding (depth, duration, and frequency) was continuously measured from June to October 1999 and May to October 2000, using wells instrumented with remote data loggers and pressure transducer units. A single well was located adjacent to each lift net (see below for description of lift nets), for a total of six wells at Piermont Marsh and twelve wells at Iona Island Marsh. Wells were constructed of 7.6 cm PVC pipe and slotted (commercial well screen with 0.325 mm slots) over the entire subsurface length (50 cm). Readings, recorded every 24 min by the pressure transducer, were converted to water table position by calibrating each instrument with measurements of actual water table level in the wells. Concurrent pressure reading and water level measurements were used to derive a linear regression for conversion of all pressure readings to water table position. Water level was manually measured with a graduated, wired rod to ensure an accurate measurement. The wires were connected to a continuity beeper to provide audible evidence of when the rod tip contacted the water. Resolution of water level measurements, after adjusting for surveyed heights of the wells, was within 1.3 cm (Osgood 2000). Flooding depth was reported as the average maximum depth/tidal cycle each month. Flooding duration was reported as the average number of hours the marsh was inundated per tidal cycle each month and flooding frequency was reported as the percentage of total tidal cycles that flooded the marsh surface monthly.

Biotic measurements

Density of resident nekton was quantified using bottomless lift nets (Rozas 1992). The lift nets enclosed a 2 × 3 m area of intertidal marsh habitat at high tide, when nekton move about and forage actively. At low tide, the walls (3 mm mesh) were buried below the marsh surface. At slack high tide, the mesh walls were rapidly pulled into place above the water surface using a system of guideposts, pulleys, and a nylon rope/PVC pipe harness assembly. Lead lines prevented the bottom edges of the net from lifting off the surface of the marsh. As the tide receded, the mesh walls of the net remained deployed. Nekton trapped within the sampling area were collected from a 3.8 L polypropylene container placed into a pit dug into the marsh surface. This provided a low tide refuge, sought by smaller size classes of nekton to avoid stranding on the marsh surface. To maximize recovery efficiency, the sampling area was visually in-

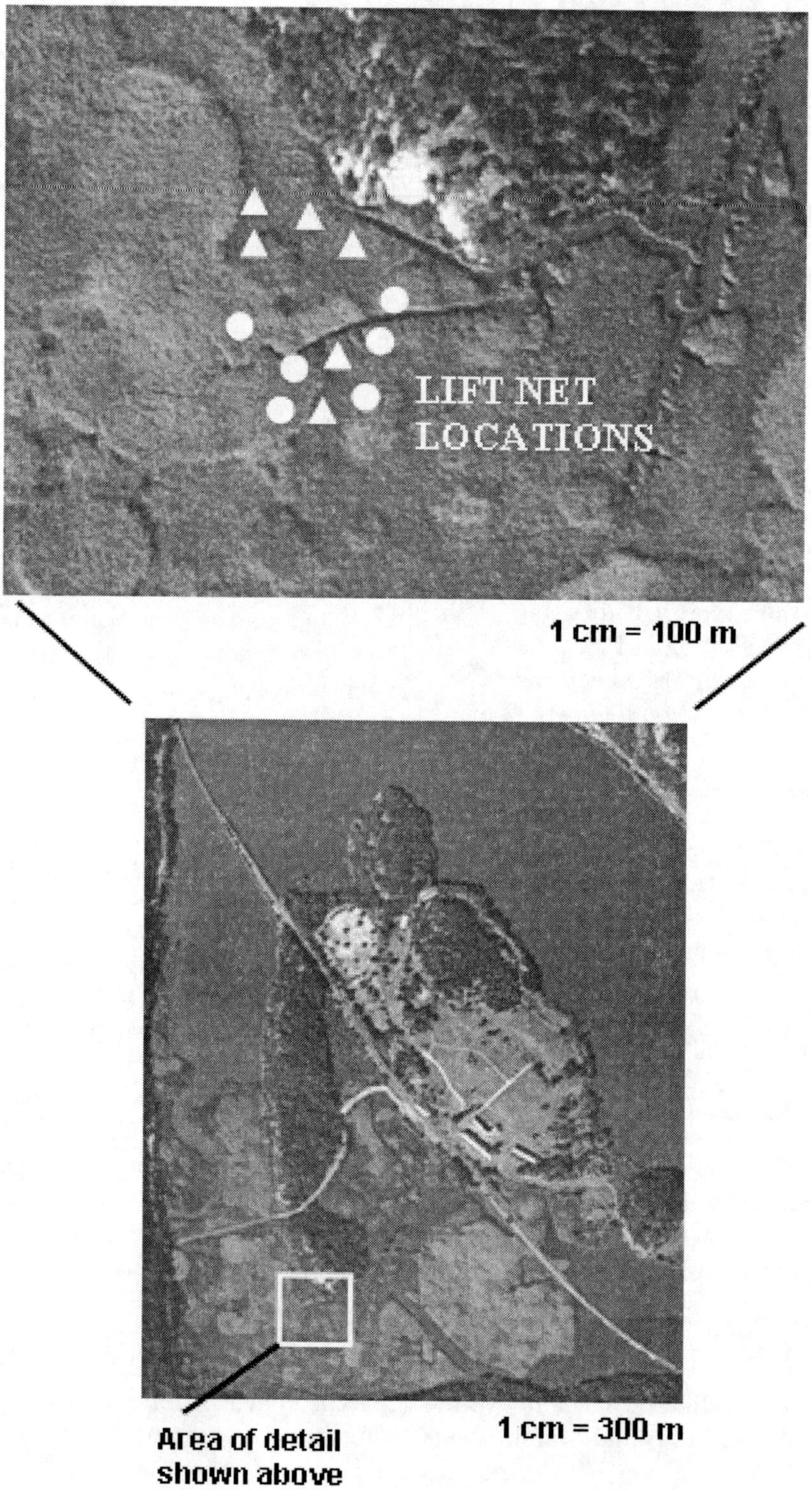

Figure 2. Iona Island site map depicting lift net sampling and well locations within narrow-leaf cattail (triangles) and common reed (circles) stands.

spected for organisms that may have been stranded on the marsh surface and the perimeter of each net was checked for nekton (Rozas 1992). Additionally, ponded water in the depressions holding the pit traps was sieved several times to capture organisms that may have sought refuge under the pit trap containers. Lift net recovery efficiencies of 32% (daggerblade grass shrimp) and 81% (Gulf killifish *Fundulus grandis*) measured in tidal salt marshes (Rozas 1992; Rozas and Minello 1997) are comparable to those of other devices designed to sample the vegetated marsh surface (Rozas and Minello 1997).

Lift nets were sampled on spring high tides every 2 weeks from July to October 1999 (Iona Island was additionally sampled on June 16 and only sampled once in August), for a total of seven sampling events for each site. During 2000, lift net sampling took place every 2 weeks during spring high tide from June to September at Piermont Marsh for a total of eight sampling events. Iona Island was similarly sampled with an additional sampling on May 06 for a total of nine sampling events. Organisms collected in the traps were preserved in the field in 5% neutral formalin and transported to the laboratory for enumeration and identification.

Benthic macroinvertebrate density was measured in June, August, and October of 1999 and 2000. Eight replicate 7.6 cm diameter cores were collected to a depth of 10 cm for each vegetation and elevation sampling location from each month of collection over both years. Individual cores were collected within 2 m of the lift net locations. Cores were sieved onto a 0.5 mm mesh and preserved in 5% neutral formalin with Rose Bengal stain. Macroinvertebrates were identified to family or lowest taxon possible. Stem density in both vegetation types was measured once in July 1999 from two 0.25 m^2 quadrats situated within 1 m of each lift net for a total of 12 and 24 quadrats at Piermont and Iona Island Marshes, respectively.

Data analysis

Repeated measures analysis of variance (ANOVA) was used to assess differences in nekton density as a function of marsh, vegetation, and elevation. Kruskal-Wallis Z tests, using a Bonferroni correction, were necessary to analyze for differences in flooding depth and duration, which did not meet the assumptions of normal distribution. Differences in mean flooding frequency were analyzed using ANOVA. Hydrologic data were pooled over the entire sampling period in any given year for the purpose of analyzing for spatial differences. Stem density was compared as a function of vegetation and elevation using ANOVA. Benthic macroinvertebrate density was normalized using a log (Y + 1) transformation and analyzed using ANOVA. The Student-Newman-Keuls multiple comparison test was used to distinguish among means when significant differences were specified. Linear regression analyses determined the relationship between nekton density, tidal hydrology, and benthic macroinvertebrate density and richness. Nekton density for each lift net collection was compared against flooding depth and duration for the tidal event coinciding with the time and date of nekton sampling. Nekton density (average for a single vegetation and elevation combination) from June, August, and October of both years was regressed against benthic macroinvertebrate density and taxonomic richness coinciding with the same months.

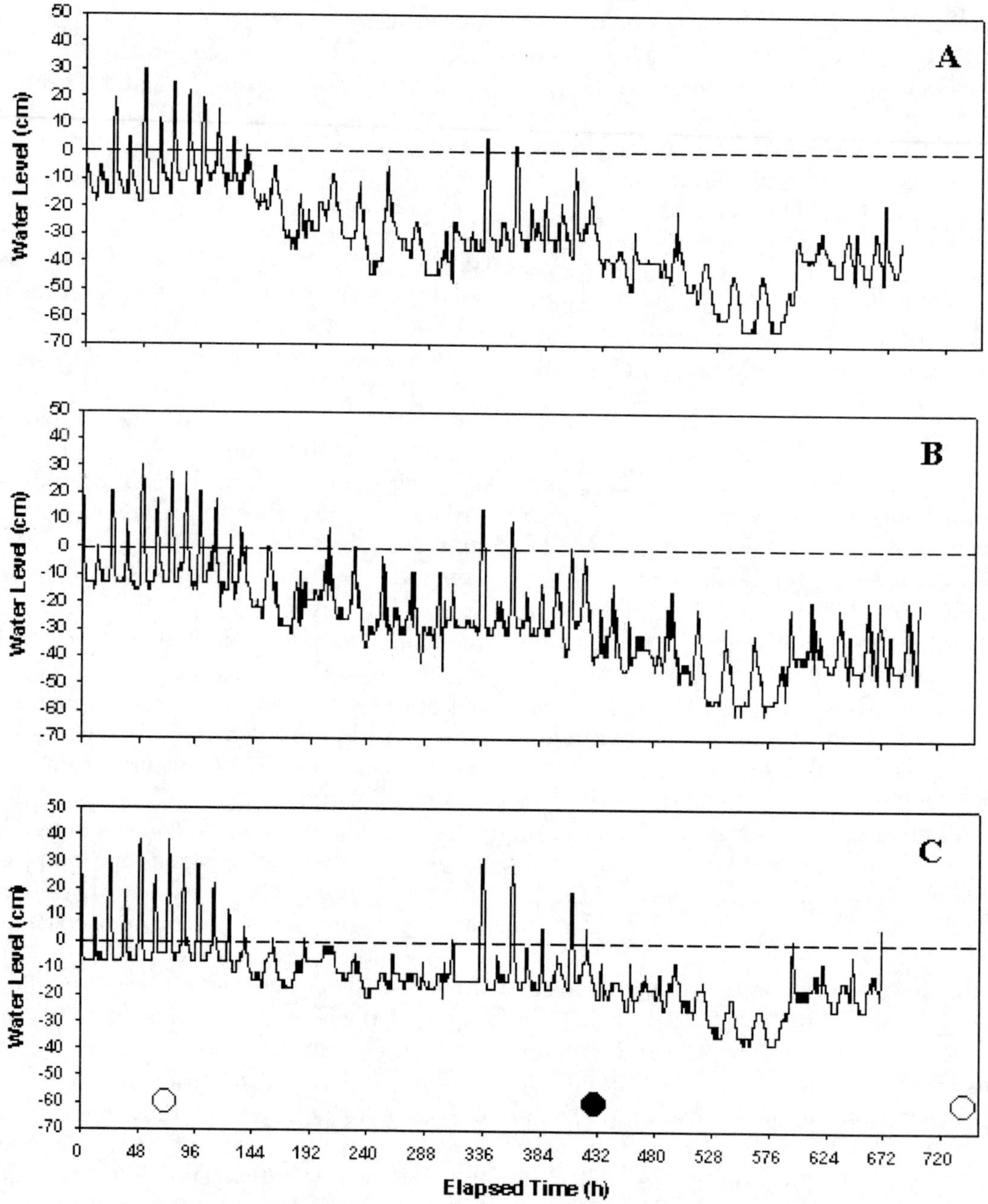

Figure 3. Water level relative to the marsh surface (dashed line) from one high elevation well each in common reed at Piermont Marsh (A), common reed at Iona Island Marsh (B), and cattail at Iona Island Marsh (C) from July 28–August 26, 2000. Data points were recorded at 24-min time intervals. Times of new moon (empty circle) and full moon (filled circle) are depicted on the graph.

Results

Hydrogeomorphology and Vegetation Structure

The water table position for August 2000 within the reed and cattail sampling locations at Piermont and Iona Island Marsh exemplifies the monthly pattern in hydrology (Figure 3). The surface of both marshes typically flooded only on spring tides. Iona Island Marsh flooded more frequently (~50–80% of high tides inundated the marsh) than Piermont Marsh where an average of 30% and 45% of high tides inundated the marsh surface at the high and low elevations, respectively, each month (Figures 3 and 4). Flooding depth was consistently higher at the low elevation station in both marshes and both vegetation types (Figure 4). Flooding duration and frequency were significantly different between elevations at Iona Island Marsh, but did not differ with elevation at Piermont Marsh (Figure 4). Flooding depth, duration, and frequency were similar between reed and cattail stands for the same elevations at Iona Island Marsh (Figure 4).

Stem density (±SE) was 82.7 (±5.8) and 70.8 (±5.2) stems m^{-2} within the reed stand at Piermont and Iona Island Marshes, respectively. Cattail stem density was 44.0 (±5.8) stems m^{-2} at Iona Island which was significantly less than the density of reed at either marsh ($P < 0.05$). The only significant difference between elevations was lower reed stem density at the low elevation relative to high elevation at Iona Island Marsh.

Marsh Surface Nekton

The 192 lift net samples (96 for each vegetation type, 48 for each elevation) from the Iona Island marsh surface yielded 489 individuals over both years (Table 1). A total of six species was represented. Mummichog comprised the majority of the individuals collected during both years. At Piermont Marsh a total of 90 lift net samples (45 for each elevation) yielded 201 individuals and six species over both years (Table 2). Daggerblade grass shrimp and mummichog dominated the marsh surface community at Piermont Marsh during both years. Both juvenile (13–35 mm total length) (size classes after Kneib 1987) and adult (>35 mm total length) size classes of mummichog were represented. No larvae (<13 mm total length) were collected during either year of the study. Given previously published lift net recovery efficiencies (see Methods), nekton density (primarily for grass shrimp caught at Piermont Marsh) should be treated as an underestimate of actual nekton density. Recovery efficiencies should, however, be comparable across sampling locations.

When pooled over both years, nekton density was not different between vegetation types at Iona Island Marsh or between vegetation types at Iona Island and Piermont Marsh. Reed at Piermont Marsh was characterized by significantly lower nekton density compared to reed and cattail at Iona Island Marsh during 2000 (Figures 5 and 6a). Otherwise there were no consistent differences among vegetation type or marshes within a single year. Total nekton density on the marsh surface was significantly lower in 2000 than in 1999 for reed at Piermont and Iona Island Marsh. Marsh surface nekton density was not different between years for cattail primarily because of an increase in density during August–September 2000 (Figure 5).

Mummichog, the dominant fish on the marsh surface, had consistently higher densities at high elevation cattail except relative to low elevation cattail and high elevation reed at Piermont Marsh (Figure 6a). Low elevation

Table 1. Total nekton collected from the marsh surface at Iona Island using lift nets.

Taxa	Common Name	Abundance 1999	Abundance 2000	Percent Abundance 1999	Percent Abundance 2000	Total length range[1]
Fundulus heteroclitus	Mummichog	226	214	84.0	97.2	14–105
Palaemonetes pugio	Dagger-blade grass shrimp	35	2	13.0	0.9	19–43
Anguilla rostrata	American eel	0	2	0.0	0.9	102–211
Callinectes sapidus	Blue crab	7	1	2.6	0.5	12–86
Cyprinus carpio	Common carp	0	1	0.0	0.5	3048
Leopomis gibbosus	Pumpkinseed	1	0	0.4	0.0	110
	Total Abundance	**269**	**220**			

Table 2. Total nekton collected from the marsh surface at Piermont Marsh using lift nets.

Taxa	Common Name	Abundance 1999	Abundance 2000	Percent Abundance 1999	Percent Abundance 2000	Total length range[1]
Palaemonetes pugio	Dagger-blade grass shrimp	72	11	47.1	22.9	15–84
Fundulus heteroclitus	Mummichog	63	21	41.2	43.7	16–105
Callinectes sapidus	Blue crab	12	7	7.8	14.6	7–50
Fundulus luciae	Spotfin killifish	0	8	0.0	16.7	21–32
Menidia menidia	Atlantic silverside	4	1	2.6	2.1	32–76
Morone americana	White perch	2	0	1.3	0	29–45
	Total Abundance	**153**	**48**			

[1]Total length for finfish, carapace length for blue crab.

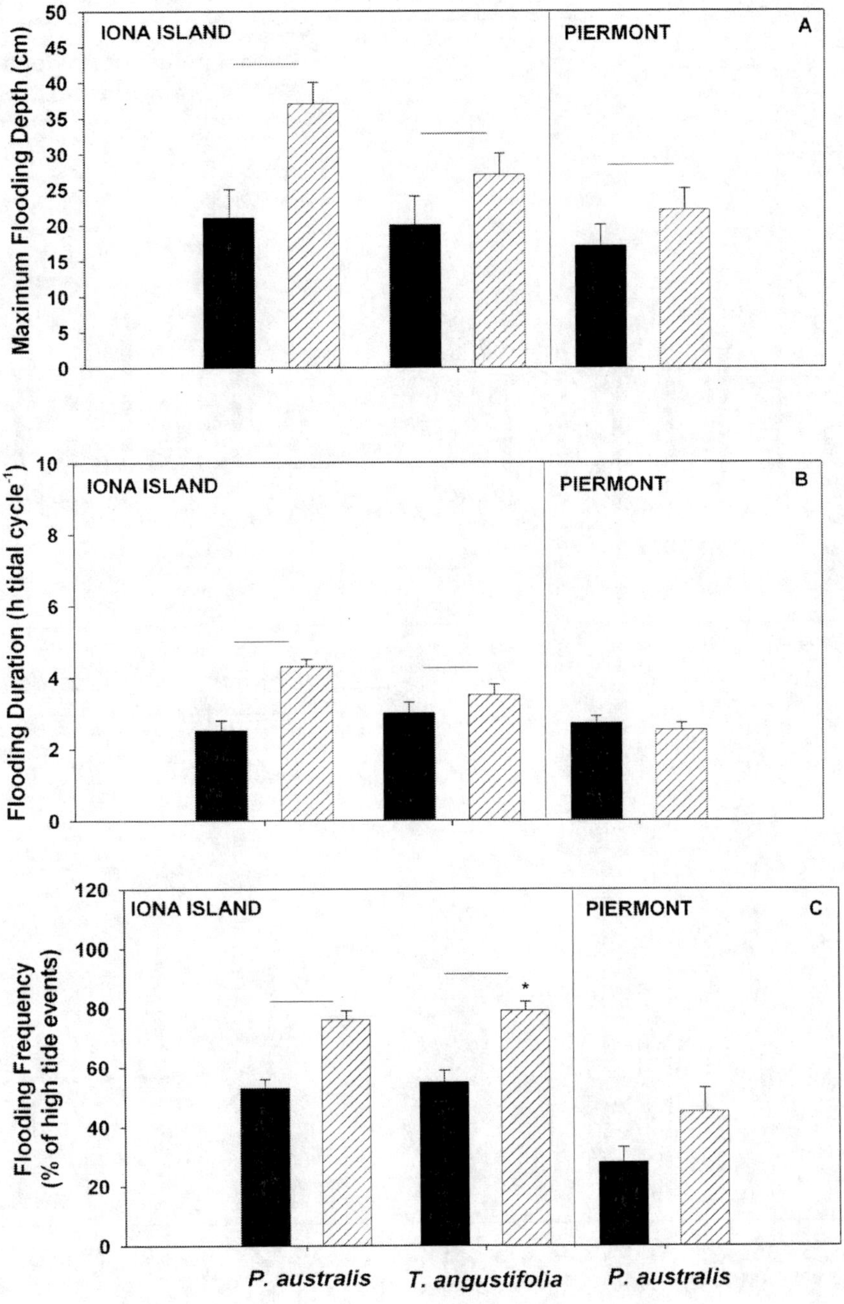

Figure 4. Mean (±SE) tidal flooding depth (A), duration (B), and frequency (C) over the 2-year study period for high (solid bars) and low (hashed bars) elevations at Piermont and Iona Island Marsh. Bars underneath the horizontal line are not significantly different ($P > 0.05$). The asterisk designates a significant difference ($P < 0.05$) relative to all other elevations and vegetation types at the same site.

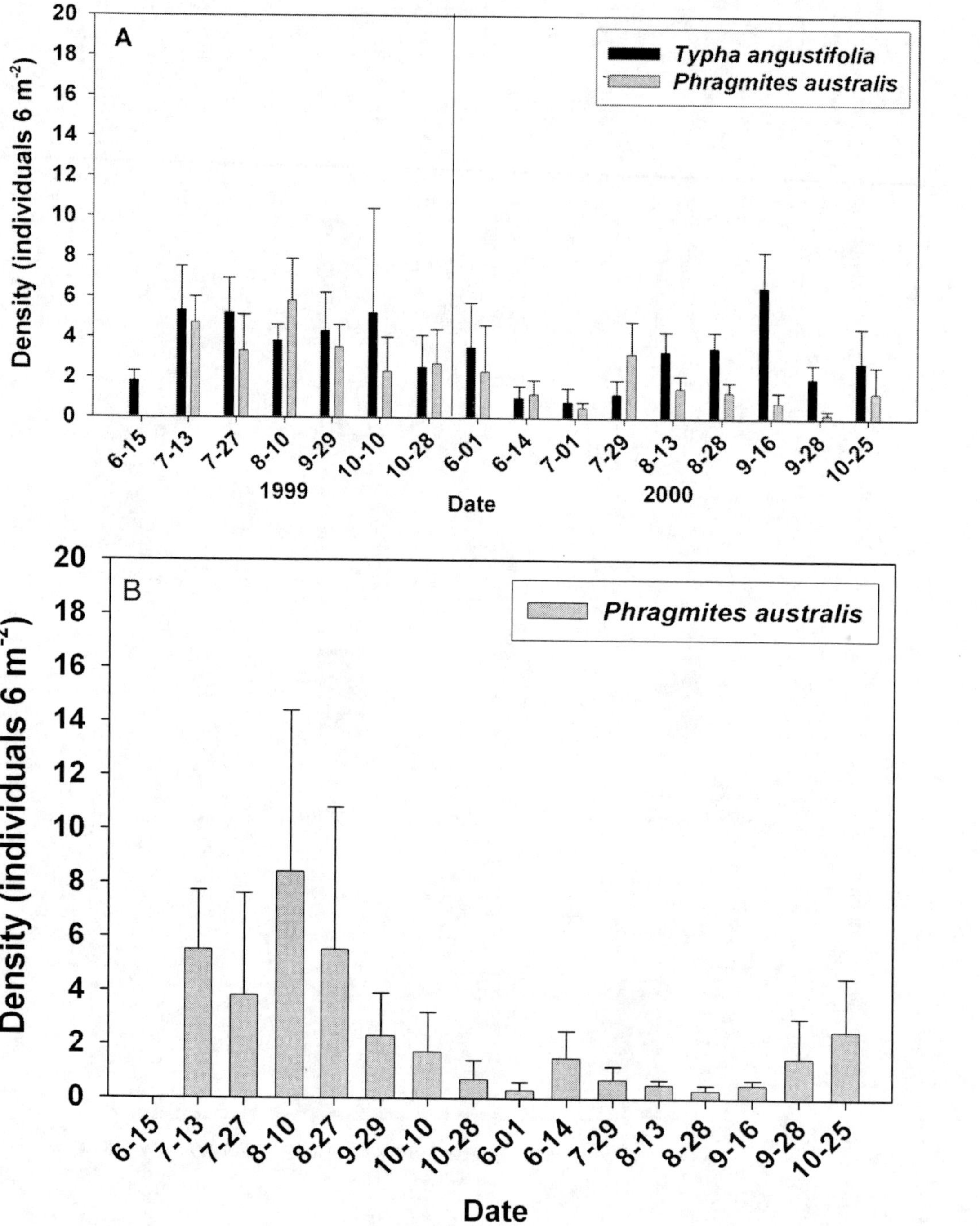

Figure 5. Temporal changes in mean density (individuals 6 m^{-2} ± SE) of total nekton species from June to October 1999 and 2000 at Iona Island (A) and Piermont Marsh (B). *N* = 6 replicate lift nets per vegetation type at each marsh.

reed at Piermont Marsh yielded the lowest mummichog density relative to all other sampling sites (Figure 6a). Otherwise, there were no consistent differences among elevations within a single vegetation type or among elevations between vegetation types.

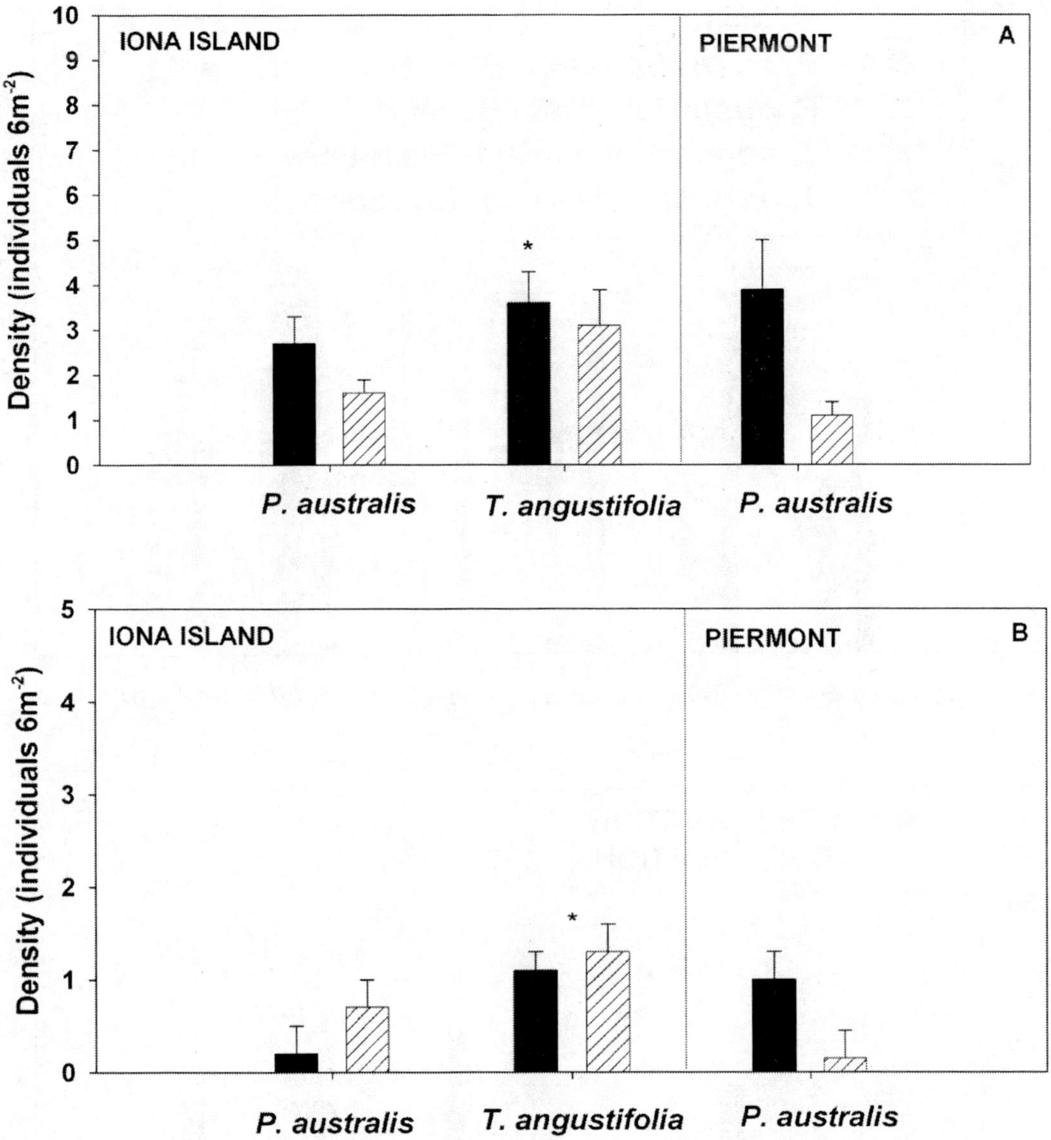

Figure 6. Mean density of mummichogs (individuals 6 m^{-2} ± SE) of adults (>30 mm total length) (A) and juveniles (15–30 mm total length) (B) for the 2-year study period. N = 3 replicate lift nets per vegetation/elevation type. High elevation bars are solid. Low elevation bars are hatched. The asterisk designates a significant difference ($P < 0.05$) relative to all other sampling sites except high elevation common reed.

In contrast with adult mummichog, juveniles (13–35 mm TL) in cattail at Iona Island (both elevations) occurred at significantly higher density relative to reed at either marsh except for high elevation reed at Piermont Marsh (Figure 6b).

Benthic Macroinvertebrates

The benthic cores yielded 6657 individuals representing 92 macroinvertebrate taxa among the six vegetation sampling sites at the two marshes. Benthic macroinvertebrate

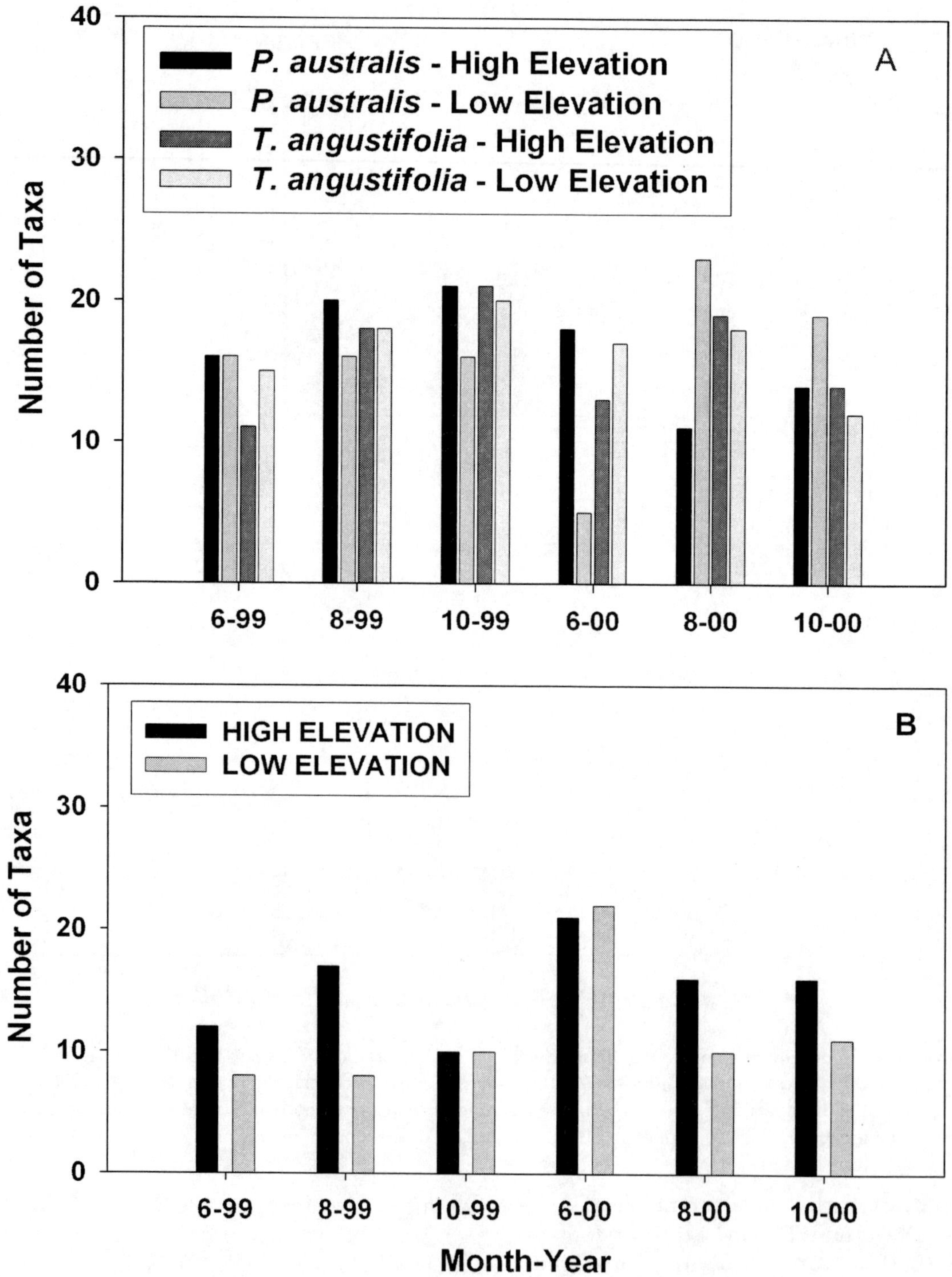

Figure 7. Taxon richness of benthic macroinvertebrates collected in June, August, and October 1999 and 2000 for high and low elevations at Iona Island (A) and Piermont Marsh (B). N = 8 replicate cores per vegetation type per month.

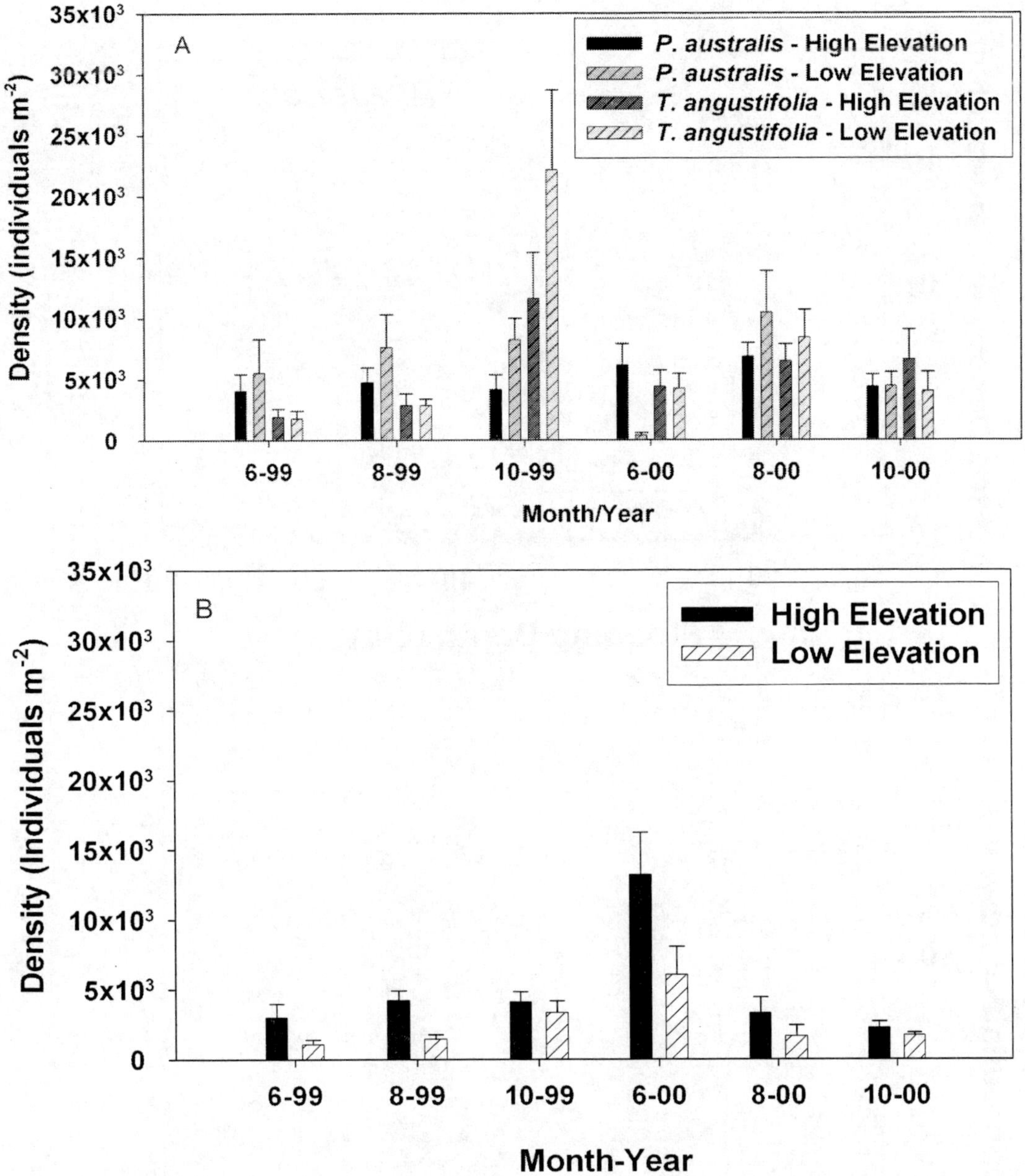

Figure 8. Mean density (individuals m^{-2} ± SE) of benthic macroinvertebrates collected in June, August, and October 1999 and 2000 for high and low elevations at Iona Island (A) and Piermont Marsh (B). N = 8 replicate cores per vegetation type per month.

communities at Iona Island Marsh and Piermont Marsh were dominated by annelids (Tubificidae, Enchytraidae), insect larvae (Diptera), and mollusks (Sphaeriidae and Hydrobiidae). Macroinvertebrate taxa richness was greater at Iona Island Marsh relative to Piermont Marsh (Figure 7). Taxonomic composition and richness was similar among vegetation types and seasons at Iona Island Marsh during 1999 (Figure 7a). In 2000, relatively lower taxa richness was observed in low elevation reed during June

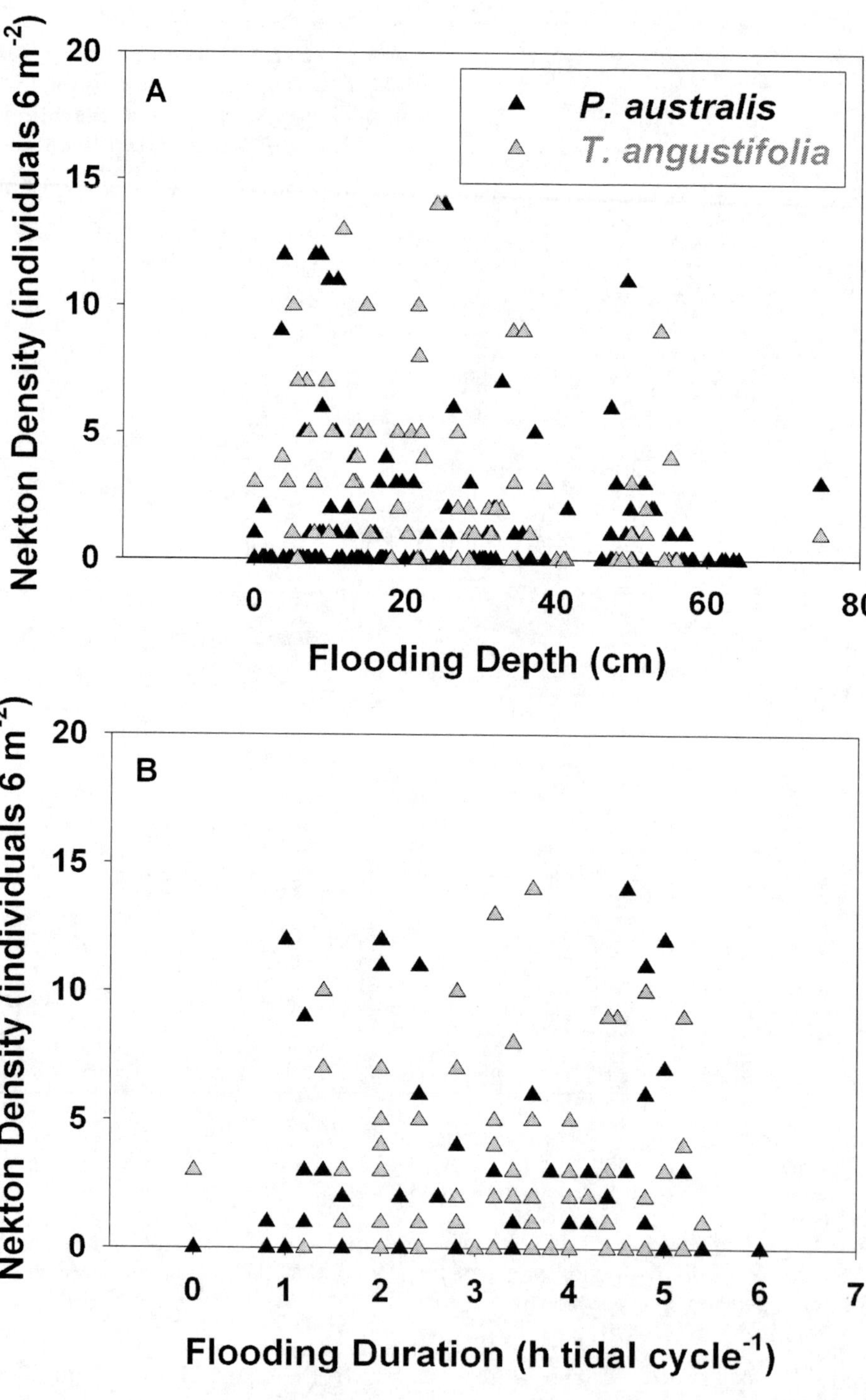

Figure 9. Plot of nekton density (individuals 6 m^{-2}) relative to flooding depth (cm) (A) and flooding duration (h tidal $cycle^{-1}$) (B) across both elevations. Individual points represent fish density from individual lift nets on individual sampling dates and the maximum flooding depth or flood duration for the tidal cycle coinciding with that sampling date. Common reed points are from both Piermont and Iona Island Marsh sites. Regression analysis revealed no significant linear trend.

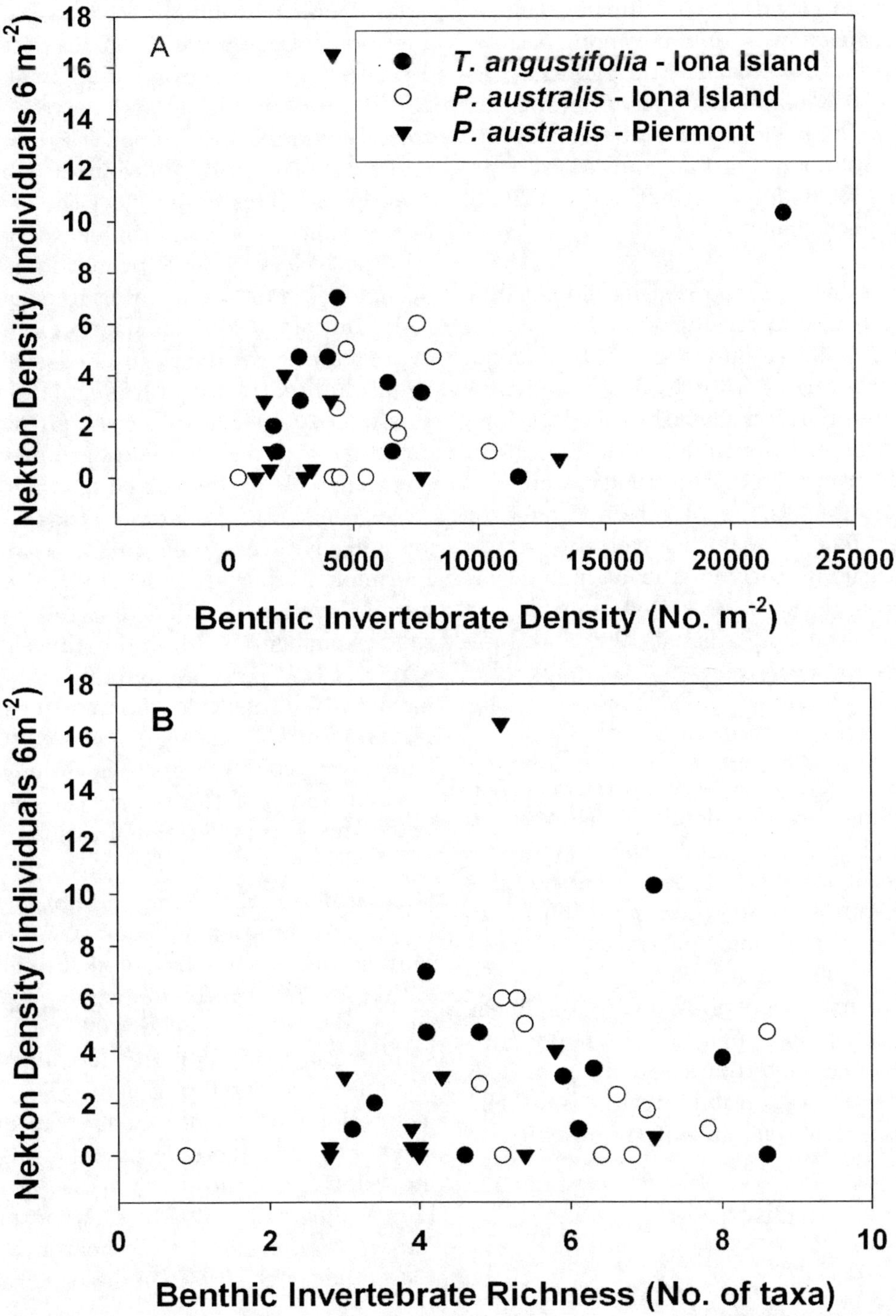

Figure 10. Plot of nekton density (individuals 6 m^{-2}) relative to benthic invertebrate density (individuals m^{-2}) (A) and richness (No. of taxa) (B) across both elevations. Individual points represent mean nekton density from June, August, and October 1999 and 2000 lift net samples and the mean benthic invertebrate density or total richness from 16 cores per month from each sampling category. Regression analysis revealed no significant linear trend.

and in high elevation reed during August. Taxic richness was similar among vegetation types at Iona Island Marsh during October 2000. Greater taxic richness was observed in high elevation reed (relative to low-elevation reed) at Piermont Marsh during June 1999, August 1999, August 2000, and October 2000.

Total macroinvertebrate density did not vary significantly with vegetation at Iona Island Marsh in 1999 (Figure 8a). In June 2000, total macroinvertebrate density was significantly lower in low elevation reed at Iona Island Marsh. No significant differences in total macroinvertebrate density were observed at Iona Island Marsh in August or October 2000. Total macroinvertebrate density was significantly greater at high elevation reed relative to low elevation reed at Piermont Marsh in August 1999 and June 2000 (Figure 8b).

Results of Regression Analyses

The tested linear relationship between fish density and flooding depth or fish density and flooding duration at either marsh site or between vegetation types was not significant (Figure 9). Low regression coefficients (r^2 values < 0.1) can be partly attributed to incidences of zero fish density across the entire range of flooding. Likewise, the linear relationship between fish density and benthic macroinvertebrate density or fish density and taxonomic richness of the benthos was not significant (Figure 10).

Discussion

Hydrogeomorphology

Piermont and Iona Island Marshes represent different stages of reed invasion and the relative age of the reed stands probably explains many of the observed patterns in tidal flooding. Common reed has occupied Piermont Marsh since the latter part of the 19th century (Winogrond and Kiviat 1997). Reed stands in this marsh have a flatter topographic relief and flood less frequently and for shorter duration than the stands at Iona Island. This may reflect the older age of the stands (i.e., more time for accretion). Alternatively, reed at Piermont Marsh may occupy areas of low flooding frequency because initial sites of colonization require elevated areas with well-drained sediments in higher salinity marshes (Bart and Hartman 2000). Lathrop et al. (2003) demonstrated that reed at Piermont Marsh preferentially colonized creek banks, which may have been elevated relative to the rest of the system. Nonreed vegetation stands at Piermont Marsh are restricted to interior, lower elevation sections that receive little tidal exchange (F. Montalto, Edesign Dynamics, LLC., personal communication) and have evidence of elevated porewater sulfide, known to inhibit the growth of reed (Chambers et al. 2002). Higher elevation across the remainder of the marsh may have favored almost complete colonization by reed.

Greater depth of flooding did not translate into longer duration of flooding at Piermont Marsh. Similar results have been found within low relief reed stands in a Connecticut marsh (Osgood et al. 2003). One possible explanation for the incongruity of flooding depth and duration may be ponding in interior sections of marsh (personal observation) that increase duration of flooding relative to the creek bank locations. Longer duration of flooding may ultimately impact the length of time nekton have to access the marsh surface in these reed stands.

In contrast to Piermont Marsh, reed at Iona Island Marsh first appeared in the early 1960s (Winogrond and Kiviat 1997) and reed stands still occupy a range of intertidal elevation and flooding similar to the older, adjacent cattail stands. Similar sediment accretion rates reported for reed and cattail stands (Leonard et al. 2002) suggest that in situ organic input may be an important contributor to increased elevation and reduced tidal flooding in older reed stands (Windham and Lathrop 1999; Rooth and Stevenson 2000). Reed stands at Iona Island may not have had sufficient time to accumulate significant organic material, or may otherwise be less productive than in other parts of the estuary. Frequency of flooding within reed stands at Iona Island is similar to that reported for reed stands in other oligohaline systems (Warren et al. 2001), suggesting that flooding differences between reed and nonreed stands may be less extreme in low salinity marshes where reed can withstand a longer duration of flooding (Chambers et al. 2002).

Marsh Surface Nekton

Previous findings in Connecticut, New Jersey, and Maryland (Fell et al. 1998; Rilling et al. 1998; Able and Hagan 2000; Meyer et al. 2001; Warren et al. 2001) show that adult finfish and macrocrustaceans are capable of utilizing reed and nonreed stands in similar abundances. The sampling techniques used in prior studies (Breder traps, open-end flume nets, pit traps) do not measure nekton from a known area (i.e., density). As a result, location-specific habitat variables (e.g., flooding dynamics, benthic prey availability) cannot be directly paired to patterns of nekton use. Our lift net sampling results at Iona Island and Piermont Marshes confirm that adult nekton density is similar between reed and cattail stands. Further, nekton density did not vary significantly across the measured range of tidal flooding for either vegetation type. The large number of lift net samples yielding zero nekton (27% of total samples) strongly suggests that nekton distribution on the marsh surface is aggregated, or very diffuse in both vegetation types. A similar distribution of nekton was found during a previous study in a polyhaline marsh in Connecticut using lift nets to quantify nekton density between reed and cattail stands (Osgood et al. 2003).

Lift net samples were collected on slack high tide and therefore only represent distribution of nekton at that point in the tidal cycle. The trend of higher nekton density caught at the high elevation stations in both vegetation types (Figure 6a) may represent the congregation of fish on the leading edge of the flood tide. This pattern would be consistent with salt marsh studies that show nekton follow the flood tide into the marsh (Kneib 1994). Higher abundance of nekton in the high marsh would also partly explain the lack of a positive relationship between nekton density and flooding depth or duration. Density of resident species, such as mummichog, can show a negative relationship with tidal duration because they spend a large portion of the tidal cycle on the marsh surface and tend to disperse through the habitat, especially at longer durations (Kneib 1997). This dilution effect may explain the absence of nekton in many of the lift nets, especially since the nets were sampled on slack high tide after maximum time for dispersal. There is no differentiation in this pattern between reed and cattail, suggesting a similar dispersal pattern for resident nekton in both habitat types under similar flooding regimes.

The collection of spotfin killifish *Fundulus luciae* at Piermont Marsh is noteworthy and represents a new and significant distribution record for the species in the northeast United States (Yozzo and Ottman 2003). This species resides on the marsh surface throughout the tidal cycle and prefers high marsh habitat (shallow flooding) with an abundance of standing water (Able et al. 1983; Kneib 1984; Kneib and Wagner 1994). Spotfin killifish also was found in reed stands in a Connecticut marsh, but was absent from adjacent salt meadow cordgrass stands (Osgood et al. 2003). Able et al. (2003) attributed an absence of spotfin killifish from reed stands in New Jersey to absence of low tide refugia (ponded depressions). A greater proportion of high marsh (low frequency of flooding) within some reed marshes, however, may favor this smaller species of fish, provided there is sufficient standing water in the vicinity (e.g., ditches, marsh ponds).

The higher density of juvenile mummichog within the cattail site at both elevations at Iona Island (Figure 6b) suggests greater spawning success within this habitat. Few larval fish were collected in either vegetation zone, which can be attributed to sampling methodology. The mesh size of the lift nets (3 mm) allowed escape by larval fish and predation may have occurred in the pit traps prior to collection. Harms et al. (2003), using pit traps, did document lower abundance of larval mummichog within large and medium stands of reed relative to cattail at Iona Island. Small, recently-developing stands of reed at Iona had similar larval abundance relative to cattail in the same study (Harms et al. 2003). Able and Hagan (2000) and Able et al. (2003) found fewer larvae and juveniles (collected in pit-traps) in reed stands than in salt meadow cordgrass stands in New Jersey salt marshes. There was also a trend of larger individuals utilizing reed relative to salt meadow cordgrass in Maryland (Meyer et al. 2001) and Connecticut (Rilling et al. 1998; Fell et al. 2003) marshes, respectively. Clearly, there is an emerging pattern of impaired spawning success by marsh resident finfish within reed marshes throughout the northeastern United States. Further, this pattern may be most evident in older, well-developed reed stands where hydrogeomorphic impacts are greatest (Weinstein and Balletto 1999; Able and Hagan 2003; Harms et al. 2003; Lathrop et al. 2003). Osgood et al. (2003) suggested that impairment of spawning and survival of larval nekton in tidal marshes in Connecticut was due to irregular flooding. In contrast, the pattern of flooding at Iona Island is regular and does not seem to affect nekton habitat use. Differences in the abundance of larval and juvenile fish have been attributed to lack of standing water (refuge) in reed stands (Able and Hagan 2000; Able and Hagan 2003). We have documented microtopographic relief within reed at Iona Island Marsh, but have observed that these features drain rapidly on ebb tide because of subsurface cavities connected to the creek (Harms et al. 2003). Raposa and Roman (2001) documented high densities and a large number of fish species, including young-of-year mummichog, in stranded pools on a tidally-restricted reed marsh surface on Cape Cod, Massachusetts. It seems likely that if such topographic features are present on a reed marsh surface they can be utilized by larval and juvenile resident nekton. This underscores the point that geomorphology and associated tidal hydrology are key factors regulating nekton use.

Taxa richness and density within the benthic macroinvertebrate community were different between vegetation stands at Iona Island and Piermont Marsh. Seasonal ponding in high reed at Piermont Marsh probably explains higher invertebrate density there. Differences in density of benthic macroinvertebrates were not, however, significantly related to nekton density on the marsh surface (Figure 10). Although not a direct indicator of prey availability to resident nekton, our regression results support findings that nekton indiscriminately acquire food they encounter from reed marsh surfaces (Fell et al. 1998; Wainwright et al. 2000). Raichel et al. (2003) documented lower benthic prey density and a greater abundance of large prey items in reed, and suggest that prey availability may regulate nekton habitat use between reed and nonreed stands. Further work, with an emphasis on experimental hypothesis testing, will be necessary to fully test for a relationship between benthic macroinvertebrate prey availability and nekton use within reed stands.

Management Implications

Similarities in nekton abundance between reed and nonreed stands from previous studies have led to the conclusion that management of reed stands may need to be reevaluated (Meyer et al. 2001). Our results show that when the marsh surface is flooded, equal density and species of nekton utilize the surfaces of reed and cattail marshes and do not vary significantly as a function of elevation within the range measured in this study. Further, benthic macroinvertebrate prey resources in both vegetation zones appear to be adequate to support resident nekton communities. Of course, any reduction in flooding frequency associated with vegetation change or other impacts will alter the length of time the marsh surface is accessible for foraging and other activity, resulting in long-term impacts on nekton populations within well-developed reed stands.

A consistent finding among nekton studies is evidence of reduced larval survival (by mummichog) within reed stands. Reduced larval density could have significant consequences in reed-dominated marshes where potential reduction in resident nekton recruitment over large areas could decrease secondary production within the system. Additionally, the relative lack of larval fundulids within reed stands could threaten to short-circuit trophic energy transfer (Kneib 1997) between the marsh surface and the open waters of the estuary. Reed stands that can be managed with respect to proper flooding and geomorphology (i.e., maintained in small, isolated stands) can be expected to provide resident nekton habitat function similar to that found in tidal marshes dominated by other plant species. This management strategy could be particularly important for low salinity and tidal freshwater marshes along the Hudson River estuary, where stands of reed are in the early stages of development and still coexist with native vegetation. Impacts of reed expansion on other aspects of resident nekton ecology (e.g., trophic transfer, recruitment rate, secondary production) still require further investigation before they can be properly factored into management decisions.

Acknowledgments

Funding for this study was provided by the Hudson River Research Grant Program of the Hudson River Foundation. The authors appreciate the logistical support of Chuck Nieder, Hudson River National Estuarine Research Reserve and the cooperation of the Palisades Interstate Park Commission. Field and laboratory assistance was provided by M. Cianciola, C. Miller, L. Vasilakos, C. Elliot, S. Matteson, and R. Murphy. The authors acknowledge the helpful comments of two anonymous reviewers.

References

Able, K. W., and S. M. Hagan. 2000. Effects of common reed (*Phragmites australis*) invasion on marsh surface macrofauna: response of fishes and decapod crustaceans. Estuaries 23:633–646.

Able, K. W., and S. M. Hagan. 2003. Impact of common reed, *Phragmites australis*, on essential fish habitat: influence on reproduction, embryological development and larval abundance of mummichog (*Fundulus heteroclitus*). Estuaries 26:40–50.

Able, K. W., S. M. Hagan, and S. A. Brown. 2003. Mechanisms of marsh habitat alteration due to *Phragmites*: response of young of the year mummichog (*Fundulus heteroclitus*) to treatment for *Phragmites* removal. Estuaries 26:484–494.

Able, K. W., C. W. Talbot, and J. K. Shishler. 1983.

The spotfin killifish (*Fundulus luciae*), is abundant in New Jersey salt marshes. Bulletin of the New Jersey Academy of Science 28:7–10.

Angradi, T. R., S. M. Hagan, and K. W. Able. 2001. Vegetation type and the intertidal macroinvertebrate fauna of a brackish marsh: *Phragmites* vs. *Spartina*. Wetlands 21:75–92.

Bart, D. J., and J. M. Hartman. 2000. Environmental determinants of *Phragmites australis* expansion in a New Jersey salt marsh: an experimental approach. Oikos 89:59–69.

Chambers, R. M., L. A. Meyerson, and K. Saltonstall. 1999. Expansion of *Phragmites australis* into tidal wetlands of North America. Aquatic Botany 64:261–273.

Chambers, R. M., D. T. Osgood, D. J. Bart, and F. Montalto. 2003. *Phragmites australis* invasion and expansion in tidal wetlands: Interactions among salinity, sulfide, and hydrology. Estuaries 26:398–406.

Chambers, R. M., D. T. Osgood, and N. Kalapasev. 2002. Hydrologic and chemical controls of *Phragmites* distribution on the Lower Housatonic River Estuary. Marine Ecology Progress Series 239:83–91.

Fell, P. E., R. S. Warren, J. K. Light, R. L. Rawson, and S. M. Fairley. 2003. Comparison of fish and macroinvertebrate use of *Typha angustifolia*, *Phragmites australis*, and treated *Phragmites* marshes along the lower Connecticut River. Estuaries 26:534–551.

Fell, P. E., S. P. Weissbach, D. A. Jones, M. A. Fallon, J. A. Zeppieri, E. K. Faison, K. A. Lennon, K. J. Newberry, and L. K. Reddington. 1998. Does invasion of oligohaline tidal marshes by reed grass, *Phragmites australis*, affect the availability of prey resources for the mummichog, *Fundulus heteroclitus*? Journal of Experimental Marine Biology and Ecology 222:59–77.

Hanson, S., D. T. Osgood, and D. J. Yozzo. 2002. Nekton use of a *Phragmites australis* marsh on the Hudson River, New York, USA. Wetlands 22:326–337.

Harms, L., E. Salak, and D. T. Osgood. 2003. Effects of *Phragmites australis* on the early life history stages of *Fundulus heteroclitus* at Iona Island Marsh, Hudson River, New York. Section IV: pages 1–34 *in* J. R. Waldman and W. C. Nieder, editors. Final Reports of the Tibor T. Polgar Fellowship Program, 2002. Hudson River Foundation, New York,

Kneib, R. T. 1984. Patterns in the utilization of the intertidal salt marsh by larvae and juveniles of *Fundulus heteroclitus* (Linnaeus) and *Fundulus luciae* (Baird). Journal of Experimental Marine Biology and Ecology 83:41–51.

Kneib, R. T. 1987. Predation risk and use of intertidal habitats by young fishes and shrimp. Ecology 68:379–386.

Kneib, R. T. 1994. Spatial pattern, spatial scale, and feeding in fishes. Pages 171–185 *in* D.J. Strouder, K. L. Fresh, and R. J. Feller, editors. Theory and application in fish feeding ecology. University of South Carolina Press, Columbia.

Kneib, R. T. 1997. Salt marsh ecoscapes and production transfers by estuarine nekton in the southeastern United States. Pages 267–292 *in* M. P. Weinstein and D. A. Kreeger, editors. Concepts and controversies in tidal marsh ecology. Kluwer Academic Publishers, Inc., Boston.

Kneib, R. T., and S. L. Wagner. 1994. Nekton use of vegetated marsh habitats at different stages of tidal inundation. Marine Ecology Progress Series 94:227–238.

Lathrop, R. G., L. Windham, and P. Montesano. 2003. Does *Phragmites* expansion alter the structure and function of marsh landscapes? Patterns and processes revisited. Estuaries 26:423–435.

Leonard, L. A., P. Ansley Wren, and R. L. Beavers. 2002. Flow dynamics and sedimentation in *Spartina alterniflora* and *Phragmites australis* marshes of the Chesapeake Bay. Wetlands 22:415–424.

Meyer, D. L., J. M. Johnson, and J. W. Gill. 2001. Comparison of nekton use of *Phragmites australis* and *S. alterniflora* marshes in the Chesapeake Bay, USA. Marine Ecology Progress Series 209:71–84.

Mills, E. L., D. L. Strayer, M. D. Scheuerell, and J. T. Carlton. 1996. Exotic species in the Hudson River basin: a history of invasions and introductions. Estuaries 19:814–823.

Orson, R. A., R. S. Warren, and W. A. Niering. 1987. Development of a New England river valley tidal marsh. Estuaries 10:20–27.

Osgood, D. T. 2000. Subsurface hydrology and nutrient export from barrier island marshes at different tidal ranges. Wetlands Ecology and Management 8:133–146.

Osgood, D. T., D. J. Yozzo, R. M. Chambers, D. Jacobson, T. Hoffman, and J. Wnek. 2003. Tidal hydrology and habitat utilization by resident nekton in *Phragmites* and non-*Phragmites* marshes. Estuaries 26:522–533.

Raichel, D. L., K. W. Able, and J. M. Hartman. 2003. The influence of *Phragmites* (common reed) on the distribution, abundance, and potential prey of a resident marsh fish in the Hackensack Meadowlands, New Jersey. Estuaries 26:511–521.

Raposa, K. B., and C. T. Roman. 2001. Seasonal habitat-use patterns of nekton in tide-restricted and unrestricted New England salt marsh. Wetlands 21:451–461.

Rilling, G. C., P. E. Fell, and R. S. Warren. 1998. Fish use of brackish tidal wetlands on the lower Connecticut River: a comparison of a *Phragmites australis*-dominated vs. a restored high marsh. Technical Report to the Connecticut Department of Environmental Protection, Office of Long Island Sound Programs, Hartford.

Rooth, J. E., and J. C. Stevenson. 2000. Sediment deposition patterns in *Phragmites australis* communities: implications for coastal areas threatened by rising sea-level. Wetlands Ecology and Management 8:173–183.

Rozas, L. P. 1992. Bottomless lift net for quantitatively sampling nekton on intertidal marshes. Marine Ecology Progress Series 89:282–292.

Rozas, L. P., and T. J. Minello. 1997. Estimating densities of small fishes and decapod crustaceans in shallow estuarine habitats: a review of sampling design with focus on gear selection. Estuaries 20:199–213.

Saltonstall, K. 2002. Cryptic invasion by a non-native genotype of the common reed, *Phragmites australis*, into North America. Proceedings of the National Academy of Science 99:2445–2449.

Wainwright, S. C., M. P. Weinstein, K. W. Able, and C. A. Currin. 2000. Relative importance of benthic microalgae, phytoplankton and the detritus of smooth cordgrass (*Spartina*) and the common reed (*Phragmites*) to brackish marsh food webs. Marine Ecology Progress Series 200:77–91.

Warren, R. S., P. E. Fell, J. L. Grimsby, E. L. Buck, G. C. Rilling, and R. A. Fertik. 2001. Rates, patterns, and impacts of *Phragmites australis* expansion and effects of experimental *Phragmites* control on vegetation, macroinvertebrates, and fish within tidelands of the lower Connecticut River. Estuaries 24:90–107.

Weinstein, M. P., and J. H. Balletto. 1999. Does the common reed, *Phragmites australis*, affect essential fish habitat? Estuaries 22:793–802.

Windham, L., and R. G. Lathrop. 1999. Effects of *Phragmites australis* (Common Reed) invasion on aboveground biomass and soil properties in brackish tidal marsh of the Mullica River, New Jersey. Estuaries 22:927–935.

Winogrond, H. G., and E. Kiviat. 1997. Invasion of *Phragmites australis* in the tidal marshes of the Hudson River. Section VI: page 29 *in* W. C. Nieder and J. R. Waldman, editors. Final Reports of the Tibor T. Polgar Fellowship Program, 1996. Hudson River Foundation, New York.

Yozzo, D. J., and F. Ottman. 2003. New distribution records for the spotfin killifish *Fundulus luciae* (Baird) in the lower Hudson River Estuary and adjacent waters. Northeastern Naturalist 10(4):399–408.

Yozzo, D. J., C. M. Way, T. R. Martin, J. M. Rhoads, and S. L. Paulus. 1999. Hudson River habitat restoration study: baseline biological data collection at project and reference sites. Final Report of Barry A. Vittor and Associates, Inc. to New York District, U. S. Army Corps of Engineers, New York.

American Fisheries Society Symposium 51:175–196, 2006

Winter Distribution and Abundance of Hudson River Fishes using Hydroacoustics

KYLE J. HARTMAN AND BRIAN W. NAGY
West Virginia University, Wildlife and Fisheries Resources Program
322 Percival Hall, Morgantown, West Virginia 26506-6125 USA
hartman@wvu.edu

Abstract.—Low water temperatures experienced by temperate fish during winter reduce muscle power, resulting in an important limitation to overwintering fishes that may explain why they often seek areas of reduced flow as winter refugia. The Hudson River is a heavily urbanized estuary, and as such, navigation and industrial influences may have particular importance to species seeking to overwinter there. In December 1995 and 1997, we used a mobile hydroacoustic survey, coupled with gill netting and midwater trawling for species identification, to identify the distribution and abundance of key species in the Hudson River. Striped bass *Morone saxatilis*, white perch *M. americana*, and bay anchovy *Anchoa mitchilli*, co-occurred in the lower 25 km of the river. Fish distributions were related to salinity. Striped bass and bay anchovy were found in highest densities in the lower river (Yonkers and Battery regions) and higher salinities. White perch dominated from Yonkers upriver. Acoustic population estimates of striped bass abundance from 1995 (576,110) compared favorably with a wintertime tagging study (949,000), but were grossly underestimated in 1997. Higher salinity in the Battery and Yonkers regions in 1997 may have pushed distributions of striped bass and white perch upriver compared to 1995 sampling and may have led to decreased precision in estimates. This study suggests that key Hudson River fish use much of the lower 61 km of the Hudson River as overwintering habitat.

Introduction

Winter represents a potentially critical time for fish in estuaries and other waterways subject to industrialization, urbanization, and commercial navigation (Bodensteiner and Lewis 1994; Lenz 2003; Lyons 1997; Raibley et al. 1997). Low water temperatures experienced by temperate fish during winter severely restrict muscle capabilities, such that efforts to maintain position or move against current are hampered (Sheehan et al. 1990). This reduction in muscle power is an important limitation to overwintering fishes that may explain why they often seek areas of reduced flow as winter refugia (Bodensteiner and Lewis 1994; Carlson 1992; Crawshaw 1984; Cunjak and Power 1987; Sheehan and Rasmussen 1999). The importance of these refugia or other overwintering locations may well act to limit carrying capacity of the system for a given species. Any alteration of key overwintering habitats from construction of new industrial sites that require cooling water, produce discharges, or alter the flow in these systems may affect overwinter survival of key fish species.

To manage riparian and aquatic use practices and potentially protect key areas from development requires an understanding of what areas are used by fish during this critical overwinter period. The objective of this study was to provide a first step towards identifying areas of importance to fishes overwintering in the Hudson River and to provide some quantification of the relative importance of overwintering areas at large geographic scales through the use of hydroacoustics. The Hudson River represents an important estuarine system that supports a diverse fish community and large populations of species of commercial and recreational importance (Beebe and Savidge 1988). We focused on three key species, striped bass *Morone saxatilis*, white perch *M. americana*, and bay anchovy *Anchoa mitchilli*. Striped bass and white perch have historically been angled or netted in the Hudson and bay anchovy are species of concern related to impingement and entrainment (Muessig et al. 1988; Boreman and Goodyear 1988). Further, the Hudson River is also a system perennially threatened by increasing demands for industrial sources of water, waste discharge, riparian development, and commercial navigation impacts due to the very high human population in the watershed.

Methods

General Methodology

We conducted winter surveys during December 1995 and 1997 to identify the distribution and abundance of fish wintering in the lower Hudson River. Surveys were restricted to these early December time periods because of conflicting demands between the need to capture low water temperature conditions and limitations of ice flows that prevented sampling on the Hudson later in the winter. Surveys consisted of mobile down-looking hydroacoutics and traditional fishery gears to identify species composition and target proximity to the bottom. In stratifying the study design we broke the river into segments, using river regions identified by utility-sponsored research to maintain continuity with other studies. The general study area was the lower 122 km (river kilometer [rkm]) of the Hudson River extending from the Battery (rkm 0) to Poughkeepsie, New York.

Fishery Gears

We used two different fishery gears to obtain ground-truthing information on the fish species in the Hudson River. Midwater trawls (6.4-m headrope, 5-cm stretch mesh upper body, 3.6-cm lower body, and 0.6-cm stretched mesh cod end liner) were used to identify pelagic targets and to obtain size-specific species composition information for species partitioning of acoustic densities. Midwater trawls were towed behind the research vessel at 2–3 knots and fished obliquely for 10–20 min. Time was varied depending upon catch rates and the variance in time was unimportant because we were interested in composition and not catch per unit effort (CPUE) from this gear. A total of 1–13 trawls were conducted per river region and year (Table 1). Unequal sample sizes of trawls between regions and years were the product of variation in catch patterns in the river and commercial navigation traffic.

We also used gill nets to determine what fraction of key species may be within our 0.5-m bottom tracking window and therefore would be inaccessible to the hydroacoustics. Experimental gill nets were 2.6 m in height and 274 m in length (stretched mesh sizes 50–152 mm in 12.5 mm increments). Gill nets were fished on the bottom during daytime slack tide and retrieved at the end of slack tide ($<$ 30 min).

Table 1. Water volumes and sampling effort (number of events) used to estimate fish distribution and abundance in the Hudson River during December 1995 and 1997. Volumes are from river areas greater than 3 m in depth and are provided by J. Young (Consolidated Edison, personal communication). Geographic regions correspond to those used by Utility-sponsored monitoring projects. Here, MWT and GN refer to midwater trawl and gillnet, respectively.

			1995		1997
Region	River km range	Volume (m^3)	MWT	GN	MWT
Battery	0–19	190,264,951	2	0	4
Yonkers	19–37	202,765,521	7	3	13
Tappan Zee	38–53	200,126,473	4	2	9
Croton-Haverstraw	54–61	93,826,649	3	2	1
Indian Point	62–74	195,688,103	3	2	0
West Point	75–88	204,807,884	2	0	
Cornwall	89–98	131,650,896	0	0	
Poughkeepsie	99–122	292,143,184	3	0	

Fish within 0.5 m of the lead line were considered too close to the bottom to detect acoustically while those above 0.5 m were considered acoustically accessible. A total of nine gill net sets were fished between Peekskill (rkm 68.8) and the Harlem River (rkm 22.4) in 1995 (Table 1).

Hydroacoustic Methodology

We used a SIMRAD EY500 120 kHz split beam hydroacoustic system to collect acoustic data on distribution and density of fish using echo integration. The acoustic system was calibrated for each cruise using a copper reference sphere and standard methods (Brandt 1996). The transducer was towed alongside the research vessel using a tow body submerged approximately 0.5 m under the water's surface. The transducer was set to pulse three times per second with a four millisecond pulse width. The minimum threshold for acoustic data during collection was –70 decibels (dB) (this was later set to –55.2 dB for processing based on minimum target strengths of key fish species). Bottom tracking threshold was set to –20 dB and all data within 0.5 m of the tracked bottom was ignored. The beam width was 7.0 degrees alongship and 7.1 degrees athwartship. Single target detection parameters were set to –60 dB minimum echo level, 0.8 minimum echo length, and 1.6 maximum echo length.

The acoustic survey design consisted of a series of spatially continuous across-river diagonal zig-zags that began at rkm 122 in 1995 and rkm 61 in 1997. Only the lower 61 km were acoustically sampled in 1997 because the primary focus of this project was on striped bass; and based upon distributions from 1995, we found no justification for repeating the upper 59 km in 1997 because striped bass were not collected there in fishery gears during either year. We also felt justified in not sampling the upper section because the lower 61 km appeared to be eminently more imperiled due to anthropogenic influences in the metropolitan New York City region. In both years the acoustic surveys were conducted during daylight after preliminary comparisons between day and night in 1995 suggested no diel effects on vertical distribution or abundance.
Processing of acoustic data were done us-

ing the DEVIS software system (Jech and Luo 2000). DEVIS divides the raw Simrad data files into absolute fish density using echo-squared integration (MacLennan and Simmonds 1992). Time varied gain was 20 $\log_{10}$ R for echo-integration and 40 $\log_{10}$ R for target strength estimates. Acoustic data were placed into a spatial array of 0.5 m vertical depth by 20 m horizontal distance. Within each cell of the array, all data on individual target strengths (TS, related to fish size) and total voltage returned were stored. Calculation of fish density within each cell was based upon the portion of total voltage returned that was attributable to fish in each size-class (see *Size Classes*, below). Within each cell the TS distribution was used to estimate proportions of the integrated echo attributable to each size-class. These proportions were calculated from total integrated echo partitioned by the proportions of total individual targets and mean target strength of each size-class. Numerical density of a size-class within a cell was derived by dividing the integrated echo by the mean backscattering coefficient for that size-class over the whole spatial array (Jech and Luo 2000). Density for each size-class within a region was estimated by averaging the density in all cells. Due to physical limitations of the hydroacoustic transducer and the depth of the transducer in the water, the first 2 m of water are eliminated from processing as well as the 0.5 m above bottom. Therefore, all reported densities do not include the top 2 m or bottom 0.5 m of the water column.

Individual targets were related to fish size from a previously-determined white perch or striped bass target strength to length regression (Hartman and Nagy 2005):

$$TS = 15.613 \log_{10} L - 25.242$$

where TS is target strength (dB) and L is length (cm). This equation was applicable to the Hudson fish populations as these two species represented the majority of fish collected in research gears and this equation provides TS values very similar to those in Love's (1971) equation for a variety of species (Hartman and Nagy 2005).

Size Classes

Using size ranges of striped bass determined in our 1995 research trawls, we partitioned the acoustic data arrays for density and abundance estimates on the basis of size to correspond to striped bass age-classes present during December. A total of four age/size classes were used. In 1995 all age-0 striped bass were greater than 90 mm in total length. Thus, size-A represented fish smaller than age-0 striped bass in the range 12 to less than or equal to 90 mm and consisted of target strengths of –55.2 to –41.6 dB. Size-B represented age-0 striped bass between 91 and 160 mm with a target strength range of –41.5 to –37.7 dB. Size-C corresponds to age-1 striped bass (160 to ≤ 300 mm) and target strengths ranged from –37.7 to –33.4 dB. The last category was size-D representing age-2 and older striped bass (300 to ≤ 1000 mm) and target strengths -33.4 to -25.2 dB. In 1997, in the Yonkers, Tappan Zee and Croton-Haverstraw regions some age-0 striped bass were captured that fell within size-A and thus, age-0 striped bass densities in 1997 included some fish from this size-group.

Fish Density and Distributions

We combined the acoustic density information with the trawl information to calculate regional and finer-scale fish density by species. Acoustic density and trawl species composition were split into the same size-classes that corresponded to age classes of the principal study species (striped bass). Catch and length information from all trawls within a geographic region were used to determine what proportion of all fish within each size-class belonged to each species. The species proportion (trawls) was multiplied by the fish density (DEVIS) to obtain species-specific density in a region. The total fish density (by species) was multiplied by the volume of the center channel of the river (areas deep enough for acoustic sampling) in each region (obtained from J. Young, ASA Analysis and Communications, personal communication) to estimate abundance of each species in a particular region (Table 1). Thus, for these estimates we assumed that the mean density for the sampled water column was representative of the unsampled portions of the water column (2.0 m from surface and within 0.5 m of bottom). We approximated 95% confidence intervals about our density and abundance values using the variance in acoustic density calculated from all cells within a particular region (mean $\pm$ 1.96 $\sigma/\sqrt{n}$). It should be noted that this method of presenting spread in density and abundance values is liberal because it uses all cells in the array in calculating sample standard deviations. Hydroacoustic data are spatially autocorrelated and our approach increases sample size and may reduce variability relative to estimates using only uncorrelated cells. Thus, our estimates of bounds on density and abundance estimates are undoubtedly smaller than would be determined from estimates following autocorrelation analysis.

Environmental Sampling

In both years a Sea-Bird CTD was used to collect vertical temperature and salinity profiles in the river. In 1995 a total of 12 vertical profiles were taken from rkm 0 to rkm 122, with at least one profile per river region. In 1997 a total of 18 vertical profiles were taken from rkm 0 to rkm 61 with a minimum of three in the Croton-Haverstraw region (the smallest region) and five in each of the other three regions surveyed. Water quality profiles were compared to vertical density profiles of the target size classes (size-A through size-D) using stepwise regression to examine any relationships between vertical fish distribution (density) and temperature or salinity profiles. Level of significance was set to a = 0.05 for all tests.

Results

A total of 13 species was collected in midwater trawls in 1995 and 11 species in 1997 (Table 2). Among these, striped bass, white perch, bay anchovy, and Atlantic croaker *Micropogonias undulatus* were the most common. Other species of importance that were collected included alewife *Alosa pseudoharengus* and blueback herring *Alosa aestivalis* (Tables 3, 4).

Striped bass were generally collected in the lower sections of the river in 1995 and 1997, although some segregation by size-class was apparent (Tables 3, 4). As the size and age classes of striped bass were used as the criteria for splitting the trawl catches and acoustic data, no striped bass were collected in size-A in 1995. In 1997 a few striped bass in size-A were collected, representing 1.1–1.8% of the trawl catch from Yonkers through Croton-Haverstraw. In the size-B, striped bass were collected primarily in Yonkers in 1995, but in 1997 striped bass represented greater than 25% of the catch from the Battery through Tappan Zee. Size-C striped bass were common in the Battery, Yonkers, and Tappan Zee in 1995, but were sporadically collected (2.7% at Croton-Haverstraw and 11.1% at Yonkers) in 1997. All fishes collected in size-D each year were striped bass. These large striped bass were found from the Battery through Croton-Haverstraw each year.

White perch was the dominant species col-

Table 2. List of species collected in midwater trawls and gillnets during December 1995 and 1997 in the Hudson River.

Common name	Scientific name
Alewife	*Alosa pseudoharengus*
Atlantic croaker	*Micropogonias undulatus*
Atlantic herring	*Clupea harengus*
Atlantic tomcod	*Microgadus tomcod*
Bay anchovy	*Anchoa mitchilli*
Blueback herring	*Alosa aestivalis*
Gizzard shad	*Dorosoma cepedianum*
Hogchoker	*Trinectes maculatus*
Naked goby	*Gobiosoma bosci*
Oyster toadfish	*Opsanus tau*
Pipefish	*Syngnathus fuscus*
Silver hake	*Merluccius bilinearis*
Shortnose sturgeon	*Acipenser brevirostrum*
Spotted hake	*Urophycis chuss*
Striped bass	*Morone saxatilis*
White catfish	*Ameiurus catus*
White perch	*Morone americana*

lected in midwater trawls in 1995 and 1997 (Tables 3 and 4). White perch were collected from the Battery through Poughkeepsie (excluding Cornwall where no fish were caught) in 1995 and from Yonkers through Croton-Haverstraw (upriver extent of sampling in 1997) in 1997. In the smallest size-class (size-A) white perch represented 66–100% of the catch from Tappan Zee up to rkm 122 in 1995. Distributions appeared pushed upriver in 1997, with white perch representing less than 3% of catches downriver of Tappan Zee and 82% of catches in Croton-Haverstraw. White perch also dominated catches in the size-B class in 1995 and 1997. In this size-class for 1995, white perch represented 43–100% of the catches in each region; increasing in the upriver regions. In 1997, no white perch were collected in the size-B class in the Battery, but as in 1995 they increased in composition upriver. Within size-C no white perch were collected in the Battery, but they represented 40.7–100% of catches from Yonkers through upriver regions in 1995 and 1997. No white perch were ever collected in size-D. However, within the three smallest size classes, white perch were the dominant species in midwater trawls in each year.

Among other notable species collected in the midwater trawl collections were bay anchovy, Atlantic croaker, and Atlantic herring *Clupea harengus* (Tables 3, 4). In 1995, bay anchovy represented 42–70% of the size-A catch, but only in the lower two river regions. In 1997, bay anchovy represented 43–61% of size-A catches in the lower three river regions. Young Atlantic croaker (size-A) were caught in trawls from Yonkers to Tappan Zee in 1995 and in all sampled regions in 1997, representing up to 53% of catches. Atlantic herring larvae were collected in 1997 in the lower three regions where they accounted for 0.7–39.0% of size-A catches.

The gillnet surveys conducted in 1995 collected five species: Atlantic tomcod *Microgadus tomcod*, gizzard shad *Dorosoma*

Table 3. Catch composition (proportion) of midwater trawls used for speciation of hydroacoustic targets from a December 1995 survey of the lower Hudson River. Species are classified as striped bass, white perch, bay anchovy, Atlantic croaker and "other." Catches are divided into size classes (size-A to size-D) that relate to age-classes of striped bass.

	Size-A (12–89 mm)							
Species	**Battery**	**Yonkers**	**Tappan Zee**	**Croton-Haverstraw**	**Indian Point**	**West Point**	**Cornwall**	**Poughkeepsie**
Atlantic croaker	0.000	0.531	0.034	0.000	0.000	0.000	—	0.000
Bay anchovy	0.700	0.423	0.000	0.000	0.000	0.000	—	0.000
White perch	0.000	0.040	0.918	0.667	1.000	1.000	—	1.000
Other	0.300	0.006	0.048	0.333	0.000	0.000	—	0.000
N	14	167	105	20	9	9	0	10

	Size-B (90–159 mm)							
Species	**Battery**	**Yonkers**	**Tappan Zee**	**Croton-Haverstraw**	**Indian Point**	**West Point**	**Cornwall**	**Poughkeepsie**
Striped bass	0.000	0.549	0.015	0.000	0.000	0.000	—	0.000
White perch	0.500	0.431	0.964	0.926	0.882	0.950	—	1.000
Other	0.500	0.020	0.021	0.074	0.118	0.050	—	0.000
N	78	100	102	54	20	25	0	17

Table 3. Continued

Species	Battery	Yonkers	Tappan Zee	Croton-Haverstraw	Indian Point	West Point	Cornwall	Poughkeepsie
Size-C (160–300 mm)								
Striped bass	1.000	0.593	0.069	0.000	0.000	0.000	—	0.000
White perch	0.000	0.407	0.917	1.000	1.000	1.000	—	1.000
Other	0.000	0.000	0.014	0.000	0.000	0.000	—	0.000
N	46	245	74	39	3	2	—	11

Species	Battery	Yonkers	Tappan Zee	Croton-Haverstraw	Indian Point	West Point	Cornwall	Poughkeepsie
Size-D (300–1000 mm)								
Striped bass	1.000	1.000	1.000	1.000	0.000	—	—	—
N	9	38	10	5	0	0	0	0

Table 4. Catch composition (proportion) of midwater trawls used for speciation of hydroacoustic targets from a December 1997 survey of the lower Hudson River. Species are classified as striped bass, white perch, bay anchovy, Atlantic croaker and "other." Catches are divided into size classes (size-A to size-D) that relate to age-classes of striped bass.

		Size-A (12–89 mm)		
Species	**Battery**	**Yonkers**	**Tappan Zee**	**Croton-Haverstraw**
Atlantic croaker	0.265	0.519	0.062	0.168
Bay anchovy	0.610	0.427	0.497	0.000
Striped bass	0.000	0.018	0.016	0.011
White perch	0.000	0.000	0.030	0.821
Atlantic herring	0.019	0.007	0.390	0.000
Other	0.106	0.029	0.005	0.000
N	210	571	400	181
		Size-B (90–159 mm)		
Species	**Battery**	**Yonkers**	**Tappan Zee**	**Croton-Haverstraw**
Striped bass	1.000	0.250	0.438	0.011
White perch	0.000	0.750	0.562	0.978
Other	0.000	0.000	0.000	0.001
N	45	216	105	91
		Size-C (160–299 mm)		
Species	**Battery**	**Yonkers**	**Tappan Zee**	**Croton-Haverstraw**
Striped bass	0.000	0.111	0.000	0.011
White perch	0.000	0.889	0.875	0.978
Other	0.000	0.000	0.125	0.000
N	0	99	105	91
		Size-D (300–1000 mm)		
Species	**Battery**	**Yonkers**	**Tappan Zee**	**Croton-Haverstraw**
Striped bass	1.000	1.000	1.000	1.000
N	1	5	2	1

cepedianum, shortnose sturgeon *Acipenser brevirostrum*, striped bass, and white perch. Due to low catch rates the catches were pooled across the nine net sets to determine what proportion of species may be too close (within 0.5 m of lead line in the gill net) to detect with the hydroacoustic gear. All Atlantic tomcod and shortnose sturgeon collected were within 0.5 m of the bottom and therefore considered undetected or underrepresented in any hydroacoustic estimates for those species. Both striped bass and gizzard shad collected were over 0.5 m off the bottom and hence, based upon this limited data were considered acoustically detectable. Of 33 white perch collected in gill nets, 67% were considered acoustically detectable. This suggests our hydroacoustic estimates of white perch are probably conservative.

Vertical distribution patterns of acoustic targets (fish) differed markedly between 1995 and 1997, likely influencing acoustic abundance estimates. Hydroacoustic surveys found the acoustic targets to be suspended

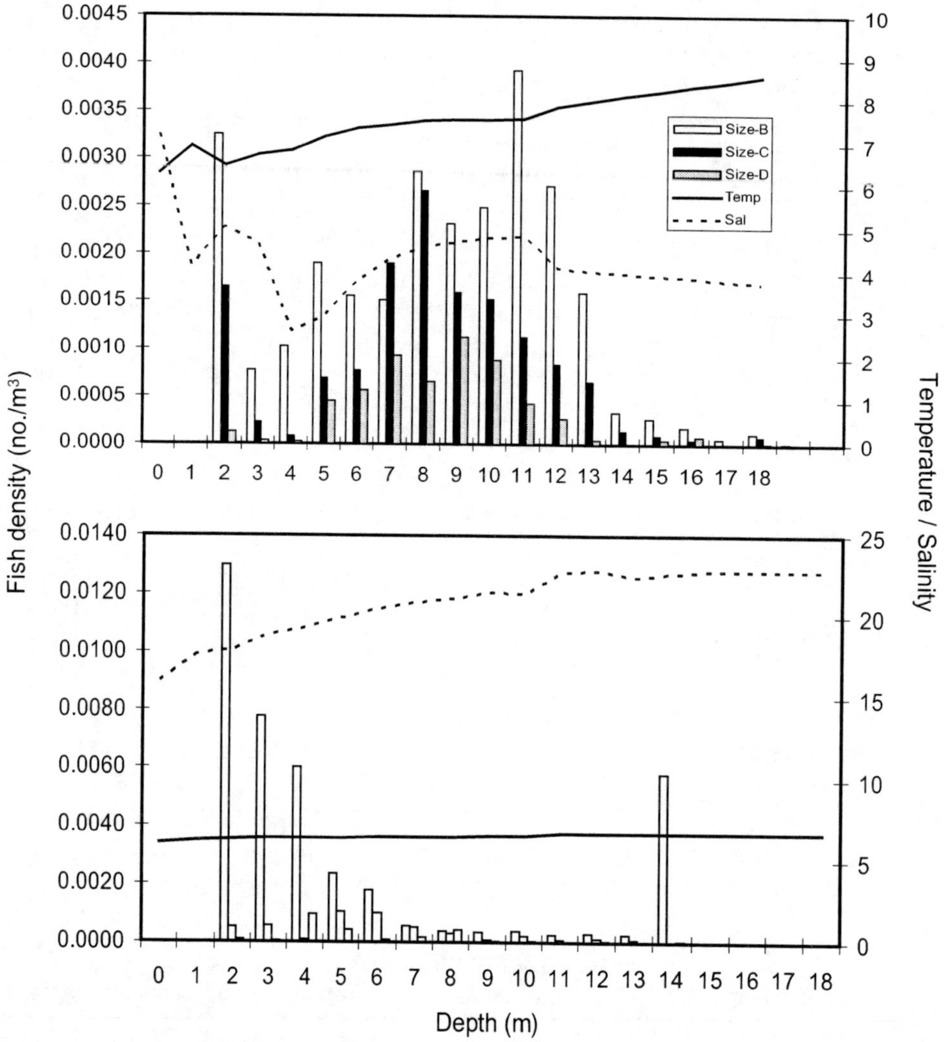

Figure 1. Vertical density distribution (fish m⁻³) of acoustic targets representative of striped bass: size-A, size-B, size-C, size-D, from across-river transects within the Battery region in December 1995 (upper panel) and 1997 (lower panel). Acoustic densities begin at 2 m depth due to transducer depth and blanking distance from the transducer eliminating the top 2 m of the water column. The 1995 densities show the typical "layering" of fish observed in the midwater region that year. In 1997, the layering of fish typical of 1995 is weaker, apparently due to a wind mixing event just prior to the 1997 survey.

in the water column in 1995, but not in 1997 (Figure 1). Salinity was the primary difference in physio-chemcial properties between the years. Salinity was much higher at the Battery in 1997 (mean 20.7‰) than 1995 (mean 4.28‰). Higher salinity levels in the Battery and Yonkers regions in 1997 (mean salinity 7.6‰ in 1997, versus 6.3‰ in 1995) may have shifted distributions observed in 1995 further upriver in 1997.

Despite the differences in vertical temperature and salinity pattern and magnitude, physio-chemical variables were frequently related to density of fish size classes (Table 5). In 1995, we found no significant relationship between physiochemical variables and size-D fish density, but temperature,

Table 5. Results of stepwise regression relating physiochemical variables to the acoustically-derived densities of fish size classes (size-A through D) in the lower Hudson River during 1995 and 1997. Variables evaluated included temperature (T), temperature-squared (T2), salinity (S), salinity-squared (S2), temperature-salinity interaction (TxS) and depth. The +/- sign shown parenthetically for each significant variable represents the nature of the relationship between that variable and density. Non-significant results for a size class are reported as NS.

		Variable	F	p	d.f.
1995					
	Size-A	S^2 (+)	5.07	0.038	1
	Size-B	T2 (-)	10.22	0.005	3
		TxS (+)	8.04	0.012	
		S (-)	4.63	0.048	
	Size-C	depth (-)	17.46	0.001	4
		T (+)	14.39	0.002	
		S (-)	6.39	0.024	
		S^2 (+)	7.71	0.015	
	Size-D	N.S.			
1997					
	Size-A	N.S.			
	Size-B	S (-)	32.00	<0.001	2
		TxS (+)	28.65	<0.001	
	Size-C	depth (-)	15.57	0.001	1
	Size-D	T^2 (-)	5.72	0.030	1

salinity, and related variables helped to explain densities of size-A through size-C fish. Salinity was negatively related to size-B and size-C vertical densities. In 1997, size-B, size-C, and size-D densities were negatively related to salinity and positively related to temperature x salinity interaction (size-B), and negatively related to depth (size-C) and temperature-squared (size-D) (Table 5).

Hydroacoustic estimates of fish density varied between years and regions within years. Total fish density (all sizes combined) was significantly higher in 1995 than 1997 (Figure 2). Highest fish densities were in the Battery and Croton-Haverstraw in 1995 (largely bay anchovy and white perch, respectively) and in the Battery in 1997. However, most regions in 1995 had densities that exceeded those in the Battery in 1997. The patterns in total fish density were driven by patterns in density of the size-A fish (Figure 3).

Within the size-B size-class (age-0 striped bass sizes) highest densities were in Poughkeepsie (0.158 fish/m^3), Croton-Haverstraw (0.058 fish/m^3) and Tappan Zee (0.043 fish/m^3) in 1995, but in 1997 Tappan Zee had the lowest densities (Figure 4). Yonkers (0.085 fish/m^3) had highest densities in 1997, but overall densities in this size-class were much lower in 1997 than 1995.

The size-C bass size-class (age-1 striped bass) had highest densities in the lower four, and the most upriver regions in 1995, but no clear pattern in 1997 (Figure 5). These patterns are probably driven by differences in the distribution of white perch and striped bass in this size-class. In 1995 white perch

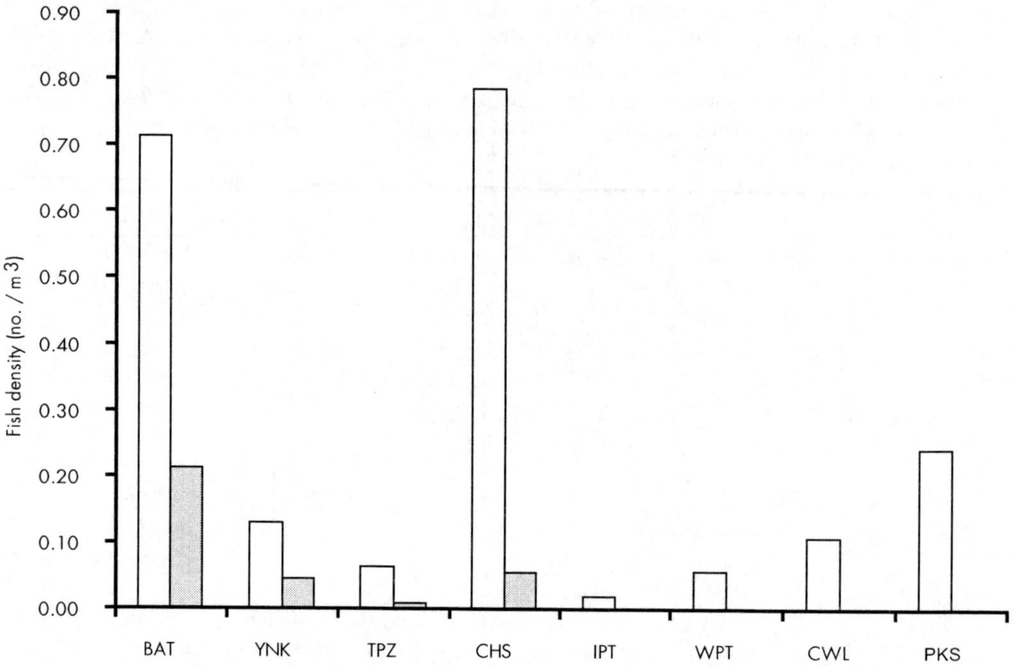

Figure 2. Mean fish density (No./m^3, all sizes) by river region determined from hydroacoustic surveys in December 1995 (open bars) and 1997 (filled bars). Fish densities do not include the top 2 m or the bottom 0.5 m of the water column due to acoustic limitations.

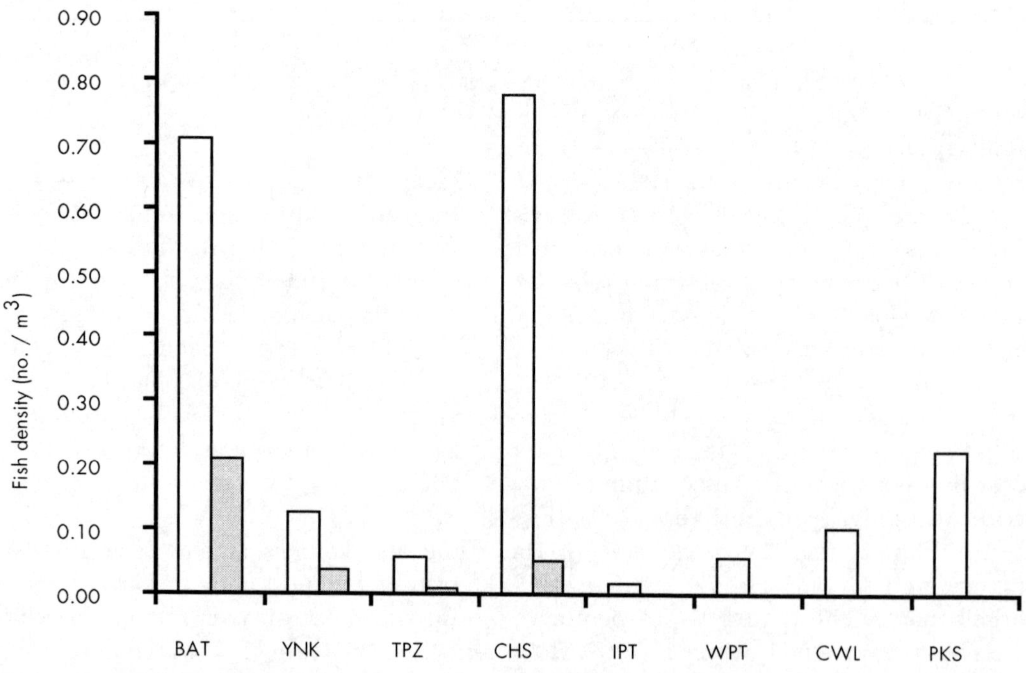

Figure 3. Mean fish density (No./m^3) of targets smaller than striped bass (Size-A) by river region determined from hydroacoustic surveys in December 1995 (open bars) and 1997 (filled bars). Fish densities do not include the top 2 m or the bottom 0.5 m of the water column due to acoustic limitations

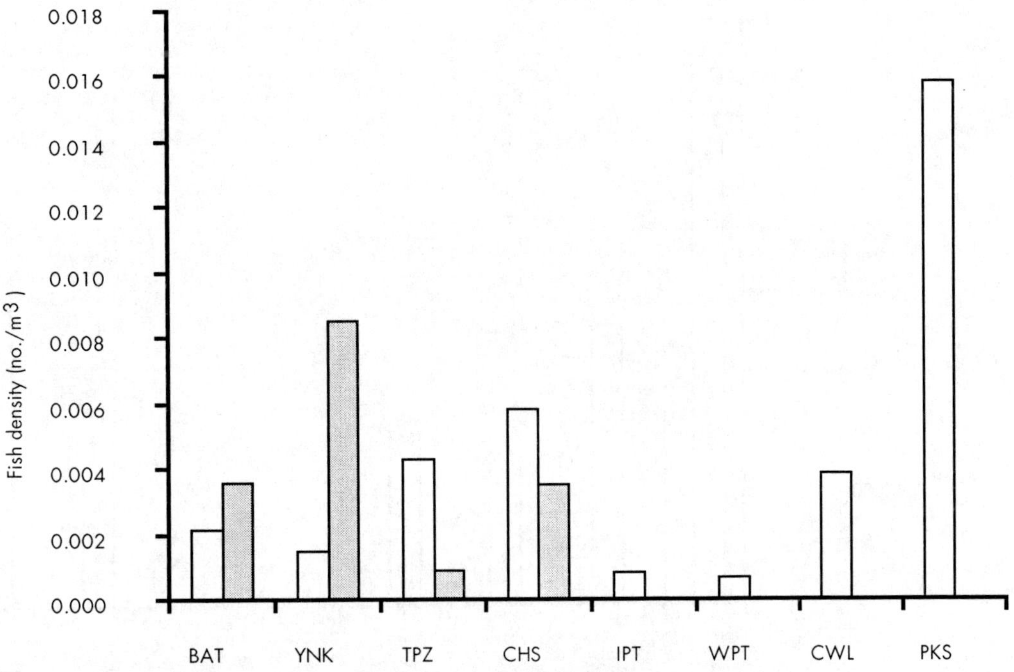

Figure 4. Mean fish density (No./m^3) of Size-B fishes by river region determined from hydroacoustic surveys in December 1995 (open bars) and 1997 (filled bars). Fish densities do not include the top 2 m or the bottom 0.5 m of the water column due to acoustic limitations.

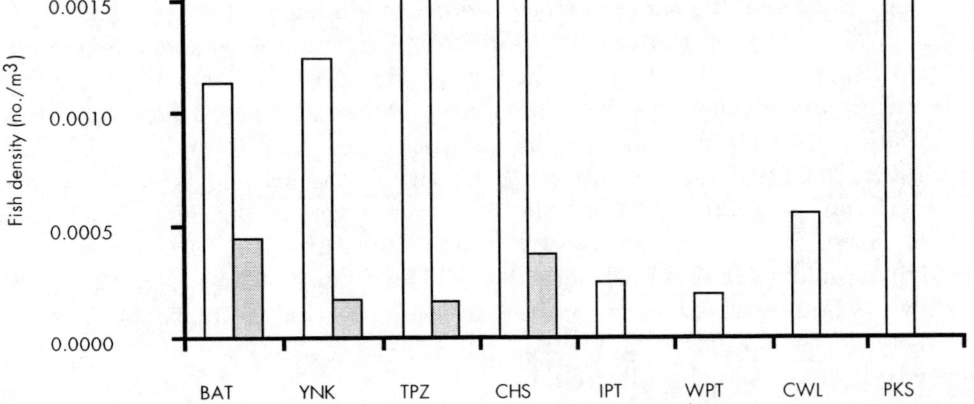

Figure 5. Mean fish density (No./m^3) of Size-C fishes by river region determined from hydroacoustic surveys in December 1995 (open bars) and 1997 (filled bars). Fish densities do not include the top 2 m or the bottom 0.5 m of the water column due to acoustic limitations.

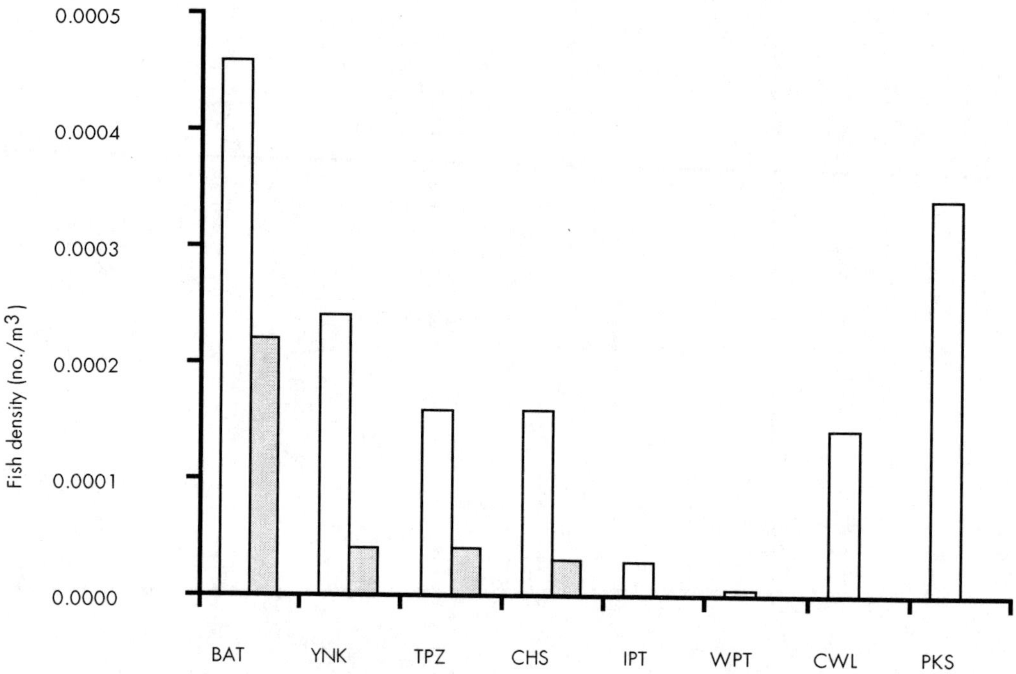

Figure 6. Mean fish density (No./m^3) of size-D fishes by river region determined from hydroacoustic surveys in December 1995 (open bars) and 1997 (filled bars). Fish densities do not include the top 2 m or the bottom 0.5 m of the water column due to acoustic limitations.

and striped bass co-occurred in the lower river, but upriver all trawl catches were white perch. In 1997, few striped bass were caught so distributions were driven even more so by white perch.

Density of the largest size-class of acoustic targets (size-D) followed the same pattern in 1995 and 1997 (Figure 6). Highest densities (0.00046 fish/m^3 in 1995) were at the Battery with gradual declines upriver, although densities were about half of 1995 levels in 1997. Densities increased again above West Point in 1995, signaling the presence of larger individuals that were unsampled in our midwater trawls. Whether these fish were large striped bass, white perch, or some other species is unknown. However, given that all fish collected in midwater trawls or gill nets (over 0.5 m from bottom) from the size-D size-class were striped bass, we considered all acoustic targets in this size range also to be striped bass.

Although spatial distributions in the abundance of striped bass were similar, we found large differences in the abundance of different cohorts between years (Figure 7). Regional abundance of striped bass was highest in Yonkers and Battery regions in each year. However, in 1995 most fish were of sizes representing age-1 and age-2 and older striped bass (size-C and size-D). In 1997, nearly all the striped bass in our acoustic estimates were size-B fish. Our estimate of total population size was 755,497 (±174,690) in 1995 and 1,473,529 (±637,200) in 1997. Of this estimate 24% were size-B in 1995 and 96% were size-B in 1997.

Comparison of the distributional abundance of white perch between years suggests that white perch experienced poor recruitment,

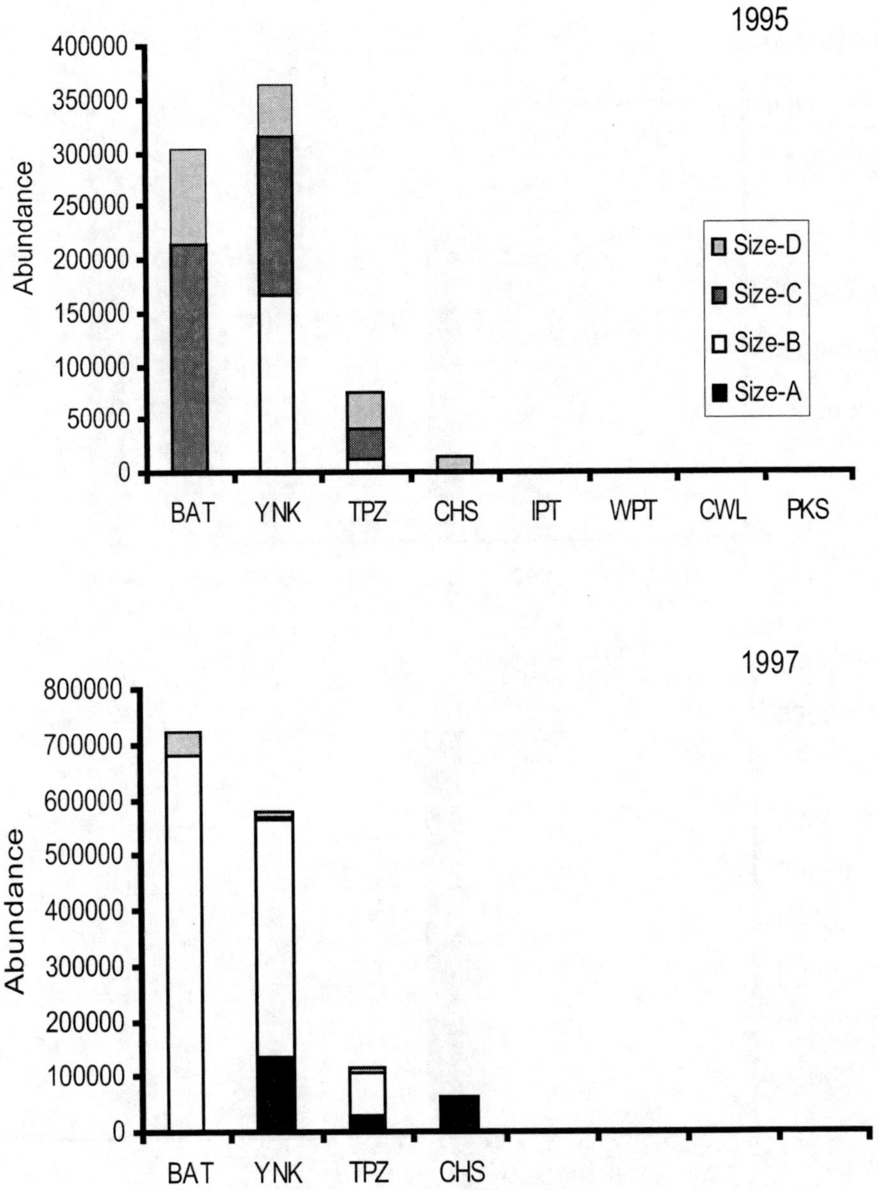

Figure 7. Regional abundance of striped bass determined by hydroacoustics in December 1995 and 1997. Note the different scales for abundance between years.

or that they were not adequately assessed in 1997 (Figure 8). Small white perch (size-A) represented the majority of the population estimate in both years, but total abundance and abundance within regions were much lower in 1997 compared with 1995. Highest regional abundance in 1995 was at Poughkeepsie (70.9 million), the most upstream region sampled, suggesting much of the population may have been upriver of this location at this time. In 1997, highest abundance was only 4.24 million fish; also occurring at the most upstream region sampled. Annual population estimates were

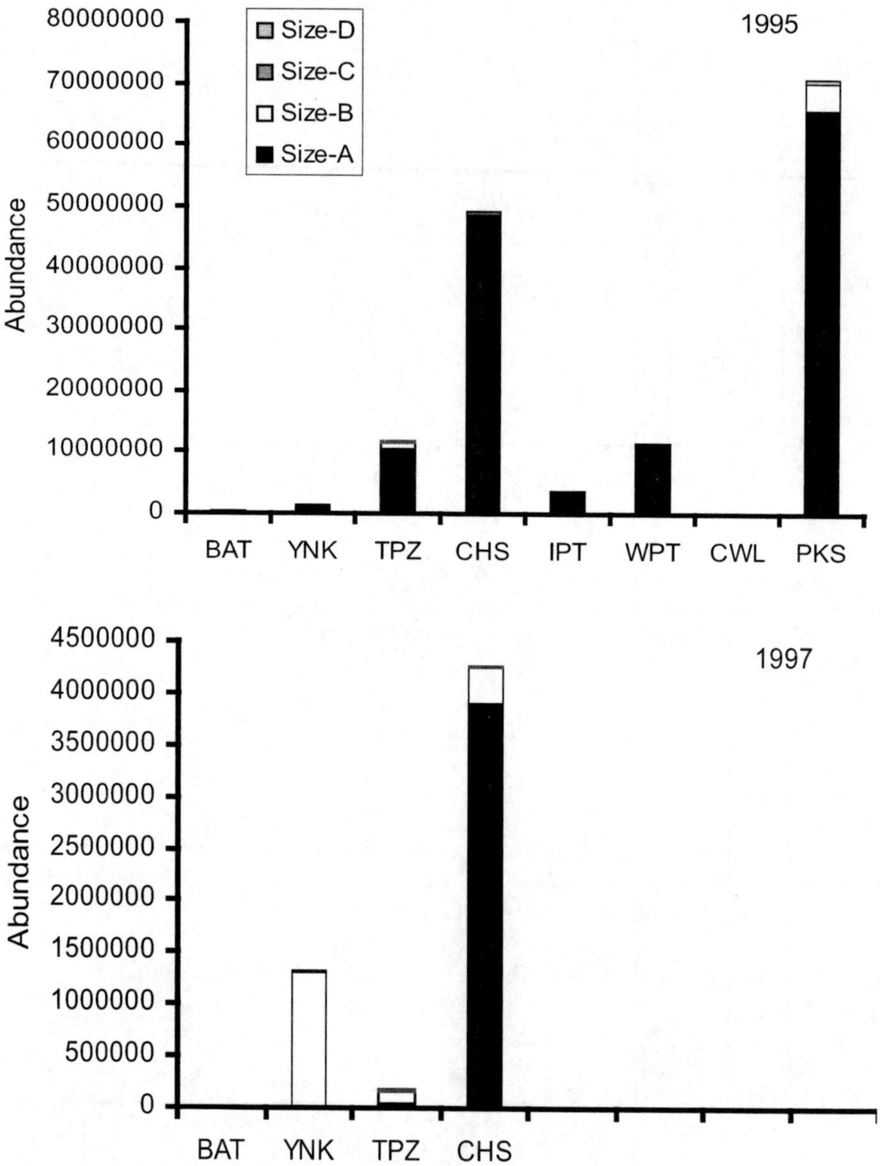

Figure 8. Regional abundance of white perch determined by hydroacoustics in December 1995 and 1997. Note the different scales for abundance between years.

148,661,400 and 5,779,500 in 1995 and 1997 respectively. Such extremely low total abundance of fish in 1997 compared with 1995 coupled with the regional abundances suggests white perch populations likely were distributed farther upriver in 1997 at the time we sampled.

Distribution of bay anchovy within the river regions was similar between years, but abundance was higher in 1995 (Figure 9). Most fish were detected in the Bat-

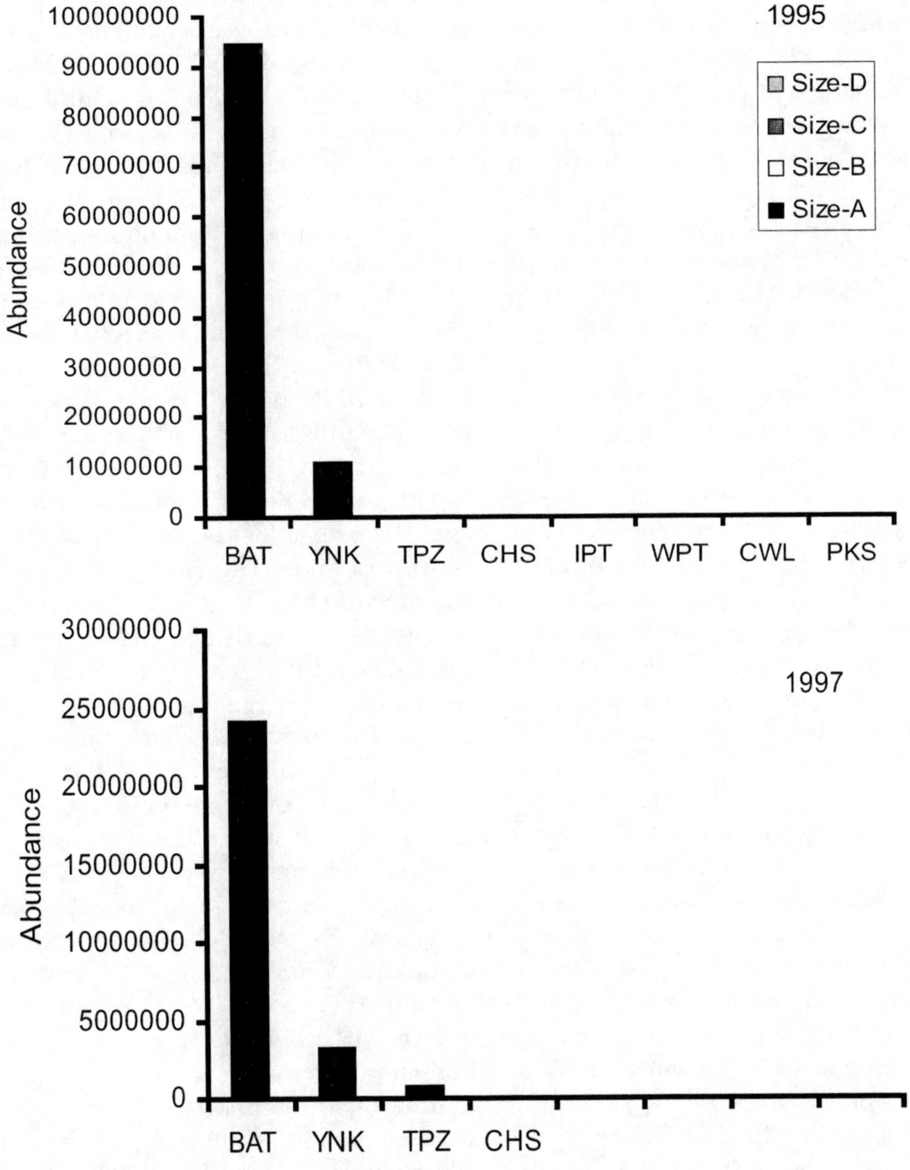

Figure 9. Regional abundance of bay anchovy determined by hydroacoustics in December 1995 and 1997. Note the different scales for abundance between years.

tery region with lesser amounts in Yonkers. In 1997, anchovy were also found in Tappan Zee. Total abundance exceeded 105 million in 1995, but was only 28.2 million in 1997. These two surveys suggest bay anchovy may overwinter in the lower Hudson River.

Discussion

This hydroacoustic study determined that several important species used the Hudson River during the winter of 1995 and 1997. White perch and bay anchovy were clearly the most abundant fish in these surveys, but Atlantic croaker and striped bass were also common inhabitants of the lower Hudson River. Each of these species is either economically (Manooch et al. 1981; Setzler-Hamilton 1991; Setzler-Hamilton and Hall 1991) or ecologically important (Houde and Zastrow 1991; Klauda et al. 1988). Thus, their use of the Hudson River during the environmentally-stressful winter period may serve to identify areas of the river as an important habitat for these species.

Management agencies have recently adopted a critical habitat approach to the protection of marine fish stocks. The "essential fish habitat" documents identify critical habitat as "those waters and substrate necessary to fish for spawning, breeding, feeding, or growth to maturity" (Magnuson-Stevens Act, 16 U.S.C. 1801 et seq). Although overwintering areas are not explicitly identified in the Act, these areas may be critical to survival and are therefore necessary to supporting a sustainable fishery and the managed species contribution to a healthy ecosystem. This study identified the lower Hudson River as an important overwintering area for striped bass, young Atlantic croaker, and bay anchovy in one or both years. White perch were also abundant in the lower Hudson River, however their high abundances upriver suggest the lower river is not necessarily more important than other areas. Based upon observations in this study it appears the lower Hudson River meets the criteria as an essential fish habitat for striped bass, bay anchovy, and Atlantic croaker during early winter.

Although consistent patterns in distribution of key fish species were noted in both 1995 and 1997, total abundance for each species differed dramatically between years. These differences could be due to differences in distributional patterns of the fish, or errors in our assumptions regarding acoustic accessibility of fish between years. Fish could have been distributed differently within the river between years (e.g., farther upriver in 1997 compared to 1995, or different vertical distributions), leading to differences in abundance estimates. The acoustic estimate of total fish abundance was much lower in 1997 compared to 1995, suggesting most of the reason for lower abundance estimates of key species in 1997 is due to the acoustic measures and not the trawl data. It is also possible that the estimates truly reflect lower abundance of key species in the lower Hudson River in 1997. Marine species may have been distributed off the mouth of the river and euryhaline species may have been further upriver in 1997 than 1995. Whatever the reason for the differences in species abundance estimates, the combined hydroacoustic and trawling surveys were successful in identifying the importance of the Hudson River for overwintering species.

The hydroacoustic population estimates for striped bass and white perch varied widely between years, but were comparable to available mark–recapture estimates for these species in the Hudson River. Young et al. (1988) used mark and recapture methods to estimate juvenile white perch and striped bass abundance from 1975 to 1979. Due to emigration, their estimates were stratified by month with their November estimates being the most similar to our December hydroacoustic estimates for comparison. As with our study, Young et al. (1988) found white perch to be much more abundant than striped bass. Striped bass abundance ranged from 1.3 to 7.0 million fish (95% confidence interval ranges from 0.7 to 14.1 million) and white perch from 12.9 to 52.8 million fish (95% confidence interval ranges from 10.9 to 65.7 million). In our hydroacoustic study we found 0.76–1.47 million striped bass and 5.8–148.6 million white perch in December surveys of Hudson River. Thus, striped bass estimates appeared on par with those previously reported. However, white perch estimates ranged more widely than those for tagging studies from the 1970s.

A comparison between an ongoing wintertime tagging study and hydroacoustic estimates of striped bass abundance also showed results that were similar in 1995, but not in 1997. Direct comparisons between the two studies are complicated by differences in population segments for which population estimates are available. The acoustic estimates are based on target strengths corresponding to observed striped bass age groups, but tagging estimates are based on fish greater than 150 mm. Because the 150 mm cut-off falls close to the upper bound of size-B, we believe the most appropriate comparison is to include only size-C and size-D in comparisons with tagging estimates. In 1995, striped bass abundance (size-C and size-D only) was estimated to be 576,110 $\pm$ 157,850 individuals with hydroacoustics. This compares to an estimate of striped bass abundance of 949,000 individuals (95% confidence limit of 745,000–1,308,000) from the wintertime tagging study (LMS 1999). The hydroacoustic estimate is within 39% of the wintertime tagging study and places it near the lower confidence interval on the tagging estimate. However, in 1997 the hydroacoustic and tagging estimates were quite different. The hydroacoustic estimate was 58,941 striped bass while the tagging study estimated the population at 446,000 (95% confidence limit of 382,000–555,000 fish, Normandeau Associates 1999). Hydroacoustic estimates appear to underestimate population sizes of striped bass overwintering in Hudson River, but it is important to consider that the acoustic surveys represent a single snapshot of fish populations while the tagging estimates are integrated over a several month marking and recapture period. Therefore direct comparisons of estimates from these studies must be viewed with caution.

This study clearly identifies white perch as the dominant species overwintering in the lower Hudson River. However, their abundance in sites further upriver suggests they may be habitat generalists during overwintering periods. The abundances of white perch reported here must also be considered an underestimate. It is important to recall that white perch catches were high in the upriver sections in 1995 suggesting they were distributed further upriver than we sampled in either year. Further, gill net sets showed about one-third of all white perch caught were too close to the bottom to detect with the acoustic system. Therefore, the white perch abundances reported here must be considered conservative measures of white perch abundance in the Hudson River for these years.

The disparity in abundance estimates of white perch between 1995 and 1997 may be accounted for by differences in distribution within the river in the two years. Lower abundance in 1997 likely arose from the smaller spatial coverage of our survey compared with 1995 coupled with a different (more upriver) distribution of white perch in 1997. Klauda et al. (1988) found white perch abundance (CPUE) increased upriver in December relative to earlier in the year. This suggests white perch may move upriver to overwinter in the river. This corroborates our suspicions that white perch were probably distributed largely upriver of our study area (rkm 61) in 1997 leading to lower estimates compared with 1995.

The finding that bay anchovy overwinter in the Hudson River adds to the evidence that this species may not migrate to warmer shelf waters during winter. Bay anchovy are believed to overwinter in Chesapeake Bay except during extremely cold winters (Houde and Zastrow 1991; Wang and Houde 1995). In more northerly latitudes bay anchovy may leave natal estuaries and overwinter off the shelf in warmer Gulf Stream waters or else migrate to warmer waters in the south. The fact that we found age-0 bay anchovy in the Hudson River in December of 1995 and 1997 suggests that significant numbers of age-0 bay anchovy in Hudson River did not move to warmer waters in these years. Bassista and Hartman (2005) found age-1 bay anchovy were virtually absent from the Hudson River in 1996 and 1997 and adult spawning populations were dominated by age-2 fish. Thus, age-0 anchovy present in December 1995 were not present in the Hudson River as age-1 fish in 1996, suggesting they either suffered complete overwinter mortality, or left the river and spent the next year in a different location. If such patterns are typical for the Hudson River, this can be a significant finding because if these fish experience mortality due to low water temperatures later in the winter, these fish may be considered a "doomed" population and therefore their impingement at regional water intake structures may have less importance to adult anchovy populations than previously considered (Boreman and Goodyear 1988).

Despite the high degree of alteration, the Hudson River supports a large fish community, many of which appear to overwinter therein. We found 17 species during December 1995 and 1997 sampling. White perch, bay anchovy, and striped bass were among the most numerous species in the river during this study. Of these, overwintering striped bass and bay anchovy appear to be distributed closest to the center of anthropogenic influence and therefore may be most impacted by further development and alterations in the river. However, white perch were still abundant in the lower river in both years so we cannot discount the potential impact of further anthropogenic stressors in the lower river upon white perch populations despite their relatively higher use of upriver areas. This study identified areas within the Hudson River that appear important to some overwintering fish species. The exact habitat attributes that these fish may be selecting were not identified in this study, but should be considered in future research (perhaps combining the recent benthic mapping surveys with hydroacoustics). Knowledge of the distribution patterns and habitat features important for overwintering fishes will be critical for conserving fish stocks in the Hudson River and other coastal areas experiencing increasing pressure due to development.

Acknowledgments

Funding for this study was provided by the Hudson River Foundation for Science and Environmental Research under contract to West Virginia University. Handling of all animals collected in this study was under protocols approved by the West Virginia University Institutional Animal Care and Use Committee.

References

Bassista, T. P., and K. J. Hartman. 2005. Reproductive biology and egg mortality of bay anchovy, *Anchoa mitchilli*, in the Hudson River Estuary. Environmental Biology of Fishes 73:49–59.

Beebe, C. A., and I. R. Savidge. 1988. Historical perspective on fish species composition and distribution in the Hudson River Estuary. Pages 25–36 *in* L. W. Barnthouse, R. J. Klauda, D. S. Vaughan, and R. L. Kendall, editors. Science, law, and Hudson River power: a case study in environmental impact assessment. American Fisheries Society, Monograph 4, Bethesda, Maryland.

Bodensteiner, L. R., and W. M. Lewis. 1994. Downstream drift of fishes in the upper Mississippi River during winter. Journal of Freshwater Ecology 9:45–56.

Boreman, J., and C. P. Goodyear. 1988. Estimates of entrainment mortality for striped bass and other fish species inhabiting the

Hudson River estuary. Pages 152–160 *in* L. W. Barnthouse, R. J. Klauda, D. S. Vaughan, and R. L. Kendall, editors. Science, law, and Hudson River power: a case study in environmental impact assessment. American Fisheries Society, Monograph 4, Bethesda, Maryland.

Brandt, S. B. 1996. Acoustic assessment of fish abundance and distribution. Pages 385–432 *in* B.R. Murphy and D.W. Willis, editors. Fisheries techniques, 2nd edition. American Fisheries Society, Bethesda, Maryland.

Carlson, D. M. 1992. Importance of wintering refugia to the largemouth bass fishery in the Hudson River estuary. Journal of Freshwater Ecology 7:173–180.

Crawshaw, L. I. 1984. Low temperature dormancy in fish. American Journal of Physiology 246:479–486.

Cunjak, R. A., and J. Power. 1987. Cover use by stream-resident trout in winter: a field experiment. North American Journal of Fisheries Management 7:539–544.

Hartman, K. J., and B. W. Nagy. 2005. A target strength and length relationship for striped bass and white perch. Transactions of the American Fisheries Society 134:375–380.

Houde, E. D., and C. E. Zastrow. 1991. Bay anchovy *Anchoa mitchilli*. Pages 8–1 to 8–14 *in* S.L. Funderburk, J.A. Mihursky, S.J. Jordan, and D. Riley, editors. Habitat requirements for Chesapeake Bay living resources. Chesapeake Research Consortium, Inc., Solomons, Maryland.

Jech, J. M., and J. Luo. 2000. Digital echo visualization and information system (DEVIS) for processing spatially-explicit fisheries acoustic data. Fisheries Research 47:115–124.

Klauda, R. J., J. B. McLaren, R. E. Schmidt, and W. P. Dey. 1988. Life history of white perch in the Hudson River estuary. Pages 69–88 *in* L. W. Barnthouse, R. J. Klauda, D. S. Vaughan, and R. L. Kendall, editors. Science, law, and Hudson River power: a case study in environmental impact assessment. American Fisheries Society, Monograph 4, Bethesda, Maryland.

Lawler, Matusky, and Skelly Engineers LLP. 1999. Hudson River Striped Bass Stock Assessment Program November 1995-April 1996. Prepared for the New York Power Authority, White Plains, New York.

Lenz, B. E. 2003. Winter habitat use of fishes in the Ohio River. Master's Thesis. West Virginia University, Morgantown, West Virginia.

Love, R. H. 1971. Measurements of fish target strength: a review. United States National Marine Fisheries Service Fishery Bulletin 69:703–715.

Lyons, J. 1997. Influence of winter starvation on the distribution of smallmouth bass among Wisconsin streams: a bioenergetic modeling assessment. Transactions of the American Fisheries Society 126:157–162.

MacLennan, D. N., and E. J. Simmonds. 1992. Fisheries acoustics. Chapman and Hall, New York.

Manooch, C. S., III, L. E. Abbas, and J. L. Ross. 1981. A biological and economic analysis of the North Carolina charter boat fishery. Marine Fisheries Review 39(8):1–11.

Muessig, P. H., J. B. Hutchinson, Jr., L. R. King, R. J. Ligotino, and M. Dailey. 1988. Survival of fishes after impingement on traveling screens at Hudson River power plants. Pages 170–181 *in* L. W. Barnthouse, R. J. Klauda, D. S. Vaughan, and R. L. Kendall, editors. Science, law, and Hudson River power: a case study in environmental impact assessment. American Fisheries Society, Monograph 4, Bethesda, Maryland.

Normandeau Associates, Inc. 1999. Hudson River Mark/Recapture Program November 1997-April 1998. Prepared for the New York Power Authority, White Plains, New York.

Raibley, P. T., K. S. Irons, T. M. O'Hara, K. D. Blodgett, and R. E. Sparks. 1997. Winter habitat used by largemouth bass in the Illinois River, a large river floodplain ecosystem. North American Journal of Fisheries Management 17:401–412.

Setzler-Hamilton, E. M. 1991. White perch *Morone americana*. Pages 12–1 to 12–20 *in* S. L. Funderburk, J. A. Mihursky, S. J. Jordan, and D. Riley, editors. Habitat requirements for Chesapeake Bay living resources. Chesapeake Research Consortium, Inc., Solomons, Maryland.

Setzler-Hamilton, E. M., and L. Hall, Jr. 1991.

Striped bass *Morone saxatilis.* Pages 13–1 to 13–31 *in* S. L. Funderburk, J. A. Mihursky, S. J. Jordan, and D. Riley, editors. Habitat requirements for Chesapeake Bay living resources. Chesapeake Research Consortium, Inc., Solomons, Maryland.

Sheehan, R. J., and J. L. Rasmussen. 1999. Large rivers. Pages 529–558 *in* W. A. Hubert and C. C. Kohler, editors. Inland Fisheries Management in North America, 2nd edition. American Fisheries Society, Bethesda, Maryland.

Sheehan, R. J., W. M. Lewis, and L. R. Bodensteiner. 1990. Winter habitat requirements and overwintering of riverine fishes. Final Report, Federal Aid in Sportfish Restoration Program, Project F-79-R.

Wang, S. B., and E. D. Houde. 1995. Distribution, relative abundance, biomass and production of bay anchovy *Anchoa mitchilli* in the Chesapeake Bay. Marine Ecology Progress Series 121:27–38.

Young, J. R., R. J. Klauda, and W. P. Dey. 1988. Population estimates for juvenile striped bass and white perch in the Hudson River estuary. Pages 89–101 *in* L. W. Barnthouse, R. J. Klauda, D. S. Vaughan, and R. L. Kendall, editors. Science, law, and Hudson River power: a case study in environmental impact assessment. American Fisheries Society, Monograph 4, Bethesda, Maryland.

American Fisheries Society Symposium 51:197–213, 2006

The Dynamics of Bay Anchovy in the Hudson River Estuary: Process-Oriented Studies and Long-Term Changes

ERIC T. SCHULTZ

Department of Ecology and Evolutionary Biology, University of Connecticut, Storrs Connecticut 06269-3043 USA
eric.schultz@uconn.edu

KAMAZIMA M. M. LWIZA

Marine Sciences Research Center, State University of New York Stony Brook, New York 11794-5000 USA

JOHN R. YOUNG

ASA Analysis & Communication, Inc.
921 Pike Street, Post Office Box 303, Lemont, Pennsylvania 16851 USA

KYLE J. HARTMAN

Wildlife and Fisheries Resources Program
West Virginia University Division of Forestry and Natural Resources
322 Percival Hall, Morgantown, West Virginia 26506-6125 USA

RONALD C. TIPTON

U.S. Fish and Wildlife Service, Maryland Fisheries Resource Office
117 Admiral Cochrane Drive, Annapolis Maryland 21401 USA

Abstract.—We review three areas of recent research on Hudson River bay anchovy *Anchoa mitchilli*. One focus has been the along-estuary movement of early life stages. A cohort analysis of samples collected in a spatiotemporally extensive monitoring program confirmed that early-stage anchovy migrate up-estuary at an estimated rate of 0.6 km/d. Complementary fine-scale field sampling found that early-stage anchovy can show preferences for depth and can conduct periodic vertical migration. To determine whether these behaviors were sufficient to produce up-estuary migration, larval flux and velocity were estimated. These estimates were consistent with local retention rather than concerted migration. High priority should be given to examining individual migration histories through analysis of otolith microchemistry. A second focus of research on Hudson anchovy has been on local population structure, permitting comparison to anchovy in other locations. Anchovy that spawn in the Hudson River are larger than those spawning in the Chesapeake Bay region and are mostly 2 years old, whereas yearlings predominate in other estuaries. Batch fecundity was lower and egg mortality higher in the Hudson River than in Chesapeake Bay. A key issue arising from these recent findings is the degree to which the Hudson anchovy pool is connected with other large anchovy pools, such as Narragansett Bay and Chesapeake Bay. A third focus of research on Hudson anchovy has been interannual variability in early-stage abundance. A more than 20-year time series of juvenile bay anchovy abundance shows that juvenile abundance has varied over one order of magnitude. There has been no significant change in abundance over the entire time series, but abundance has declined 10-fold since a peak in the late 1980s. Anchovy abundance was negatively associated with the abundance of early-stage striped bass *Morone saxatilis* and positively associated with the abundance of early-stage Atlantic tomcod *Microgadus tomcod*. We suggest that these associations reflect direct interactions among the species and urge further work on the ecological role of striped bass in the estuary.

Introduction

The bay anchovy *Anchoa mitchilli* (hereafter referred to as anchovy) is an abundant marine-estuarine fish of the eastern seaboard. It is a predominant member of the estuarine fish assemblage, ranging from tropical and subtropical estuaries in the Gulf of Mexico and the Atlantic coast of the southeastern U.S. (Deegan and Thompson 1985; Castillo-Rivera et al. 1994; Snelson and Johnson 1995; Ayala Perez et al. 1998) to the temperate estuaries of the North Atlantic Bight and southern New England (Dovel 1981; Monteleone 1992; Keller et al. 1999; Dorfman 2000). Early-stage (larvae and juvenile) anchovy are often the most abundant members of the summer ichthyoplankton assemblage (Olney 1983; Setzler-Hamilton 1987; Monteleone 1992; Keller et al. 1999; Schultz et al. 2003). In northern latitudes, anchovy migrate out of estuaries to coastal waters in the autumn, returning in spring as reproduction commences (Dovel 1981; Vouglitois et al. 1987; Schmidt 1992; Wang and Houde 1995). In contrast, in the Gulf of Mexico (Hackney and de la Cruz 1981; Dokken et al. 1984) and Florida (Tremain and Adams 1995) anchovy are estuarine residents for the entire year. Anchovy are found in a wide range of habitats, including salt-marsh creeks (Rozas and Hackney 1984; Allen et al. 1995; Able et al. 2001), marsh tide pools (Crabtree and Dean 1982), marsh lakes (Griffith and Bechler 1995), eelgrass beds (Orth and Heck 1980), the surf zone (DeLancey 1989), and the open water of small to large estuaries. Anchovy are pelagic with no preference for structure; they are most abundant in unvegetated areas (Rozas and Minello 1998; Thayer et al. 1999; Rozas and Zimmerman 2000). Older studies of bay anchovy habitat use were reviewed by Morton and Moran (1989).

Anchovy are an important part of estuarine food webs. Anchovy are planktivorous; they prey on crustaceans (Johnson et al. 1990; Peebles et al. 1996; Peebles 2002) and fish eggs and larvae (Cowan and Houde 1992). There are indications of competitive interactions and resource partitioning with other zooplanktivorous fish (Ogburn-Matthews and Allen 1993; Allen et al. 1995). Anchovy are major prey of piscivorous fish such as bluefish, striped bass, and weakfish (Manooch 1973; Richards 1976; Baird and Ulanowicz 1989; Juanes et al. 1993; Hartman and Brandt 1995). Anchovy production is also important to ctenophores, which feed on eggs and small larvae (Monteleone and Duguay 1988; Govoni and Olney 1991; Purcell et al. 1994).

The anchovy has been subject to annual monitoring in the Hudson River for almost three decades. An extensive sampling program was designed to assess the effects of power plant operations on fish populations and communities (Barnthouse et al. 1988). Data on distribution and abundance of anchovy and other target species are disseminated in annual reports (e.g., ASA Analysis & Communication 2004) and in research publications (e.g., Schmidt 1992 and other papers in this volume).

Here, we review three areas of recent research on Hudson River anchovy. One focus has been the up-estuary migration of early life stages. A second focus has been population structure and vital rates of Hudson River anchovy, in comparison to anchovy in other locations. Careful attention to interpopulation variability in these vital rates is warranted, given the ecological importance of this species. A third focus has been interannual variability in anchovy abundance. One important application of long-term monitoring program data is exploration of local factors that may influence the abundance of 'target species' in the estuary. Hypothesis-based explorations, combined with tightly-focused research on processes, are essential to understand the Hudson River estuary's ecological dynamics.

Analyses of Migration by Young Anchovy

A conspicuous feature of anchovy estuarine residency is the apparent up-estuary migration of early life stages. Seasonal redistribution of early-stage anchovy to upper estuary areas has been reported for the Hudson River (Dovel 1981; Schmidt 1992), tributaries of Chesapeake Bay such as the Patuxent River (Dovel 1971; Loos and Perry 1991), and Chesapeake Bay proper (Wang and Houde 1995), but has not been found in some locations (e.g., Narragansett Bay, Bourne and Govoni 1988). An instance of

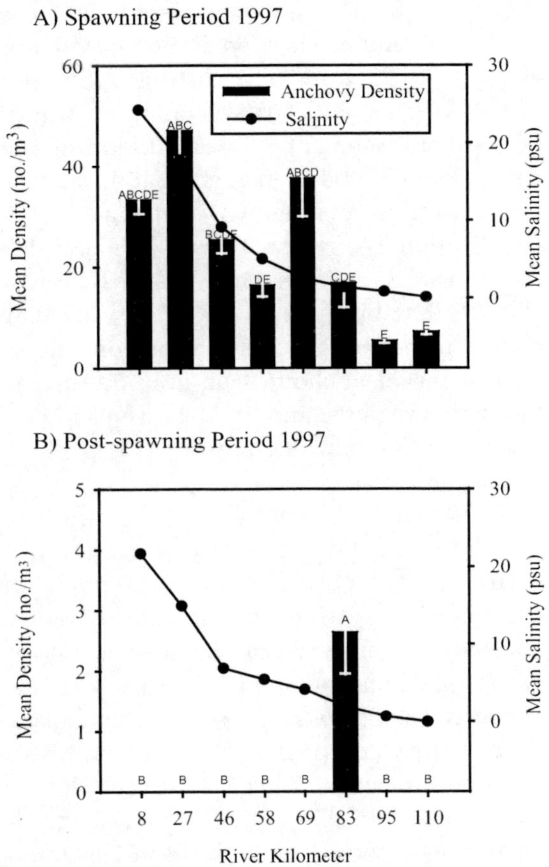

Figure 1. Along-river distribution of early-stage anchovy and salinity for two different time periods in 1997. Concentration of anchovy (mean ± standard error) that were 10-22 mm (total length) and salinity are plotted against location as the mean along-river distance of eight regions used for the Longitudinal River Survey (LRS). Concentration of anchovy was determined in hydroacoustic surveys; salinity was determined from the mean of two CTD deployments per LRS region. Letters above the bars signify assignment to groups that are not statistically distinguishable by Tukey's Studentized Range Test on LS means. Note that the scales for anchovy densities differ between the two panels. A) During the spawning season, June – early August. B) After the spawning season, September. From Tipton (2003).

the apparent seasonal upriver shift in the Hudson River is provided by hydroacoustic data collected during 1997. Mobile, down-looking hydroacoustic surveys with a

Simrad, 120 kHz split-beam sounder were used to estimate the density of larval anchovy (10–22 mm TL) during spawning (June-August) and postspawning (September) periods along the lower 122 km of the river (Tipton 2003). During the 1997 spawning season, spawning was most active in the Tappan Zee region (mean river km 46) and in the East River and Arthur Kill/Kill Van Kull tributaries of the Hudson (near river km 0; Bassista 2000). The concentration of larval anchovy was greatest in the lower two regions, and at mean river km 69 upriver of the spawning patches (Figure 1A). Subsequently there was a pronounced upriver shift in larval distribution; during the postspawning period larvae were present entirely in the West Point region (river km 83; Figure 1B). The distribution of eggs, larvae and young-of-year anchovy sampled in the 1997 Longitudinal River Survey (LRS) of ichthyoplankton conducted by the Hudson River utility companies (see ASA Analysis & Communication 2004 for further details on this sampling program) is broadly similar to that found in the hydroacoustic analysis.

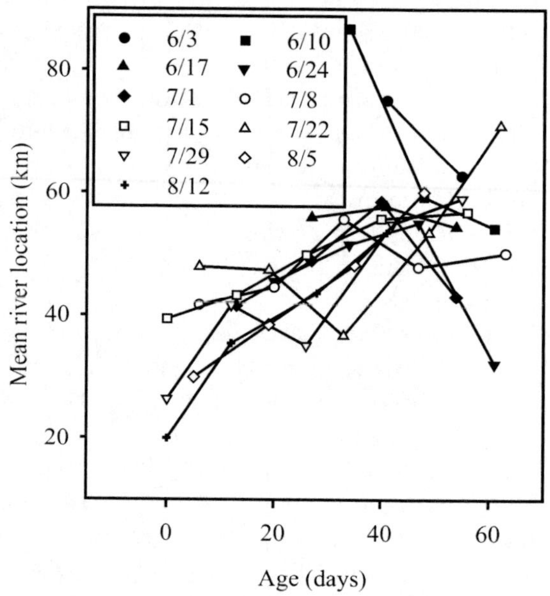

Figure 2. Net cohort transport of early-stage bay anchovy. Eleven cohorts were sampled multiple times at biweekly intervals in 1998. The mean river location of each cohort is plotted against the age of the cohort. Identity of each cohort is indicated by symbol, as represented in the legend. From Schultz et al. (2005).

Three general processes could produce up-estuary shifts, alone or in combination: 1) migration by larvae and juveniles (hypothesized by Dovel 1981); 2) migration by spawning adults (hypothesized by Schmidt 1992); and, 3) mortality gradients. Thus, if larvae in lower portions of the estuary are more likely to die than those in upper portions, redistribution will occur that mimics migration. MacGregor and Houde (1996) concluded that mortality gradients had a pronounced effect on redistribution of anchovy larvae from offshore to inshore areas of Chesapeake Bay. In contrast, Loos and Perry (1991) refuted the mortality hypothesis in favor of the early-stage migration hypothesis. Here, we first present analyses demonstrating that neither adult migration nor gradients in early-stage mortality are sufficient to explain the up-estuary redistribution of anchovy in the Hudson River. Then we describe efforts to determine the mechanisms of migration.

The hypothesis that early-stage anchovy migrate up-estuary has been tested with a cohort approach (Schultz et al. 2005). Cohort identification involves assigning individuals to birth-date classes (Limburg 2002), which is possible in many fish species via analysis of the daily age record in otoliths (Jones 2002). The cohort analysis of anchovy migration focused on the summer and autumn portion of the 1998 LRS (ASA Analysis and Communication 2001). Briefly, the steps in this analysis were: 1) determining daily age of an LRS subsample representative of LRS dates and locations; 2) estimating birth-date for all anchovy collected in the 1998 LRS based on the strong relationship ($R^2 = 0.82$; Schultz et al. 2005) between size and age found in step 1, and pooling the birth dates into weekly cohorts; 3) mapping the distribution of 11 different cohorts as they aged.

Cohorts shifted up-estuary, favoring the migration hypothesis over the alternative hypotheses of adult migration and mortality gradients (Schultz et al. 2005). We summarized our data on net cohort movements as the relationship between mean river location of a cohort and its age (Figure 2). Cohorts early in their life showed a net movement up-estuary at a mean rate of 0.6 km/d. However, once they aged to 35–40 d there was no further up-estuary movement. The overall relationship between cohort position and cohort age was strong ($R = 0.58$, $P = 0.0001$). This test demonstrated that the apparent movement of early-stage anchovy was not the result of seasonal movements of adult spawners. Another prediction we tested was that migration should entail an upriver increase and a downriver decrease in cohort regional abundance. We estimated regional standing crop (the number of individuals in a specified portion of the river), scored whether a cohort's regional standing crop decreased from one sampling period to the next, and then tested whether the probability that regional standing crop decreased varied with region. The results were clear; in downriver regions, standing crop usually decreased, while in upriver regions they usually increased (Figure 3). This test demonstrated that the apparent movement of early-stage anchovy was not entirely the result of spatial variation in mortality.

By what means might larvae migrate up-estuary? Because net flow in a river-fed estuary is seaward, a pool of passive larvae distributed randomly in the water column would on average move down-estuary (for an empirical demonstration of passive particle movement in the Hudson estuary see Hellweger et al. 2004). Of course larvae are not passive and could exhibit behaviors that function to promote up-estuary migration, such as 1) directed swimming away from the sea and 2) lateral preference (e.g., Forward et al. 1999), because along-river flow can have pronounced lateral variability; 3) preference for deeper water, where in many systems there is residual up-estuary flow (e.g., Melville-Smith et al. 1981); 4) periodic vertical migration. A well-documented form of the last strategy is synchronization of vertical migration and tidal flow, termed 'selective tidal stream transport' (reviewed by Forward and Tankersley 2001). Vertical migration that is entrained to the day/night cycle is also common in marine organisms and can in theory affect along-estuary movement (Hill 1991).

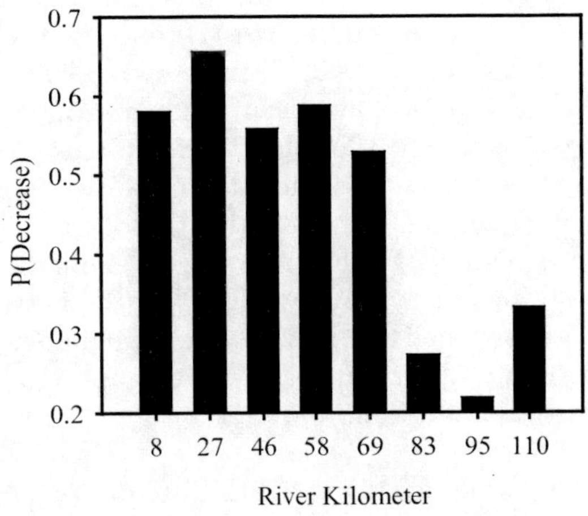

Figure 3. Along-river shifts in cohort distributions, as indicated by changes in regional standing crop. Eleven cohorts were sampled multiple times at biweekly intervals in 1998 (see Fig. 2). Changes in regional standing crop from one sample to the next are summarized as the proportion of intervals that cohorts decreased in regional standing crop against location, as the mean along-river distance of eight regions used for the LRS. From Schultz et al. 2005.

The behavioral mechanisms underlying migration of early-stage anchovy have been recently studied in the Hudson River (Schultz et al. 2000; Schultz et al. 2003). These projects have been conducted in the field, sampling the ichthyoplankton to reveal behaviors through patterns of distribution. Sampling was precisely controlled with respect to depth and tidal stage, permitting tests of temporally-averaged depth distribution as well as tests of temporal changes in depth distribution indicating periodic vertical migration.

Temporally-averaged depth distributions of early-stage anchovy have often been non-uniform but have not been consistent. In some circumstances anchovy exhibited preference for greater depth (see also Loos and Perry 1991; Schultz et al. 2003). During one three-day period of sampling in neap tide conditions at Croton Point (Figure 4B) anchovy were two to four times more concentrated at depth in the channel than at the surface. Yet under spring tide conditions a week earlier at the same site no depth preference was evident (Figure 4A). The difference in vertical stratification between these periods may be attributable to changes in vertical mixing: larval preference for greater depth was evident during neap tide but may have been overwhelmed by increased turbulence associated with more vigorous tidal flows during spring tide. Another sampling effort attempted to assess vertical stratification of anchovy at multiple sites and over longer periods to smooth over neap-spring variability (Schultz et al. 2000). Vertical stratification was often significant but no consistent pattern emerged: at some sites anchovy were most concentrated at middle depths, at other sites the most anchovy were found in the shallowest water, and at yet other sites the greatest concentrations were in the shallows and at depth.

Anchovy also appear to exhibit vertical migration in some situations (Schultz et al.

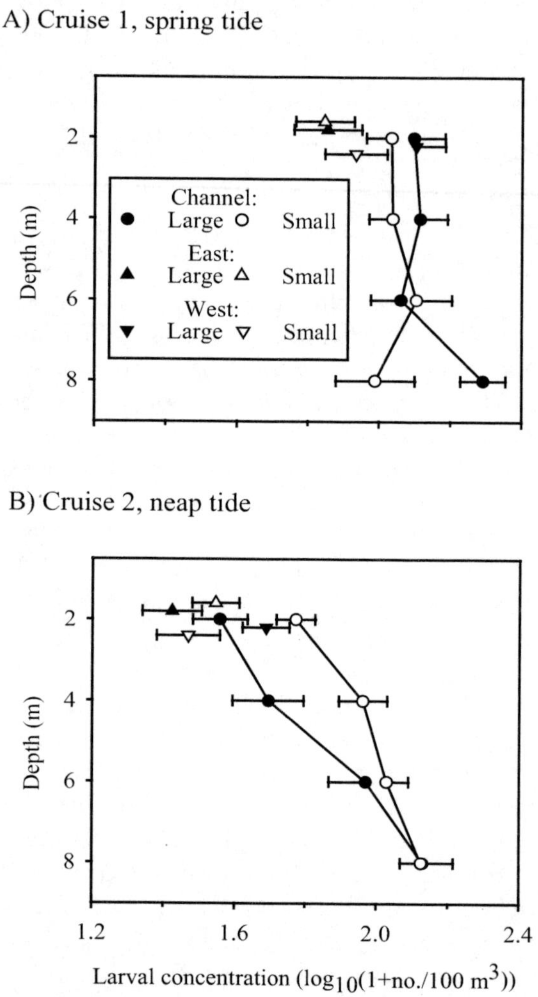

Figure 4. Lateral and vertical distribution of early-stage anchovy at Croton Point, late July 1998. Log-transformed larval concentration (mean ± standard error) is plotted for multiple depths in the river channel and for the uppermost 2 m over the east and west shoals. For clarity the depth values for east and west shoal stations have been slightly offset from 2 m and only one side of the error bar is displayed for some points. Small larvae are ≤11 mm standard length and large larvae and juveniles are ≥13 mm SL. The legend applies to both panels. A) Cruise 1, during spring tide conditions. B) Cruise 2, during neap tide conditions. From Schultz et al. (2003).

2003). Harmonic regression (Lorda and Saila 1986) was used to test for periodic change in abundance and mean depth of anchovy in synchrony with predominant tidal cycles (i.e., the M2 tide, 12.4 h, and the M4 tide, 6.2 h) and/or the diel cycle (24 h). Periodic changes in mean abundance (Figure 5) were consistent with diel migration into the water column at night. However small larvae during spring tide displayed no periodic changes in abundance or depth distribution. Other studies have also found that early-stage anchovy migrate vertically in a diel cycle, but while anchovy were shallowest at night in one case(Bourne and Govoni 1988), anchovy were deepest at night in another case (Loos and Perry 1991).

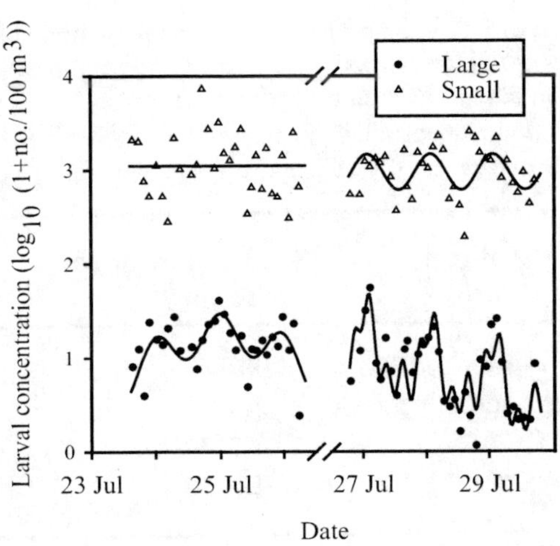

Figure 5. Temporal variability in abundance of early-stage anchovy at Croton Point, late July 1998. The symbols represent observed values of depth-averaged concentration (log-transformed) against the time the sample was collected. The lines represent the fitted relationship based on harmonic regression. Results for small ($\leq$ 11 mm) and large ($\geq$13 mm) are presented separately; for clarity the values for small larvae have been shifted up one increment on the y-axis and the values for large larvae have been shifted down one increment. The time series for cruise 1, during spring tide conditions, are to the left, and the time series for cruise 2 (neap tide conditions) are to the right. From Schultz et al. (2003).

Further work on depth preferences and vertical migrations is needed. There is as yet no coherent picture of how young anchovy are distributed in the water column. An important issue that has not been adequately addressed is the need for a complete depth profile, including a sample just above the bottom. The midwater collecting gear used in previous studies (Schultz et al. 2000; Schultz et al. 2003) was kept above the bottom to avoid fouling. Depth distributions such as those shown in Figure 4B suggest that such collecting methods may be missing the highest concentrations of young anchovy at times.

The patterns of depth preference and vertical migration observed off Croton Point (Schultz et al. 2003) were consistent with retention in an area rather than concerted migration upriver. Estimates of transport rate were derived by combining the data on early-stage anchovy distribution with contemporaneous data on flow collected with a 600 kHz Acoustic Doppler Current Profiler (RD Instruments, San Diego CA). We estimated flux and transport velocity (Dittel et al. 1991; Rowe and Epifanio 1994; Jager and Mulder 1999) of early-stage anchovy during spring and neap tides. These analyses indicate weak movement that was predominantly downriver (Lwiza and Schultz, unpubl. data). During spring tide, small and large early-stage anchovy were moving downriver at roughly 0.05 km/d. During neap tide, small anchovy were moving downriver at about one-third that rate, while large early-stage anchovy were virtually motionless. This result is quite inconsistent with the upstream migration scenario suggested by larval survey data (Dovel 1981; Tipton 2003) and our cohort analysis. A likely explanation for this discrepancy is that flux and transport calculations were biased; no data were available on anchovy concentrations in the deepest part of the water column where net flow tends to be up-estuary.

Table 1. Comparison of empirical methods of transport/migration analysis. For each method we list the scale of the analysis, precision of migration estimates, whether the method provides any direct information on the mechanism by which the transport occurs, and whether the method provides any information on individual variability in transport history. The fine-scale analyses of larval distribution represented by Schultz et al. (2003), and of larval flux (Lwiza and Schultz, unpublished data) are combined into one row.

Method	Spatio–temporal scale	Precision	Mechanistic	Individual variability
Behavior/Flux	Fine	Moderate	yes	no
Microchemistry	Fine	High	no	yes
Survey	Broad	Low	no	no
Cohort	Broad	Moderate	no	no

We conclude this section with further thoughts on approaches to analyzing the estuarine redistribution of young fish. The research to date on early-stage anchovy has been conducted on two spatiotemporal scales, with fine-scale analyses of larval distributions and fluxes and broad-scale analyses of net transport. Both approaches can yield estimates of migration rate. We advocate such complementary multi-scale research programs (Table 1). For instance, the cohort-level analysis yields insight into the broad patterns of redistribution but not the mechanism of migration. None of the methods so far employed to assess anchovy migration in the Hudson River can address individual variability. Fortunately there are methods for retrospective analysis of movement via assays for tracers of environmental conditions. Analysis of strontium-calcium ratios in fish otoliths has yielded important insights into estuarine migration (Limburg 1995; Secor and Piccoli 1996; Jessop et al. 2002). The microchemistry technique was recently applied to quantify the timing and rate of young anchovy migration in Chesapeake Bay (Kimura et al. 2000). Kimura et al.'s (2000) analyses indicated that anchovy juveniles did indeed migrate up-estuary. The microchemical signal of migration did not appear until individuals were relatively large (25 mm and larger), in contrast to the cohort analysis of Hudson River anchovy indicating that net migration occurs only among smaller individuals. There are several possible reasons for this discrepancy, including the possibility that it arises from differences between estuaries in flow patterns and larval behavior. Clearly conducting parallel cohort-level and individual-level analyses of migration in Chesapeake Bay and/or the Hudson River would be enlightening.

Demography and Vital Rates of Hudson River Anchovy

Extensive sampling of adult anchovy permits demographic comparisons between anchovy in the Hudson River and other locations. Anchovy size structure varies geographically. In the Hudson River in July 1997, yearling-and-older (mature) anchovy averaged about 70 mm fork length (Figure 6A; Bassista and Hartman 2005). A similar size structure was observed in Narragansett Bay in July 1997 (Figure 6B). In contrast, mature anchovy in July in Chesapeake Bay were mostly about 50 mm fork length (Newberger and Houde 1995). There is also variability in age structure among estuaries. In Narragansett Bay (Figure 6B; Lapolla 2001) and Chesapeake Bay (Newberger and

Houde 1995) most of the mature anchovy were yearlings, but the mature anchovy in the Hudson River were predominantly age-2 and older (Figure 6A). Some reproductive parameters vary across estuaries (Table 2). In particular, batch fecundity was lower and egg mortality higher in Hudson River than in Chesapeake Bay. There also may be differences between Hudson River and Chesapeake spawning frequency and sex ratios: in comparison to the results of some Chesapeake studies, Hudson River anchovy spawn more frequently, in more equitable sex ratios. Such differences in reproductive parameters may be related to the greater size and age of the Hudson River adults, environmental differences, or to geographic genetic variation.

The demographic differences among estuaries are remarkable, yet difficult to interpret. Lapolla (2001) suggested that the large size of yearling anchovy in Narragansett Bay reflects faster growth of fish in northern populations (see also Castro and Cowen 1991) and relatively little mixing with Chesapeake Bay. However, Hudson River anchovy are comparable in size-at-age to Chesapeake Bay (cf. Figure 6 and Newberger and Houde 1995). The absence of yearling spawners in the Hudson River could be the result of delayed maturation or an age-specific exchange with other locations, such that Hudson River origin fish spawn as yearlings elsewhere and only older fish use the Hudson River estuary. The best resolution of these issues would involve a comprehensive sampling program of adults during the spawning season from multiple major estuaries, to clarify geographic patterns in size at age, age structure, and genetic population structure.

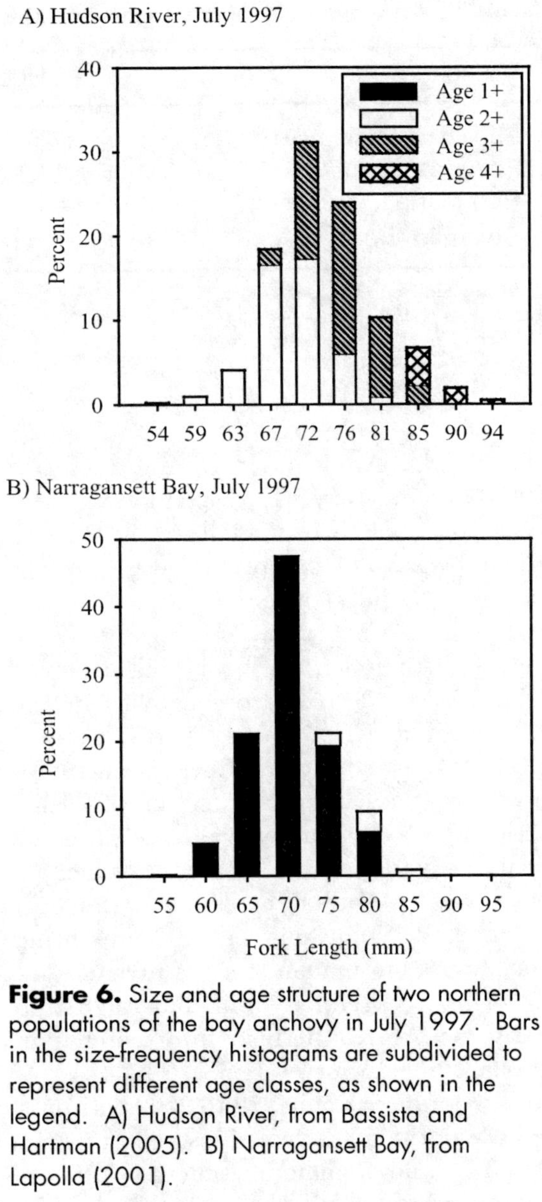

Figure 6. Size and age structure of two northern populations of the bay anchovy in July 1997. Bars in the size-frequency histograms are subdivided to represent different age classes, as shown in the legend. A) Hudson River, from Bassista and Hartman (2005). B) Narragansett Bay, from Lapolla (2001).

Year-class Strength: Long-term Indices and their Predictors

A time series of juvenile bay anchovy abundance in the Hudson Estuary was collected through the utility monitoring program (Central Hudson Gas and Electric Corp. et al.

Table 2. Demographic comparisons between bay anchovy in the Hudson River and Chesapeake Bay.

Variable	Hudson	Chesapeake
Spawning frequency (%/d)	73 – 89%[a]	67 - 100%[b]; 54 – 78%[b]
Fecundity (eggs/g)	506 ± 17.3 (SE)[a]	687 ± 55 (SE)[b]; 740 – 803[c]
Egg mortality (%/d)	72 - 99% (mean 93%)[a]	73 - 78%[d]
Sex ratio (%)	50%[a]	>50%[e]; 45% - 57%[c]

[a] Bassista and Hartman (2005)
[b] Zastrow et al. (1991), June and July
[c] Luo and Musick (1991), July and August
[d] Dorsey et al. (1996)
[e] Newberger (1989)

1999; ASA Analysis & Communication 2004). No index was developed for anchovy eggs, yolk sac larvae, or adults in the Hudson River because of inappropriate gear (yolk sac larvae, adults) or sampling interval mismatched to stage duration (eggs, yolk sac larvae). An index for postyolksac larvae (PYSL) was derived from catches in the Longitudinal River Survey (LRS; Central Hudson Gas and Electric Corp. et al. 1999); however, the earlier portions of this time series may be of limited value because the spatial coverage of the LRS did not include the lowest region of the estuary (Battery) until 1988. The juvenile index, representing the annual mean abundance of juveniles for an eight-week period from mid-August to early October, is derived from catches in the Fall Shoals Survey (FSS). The series of juvenile indices from 1979 to 2002 was published in the 2002 year-class report (ASA Analysis and Communication 2004).

Juvenile abundance varied over one order of magnitude in the 23-year time series (Figure 7A). Maximal abundance was recorded at 34.1 individuals/100 m^3 in 1988, while lowest abundance was 3.7 individuals/100 m^3 in 2000. There was no significant change in abundance over the entire time series (bivariate regression of log-transformed abundance against year, $P = 0.10$), and there was no evident temporal autocorrelation in the time series (detrended time series, test for effects of one- and two-year lagged abundance $P = 0.35$ and 0.88 respectively).

What factors contributed to fluctuations in abundance of juvenile anchovy within this time period? Among the possible ecological relationships of importance, work has focused on three potentially interacting fish species: striped bass *Morone saxatilis* (a potential competitor and predator), Atlantic tomcod *Microgadus tomcod* (a potential competitor), and bluefish *Pomatomus saltatrix* (a predator). Young striped bass abundance was estimated via two indices, of post yolk sac larvae (from the LRS, as mean abundance in a seven-week period around the period of peak abundance) and juveniles (from the Beach Seine Survey (BSS), as average number in a 100 ft haul from mid-August to early October). Other indices of early-stage striped bass (for eggs and yolk sac larvae) may be less reliable because the duration of these stages is short relative to the one-week sampling interval of the LRS (Central Hudson Gas and Electric Corp. et al. 1999). Bluefish abundance was indexed from the BSS as average number of juveniles in a 100 ft haul from mid-August to early October. Tomcod abundance was indexed from the LRS as mean abundance of PYSL and juveniles in the LRS in May.

Analysis of the relationships among the indices supports the hypothesis that striped bass abundance is affecting anchovy abundance but does not support the hypothesis of tomcod-anchovy competition. Over the entire time series, striped bass PYSL was negatively correlated with anchovy abundance and tomcod juveniles were positively correlated with anchovy abundance; the strength of the relationships depends on whether the abundance indices are transformed (Table 3A). A multiple regression indicated that both striped bass and tomcod were independently associated with anchovy abundance (all indices log-transformed; $P = 0.05$ for striped bass, $P = 0.007$ for tomcod). We repeated the analyses for the latter portion of the time series (1988–2002) because: 1) the increased spatial coverage of the LRS in those years should yield more reliable estimates of striped bass and tomcod PYSL abundance; 2) striped bass abundance in all stages has been consistently higher in recent years. The negative association between striped bass and anchovy was pronounced in the last 14 years, and the positive association between tomcod and anchovy persisted (Table 3B; multiple regression of log-transformed indices, $P < 0.0001$ for striped bass, $P = 0.01$ for tomcod). In the first 9 years of the time series the abundance of tomcod was the only significant predictor of anchovy abundance ($r = 0.72$, $P = 0.03$).

Abiotic factors that may influence anchovy abundance were also tested for the 1988–2002 time series. The amount of freshwater flow, which has demonstrable effects on Hudson River fish community structure (Hurst et al. 2004) and has been observed to influence the distribution of early-stage bay anchovy (Schultz et al. 2000), was quantified as USGS flow at Poughkeepsie (http://ny.water.usgs.gov/projects/dialer_plots/Hudson_R_at_Poughkeepsie_Freshwater_Discharge.htm). We also tested for the effect of temperature; mean regional temperatures for every sampling week of the LRS (drawn from each Year Class Report, May -

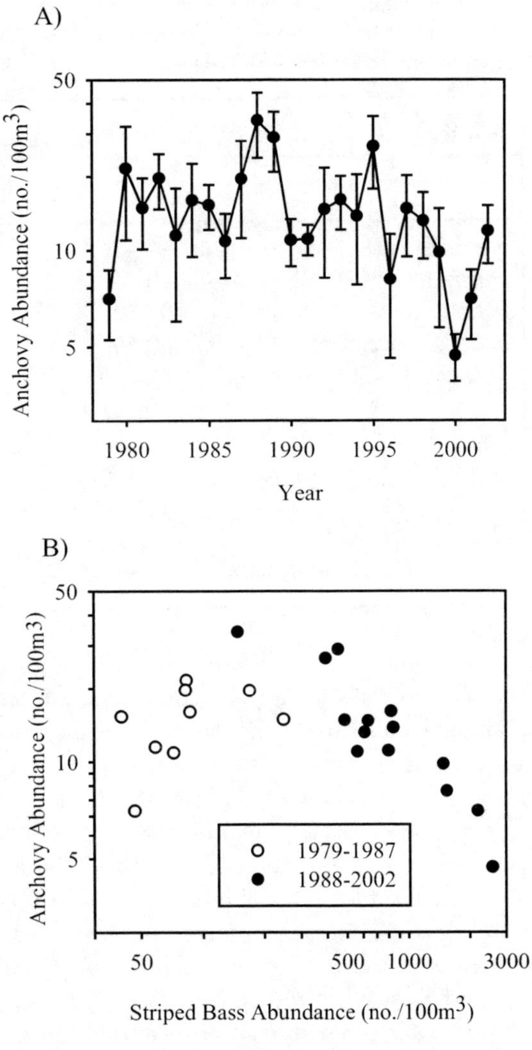

Figure 7. Long-term record of early-stage bay anchovy abundance in the Hudson River. A) The year-class index for juvenile anchovy for 1979 - 2001. Error bars represent ± 2 standard errors. B) A long-term inverse relationship between bay anchovy abundance and striped bass abundance, 1979-2001. The index of abundance for juvenile bay anchovy is plotted against the index for post-yolksac striped bass larvae. The first 9 years of data are plotted in a different symbol from the subsequent 14 years.

Table 3. Interannual variation in anchovy abundance: correlation of juvenile bay anchovy abundance with abundance of potential predators and competitors. The correlations are tested for untransformed and log-transformed indices, and over two intervals.

Factor	*Untransformed*		*Log-transformed*	
	r	*P*	*r*	*P*
A) 1979-2002				
Striped bass PYSL	–0.52	0.0091	–0.39	0.062
Striped bass juveniles	–0.051	0.81	0.14	0.51
Bluefish juveniles	–0.17	0.4300	0.05	0.81
Tomcod PYSL/ juveniles	0.48	0.018	0.53	0.0075
B) 1988-2002				
Striped bass PYSL	–0.70	0.0034	–0.84	< 0.0001
Striped bass juveniles	–0.081	0.77	0.24	0.38
Bluefish juveniles	–0.20	0.50	0.014	0.96
Tomcod PYSL/ juveniles	0.42	0.12	0.46	0.087

September) were condensed into mean monthly temperatures for the estuary from Poughkeepsie to the Battery. The anchovy juvenile index was not correlated with any monthly mean flow or temperature, nor was it correlated with seasonal average (May-September) flow or temperature (results not shown). We tested combined biotic and abiotic factors via a stepwise regression. In addition to the negative effect of PYSL striped bass abundance ($P < 0.0001$), the flow during August also entered the model as a significant regressor with a negative slope ($P = 0.04$). Given the number of regressors that were tested, it seems likely that this is a spuriously significant result.

What are the mechanisms driving the negative association between anchovy and striped bass, and the positive association between anchovy and tomcod? Based on the nonsignificance of abiotic factors, we suggest that the associations are not the result of independent responses to the same environmental drivers. Direct interactions among these species are more likely. One possibility that explains both associations is that striped bass have a dominant negative influence on both anchovy and tomcod. Further analysis of the ecological role of striped bass should be fruitful as the species continues to recover from overexploitation.

Summary and Prospects for Further Work

Recent process-oriented studies of bay anchovy in the Hudson River include research into the redistribution or migration of early life stages. We have learned that young anchovy do indeed migrate into the estuary, and we are working towards agreement among different methods of estimating migration rate. Future studies of the migration process will need to assess interannual variability in migration patterns and should include an individual-level approach via otolith microchemistry. The function of the migration is poorly understood; what benefit might there be for individuals who move up-estuary? Are these individuals substantially more successful in growth and/or survival than those that remain in coastal waters? Is there a continuous and substantial exchange of early-stage anchovy between the estuary and coastal waters or do these pools of anchovy remain distinct?

The demography of the Hudson River anchovy may be unique, even among the higher-latitude locations; certainly the anchovy in higher-latitude estuaries have a distinctly different size structure than others to the south. How these differences arise is not known. A key question in this regard is: to what extent is the Hudson anchovy pool connected with other pools in the area (e.g., Narragansett Bay) and elsewhere in the range (e.g., Chesapeake Bay)?

The accumulation of long-term monitoring data offers a rare opportunity to see how processes play out over successive year classes. Analysis of the anchovy data has shown that (unsurprisingly) there is interannual variability in juvenile abundance. Anchovy abundance has declined 10-fold since a peak in the late 1980s, but no overall change in abundance is evident over the entire time series. The analysis further suggested that one cause for the decline has been a rebound in striped bass abundance. It is not clear whether this negative interaction is competitive or predatory. There was a positive association between anchovy and tomcod, contrary to the expectation of a competitive negative association. A missing element in the analysis of ecological relationships is gelatinous zooplankton. Jellies are known to be important predators elsewhere and are certainly abundant in the lower reaches of the estuary during the summer (E. Schultz, personal observation). The lack of temperature and flow effects on the estuarine population of juvenile anchovy may be attributable to connections with coastal waters. Perhaps local perturbations disappear because their effect is spread out well beyond the estuary, as young anchovy in the river move out to the coastal waters and vice versa. Revealing the various connections between the Hudson Estuary anchovies, those in other local waters, and those of other major estuaries, is clearly an important and challenging task for the future.

Acknowledgments

Funding for the work described here was provided through grants, contracts and subcontracts from the Hudson River Foundation and the Hudson River Utilities. The field and laboratory workers that turned reality into data were legion and can only be thanked in a blanket expression of appreciation. We are also grateful to those who helped to bring the Hudson River Fishes conference and the proceedings volume together, including K. Limburg, M. Mattson, D. Strayer, and J. Waldman.

References

Able, K. W., D. M. Nemerson, R. Bush, and P. Light. 2001. Spatial variation in Delaware Bay (U.S.A.) marsh creek fish assemblages. Estuaries 24:441–452.

Allen, D. M., W. S. Johnson, and V. Ogburn-Matthews. 1995. Trophic relationships and seasonal utilization of salt-marsh creeks by zooplanktivorous fishes. Environmental Biology of Fishes 42:37–50.

ASA Analysis & Communication. 2001. 1998 year class report for the Hudson River Estuary Monitoring Program, New Hampton, NY.

ASA Analysis & Communication. 2004. 2002 year class report for the Hudson River Estuary Monitoring Program, Washingtonville, NY.

Ayala Perez, L. A., O. A. Aviles Alatriste, and J. L. Rojas Galaviz. 1998. Community composition of fish in the Candelaria-Panlau, Campeche, Mexico. Revista de Biologia Tropical 46:763–774.

Baird, D., and R. E. Ulanowicz. 1989. The seasonal dynamics of the Chesapeake Bay ecosystem. Ecological Monographs 59:329–364.

Barnthouse, L. W., J. Boreman, T. L. Englert, W. L. Kirk, and E. G. Horn. 1988. Hudson River settlement agreement: technical rationale and cost considerations. Pages 267–273 *in* L. W. Barnthouse, R. J. Klauda, D. S. Vaughn, and R. L. Kendall, editors. Science, Law, and Hudson River Power Plants. American Fisheries Society Monograph 4. American Fisheries Society, Bethesda, Maryland.

Bassista, T. P. 2000. Reproductive biology and early life stage abundance and distribution of bay anchovy, (*Anchoa mitchilli*) in the Hudson River Estuary, USA. M.S. thesis. West Virginia University, Morgantown, WV.

Bassista, T. P., and K. J. Hartman. 2005. Reproductive biology and egg mortality of bay anchovy (*Anchoa mitchilli*) in the Hudson River Estuary. Environmental Biology of Fishes 73:49–59.

Bourne, D. W., and J. J. Govoni. 1988. Distribution of fish eggs and larvae and patterns of water circulation in Narragansett Bay, 1972–1973. Pages 132–148 *in* M. P. Weinstein, editor. Larval fish and shellfish transport through inlets. American Fisheries Society Symposium 3. American Fisheries Society, Bethesda, Maryland.

Castillo-Rivera, M., G. Moreno, and R. Iniestra. 1994. Spatial, seasonal, and diel variation in abundance of the bay anchovy, *Anchoa mitchilli* (Teleostei: Engraulidae), in a tropical coastal lagoon of Mexico Southwestern Naturalist 39:263–268.

Castro, L. R., and R. K. Cowen. 1991. Environmental factors affecting the early life history of bay anchovy *Anchoa mitchilli* in Great South Bay, New York. Marine Ecology Progress Series 76:235–247.

Central Hudson Gas and Electric Corp., Consolidated Edison Company of New York Inc., New York Power Authority, and Southern Energy New York. 1999. Draft environmental impact statement for state pollutant discharge elimination system permits for Bowline Point, Indian Point 2 & 3, and Roseton steam electric generating stations.

Cowan, J. H., Jr., and E. D. Houde. 1992. Size-dependent predation on marine fish larvae by ctenophores, Scyphomedusae, and planktivorous fish. Fisheries Oceanography 1:113–126.

Crabtree, R. E., and J. M. Dean. 1982. The structure of two South Carolina estuarine tide pool fish assemblages. Estuaries 5:2–9.

Deegan, L. A., and B. A. Thompson. 1985. The ecology of fish communities in the Mississippi River deltaic plain. Pages 35–56 *in* A. Yáñez-Arancibia, editor. Fish community ecology in estuaries and coastal lagoons. Towards an ecosystem integration. UNAM Press, Mexico City.

DeLancey, L. B. 1989. Trophic relationship in the surf zone during the summer at Folly Beach, South Carolina. Journal of Coastal Research 5:477–488.

Dittel, A. I., C. E. Epifanio, and O. Lizano. 1991. Flux of crab larvae in a mangrove creek in the Gulf of Nicoya, Costa Rica. Estuarine, Coastal and Shelf Science 32:129–140.

Dokken, Q. R., G. C. Matlock, and S. Cornelius. 1984. Distribution and composition of larval fish populations within Alazan Bay, Texas. Contributions in Marine Science, University of Texas 27:205–222.

Dorfman, D. 2000. Fish collections from the Hudson River in and around Port Liberte, New Jersey. Underwater Naturalist 25:23–24.

Dorsey, S. E., E. D. Houde, and J. C. Gamble. 1996. Cohort abundances and daily variability in mortality of eggs and yolk-sac larvae of bay anchovy, *Anchoa mitchilli*, in Chesapeake Bay. Fishery Bulletin 94:257–267.

Dovel, W. L. 1971. Fish eggs and larvae of the upper Chesapeake Bay. University of Maryland, Natural Resources Institute Special Report 4.

Dovel, W. L. 1981. Ichthyoplankton of the Lower Hudson Estuary, New York. New York Fish and Game Journal 28:21–39.

Forward, R. B., Jr., K.A.Reinsel, D.S.Peters, R.A. Tankersley, J.H. Churchill, L.B. Crowder, W.F. Hettler, S.M. Warlen, and M.D. Green. 1999. Transport of fish larvae through a tidal inlet. Fisheries Oceanography 8(Supplement 2):153–172.

Forward, R. B., Jr., and R. A. Tankersley. 2001. Selective tidal-stream transport of marine animals. Oceanography and Marine Biology: An Annual Review 39:305–353.

Govoni, J. J., and J. E. Olney. 1991. Potential predation on fish eggs by the lobate ctenophore *Mnemiopsis leidyi* within and outside the Chesapeake Bay plume. Fishery Bulletin 89:181–186.

Griffith, S. A., and D. L. Bechler. 1995. The distribution and abundance of the bay anchovy, *Anchoa mitchilli*, in a southeast Texas marsh lake system. Gulf Research Reports 9:117–122.

Hackney, C. T., and A. A. de la Cruz. 1981. Some notes on the macrofauna of an oligohaline tidal creek in Mississippi. Bulletin of Marine Science 31:658–661.

Hartman, K. J., and S. B. Brandt. 1995. Predatory demand and impact of striped bass, bluefish, and weakfish in the Chesapeake Bay: Applications of bioenergetics models. Canadian Journal of Fisheries and Aquatic Sciences 52:1667–1687.

Hellweger, F. L., A. F. Blumberg, P. Schlosser, D.T. Ho, T. Caplow, U. Lall, and H. Li. 2004. Transport in the Hudson Estuary: a modeling study of estuarine circulation and tidal trapping. Estuaries 27:527–538.

Hill, A. E. 1991. A mechanism for horizontal zooplankton transport by vertical migration in tidal currents. Marine Biology 111:485–492.

Hurst, T. P., K. A. McKown, and D. O. Conover. 2004. Interannual and long-term variation in the nearshore fish community of the mesohaline Hudson River Estuary. Estuaries 27:659–669.

Jager, Z., and H. P. J. Mulder. 1999. Transport velocity of flounder larvae (*Platichthys flesus* L.) in the Dollard (Ems Estuary). Estuarine, Coastal and Shelf Science 49:327–346.

Jessop, B. M., J.-C. Shiao, Y. Iizuka, and W.-N. Tzeng. 2002. Migratory behavior and habitat use by American eels *Anguilla rostrata* as revealed by otolith microchemistry. Marine Ecology Progress Series 233:217–229.

Johnson, W. S., D. M. Allen, M. V. Ogburn, and S. E. Stancyk. 1990. Short-term predation responses of adult bay anchovies *Anchoa mitchilli* to estuarine zooplankton availability. Marine Ecology Progress Series 64:55–68.

Jones, C. M. 2002. Age and growth. Pages 33–63 *in* L. A. Fuiman, and R. G. Werner, editors. Fishery science: The unique contributions of early life stages. Blackwell Scientific Publications, Oxford.

Juanes, F., R. E. Marks, K. A. McKown, and D. O. Conover. 1993. Predation by age-0 bluefish on age-0 anadromous fishes in the Hudson River Estuary. Transactions of the American Fisheries Society 122:348–356.

Keller, A. A., G. Klein-MacPhee, and J. S. O. Burns. 1999. Abundance and distribution of ichthyoplankton in Narragansett Bay, Rhode Island, 1989–1990. Estuaries 22:149–163.

Kimura, R., D. H. Secor, E. D. Houde, and P. M. Piccoli. 2000. Up-estuary dispersal of young-of-the-year bay anchovy *Anchoa mitchilli* in the Chesapeake Bay: Inferences from microprobe analysis of strontium in otoliths. Marine Ecology Progress Series 208:217–227.

Lapolla, A. E. 2001. Bay anchovy *Anchoa mitchilli* in Narragansett Bay, Rhode Island. 1. Population structure, growth and mortality. Marine Ecology Progress Series 217:93–102.

Limburg, K. E. 1995. Otolith strontium traces environmental history of subyearling American shad *Alosa sapidissima*. Marine Ecology Progress Series 119:25–35.

Limburg, K. E. 2002. Cohort identification. Pages 143–160 *in* L. A. Fuiman, and R. G. Werner, editors. Fishery science: the unique contribution of early life stages. Blackwell Scientific Publications, Oxford, UK.

Loos, J. J., and E. S. Perry. 1991. Larval migration and mortality rates of bay anchovy in the Patuxent River. Pages 65–76 *in* R. D. Hoyt, editor. Larval fish recruitment and research in the Americas. NOAA Technical Report NMFS 95.

Lorda, E., and S. B. Saila. 1986. A statistical technique for analysis of environmental data containing periodic variance components. Ecological Modelling 32:59–69.

Luo, J., and J. A. Musick. 1991. Reproductive biology of the bay anchovy in Cheseapeake Bay. Transactions of the American Fisheries Society 120:701–710.

MacGregor, J. M., and E. D. Houde. 1996. Onshore-offshore pattern and variability in distribution and abundance of bay anchovy *Anchoa mitchilli* eggs and larvae in Chesapeake Bay. Marine Ecology Progress Series 138:15–25.

Manooch, C. S. 1973. Food habits of yearling and adult striped bass, *Morone saxatilis* (Walbaum), from Albemarle Sound, North Carolina. Chesapeake Science 14:73–86.

Melville-Smith, R., D. Baird, and T. Wooldridge. 1981. The utilization of tidal currents by the larvae of an estuarine fish. South African Journal of Zoology 16:10–13.

Monteleone, D. M. 1992. Seasonality and abundance of ichthyoplankton in Great South Bay, New York. Estuaries 15:230–238.

Monteleone, D. M., and L. E. Duguay. 1988. Laboratory studies of predation by the ctenophore *Mnemiopsis leidyi* on the early stages in the life history of the bay anchovy, *Anchoa mitchilli*. Journal of Plankton Research 10:359–372.

Morton, T., and D. Moran. 1989. Species profiles: Life histories and environmental requirements of coastal fishes and invertebrates (Mid-Atlantic). Bay anchovy. U.S. Fish and Wildlfe Service

Biological Reports, FWS-82/11.97.

Newberger, T. A. 1989. Relative abundance, age, growth and mortality of bay anchovy in the Mid-Chesapeake Bay. M.S. Thesis. University of Maryland, College Park.

Newberger, T. A., and E. D. Houde. 1995. Population biology of bay anchovy *Anchoa mitchilli* in the mid Chesapeake Bay. Marine Ecology Progress Series 116:25–37.

Ogburn-Matthews, M., and D. M. Allen. 1993. Interactions among some dominant estuarine nekton species. Estuaries 16:840–850.

Olney, J. E. 1983. Eggs and early larvae of the bay anchovy, *Anchoa mitchilli*, and the weakfish, *Cynoscion regalis*, in lower Chesapeake Bay with notes on associated ichthyoplankton. Estuaries 6:20–35.

Orth, R. J., and K. L. Heck, Jr. 1980. Structural components of eelgrass (*Zostera marina*) meadows in the lower Chesapeake Bay - fishes. Estuaries 3:278–288.

Peebles, E. B. 2002. Temporal resolution of biological and physical influences on bay anchovy *Anchoa mitchilli* egg abundance near a river-plume frontal zone. Marine Ecology Progress Series 237:257–269.

Peebles, E. B., J. R. Hall, and S. G. Tolley. 1996. Egg production by the bay anchovy *Anchoa mitchilli* in relation to adult and larval prey fields. Marine Ecology Progress Series 131:61–73.

Purcell, J. E., D. A. Nemazie, S. E. Dorsey, E. D. Houde, and J. C. Gamble. 1994. Predation mortality of bay anchovy *Anchoa mitchilli* eggs and larvae due to scyphomedusae and ctenophores in Chesapeake Bay. Marine Ecology Progress Series 114:47–58.

Richards, S. W. 1976. Age, growth, and food of bluefish (*Pomatomus saltatrix*) from east-central Long Island Sound from July through November 1975. Transactions of the American Fisheries Society 105:523–525.

Rowe, P. M., and C. E. Epifanio. 1994. Flux and transport of larval weakfish in Delaware Bay, USA. Marine Ecology Progress Series 110:115–120.

Rozas, L. P., and C. Hackney. 1984. Use of oligohaline marshes by fishes and macrofaunal crustaceans in North Carolina. Estuaries 7:213–224.

Rozas, L. P., and T. J. Minello. 1998. Nekton use of salt marsh, seagrass, and nonvegetated habitats in a south Texas (USA) estuary. Bulletin of Marine Science 63:481–501.

Rozas, L. P., and R. J. Zimmerman. 2000. Small-scale patterns of nekton use among marsh and adjacent shallow nonvegetated areas of the Galveston Bay Estuary, Texas (USA). Marine Ecology Progress Series 193:217–239.

Schmidt, R. E. 1992. Temporal and spatial distribution of bay anchovy eggs through adults in the Hudson River. Pages 228–241 *in* C. L. Smith, editor. Estuarine research in the 1980's. State University of New York Press, Albany, NY.

Schultz, E. T., R. K. Cowen, K. M. M. Lwiza, and A. M. Gospodarek. 2000. Explaining advection: do larval bay anchovy (*Anchoa mitchilli*) show selective tidal stream transport? ICES Journal of Marine Science 57:360–371.

Schultz, E. T., K. M. M. L. Lwiza, M. C. Fencil, and J. M. Martin. 2003. Mechanisms promoting upriver transport of two species of larval fish in the Hudson River Estuary (USA). Marine Ecology Progress Series 251:263–277.

Schultz, E. T., J. Young, J. M. Martin, and K. M. M. Lwiza. 2005. Tracking cohorts: analysis of migration in early life stages of an estuarine fish. Estuaries 28:394–405.

Secor, D. H., and P. M. Piccoli. 1996. Age- and sex-dependent migrations of striped bass in the Hudson River as determined by chemical microanalysis of otoliths. Estuaries 19:778–793.

Setzler-Hamilton, E. M. 1987. Utilization of Chesapeake Bay by early life history stages of fishes. Pages 63–93 *in* S. K. Majumdar, L. W. Hall, Jr., and H. M. Austin, editors. Contaminant problems and management of living Chesapeake Bay resources. Pennsylvania Academy of Science, Easton, PA.

Snelson, F. F., Jr., and M. R. Johnson. 1995. Epibenthic fish diversity in Mosquito Lagoon: A decade of relative stability. Bulletin of Marine Science 57:284–285.

Thayer, G. W., A. B. Powell, and D. E. Hoss. 1999. Composition of larval, juvenile, and small adult fishes relative to changes in environmental conditions in Florida Bay. Estuaries 22:518–533.

Tipton, R. C. 2003. Distributional ecology of bay anchovy *Anchoa mitchilli* in the Hudson River Estuary, New York. Doctoral dissertation. West Virginia University, Morgantown, WV.

Tremain, D. M., and D. H. Adams. 1995. Seasonal variations in species diversity, abundance, and composition of fish communities in the northern Indian River Lagoon, Florida. Bulletin of

Marine Science 57:171–192.

Vouglitois, J. J., K. W. Able, R. J. Kurtz, and K. A. Tighe. 1987. Life history and population dynamics of the bay anchovy in New Jersey. Transactions of the American Fisheries Society 116:141–153.

Wang, S. B., and E. D. Houde. 1995. Distribution, relative abundance, biomass and production of bay anchovy *Anchoa mitchilli* in the Chesapeake Bay. Marine Ecology Progress Series 121:27–38.

Zastrow, C. E., E. D. Houde, and L. G. Morin. 1991. Spawning, fecundity, hatch-date frequency and young-of-the-year growth of bay anchovy *Anchoa mitchilli* in mid-Chesapeake Bay. Marine Ecology Progress Series 73:161–171.

American Fisheries Society Symposium 51:215–226, 2006

Distribution and Abundance of Bay Anchovy Eggs and Larvae in the Hudson River and Nearby Waterways

DENNIS J. DUNNING
New York Power Authority
123 Main Street, White Plains, New York 10601 USA
Dennis.Dunning@nypa.gov

QUENTIN E. ROSS
348 County Highway 2, Mount Upton, New York 13809 USA

MARK T. MATTSON
Normandeau Associates, Inc.
25 Nashua Road, Bedford, New Hampshire 03110 USA

DOUGLAS G. HEIMBUCH
AKRF, Inc.
7250 Parkway Drive, Suite 210, Hanover, Maryland 21076 USA

Abstract.—Our objectives were to examine the distribution and abundance of bay anchovy *Anchoa mitchilli* eggs and larvae in the Hudson River and nearby waterways and to determine if past conditional mortality rate (CMR) estimates for bay anchovy entrained at Hudson River power plants may be substantially biased because they were based solely on sampling in the Hudson River. We addressed these objectives by comparing ichthyoplankton samples collected in the Hudson River with those collected in New York Harbor, the East River, and Long Island Sound using the same gear during 2002. Bay anchovy eggs were collected from late April through the end of sampling in the Hudson River (early October) and through the end of sampling in nearby waterways (late July). Bay anchovy larvae were collected from early June through end of sampling in both the Hudson River and nearby waterways. The highest densities of bay anchovy eggs and larvae in nearby waterways were about 13 and 14 times greater, respectively, than the highest densities in the lower Hudson River. The peak standing crops of bay anchovy eggs and larvae in nearby waterways were about eight times larger than those in the Hudson River. Therefore, past CMR estimates for bay anchovy entrained at Hudson River power plants may be substantially biased if the bay anchovy eggs and larvae collected in the Hudson River and nearby waterways during 2002 belonged to one population, as it appears they did, and 2002 was representative of other years.

Introduction

The bay anchovy *Anchoa mitchilli* is abundant in estuaries along the Atlantic coast and serves as prey for important commercial and recreational species including striped bass *Morone saxatilis* and bluefish *Pomatomus saltatrix* (Bassista and Hartman 2005). In New York, bay anchovy are found in the New York Harbor, Long Island Sound, and the Hudson River (Smith 1985) and along the south shore of Long Island (Monteleone 1992). Considerably more is known about bay anchovy in the Hudson River than elsewhere in New York. Bay anchovy migrate into the Hudson River starting in late April and begin to spawn in May or June, depending upon water temperature (Schmidt 1992). They continue to spawn into September (Bassista and Hartman 2005) when adults are thought to leave the Hudson River or die (Schmidt 1992). Spawning generally occurs in portions of the Hudson River where the salinity is greater than or equal to 10.0 practical salinity unit (psu) (Dovel 1981). However, eggs and larvae are found throughout the Hudson River (Schmidt 1992). This may be due to adults that spawn in lower salinity water, clines in larval mortality, tidal transport upriver, or larval migration upriver (Schultz et al. 2003; Schultz et al. 2005).

The range of bay anchovy in the Hudson River is from the southern tip of Manhattan Island (river kilometer [rkm] 0) to the federal dam at Troy, New York (rkm 245). It includes the location of the Bowline (rkm 59), Indian Point (rkm 67), and Roseton (rkm 104) power plants. These power plants operate with once-through cooling systems that withdraw water directly from the Hudson River (Hutchison 1988). As a result, bay anchovy eggs, larvae, and young of year (age-0) are entrained through their cooling systems and older bay anchovy are impinged on their intake screens.

The effect of entrainment and impingement of bay anchovy at power plants on the Hudson River has been quantified as a conditional mortality rate (CMR). CMR is a measure of the reduction in year-class abundance due to entrainment and impingement if no other sources of mortality occur (Barnthouse 2000). It is analogous to the conditional fishing mortality rate used in classical fisheries science (Ricker 1975). Expressing the effect of entrainment and impingement as a CMR is useful because CMRs from multiple power plants and life stages of fish can be combined and compared and because CMRs can be entered directly into life cycle models for assessing long-term effects (Vaughn 1988). Recently, the U.S. Environmental Protection Agency used CMRs to characterize the effect of entrainment and impingement at cooling water intake structures (USEPA 2004). Nonetheless, concerns have been expressed that such system-wide CMR estimates for bay anchovy entrained and impinged at Hudson River power plants do not adequately represent the potential effects of localized depletion of bay anchovy on predators (NYSDEC 2003).

Boreman and Goodyear (1988) calculated CMR estimates for bay anchovy entrained at Hudson River power plants using the Empirical Transport Model (ETM), widely adopted by the electric utility industry (EPRI 1999). The ETM relies on the spatial and temporal distribution of organisms and the rate of water withdrawal at a cooling water intake (Boreman et al. 1981). If ETM calculations are based on only a portion of the spatial and temporal distribution of organisms in a population, the resulting CMR estimate will be biased high (Barnthouse 2000). The CMR for bay anchovy entrained collectively at the Bowline, Indian Point, and Roseton power plants was initially estimated to range between 25.9% and 65.1% annually and projected to range between

38.1% and 75.3% annually in the future, based on ichthyoplankton sampling done during 1974 and 1975 (Boreman and Goodyear 1988). However, Boreman and Goodyear (1988) acknowledged that their initial estimates were biased high because ichthyoplankton sampling was not done in all regions of the Hudson River where bay anchovy eggs and larvae were likely to be present.

The degree of bias in the Boreman and Goodyear (1988) estimates only became apparent when ichthyoplankton sampling was done throughout the tidal portion of the Hudson River starting in 1988. The sum of CMR estimates for bay anchovy entrained at the Bowline, Indian Point, and Roseton power plants each year between 1988 and 1997 was substantially lower than the Boreman and Goodyear (1988) estimates, ranging from 8.7% to 27.2% annually (Central Hudson et al. 1999). Although the estimates from 1988 through 1997 are less biased than those calculated by Boreman and Goodyear (1988), they are based solely on the distribution and abundance of bay anchovy eggs, larvae, and young of year within the Hudson River. However, bay anchovy eggs and larvae have been collected in nearby New York Harbor (Dovel 1981), the East River (PBSJ/LMS 2003), and Long Island Sound (Wheatland 1956; Richards 1959). If bay anchovy eggs and larvae collected in the Hudson River, New York Harbor, the East River, and Long Island Sound belong to one population, then the revised CMR estimates for bay anchovy entrained at the Bowline, Indian Point, and Roseton power plants are still biased high.

During 2002, ichthyoplankton sampling was done concurrently in the Hudson River, New York Harbor, the East River, and Long Island Sound using the same gear. We examined the distribution and abundance of bay anchovy eggs and larvae from that ichthyoplankton sampling to determine if past CMR estimates for bay anchovy entrained at Hudson River power plants may be substantially biased because they were based solely on sampling in the Hudson River.

Methods

Two ichthyoplankton programs were conducted independently during 2002: one to sample the estuarine portion of the Hudson River from rkm 0 to rkm 245, and the other to sample principally outside the Hudson River from lower New York Harbor through the east end of Long Island Sound, referred to as the New York Harbor/Long Island Sound (NYH/LIS) program. In the Hudson River program, ichthyoplankton sampling was done during 23 weeks from the week of March 4 through the week of October 7. We term each week of sampling a river run. Sampling was allocated among 13 regions of the Hudson River and 3 strata (shoal, bottom, and channel) (ASA 2006). We combined the 13 regions of the Hudson River into 3: the lower, middle, and upper regions. In the lower region (rkms 0–74) and the middle region (rkms 75–122), a river run was done biweekly from the weeks of March 4 through March 31, from the weeks of July 29 through October 7, and weekly during the remaining period. In the upper region (rkms 123–245), river runs were done weekly from the weeks of April 8 through July 8. The gear, 1.0-m^2 Tucker trawls and 1.0-m^2 epibenthic sleds fitted with 500-μm mesh nets, were deployed during the day except for 10 weeks from the week of May 6 through week of July 8, when they were deployed at night. Both gear types were towed against the current for 5 min. A digital flowmeter was used to measure the volume of water sampled. After sampling, the contents of the plankton net were concentrated and preserved in 10% buffered for-

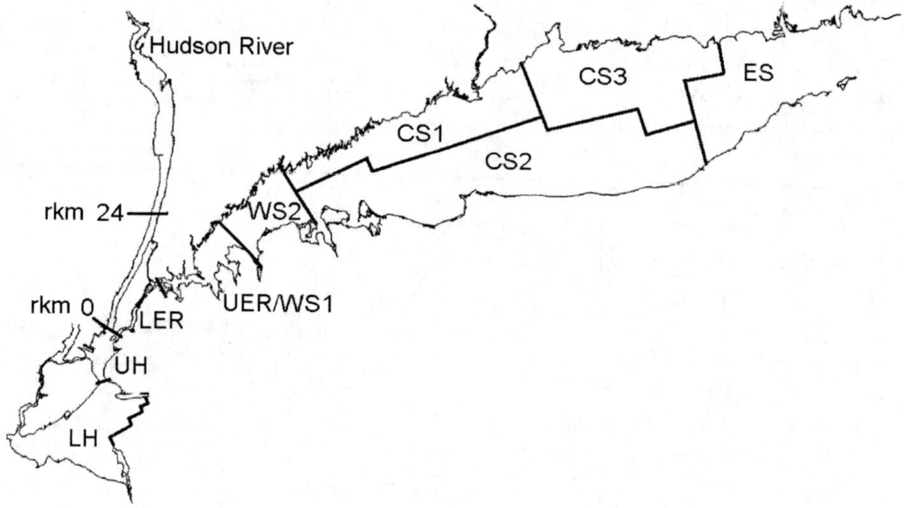

Figure 1. Regions of the New York Harbor/Long Island Sound program where ichthyoplankton sampling was done during 2002: lower New York Harbor (LH), upper New York Harbor (UH), the lower East River (LER), the upper East River and adjacent Long Island Sound (UER/WS1), western Long Island Sound (WS2), central Long Island Sound (CS1, CS2, and CS3), and eastern Long Island Sound (ES).

malin. Preserved contents were examined and all fish eggs and larvae were identified to the lowest distinguishable taxon and life stage and then counted, unless a sample contained large numbers of eggs or larvae, in which case the sample was split and a randomly selected subsample was analyzed. Bay anchovy larvae spawned during 2001 were considered juveniles on January 1, 2002, which follows the convention used to assign the life stage of fishes caught in the Hudson River Long River Survey (Heimbuch et al. 1992). Therefore, bay anchovy reported as larvae in this paper were spawned during 2002.

In the NYH/LIS program, ichthyoplankton sampling was conducted during 11 biweekly periods, each of which we term a survey, from the week of March 4 through the week of July 22 in eight out of nine regions outside Hudson River: lower New York Harbor (LH); upper New York Harbor (UH); the upper East River and adjacent western Long Island Sound (UER/WS1); western Long Island Sound (WS2); central Long Island Sound (CS1, CS2, and CS3); eastern Long Island Sound (ES); and a region of the Hudson River between rkms 0 and 24 (Figure 1). No ichthyoplankton sampling was done in the lower East River (LER) because strong tidal currents and the presence of exclusion zones established after the attack on the World Trade Center on September 11, 2001 made ichthyoplankton sampling in that region either dangerous or prohibited.

The ichthyoplankton sampling in the NYH/LIS program was conducted to help generate CMR estimates for ten representative species of fish entrained and impinged at the Charles Poletti Power Project. If bay anchovy had been among the ten, a 300-μm mesh net would have been used because bay anchovy eggs and larvae are smaller than those of the representative species. Also, ichthyoplankton sampling would have been extended beyond July because bay anchovy eggs have been collected in

Long Island Sound during August, and larvae have been collected there as late as November (Wheatland 1956).

The regions of the NYH/LIS program were divided into subareas based on bathymetry (3–6 m, >6–30 m, and >30 m) and depth strata based on location within the water column (midwater and bottom). Sampling was allocated based on a stratified random design in which region, sub-area, and depth strata defined the statistical strata. About 100 samples were generally collected during each survey. The Tucker trawls and the epibenthic sleds used in the NYH/LIS program were the same as those used for the Hudson River program and were deployed in the same way except that all tows were done during the day. Samples from each tow in the NYH/LIS program were handled like those collected in the Hudson River program except that they were split. Half of each sample was preserved separately from other samples and archived. The other half was used to create a composite sample for each region by combining samples from the region's subareas and depth strata. Composite samples were created for each survey.

The weekly densities and standing crops of bay anchovy eggs and larvae spawned during 2002 in the Hudson River were estimated using approaches previously developed (ASA 2006). The same approaches were used for estimating the densities and standing crops of bay anchovy eggs and larvae in the NYH/LIS program as in the Hudson River program except that the estimates were for a two-week survey period instead of a weekly period used in the Hudson River program.

Differences in collection efficiency for mobile ichthyoplankton collected during the day and at night, due to gear avoidance, have been reported (Clutter and Anraku 1968). Therefore, we examined density estimates for bay anchovy post-yolk-sac larvae (PYSL) collected during the day with PYSL collected at night in the Hudson River between rkms 0 and 24 to determine if they were directly comparable. We used the PYSL life stage because it was likely to be more mobile and more abundant than yolk sac larvae and thus more likely to avoid the ichthyoplankton gear. We also examined the densities of bay anchovy eggs collected during the day with eggs collected at night. Density estimates for bay anchovy eggs and PYSL in the Hudson River were calculated for the same week, gear, and depth stratum where sampling was done both during the day in the NYH/LIS program and at night during the Hudson River program. The density estimates were also pooled to produce a mean density estimate for collections made during the day and at night. The difference in the densities of eggs and of PYSL collected during the day and at night was tested using a nonparametric Wilcoxon signed-rank test in SAS (SAS 1990), based on the expectation that the density estimates would not be normally distributed.

Results

Day–Night Comparison

In the Hudson River between rkms 0 and 24, the mean density of bay anchovy eggs was 6,923 eggs/1,000 m^3 during the day and 10,462 eggs/1,000 m^3 at night, while the mean density of bay anchovy PYSL was 1,352 PYSL/1,000 m^3 during the day and 2,096 PYSL/1,000 m^3 at night. The densities of bay anchovy eggs and PYSL were not normally distributed even when a log-transformation was used. There was no significant difference in the densities of either bay anchovy eggs or PYSL collected during the day compared with those collected at night ($P > 0.05$).

Distribution and Abundance of Eggs

Bay anchovy eggs were collected in the Hudson River during every river run from the week of April 29 through the week of October 7 except during the weeks of May 13 and September 23 (Table 1). They were collected in every region of every river run from the week of June 3 through the week of June 17. The density of bay anchovy eggs was considerably higher in the lower region than in the middle and upper regions of the Hudson River during all river runs. The highest density of bay anchovy eggs in the Hudson River occurred during the July 15 river run in the lower region.

Bay anchovy eggs were collected in waterways near the Hudson River during every survey from the biweekly period starting April 29 through the biweekly period starting July 22, the last survey (Table 1). They were collected in every region of every survey from the biweekly period starting May 27 through the biweekly period starting July 22. The density of bay anchovy eggs in nearby waterways was highest in LH during the surveys of May 13, May 27, and July 22 and in UER/WS1 during the surveys of June 10, June 24, and July 8. The peak density of bay anchovy eggs in nearby waterways occurred during the survey of May 27 in LH; a nearly equal density occurred during the survey of June 10 in UER/WS1. The highest density of bay anchovy eggs in waterways nearby the Hudson River was about 13 times greater than the highest density in the lower region of the Hudson River.

The standing crop of bay anchovy eggs in the Hudson River peaked during the river run of July 15 and was greater than or equal to 2.2 billion during every river run from the week of June 3 through August 12 (Table 2). The standing crop of bay anchovy eggs in nearby waterways peaked during the survey of June 10 and was greater than or equal to 13.1 billion during every survey from the week of May 27 through the week of July 22. The peak standing crop of bay anchovy eggs in nearby waterways was 7.9 times larger than in the Hudson River.

Distribution and Abundance of Larvae

Bay anchovy larvae were collected in the Hudson River during every river run from the week of June 3 through the week of October 7 (Table 3). The density of bay anchovy larvae was always higher in the lower region than in the middle region; none were collected in the upper region. The highest density of bay anchovy larvae in the Hudson River occurred during the river run of July 15 in the lower region.

Bay anchovy larvae were collected in waterways near the Hudson River in every region of every survey from the biweekly period starting June 24 through the biweekly period starting July 22. The density of bay anchovy larvae was highest in LH during the surveys of June 10 and in UER/WS1, and WS2 during the survey of July 8. The highest density of bay anchovy larvae in waterways near the Hudson River was about 14 times greater than the highest density in the lower region of the Hudson River.

The standing crop of bay anchovy larvae in the Hudson River peaked at 4.8 billion during the river run of July 15 and was greater than or equal to 2.2 billion from the river runs of July 15 through August 12 (Table 2). The standing crop of bay anchovy larvae in waterways near the Hudson River peaked at 35.2 billion during the survey of July 8 and was greater than of equal to 7.9 billion from the surveys of June 10 through July 22. The peak standing crop of bay anchovy larvae in nearby waterways was 7.7 times larger than that in the Hudson River.

Table 1. Density of bay anchovy eggs (eggs/1,000 m^3) by week in regions of the Hudson River (lower, middle, and upper) and in regions of nearby waterways: the lower New York Harbor (LH), upper New York Harbor (UH), the upper East River and western Long Island Sound Region 1 (UER/WS1), western Long Island Sound Region 2 (WS2), central Long Island Sound Region 1 (CS1), central Long Island Sound Region 2 (CS 2), central Long Island Sound Region 3 (CS 3), and eastern Long Island Sound (ES) during 2002. A blank means no sampling was done.

	Hudson River regions			Nearby waterway regions							
Week	Lower	Middle	Upper	LH	UH	UER/WS1	WS2	CS1	CS2	CS3	ES
Mar 4	0	0		0	0	0	0	0	0	0	0
Mar 18	0	0		0	0	0	0	0	0	0	0
Apr 1	0	0		0	0	0	0	0	0	0	0
Apr 8	0	0	0								
Apr 15	0	0	0	0	0	0	0	0	0	0	0
Apr 22	0	0	0								
Apr 29	<1	0	0	0	0	0	0	13	0	0	0
May 6	1	0	0								
May 13	0	0	0	13,840	35	4	0	0	0	0	0
May 20	3	0	0								
May 27	62	<1	0	253,606	11,510	6,253	12,263	345	13	77	157
Jun 3	4,736	<1	1								
Jun 10	3,701	<1	<1	120,434	8,801	248,093	154,798	6,462	8,529	316	271
Jun 17	2,112	<1	<1								
Jun 24	2,691	<1	0	22,408	11,291	108,975	64,075	2,894	619	795	189
Jul 1	6,863	0	<1								
Jul 8				1,329	963	36,565	3,1134	72	18	2,548	98
Jul 15	19,460	<1									
Jul 22				29,504	22,112	69	418	156	171	2,075	30
Jul 29	4,372	1									
Aug 12	3,125	20									
Aug 26	11	0									
Sep 9	18	0									
Sep 23	0	0									
Oct 7	<1	0									

Table 2. Standing crop of bay anchovy eggs and larvae (in thousands) in the Hudson River and in the nearby waterways of New York Harbor, the East River, and Long Island Sound by week during 2002. A blank means no sampling was done.

Week	Eggs		Larvae	
	Hudson River	Nearby waterways	Hudson River	Nearby waterways
Mar 4	0	0	0	0
Mar 18	0	0	0	0
Apr 1	0	0	0	0
Apr 8	0		0	
Apr 15	0	0	0	0
Apr 22	0		0	
Apr 29	58	51,920	0	0
May 6	755		0	
May 13	0	4,841,700	0	0
May 20	2,931		0	
May 27	67,124	70,696,880	0	0
Jun 3	5,237,831		3,135	
Jun 10	4,044,377	183,288,480	138,773	7,920,320
Jun 17	2,242,283		226,266	
Jun 24	2,849,047	47,293,580	981,958	33,742,530
Jun 30	7,294,663		884,879	
Jul 8		13,145,030		35,168,690
Jul 15	23,349,745		4,813,293	
Jul 22		26,690,710		8,930,160
Jul 29	4,890,202		4,583,182	
Aug 12	4,189,310		2,163,038	
Aug 26	12,256		551,145	
Sep 9	20,268		220,122	
Sep 23	0		72,756	
Oct 7	117		56,153	

Discussion

The larger standing crops of bay anchovy eggs and larvae in waterways near the Hudson River were due to the greater volume of water in New York Harbor, the East River, and Long Island Sound than in the Hudson River, and higher densities of bay anchovy eggs and larvae in nearby waterways. The higher densities of bay anchovy eggs and larvae in nearby waterways and the higher densities in the lower region of the Hudson River than in the middle and upper regions may be due to a preference by spawning bay anchovy for higher salinities, higher survival of eggs and larvae at higher salinities, or both.

The spring migration of spawning bay anchovy into the Hudson River during 2002, based on the collection of eggs, appeared to begin at about the same time as the migration into Long Island Sound. However, the peak density of bay anchovy eggs in the Hudson River occurred during mid-July, about a month later than the peak density in Long Island Sound. It seems unlikely that the peak density of bay anchovy eggs

Table 3. Density of bay anchovy larvae (larvae/1,000 m^3) by week in regions of the Hudson River (lower, middle, and upper) and in regions of nearby waterways: lower New York Harbor (LH), upper New York Harbor (UH), the upper East River and western Long Island Sound Region 1 (UER/WS1), western Long Island Sound Region 2 (WS2), central Long Island Sound Region 1 (CS1), central Long Island Sound Region 2 (CS 2), central Long Island Sound Region 3 (CS 3), and eastern Long Island Sound (ES) during 2002. A blank means no sampling was done.

	Hudson River regions			Nearby waterway regions							
Week	Lower	Middle	Upper	LH	UH	UER/WS1	WS2	CS1	CS2	CS3	ES
Mar 4	0	0		0	0	0	0	0	0	0	0
Mar 18	0	0		0	0	0	0	0	0	0	0
Apr 1	0	0		0	0	0	0	0	0	0	0
Apr 8	0	0	0								
Apr 15	0	0	0	0	0	0	0	0	0	0	0
Apr 22	0	0	0								
Apr 29	0	0	0	0	0	0	0	0	0	0	0
May 6	0	0	0								
May 13	0	0	0	0	0	0	0	0	0	0	0
May 20	0	0	0								
May 27	0	0	0	0	0	0	0	0	0	0	0
Jun 3	3	0	0								
Jun 10	109	0	0	14,747	3,622	1,374	11,578	18	552	58	77
Jun 17	197	0	0								
Jun 24	912	<1	0	15,802	2,633	31,332	15,865	3,034	1,854	1,083	336
Jul 1	738	<1	0								
Jul 8			0	551	1,143	34,348	54,532	17,955	1,311	6,802	1,133
Jul 15	3,911	296									
Jul 22				10,378	1,162	9,268	2,495	627	236	2,031	1,127
Jul 29	3,453	767									
Aug 12	1,624	290									
Aug 26	330	244									
Sep 9	172	47									
Sep 23	39	33									
Oct 7	45	9									

in Long Island Sound would have occurred after sampling ended during July 2002 because the density of eggs declined during three consecutive surveys after the peak. It also seems unlikely that peak spawning occurred later in the Hudson River than in Long Island Sound because optimum water temperatures for bay anchovy spawning would occur before mid-July. Furthermore, water temperatures in the lower region of the Hudson River were generally warmer than those outside the river. Therefore, it seems more likely that the peak density of bay anchovy eggs in the Hudson River during mid-July spawning was due to an influx of spawning bay anchovy from New York Harbor to the Hudson River, perhaps due to a seasonal decline in freshwater flow.

The peak densities of bay anchovy eggs and larvae collected in Long Island Sound during 2002 occurred earlier than the peak density of bay anchovy eggs during July and the peak density of larvae during August reported by Wheatland (1956) and Richards (1959). Although it is possible that the peaks during 2002 could have occurred after ichthyoplankton sampling during July, it seems more likely that the differences between 2002 and previous years were real. This uncertainty could be eliminated if future ichthyoplankton sampling in Long Island Sound continued beyond July.

Small bay anchovy, ranging from 19.1 mm to 41.0 mm total length (TL), were collected in the Hudson River from March through May 2002. Based on developmental characteristics, these small bay anchovy would be classified as larvae, although based on the criteria used in the Hudson River Long River Survey and in our paper, which increments age up by one year on January 1, they would be classified as yearlings. These small bay anchovy were considerably larger than the first bay anchovy yolk-sac larvae collected during early June 2002 (maximum TL 2.6 mm) and the first PYSL collected during the second week of June 2002 (2.8 mm–10.5 mm TL). Although it is possible that the small bay anchovy collected during March through May were spawned very early in 2002, when the water temperature in the Hudson River and nearby waterways was well below that previously reported for spawning, it is more likely that they were larvae spawned during 2001 and survived the winter. We could not ascertain how frequently bay anchovy larvae survive during the winter because in all years except 2002 bay anchovy larvae collected after December 31 were reported as yearlings.

The appearance of substantial numbers of bay anchovy eggs and larvae during the spring of 2002 first in New York Harbor and then, at about the same time, in the Hudson River, the East River, and Long Island Sound could indicate migration into New York estuaries by one or several populations. Vouglitois et al. (1987) believed that bay anchovy migrating into estuaries along the coast of New Jersey during the spring came from one population. Furthermore, Munroe (2002) reported that bay anchovy exhibit seasonal onshore-offshore migrations in the temperate parts of its range but not substantial north-south seasonal migration. Therefore, it is reasonable to believe that the bay anchovy migrating into New York Harbor, the Hudson River, the East River, and Long Island Sound during the spring belong to one population. Stock identification techniques, such as mitochondrial DNA (Waldman et al. 1997), could provide empirical proof that the bay anchovy eggs and larvae collected in the Hudson River and nearby waterways belong to one population. These techniques could also be used to determine if the bay anchovy entering New York and New Jersey estuaries belong to one population.

If the bay anchovy in the Hudson River

and nearby waterways belong to one population, past CMR estimates for bay anchovy entrained at the Bowline, Indian Point, and Roseton power plants may be substantially biased because: 1) the peak standing crops of bay anchovy eggs and larvae in nearby waterways were 7.9 and 7.7 times greater, respectively, than those in the river, and 2) bay anchovy eggs and larvae in nearby waterways were not considered in calculating past CMR estimates. This conclusion assumes that the standing crop estimates for bay anchovy eggs and larvae in the Hudson River and nearby waterways are directly comparable. The peak standing crops of bay anchovy eggs and larvae from ichthyoplankton samples collected during 2002 in the Hudson River should be directly comparable to those collected in nearby waterways without adjustment for potential diel differences in gear avoidance and collection efficiency because the peaks occurred during river runs and surveys when ichthyoplankton sampling was done only during the day. Even when the surveys and river runs were conducted during different diel periods, the standing crops from ichthyoplankton samples should still be directly comparable because we did not detect a significant difference in the densities of either bay anchovy eggs or PYSL collected during the day and at night in the lower Hudson River. This result is consistent with the findings of Vouglitois et al. (1987) who did not detect a significant difference in the densities of bay anchovy larvae collected during the day and at night in Barnegat Bay, New Jersey.

Acknowledgments

We thank Jim Morrison and the field crews of Lawler, Matusky & Skelly Engineers LLP (LMS) and Mike Ricci and the field crews of Normandeau Associates, Inc. (NAI) for completing the ichthyoplankton sampling; Paul Moccio and the laboratory technicians of LMS and Joe Strube and the laboratory technicians of NAI for identifying the bay anchovy eggs and larvae; Paul Lindsay for performing the statistical analysis of day–night sampling; John Young and the staff of ASA Analysis and Communications for calculating and providing the estimates of density and standing crop for bay anchovy eggs and larvae in the Hudson River; and Dennis Logan and the staff of Post, Buckley, Schuh, and Jernigan, Inc. for calculating estimates of density and standing crop for bay anchovy eggs and larvae outside the Hudson River. We also thank Eric Schultz and Kim McKown for reviewing the paper and providing thoughtful and constructive comments which were very useful in making revisions. The Hudson River program during 2002 was funded by Dynegy Roseton LLC, Entergy Nuclear Indian Point 2 LLC, Entergy Nuclear Indian Point 3 LLC, and Mirant Bowline LLC. The NYH/LIS program was funded by the New York Power Authority.

References

ASA (ASA Analysis and Communication). 2006. 2004 Year Class Report for the Hudson River Estuary Monitoring Program. Prepared for Central Hudson Gas and Electric Corporation, New York.

Barnthouse, L. W. 2000. Impacts of power plant cooling systems on estuarine fish populations: the Hudson River after 25 years. Environmental Science and Policy 3:S341–S348.

Bassista, T. P., and K. J. Hartman. 2005. Reproductive biology and egg mortality of bay anchovy, *Anchoa mitchilli*, in the Hudson River estuary. Environmental Biology of Fishes 73:49–59.

Boreman, J., C. P. Goodyear, and S. W. Christensen. 1981. An empirical methodology for estimating entrainment losses at power plants sited on estuaries. Transactions of the American Fisheries Society 110:253-260.

Boreman, J., and C. P. Goodyear. 1988. Estimates of entrainment mortality for striped bass and other fish species inhabiting the Hudson River Estuary. Pages 152–160 *in* L. W. Barnthouse, R. J. Klauda, D. S. Vaughan, and R. L. Kendall, editors. Science, law, and Hudson River Power: a case study in environmental impact assessment. American Fisheries Society, Monograph 4, Bethesda, Maryland.

Central Hudson Gas & Electric Corp., Consolidated Edison Co. of New York, Inc., New York Power Authority, and Southern Energy New York. 1999. Draft Environmental Impact Statement for State Pollutant Discharge Elimination System Permits for Bowline Point, Indian Point 2&3, and Roseton Steam Electric Generating Stations. New York State Department of Environmental Conservation, Albany.

Clutter, R. I., and M. Anraku. 1968. Avoidance of samplers. Pages 57–76 *in* D. J. Tanter, editor. Monographs on oceanographic methodology 2. Zooplankton sampling. UNESCO, Paris.

Dovel, W. L. 1981. Ichthyoplankton of the lower Hudson Estuary, New York. New York Fish and Game Journal 28:21–39.

Electric Power Research Institute (EPRI). 1999. Catalog of Assessment Methods for Evaluating the Effects of Power Plant Operations on Aquatic Communities. Palo Alto, California.

Heimbuch, D. G., D. J. Dunning, and J. R. Young. 1992. Post-yolk-sac larvae abundance as an index of year class strength of striped bass in the Hudson River. Pages 376–391 *in* C. L. Smith, editor. Estuarine research in the 1980s. State University of New York Press, Albany.

Hutchison, J. B. 1988. Technical descriptions of Hudson River electricity generating stations. Pages 113–120 *in* L. W. Barnthouse, R. J. Klauda, D. S. Vaughan, and R. L. Kendall, editors. Science, law, and Hudson River Power: a case study in environmental impact assessment. American Fisheries Society, Monograph 4, Bethesda, Maryland.

Monteleone, D. M. 1992. Seasonality and abundance of ichthyoplankton in Great South Bay, New York. Estuaries 15:230–238.

Munroe, T. A. 2002. Anchovies. Family Engraulidae. Pages 105–111 *in* B. B. Collette and G. Klein-MacPhee, editors. Bigelow and Schroeder's fishes of the Gulf of Maine. Smithsonian Institution Press, Washington, D.C.

PBSJ/LMS (Post, Buckley, Schuh and Jernigan/Lawler, Matusky, and Skelly Engineers Jojnt Venture). 2003. Charles Poletti Power Project Studies to Determine the Effects of Entrainment and Impingement. Prepared for the New York Power Authority, White Plains, New York.

Richards, S. W. 1959. Pelagic fish eggs and larvae of Long Island Sound. Bulletin of the Bingham Oceanographic Collection 17:95–124.

Ricker, W. E. 1975. Computation and interpretation of biological statistics of fish populations. Fisheries Research Board of Canada Bulletin 191.

SAS Institute Inc. (SAS). 1990. SAS/STAT User's Guide. SAS Institute Inc., Cary, NC.

Schmidt, R. E. 1992. Temporal and spatial distribution of bay anchovy eggs through adults in the Hudson River Estuary. Pages 228–241 *in* C. L. Smith, editor. Estuarine Research in the 1980s. State University of New York Press, Albany.

Schultz, E. T., K. M. M. Lwiza, M. C. Fencil, and J. M. Martin. 2003. Mechanisms promoting upriver transport of larvae of two fish species in the Hudson River estuary. Marine Ecology Progress Series 251:263–277.

Schultz, E. T., J. Young, J. M. Martin, and K. M. M. Lwiza. 2005. Tracking cohorts: analysis of migration in the early life stages of an estuarine fish. Estuaries 29:394–405.

Smith. C. L. 1985. The inland fishes of New York State. The New York State Department of Environmental Conservation, Albany.

USEPA (U.S. Environmental Protection Agency). 2004. National pollutant discharge elimination system—Final regulations to establish requirements for cooling water intake structures at Phase II existing facilities, final rule. Federal Register 69 (9 July 2004):41575–41693.

Vaughn, D. S. 1988. Introduction. Pages 121–123 *in* L. W. Barnthouse, R. J. Klauda, D. S. Vaughan, and R. L. Kendall, editors. Science, law, and Hudson River Power: a case study in environmental impact assessment. American Fisheries Society, Monograph 4, Bethesda, Maryland.

Vouglitois, J. J., K. W. Able, R. J. Kurtz, and K. A. Tighe. 1987. Life history and population dynamics of the bay anchovy in New Jersey. Transactions of the American Fisheries Society 116:141–153.

Waldman, J. R., R. A. Richards, W. B. Schill, I. Wirgin, and M. C. Fabrizio. 1997. An empirical comparison of stock identification techniques applied to striped bass. Transactions of the American Fisheries Society 126:369–385.

Wheatland, S. B. 1956. Oceanography of Long Island Sound, 1952–1954. VII. Pelagic fish eggs and larvae. Bulletin of the Bingham Oceanographic Collection 15:234–314.

American Fisheries Society Symposium 51:227–249, 2006

Distributional Ecology of Bay Anchovy *Anchoa mitchilli* in the Hudson River Estuary

Ronald C. Tipton
United States Fish and Wildlife Service, Maryland Fishery Resources Office
177 Admiral Cochrane Drive, Annapolis, Maryland 21401 USA
clif_tipton@fws.gov

Kyle J. Hartman
West Virginia University, Wildlife and Fisheries Resources Program
322 Percival Hall, Morgantown, West Virginia 26506-6125 USA

Abstract.–Bay anchovy are the most abundant fish found along the U.S. Atlantic coast. This coupled with trophic position and importance to piscivores makes them ecologically vital. However, the anchovy's small size renders them vulnerable to impingement and entrainment at industrial water intakes. Therefore, knowledge of anchovy spatial distributions relative to such intakes is critical to understanding anthropogenic effects on this species. To evaluate the distributional ecology of bay anchovy in the Hudson River Estuary we conducted seasonal hydroacoustic surveys to define regional density and total abundance of larval, juvenile, and adult anchovy in the river from 1996–1998. Adults concentrated in the lower river during prespawning periods, utilizing the lower 50 km during spawning. Adults dispersed within the river and then emigrated during fall and winter. Larvae and juveniles were in highest densities upriver of spawning locations, with a downriver shift in density distribution during fall. Few juveniles overwintered in the river. Abundance of all stages peaked in spawning season, 1997, when abundance of adults, juveniles, and larvae were estimated at 150 million, 3 billion, and 5.5 billion individuals, respectively. Distributional patterns suggest a panmictic population that includes the Hudson River Estuary and nearby coastal areas, but with possible age segregation. These findings suggest population impacts to anchovy within the River may be distributed on a more regional than a local scale.

Introduction

Bay anchovy *Anchoa mitchilli* is the numerically dominant species in fisheries research collections in Chesapeake Bay and Hudson River estuaries (Dovel 1981; Byrne 1982; Gallagher et al. 1983; Horwitz 1987; Houde and Zastrow 1991; Brandt 1992). In estuaries, bay anchovy are a significant component of the pelagic food web, serving as prey to larger, commercially and recreationally important piscivores such as striped bass *Morone saxatilis*, bluefish *Pomatomus saltatrix*, and weakfish *Cynoscion regalis* (Schaefer 1970; Merriner 1975; Texas Instruments 1976; Gardinier and Hoff 1982; Friedland et al. 1988; Hartman and Brandt 1995). Bay anchovies are also major consumers of zooplankton (Klebasko 1991). In the Hudson River it has been suggested that bay anchovy act as a buffer species and when densities are low, piscivores may switch to feeding on the young of each other (Texas Instruments 1976). Thus, reductions in anchovy densities may be indirectly responsible for increased predation on young striped bass, bluefish, or other important species.

Bay anchovy abundance may significantly affect fish community structure as well as the quality and quantity of secondary production in the Chesapeake Bay (Wang and Houde 1994). Wang (1992) demonstrated that variability in future spawning biomass of the bay anchovy in Chesapeake Bay is most sensitive to factors operating during two periods of its annual cycle: (1) mid-July to October, the period of spawning to recruitment, when predation on bay anchovy is believed to be high (Baird and Ulanowicz 1989), and (2) winter, when mortality may occur from predation or starvation. Size selective overwintering mortality may kill smaller fish in the northern part of their range, although it is difficult to conclusively separate possible size-selective mortality from migration (Wang 1992). Bay anchovy may act as a buffer to production in both a bottom-up and top-down manner (Luo and Brandt 1993), making it a key species in the pathway from plankton production to harvestable fish (Baird and Ulanowicz 1989).

Bay anchovy distribution and abundance varies temporally (Vouglitois et al. 1987; EA 1991; Newberger and Houde 1995; Wang and Houde 1995; Houde 1996; Bassista 2000), with recruitment success linked to food availability (patchiness) and predators (McGurk 1986; Luo 1993; Houde 1996), variable physical factors of the environment, such as temperature, salinity, dissolved oxygen, discharge (Vouglitois et al. 1987; Limburg and Schmidt 1990; EA 1991; LMS 1992; Schmidt 1992; Bassista 2000) and entrainment (Boreman and Goodyear 1988; Morgan et al. 1995). Therefore, studies aimed at determining distribution and abundance trends of bay anchovy are essential to the management and understanding of estuarine systems. In the Hudson River Estuary, describing some of the variability in spatial distribution of bay anchovy can identify areas that are consistently used by them. Subsequent consideration of these areas in management decisions concerning human activity, with the potential to negatively affect survival and recruitment of bay anchovy, could benefit the entire fish community.

The purpose of this study was to gain insight into seasonal patterns of bay anchovy presence and density over 3 years (1996–1998), in an effort to describe interannual variability. Goals of this study are:

1) To identify areas within the estuary that were consistently used by bay anchovy, with emphasis on determining seasonal patterns;

2) To determine if bay anchovy are continuously distributed between the Hudson River and adjacent waters such as Long Island Sound and Newark Bay;

3) To compare Hudson River total abundance estimates among years to capture some of the year-to-year variability in bay anchovy production; and

4) To validate acoustic performance through comparison with results from a study using ichthyoplankton survey techniques.

Methods

Hydroacoustics

We used hydroacoustics to estimate bay anchovy abundance in the Hudson River (May 1996–September 1998). Acoustic data were collected continuously over zig-zag cross-river transects (Brandt 1996) using a Simrad EY 500 echosounder equipped with a 120 kHz split-beam transducer. The echosounder digitally controls pulse length and pulse repetition rate. Time varied gain is also controlled to correct for signal spreading and absorption through the water column. The transducer was towed near-surface via a vessel mounted boom and tow body. A blanking distance 1 m from the transducer face was necessary to allow acoustic beams, from multiple pressure sources within the transducer, to coalesce into a single sound wave. Because of this, targets in the near-field of the transducer cannot be detected. Transducer blanking distance coupled with a tow body depth of approximately 0.5 m effectively eliminates the top 1.5 m of water from sampling. An off-bottom integration value of 0.5 m was set to prevent integration of the bottom during postprocessing of digital acoustic data. This combined with losses at the surface eliminates approximately 2.0 m of total water column. Within these boundaries, fish densities were averaged over constant time intervals throughout the water column using 1 m depth intervals.

Echo-squared integration and target strength analysis (Foote et al. 1986; Foote 1987) of acoustic data were performed using Digital Echo Visualization and Information System (DEVIS) software (Jech and Luo 2000) to determine density of acoustic scatterers (fish). Fish lengths were initially estimated from measured target strengths using equations for bay anchovy established in the Chesapeake Bay (Brandt 1992) and were refined based on target strength for bay anchovies sampled within the Hudson River. The result was a modified target strength–length equation for bay anchovy in the Hudson River: $TS = 14.7 \, Log_{10}L_{cm} - 63.5$, where TS = target strength (dB) and L_{cm} = fish total length (cm). Bay anchovy densities were estimated on the basis of size classes corresponding to prejuvenile anchovy, hereafter referred to as larval anchovy, 10–21 mm TL (–63.5 to –58.76 dB, size 1); juvenile anchovy 22–49 mm TL (–58.75 to –53.35 dB, size 2); and adult anchovy 50–99 mm TL (–53.34 to –48.86 dB, size 3). Species generally too large to be bay anchovy were included in the 100–1000 mm TL size-class (>–48.8 dB, size 4). Larval anchovy abundances were restricted to a minimum of 10 mm TL due to insufficient resolution of targets below that size with a 120 kHz echosounder.

Density estimates for bay anchovy were determined using a correction factor based on species composition from research trawls. We conducted trawls and hydroacoustics during night when bay anchovy schools disperse (Luo 1993), making them more susceptible to hydroacoustic methods. This behavior is beneficial to hydroacoustic surveys as it targets most species at a time when they are likely to be pelagic and thus susceptible to acoustic gear. Bay anchovy density estimates ($\#/m^2$) were calculated for each geographic region sampled (1996–1998).

Acoustic surveys were performed along individual transects from the near-shore coastal area, east of the Lower New York Bay, and within Lower New York Bay (Lower Bay) itself. Individual transects were also surveyed in Jamaica Bay. Continuous zig-zag transects were surveyed in the East River (to Execution Rock in western Long Island Sound), Arthur Kill, Kill Van Kull, Upper

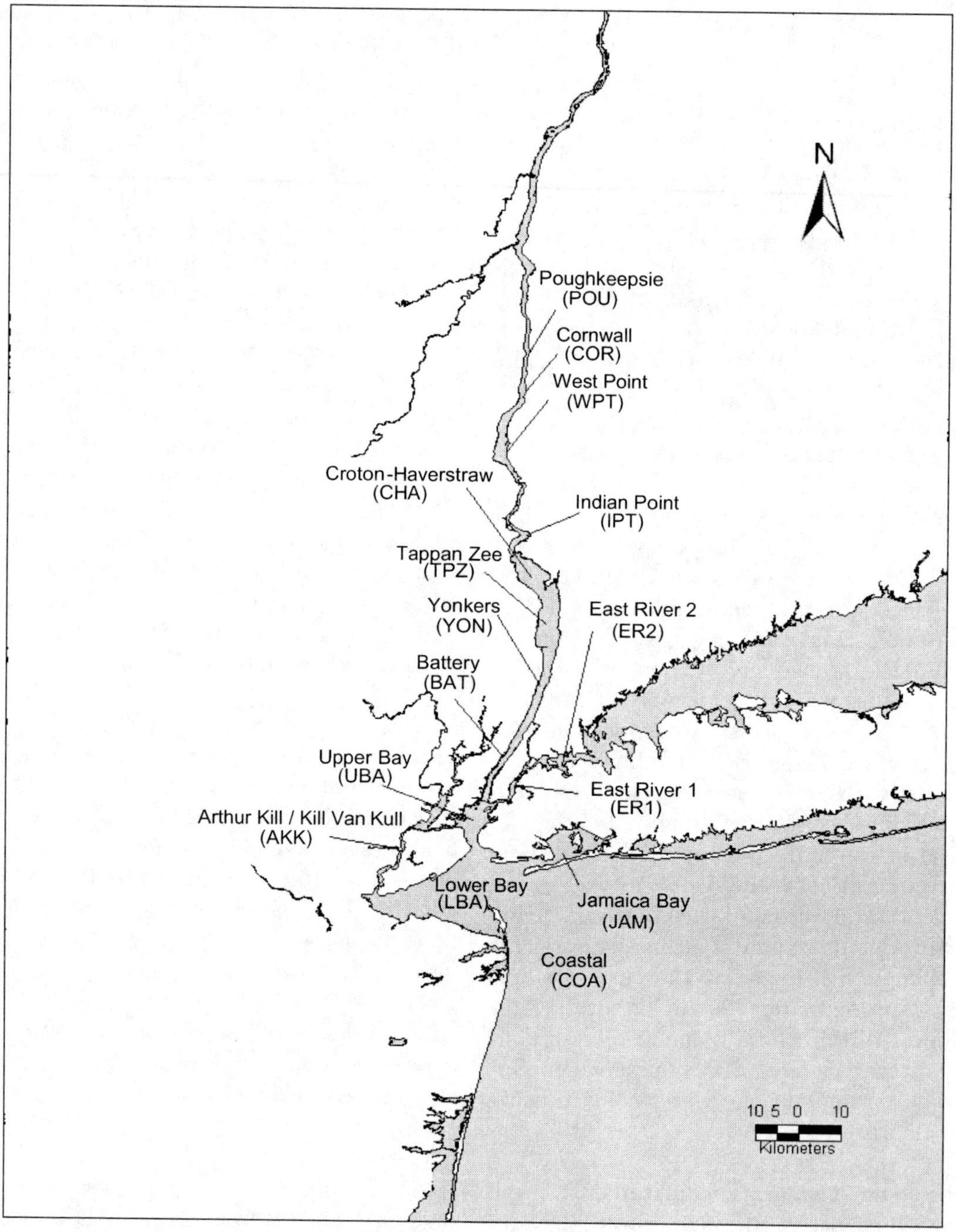

Figure 1. Map of regions surveyed by hydroacoustics and midwater trawls within the Hudson River Estuary and adjacent coastal areas during 1996-1998.

New York Bay (Upper Bay), and in the Hudson River proper (Figure 1). We designated these areas as peripheral regions because they lie outside the Hudson River. Separation of peripheral regions from Hudson River regions occurs at the Upper

Bay and Battery region border. By convention (Barnthouse et al. 1988), the Hudson River proper begins off the tip of Battery Park (Manhattan, New York) (Battery region, river kilometer [rkm] 0–18) and continues north to the Albany region (rkm 199–243). We adhered to these predefined regions up to and including the Poughkeepsie region (rkm 99–122), which was the northern (upstream) boundary for this study. Survey effort continued north in the river to the distance at which bay anchovy were no longer present in great number and never exceeded the Poughkeepsie region.

We used trawl data in size-class bins corresponding to bay anchovy functional size groups, listed above, to correct areal (fish/m^2) density estimates to bay anchovy density estimates. We assumed a correction factor of 1.00 for larval bay anchovy. Our decision was based on the dominance of bay anchovy in this size range from July to August within the Hudson River Estuary (Bassista 2000). Juvenile and adult bay anchovy estimates were corrected via trawl-based percentages for each region and season. This designation of the 100–1000 mm TL size bin relates to fish often too large to include bay anchovy within the Hudson River Estuary and sizes for which the small trawl did not sample efficiently. For this reason we did not include this size range in acoustic results for bay anchovy.

Surveys were conducted once a month May–December 1996 and were conducted twice monthly May–September 1997. During 1998, surveys were restricted to once a month during May, July, August, and September, respectively. In 1998, logistical and equipment problems prevented the June survey as well as duplication of the fine-scale, bimonthly, sampling effort employed in 1997. To facilitate recruitment of bay anchovy to sampling gear all surveys were conducted between 0.5 h after sunset to 0.5 h before sunrise.

Trawling

Since the use of acoustics did not distinguish species, midwater trawls were used to estimate species composition for each region sampled acoustically. Due to logistical limitations trawls and acoustics were usually not conducted simultaneously. Trawling was performed within 0–2 d of an acoustic survey in an attempt to obtain samples adequately representing fish populations censused by the acoustics. Because we needed a large number of trawls to correct acoustic results for bay anchovy, trawling effort (time) was recorded but not standardized to a specific period.

Two trawl nets were used during sampling depending on the size and availability of chartered research vessels. The larger midwater trawl was deployed from the 16 m long R/V Walford (New Jersey Marine Sciences Consortium, Sandy Hook, New Jersey). The smaller otter trawl was fished primarily from the 9 m long R/V Victoria (New Jersey Marine Sciences Consortium, Sandy Hook, New Jersey). The large midwater trawl had a 6.40 m headrope, 6.40 m footrope, 7.62 cm stretch (mouth), 5.08 cm stretch mesh (lower body), and 3.00 mm stretch in the middle and cod end liner. The smaller net was a 4.88 m semiballoon otter trawl with a 5.18 m headrope, 6.40 m footrope, 3.81 cm stretch mesh (mouth), and 3.00 mm stretch mesh in the middle and cod end liner. Both trawls were fished obliquely from surface to near bottom, in timed steps, to effectively sample the water column. Trawl times varied from 5 to 15 min total fishing time per trawl. Trawl time varied depending on water depth, which increased or decreased the number of trawl depth settings from the surface to the bottom.

For purposes of this study, estimates of fish density are split geographically to reflect temporal density and distribution of bay anchovy within the Hudson River (river proper regions) and areas outside the river (peripheral regions) on a finer scale. We conducted midwater trawling in 15 regions of the Hudson River Estuary. Eight regions were surveyed from the Battery region north (river proper). Seven regions were sampled outside the river (peripheral regions). Within these two categories, regions consistent with those established in earlier, utility-sponsored studies were used when possible, and are the scale basis for comparisons of bay anchovy mean regional density and distribution in this study. References herein to the lower, middle, and upper river apply to: Battery, Yonkers, Tappan Zee (lower); Croton-Haverstraw, Indian Point (middle); and West Point, Cornwall, Poughkeepsie (upper) river regions, respectively.

We formed seasonal groupings based on the reproductive cycle for bay anchovy in the Hudson River (Bassista 2000; Schmidt 1992) which correspond to prespawning, peak-spawning, and postspawning. The presence of eggs in ichthyoplankton samples suggested that in 1996 and 1998, prespawning included May and June samples, peak-spawning included July and August, and postspawning included samples from September through November (Bassista 2000). The presence of eggs (and therefore spawning) occurred earlier in 1997 (Bassista 2000) and for this reason data were processed using May as prespawn, June through August as peakspawn, and September or later as postspawn seasons for 1997.

Our trawl was equipped with a 3 mm mesh cod end liner that likely did not sample larval anchovy effectively. For this reason, their contribution to trawl catch was considered suspect, resulting in their omission here. However, Bassista (2000) found that from June through September 1996–97, greater than 99% of all larval sized fish in larval net tows were bay anchovy. For this reason, we present acoustically derived larval bay anchovy density estimates without trawl correction. Tomcod *Microgadus tomcod* and alosines occur sympatrically, but at much lower numbers than bay anchovy. Catches of these two species were included in the trawl catch composition figures and all other species encountered during the study were placed in an "other" category.

Specimens collected in trawls were immediately identified, enumerated, and measured to the nearest mm TL, time permitting. Otherwise, species were enumerated, weighed, and frozen. Frozen samples were returned to the laboratory where a subsample, not exceeding one hundred randomly chosen individuals of each species, was measured to the nearest mm TL.

Data Analysis

Post processing of acoustic data (using DEVIS) resulted in size-specific density estimates (fish/m^3) for bay anchovy in 1-m depth intervals by approximately 17.14 m of horizontal distance along the survey route. Bins of this resolution result from an approximate average ship speed of 5.00 knots (2.57 m/s), an acoustic sampling rate of 3.00 pings/s, and raw data storage at 20.00 ping intervals (6.67 s/data storage). To facilitate interpretation and statistical analysis, fish density (fish/m^3) estimates within data bins were summed vertically (among sampled depth strata) yielding density estimates of mean fish/m^2 of surface area. Areal densities were then corrected proportionally for bay anchovy from trawl data. A geographic information system (ArcView software) was used to map the study area, set regional boundaries, and calculate regional surface area (m^2) statistics for use in converting mean density estimates to total abundance estimates. Map data covering the study area were obtained using a coastline extractor tool from the National Oceanic and Atmospheric Administration's web site.

Given the nearly continuous nature of hydroacoustic sampling and the large geographic scale of this study, there exists the potential for spatial autocorrelation within postprocessed data. To obtain means from independent samples we fit semivariograms to a subset of data using various nonlinear models with Statistical Analytical Software (SAS 1999). Data used in variogram modeling originated from two regions (Tappan Zee and Croton-Haverstraw) and focused on adult bay anchovy (50–99 mm TL) during the 1997 spawning season (July and August). This subset of data were chosen because trawls and egg collections (Bassista and Hartman 2005) showed adult bay anchovy were concentrated in this area at the time, increasing their susceptibility to sampling effort. Variogram analysis was performed in an effort to obtain an unbiased estimate of the distance (observations) at which samples may be considered spatially independent (Hohn 1988; Rivoirard et al. 2000). We fit nugget and nonnugget versions of Gaussian, spherical, and exponential models to variogram data (Hohn 1988; Rivoirard et al. 2000). The spherical model with nugget provided the best-fit model for sub-sampling acoustic data. Sub-sampling, without replacement, was initiated from a randomly selected location within the data set and progressed bidirectionally throughout.

We tested for significant ($p < 0.05$) differences in mean areal density (fish/m^2), from sub-sampled data, within Hudson River regions and within peripheral regions using general linear models (Proc GLM) in PC SAS. Under each of the broad geographic classifications tests were run within a given year, season, and size-class. We used Tukey's studentized range (HSD) post comparison test on LS means to determine significant differences among bay anchovy regional mean densities. Ninety-five percent confidence intervals were calculated for means using

$$\mu = \bar{x} \pm 1.96\left(\frac{\sigma}{\sqrt{n}}\right)$$

where μ = the population mean, σ = the sample mean, s = sample variance, and n = the number of observations.

Results

We completed a total of 71, 227, and 90 trawls in 1996, 1997, and 1998, respectively. Bay anchovy were the dominant species encountered, constituting 72.23% of the total catch. Atlantic tomcod and *Alosa* spp. were next in abundance and accounted for 12.94 and 4.54% of the total catch, respectively. Trawls were used to partition species observed acoustically and their results are reported in Figures 2–4.

Seasonal Population Abundance

Population estimates of adult bay anchovy within the Hudson River showed seasonal as well as annual patterns (Figure 5). Adult numbers were lowest in the prespawn period, peaked during spawning season at 5–15 times prespawn levels, and declined following spawning (postspawn) each year. River-wide prespawn total abundance estimates indicate that more adult bay anchovy had entered the river in 1997 compared to 1996 and 1998, though the 1996 estimate did not differ significantly from 1997. River-wide adult total abundance estimates during the spawning seasons were similar between 1996 and 1998, but during 1997 the amounts of adult bay anchovy in the river were substantially greater. This trend persisted to the postspawn season with total river adult abundance being far

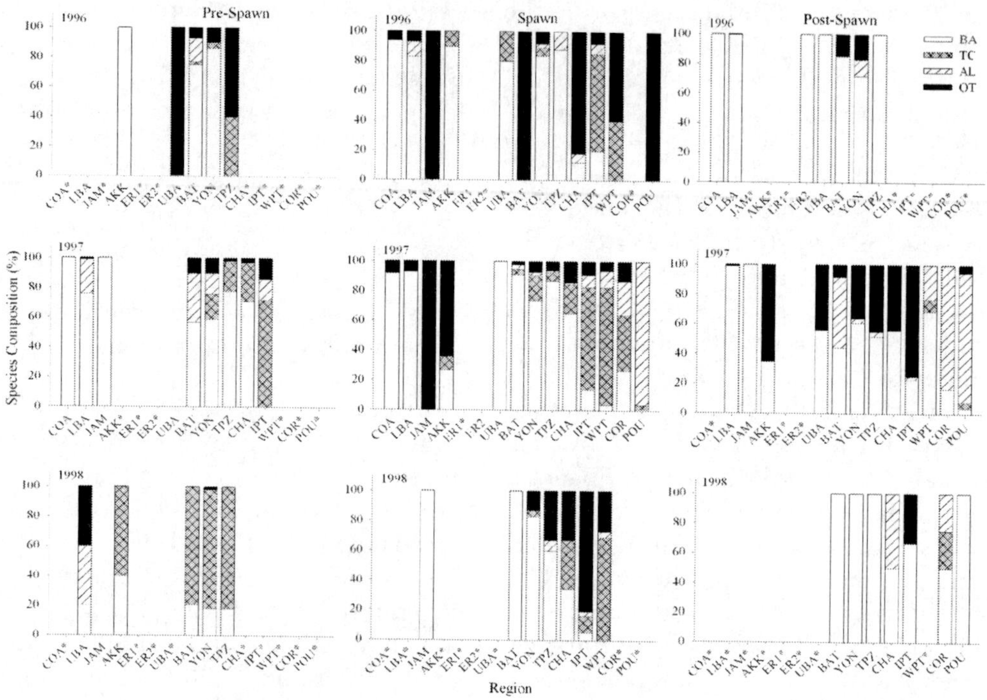

Figure 2. Percent trawl composition of adult anchovy-sized species (50 – 99.9 mm) from the Hudson River Estuary. BA = bay anchovy, TC = tomcod, AL = alosines, OT = other species. See Figure 1 for definitions of region abbreviations, * indicate unsampled regions.

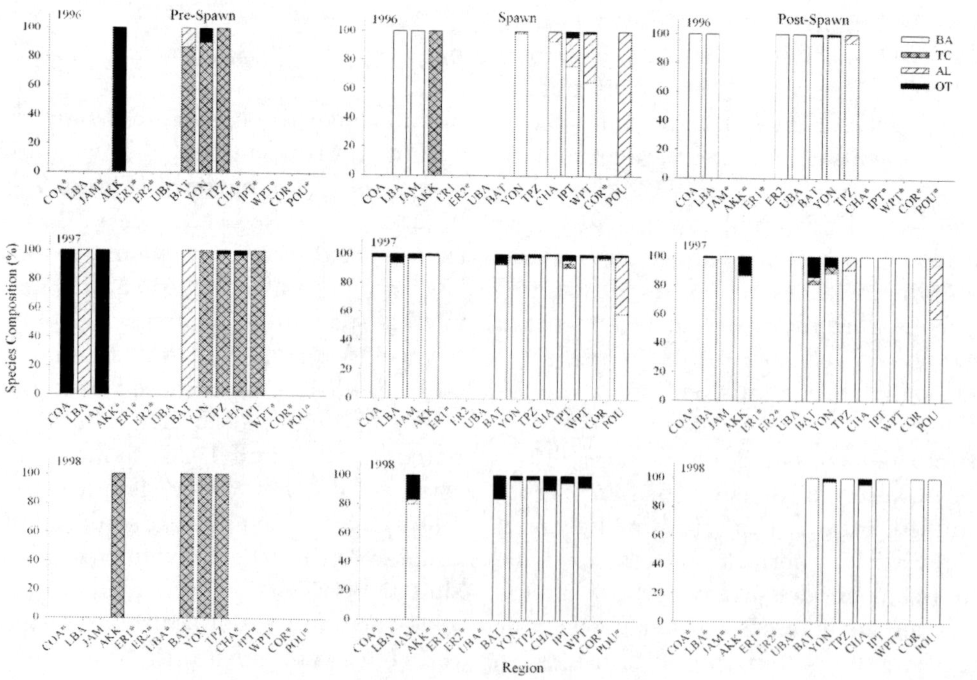

Figure 3. Percent trawl composition of juvenile anchovy-sized species (22 – 49.9 mm) from the Hudson River Estuary. BA = bay anchovy, TC = tomcod, AL = alosines, OT = other species. See Figure 1 for definitions of region abbreviations, * indicate unsampled regions.

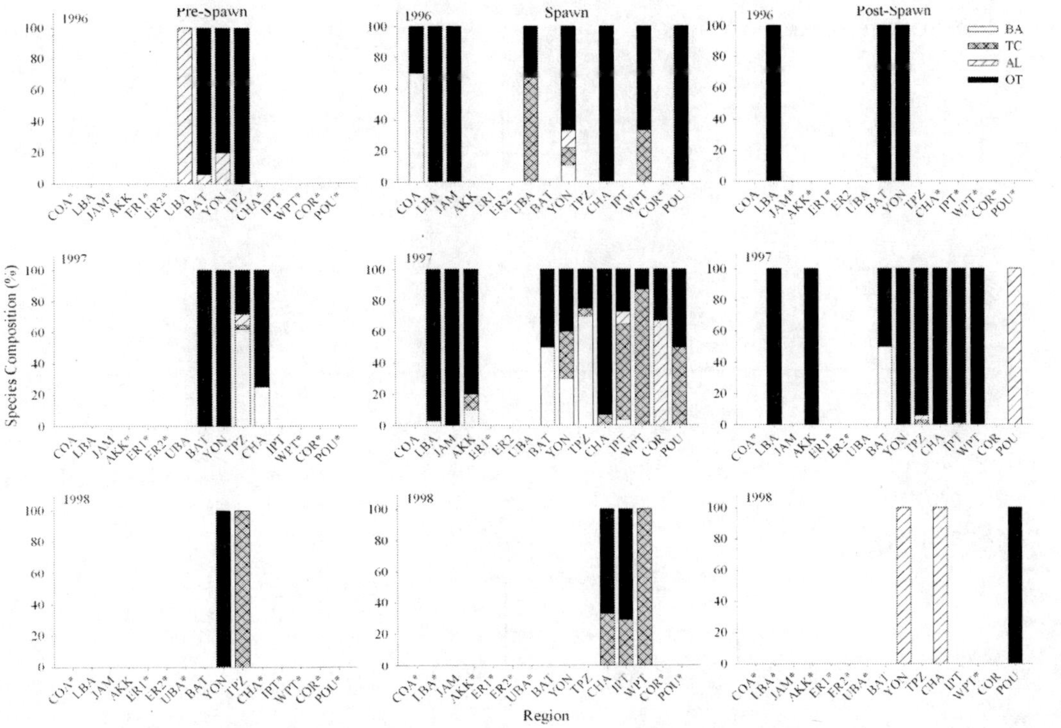

Figure 4. Percent trawl composition of fish larger than adult-sized bay anchovy (100 – 1000 mm) from the Hudson River Estuary. BA = bay anchovy, TC = tomcod, AL = alosines, OT = other species. See Figure 1 for definitions of region abbreviations, * indicate unsampled regions.

greater during 1997 than in 1996 and 1998.

Larval and juvenile anchovy abundance also showed seasonal and annual differences. No larval or juvenile anchovy were collected during prespawning season leading to abundance estimates of 0. Spawning season trends in bay anchovy production of larvae and juvenile within the river mirrored those for adults, with 1997 yielding the greatest numbers of each size-class (Figures 6–7). Larvae and juvenile production from within the river was lower during the 1996 spawning season than in 1998, despite their similarity in adult abundance. By the postspawn season production of larval bay anchovy had ceased in 1996 and persisted at a low level in 1997 compared to 1998 (Figure 6). By contrast, juveniles remaining in the river during the postspawn season were markedly higher for 1997 compared to 1996 and 1998, respectively (Figure 7).

Anchovy Distributional Density

Adults

Although adult anchovy were typically collected in all locations and seasons, density estimates suggest certain areas represented population focal points. Prespawn season trends in adult bay anchovy density were variable over the course of the study, but often were highest from the southern Lower Bay and lower Hudson River (Figure 8). During the spawning season, high densities of adults were found in the Lower Bay (1996–97) and Jamaica Bay (1998), while within the Hudson River, the Yonkers, Tappan Zee, and Battery regions accounted

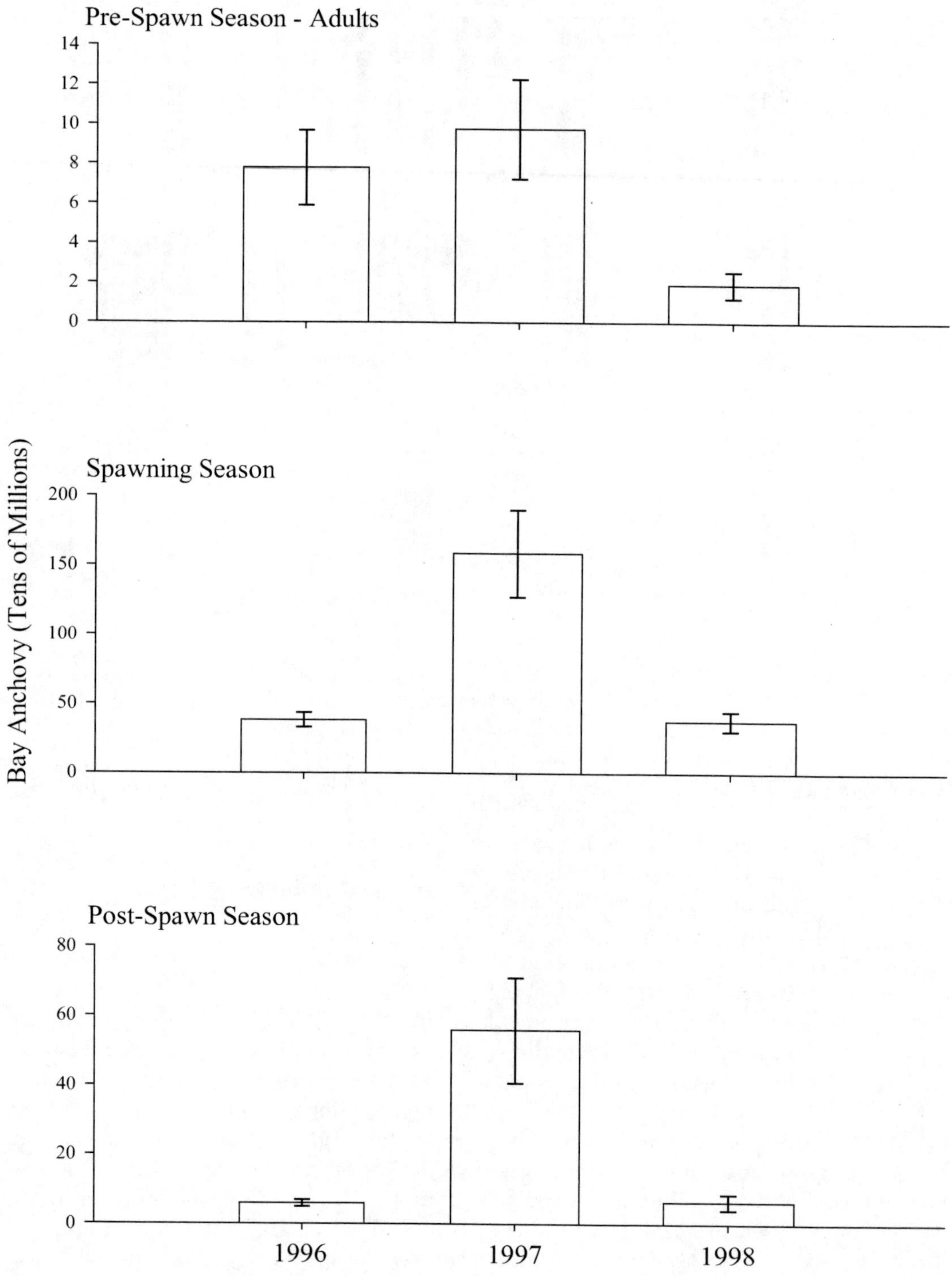

Figure 5. Total abundance estimates of adult bay anchovy from within the Hudson River by year and season. Error bars are 95% confidence intervals. Y-axis differs between seasons.

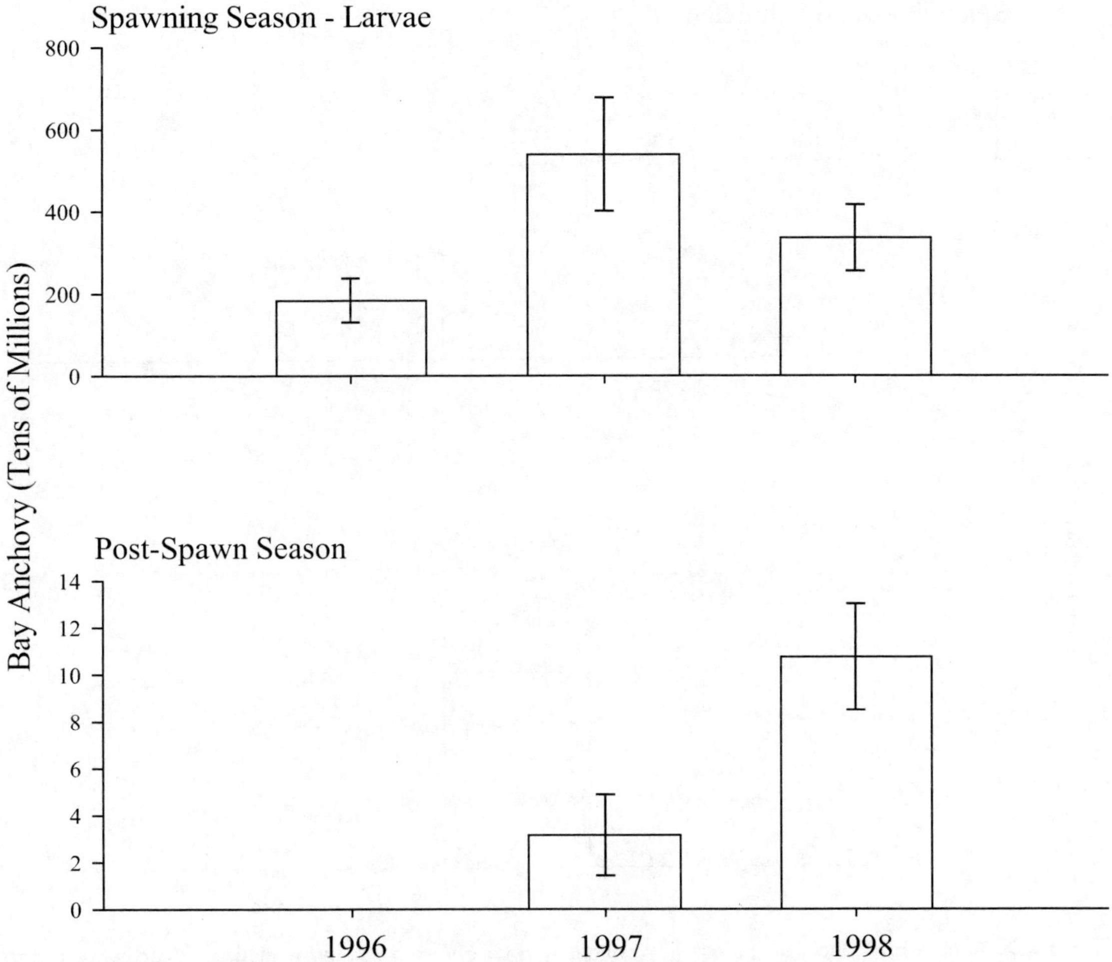

Figure 6. Total abundance estimates of larval bay anchovy from within the Hudson River by year and season. Error bars are 95% confidence intervals. Y-axis differs between seasons.

for the highest densities of adults (Figure 9). Post spawning distribution of adult anchovy was more equivocal with peaks in the lower bay in 1996 and 1997 and smaller peaks in the lower river each year (Figure 10).

Larvae

During the spawning season larval bay anchovy were detected in all surveyed locations, but patterns in their distribution were evident. When surveyed, coastal areas and the Lower Bay held higher densities of larval anchovy than Jamaica Bay (Figure 11). Other peripheral areas in the East River, Arthur Kill and Kill Van Kull contained larval anchovy, although the density in these areas was lower than in many of the Hudson River regions. River distributions of larvae differed among years. In 1996, highest densities were between Yonkers and Croton-Haverstraw. However, in 1997 and 1998 peak larval densities occurred in Yonkers and Indian Point (Figure 11).

During postspawning season few larval anchovy were detected in the Hudson River (Figure 12). However, a peak in the Lower

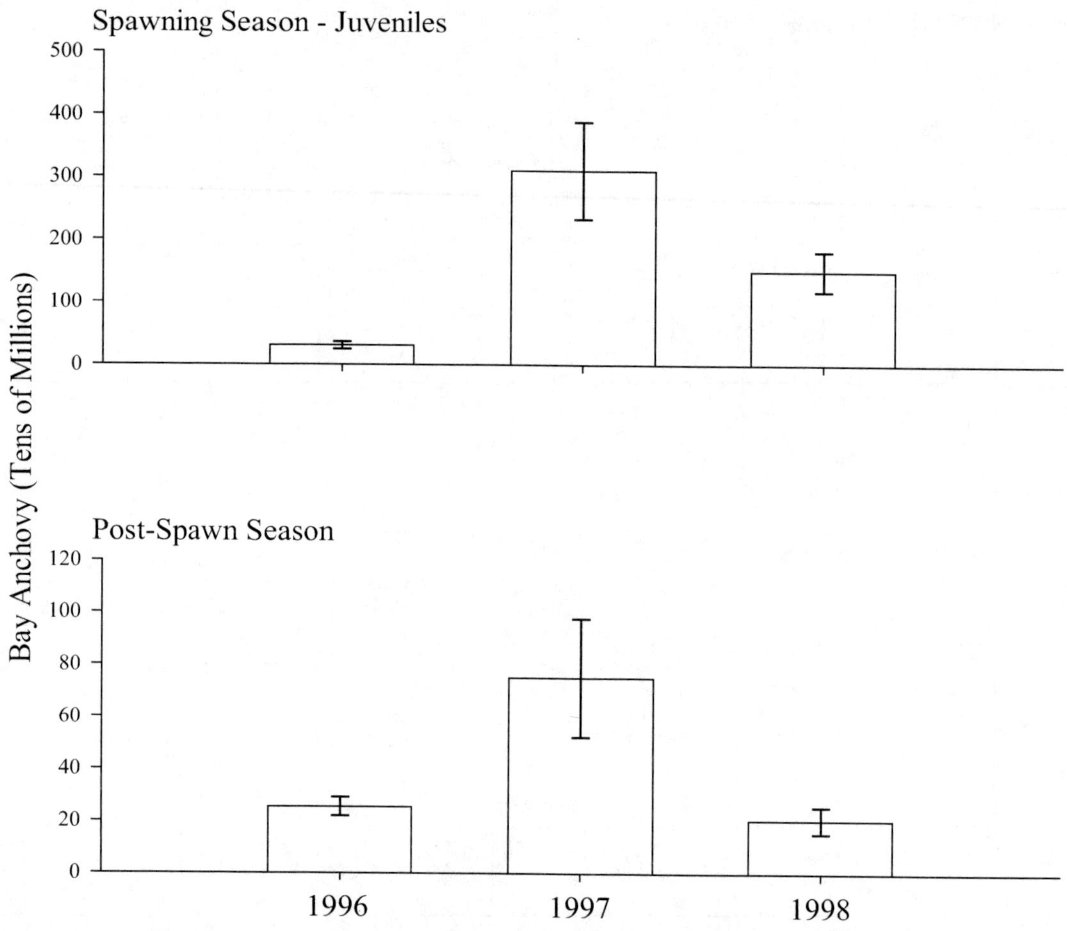

Figure 7. Total abundance estimates of juvenile bay anchovy from within the Hudson River by year and season. Error bars are 95% confidence intervals. Y-axis differs between seasons.

Bay, Jamaica Bay, Upper Bay, and West Point in 1997 suggested either a protracted spawning in 1997 or that these were focal locations in which metamorphosis to the juvenile stage had not yet been completed.

Juveniles

Densities of juvenile bay anchovy were generally higher during spawning season than during postspawning season (Figures 13 and 14), but no juveniles were detected during prespawning season. Highest juvenile densities occurred in 1997. In the spawning season peaks in juvenile density occurred in the coastal, Lower Bay and Jamaica Bay areas. Densities declined in the Arthur Kill/Kill Van Kull, East River, and Upper Bay before increasing within the Hudson River. Highest densities within the river were focused around the Yonkers region with secondary peaks upriver: around Croton-Haverstraw and Indian Point in 1996, Indian Point and West Point in 1997, and Indian Point in 1998. Densities tended to decline significantly in regions between these peaks.

During the postspawn season juvenile anchovy were densest in the Lower Bay, but generally widely distributed in the Hudson River (Figure 14). In 1996, very high densities (~10–25 fish m^2) were found in Lower

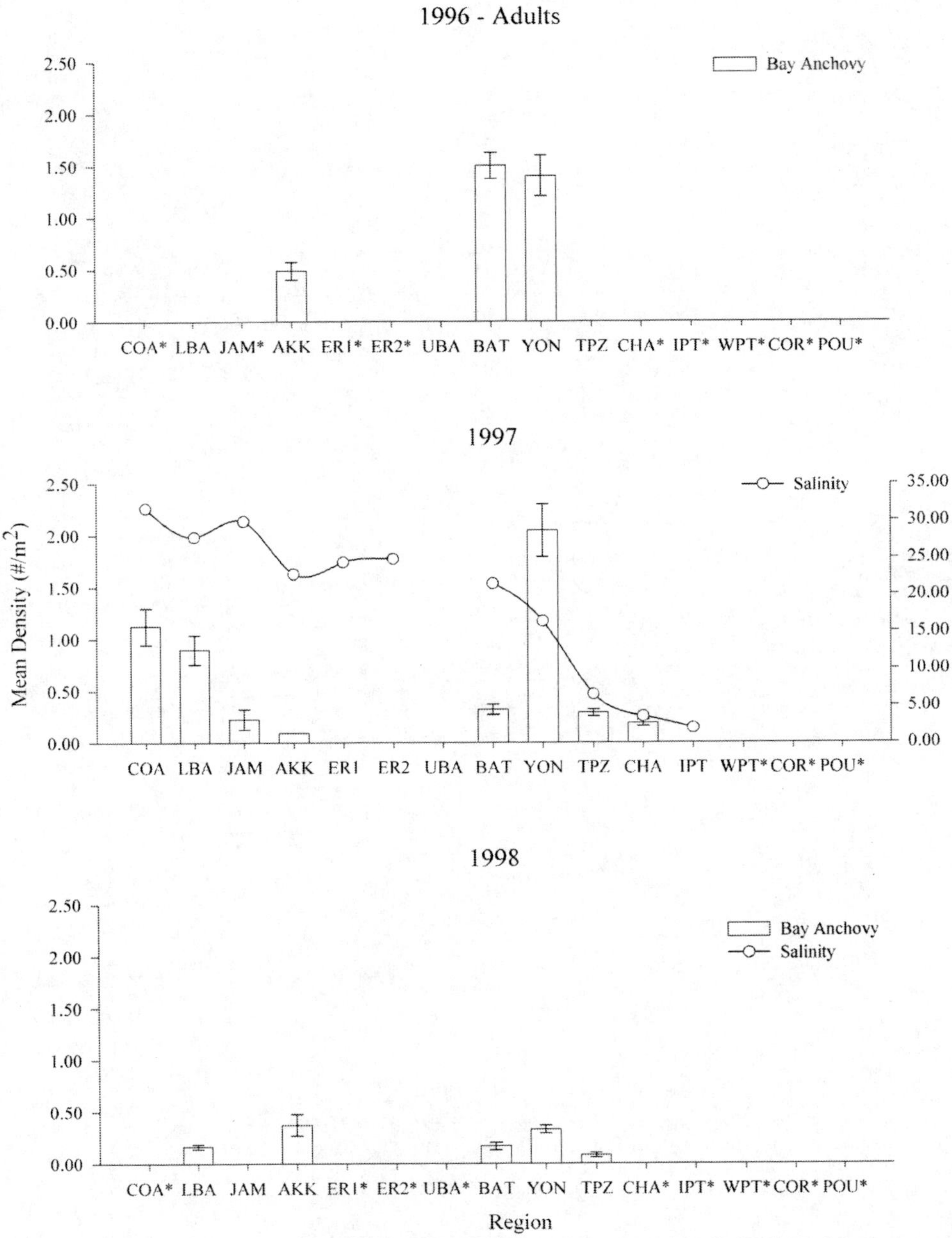

Figure 8. Pre-spawn season adult bay anchovy regional density estimates (+/- SE). See Figure 1 for definitions of region abbreviations, * indicate unsampled regions. Salinity data not available in 1996 and 1998 due to equipment malfunction.

Bay and coastal areas. Similar densities were detected in 1997 when Lower Bay reached greater than 17 fish per m^2. No peripheral sampling was completed in 1998 during postspawning season. In the Hudson River very few juvenile anchovy were detected in 1996: only in Tappan Zee and the Battery. Although not as dense as reported for the coastal areas, juvenile anchovy were commonly found in densities greater than 2.5 fish m^2 between the Battery and Croton-Haverstraw. In 1997, when anchovy abun-

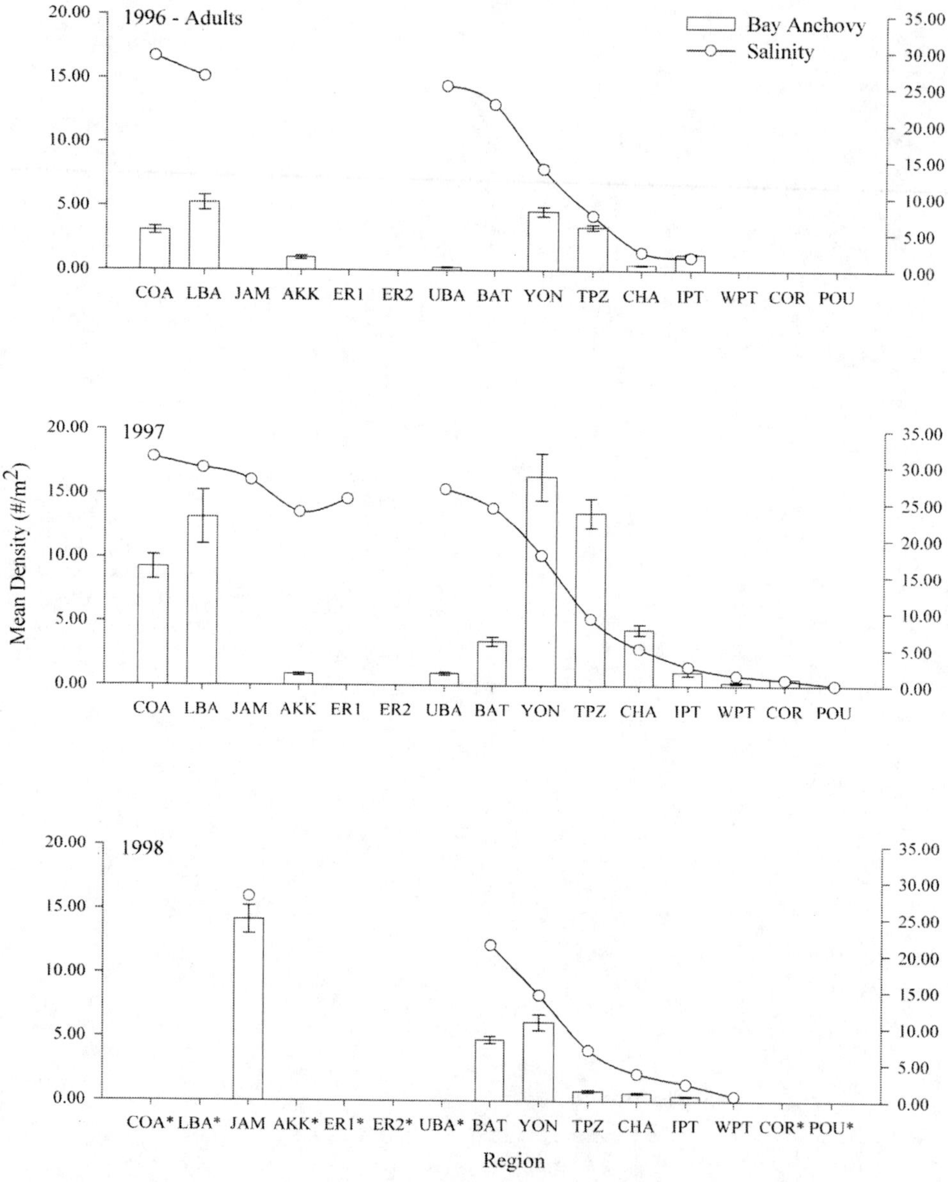

Figure 9. Spawning season adult bay anchovy regional density estimates (+/- SE). See Figure 1 for definitions of region abbreviations, * indicate unsampled regions.

dances were highest of the 3 years, juveniles were distributed upriver (Figure 14) throughout the study area (Poughkeepsie) .

Discussion

The results of this study provide two important pieces to the puzzle involving the distributional ecology of bay anchovy in the Hudson River estuary. Our surveys of the Hudson River and adjoining coastal areas in 1996–1998 suggest the Hudson River population of bay anchovy is not a discrete population, but rather it appears to be part of a larger Long Island Sound/New York Bight population. Further, our study suggests that the area around Indian Point is

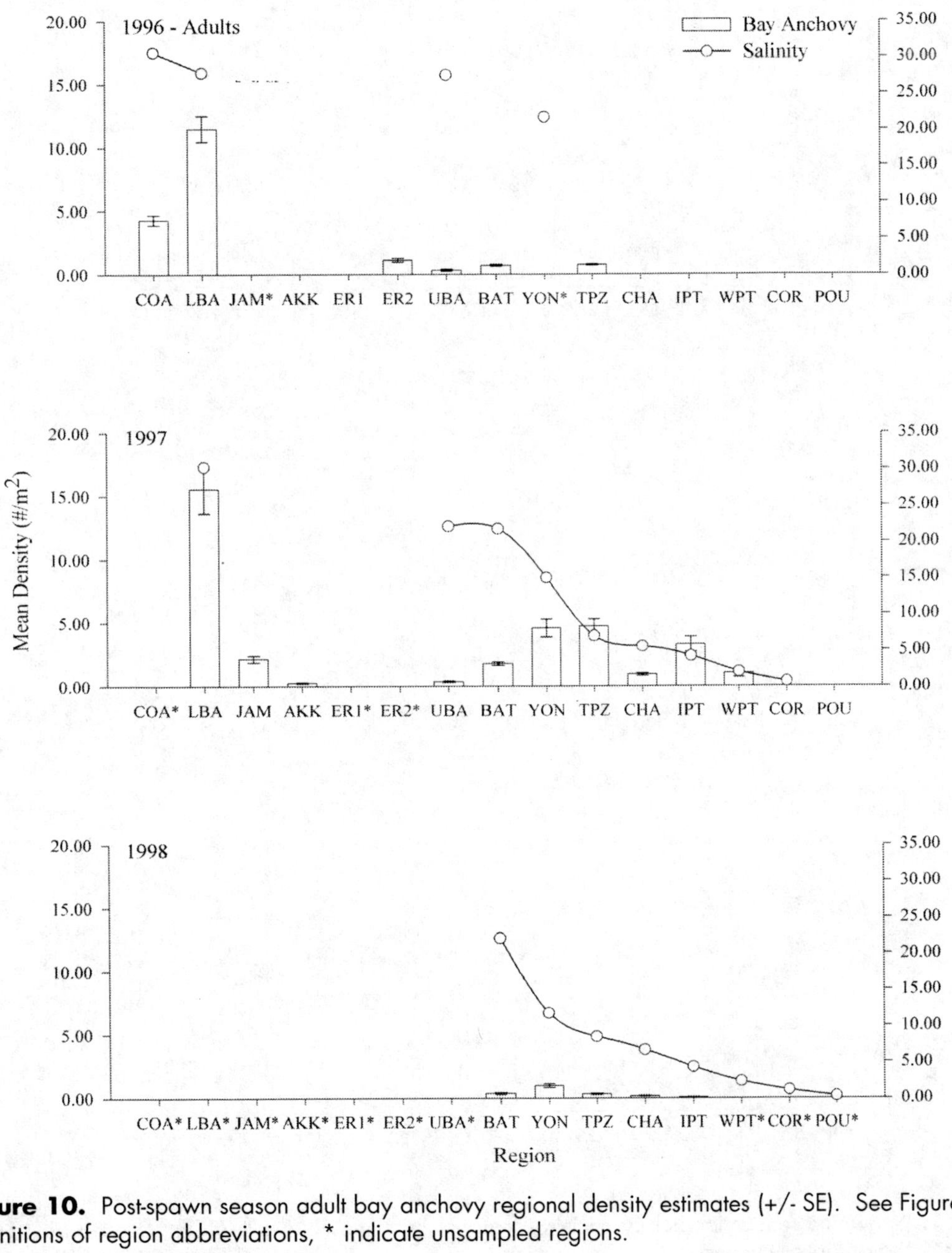

Figure 10. Post-spawn season adult bay anchovy regional density estimates (+/- SE). See Figure 1 for definitions of region abbreviations, * indicate unsampled regions.

a transitional area from a fish community dominated by bay anchovy downriver, to one dominated by other species upriver with an apparent disconnect between adult and larval bay anchovy at this point. These findings have great importance to the debate regarding the impacts of Hudson River Power plants in the area.

Evidence that Hudson River bay anchovy are part of a larger New York Bight population is provided by the seasonal distribution data. Bay anchovy spawn nightly during the spawning season (Luo and Music 1991; Bassista and Hartman 2005). At this time we found bay anchovy adults at least as far upriver as Indian Point and extend-

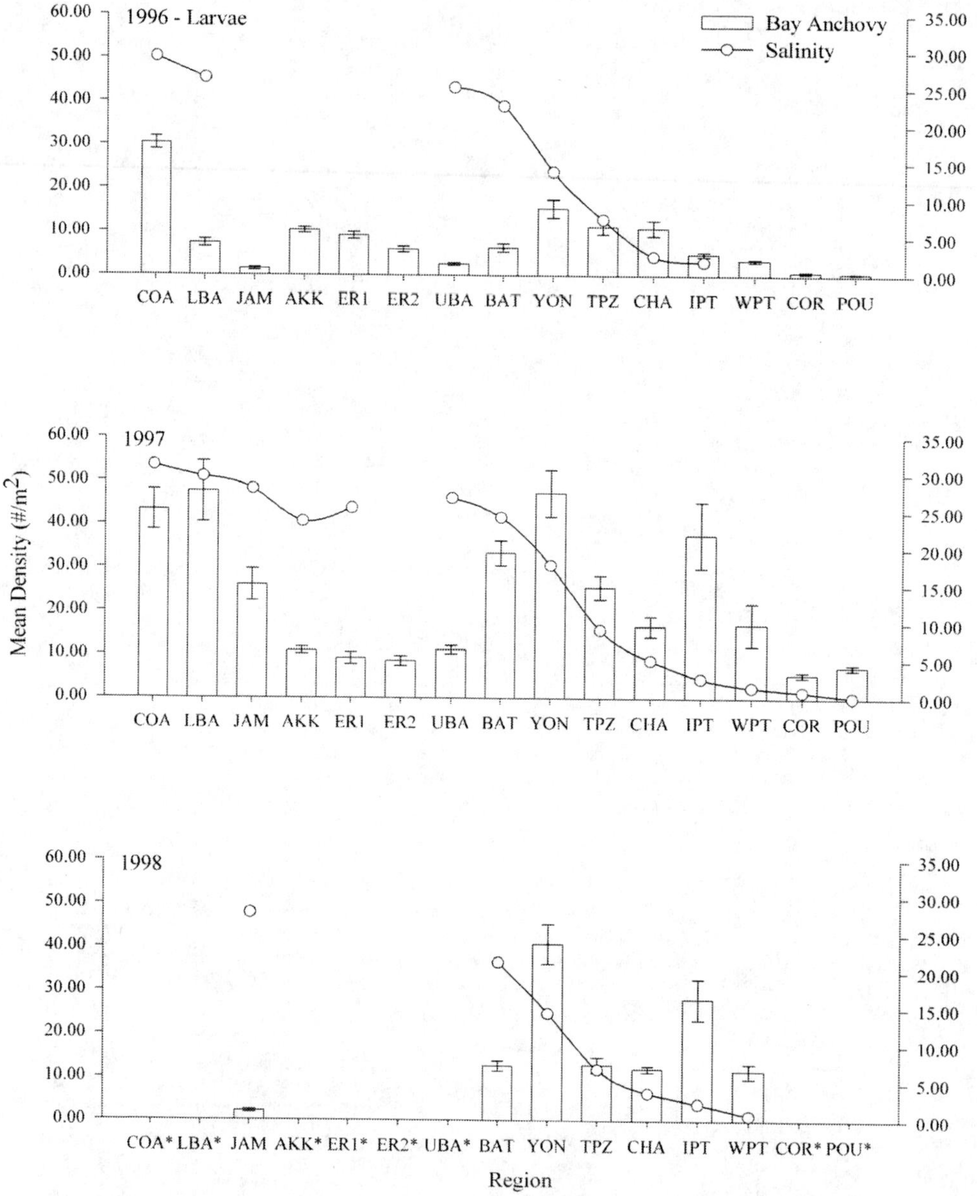

Figure 11. Spawning season larval bay anchovy regional density estimates (+/- SE). See Figure 1 for definitions of region abbreviations, * indicate unsampled regions.

ing downriver to the Battery. Sampling of areas outside of the Hudson River was sporadic, but when sampled we also found spawning adults in the upper bay, Arthur Kill and Kill Van Kull, lower bay, Jamaica Bay, and coastal areas. Hartman and Nagy (2006, this volume) found bay anchovy juveniles present in the river during December, but they were absent from our surveys in spring. Further, Schultz et al. (2006, this volume) found primarily older (age-2) bay anchovy in the spawning stock in the Hudson, and Bassista and Hartman (2005) suggested the pres-

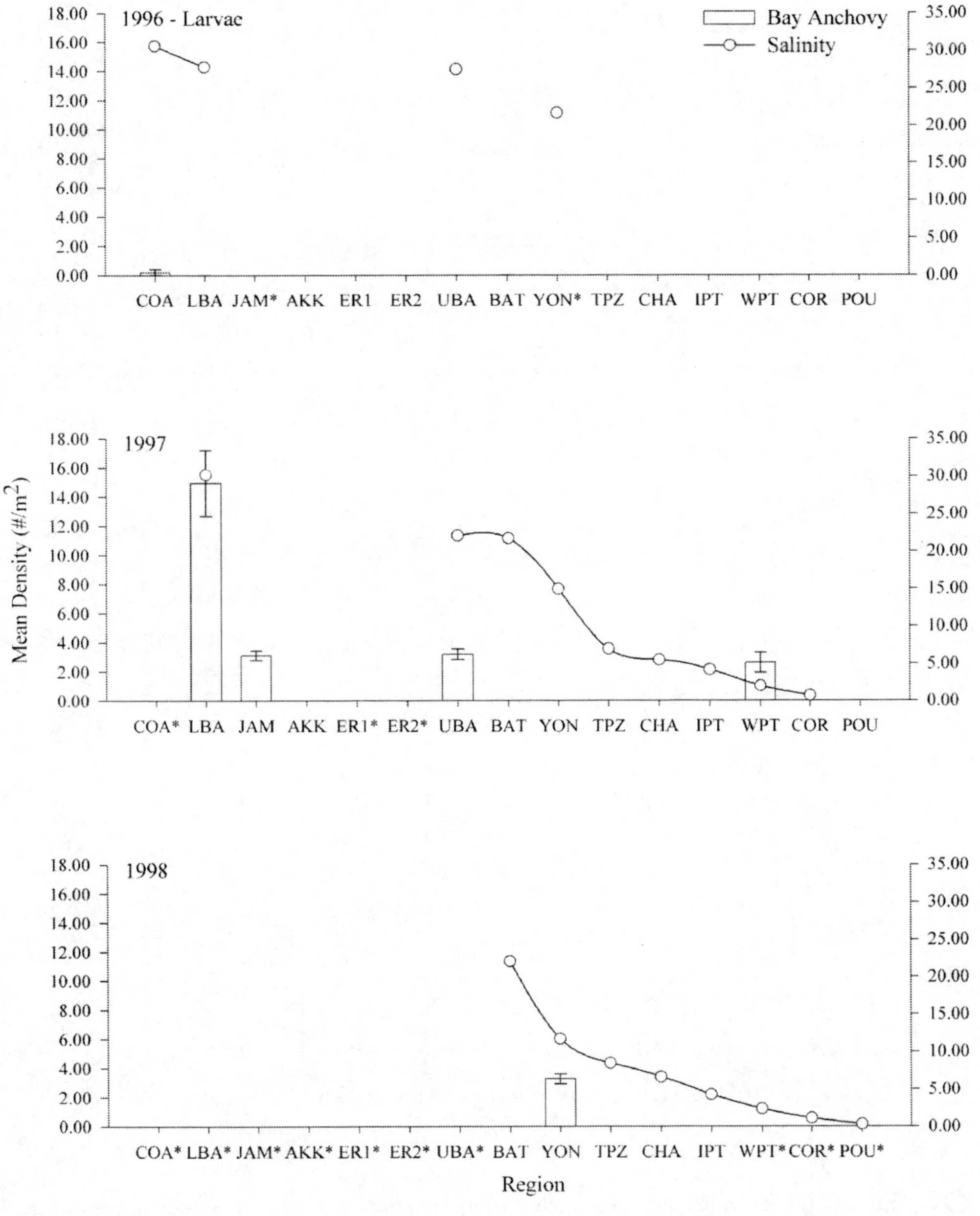

Figure 12. Post-spawn season larval bay anchovy regional density estimates (+/- SE). See Figure 1 for definitions of region abbreviations, * indicate unsampled regions.

ence of age-1 adult anchovy in nearby coastal areas such as Jamaica Bay may indicate age-related segregation of spawning stocks in the greater New York Bight area. The continuous presence of adult anchovy in these regions, coupled with the age-distribution of adults in the Hudson River strongly suggests that bay anchovy in the Hudson River are part of a panmictic population that extends into Long Island Sound and the New York Bight.

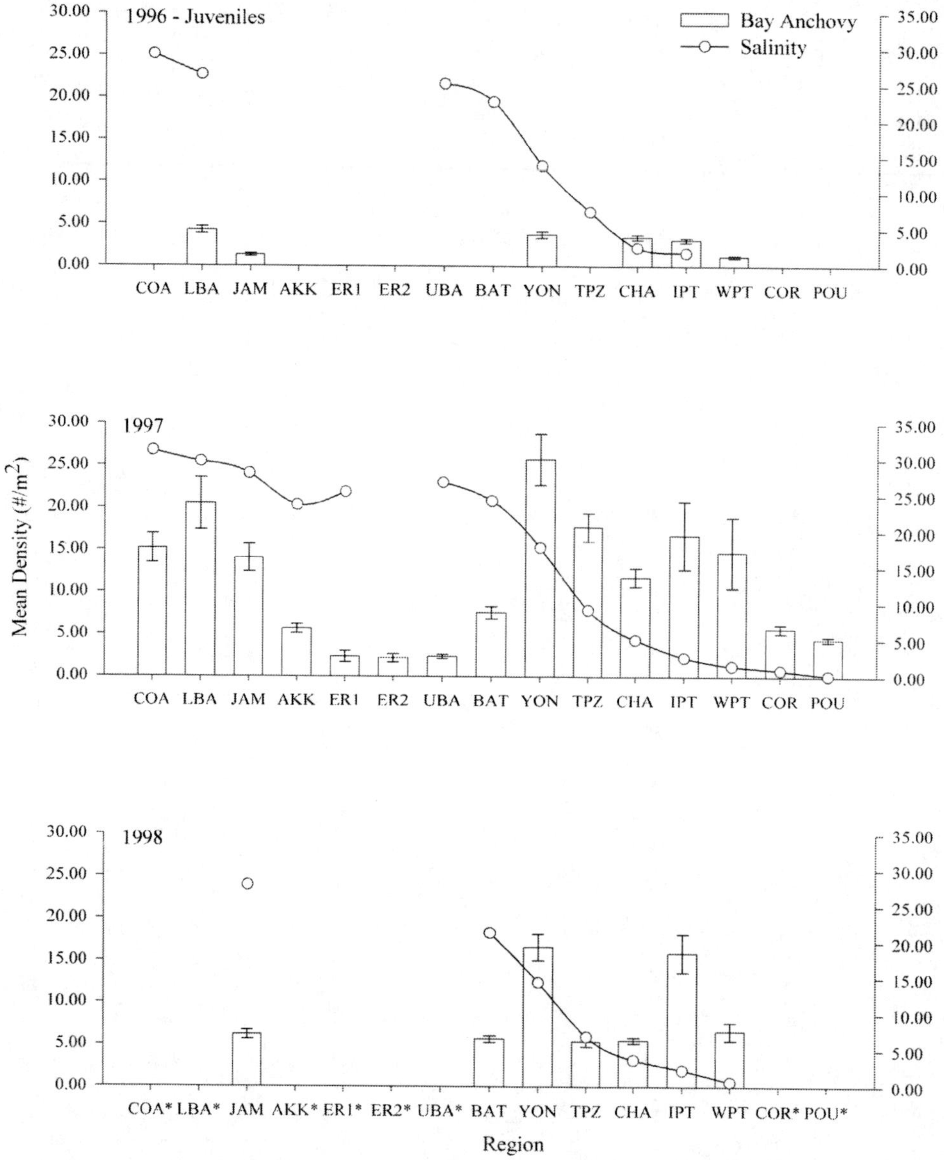

Figure 13. Spawning season juvenile bay anchovy regional density estimates (+/- SE). See Figure 1 for definitions of region abbreviations, * indicate unsampled regions.

Further evidence for Hudson River bay anchovy membership in a larger New York Bight stock comes from the distribution of early life stages. During the spawning season larval bay anchovy distribution was complete throughout sampled peripheral and river proper regions, with larval-sized targets being observed in each region surveyed. This is consistent with findings from previous studies in the Hudson River Estuary in which eggs and or adults were found to be well dispersed (Dunning et al. 2006, this volume; EA 1991; LMS 1992; Schmidt 1992). Bassista (2000) found high densities of bay anchovy eggs throughout the lower Hudson River, Upper and Lower Bay, Kill

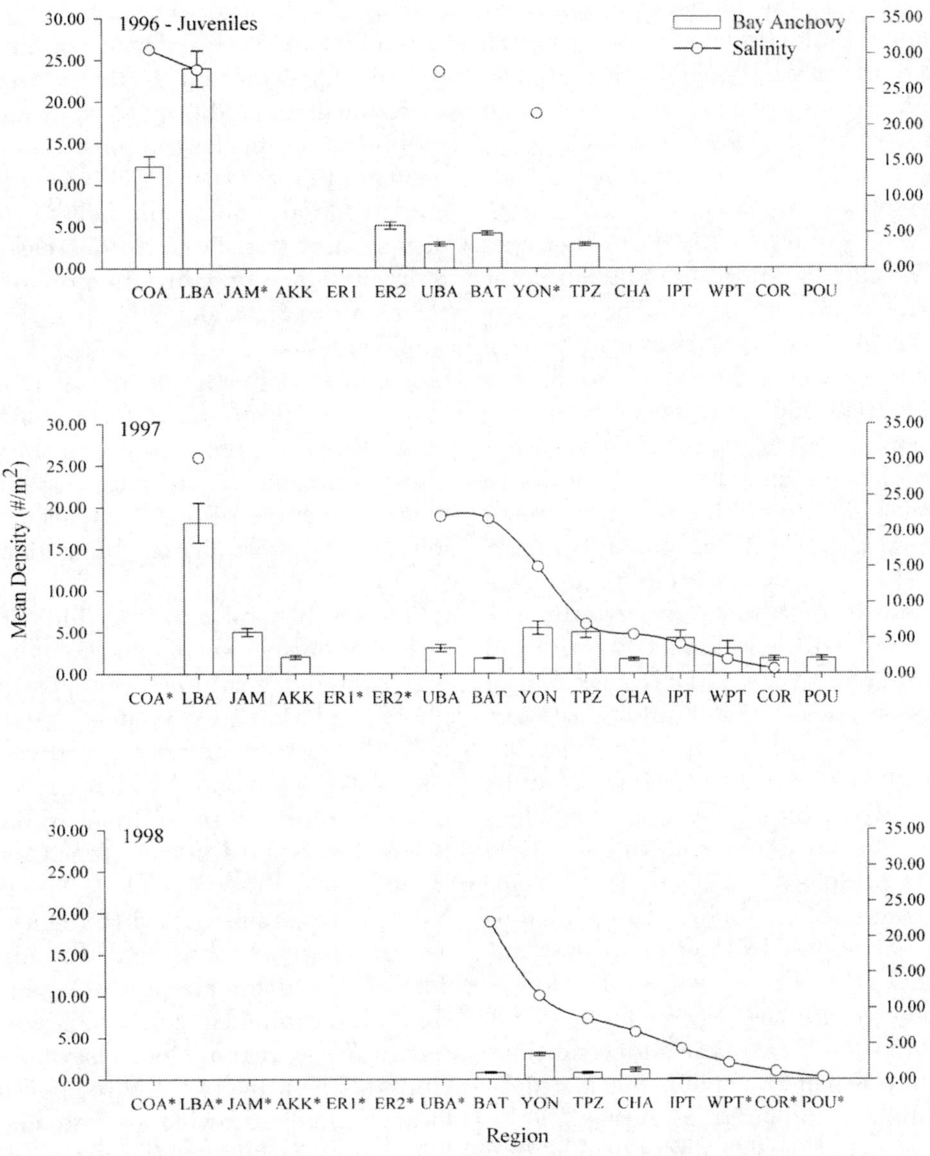

Figure 14. Post-spawn season juvenile bay anchovy regional density estimates (+/- SE). See Figure 1 for definitions of region abbreviations, * indicate unsampled regions.

Van Kull/Arthur Kill, East River, and coastal areas. Our findings support those of Dunning et al. (2006, this volume) which show that anchovy are continuously distributed between the Hudson River and adjacent coastal areas during spawning. Using genetic markers, Morgan et al. (1995) found bay anchovy to be panmictic in the Chesapeake Bay. We suggest that the lack of anchovy spawning isolation, especially in the corridor connecting the lower Hudson River and Upper Bay regions to the Long Island Sound, suggests that bay anchovy are likely panmictic within the Hudson River Estuary and adjacent coastal areas.

Our study suggests that the Indian Point region is generally the upriver extent of adult bay anchovy in the Hudson River, although larvae and juveniles are found further upriver. Adult bay anchovy density from the Indian Point region was among the lowest observed from Hudson River proper regions. Adult bay anchovy were replaced by Atlantic tomcod (65%—1996, 62%—1997) or other species (81%—1998) in trawls from Indian Point during the spawning season. These results suggest the Indian Point region may be a region of transition to the dominance of species other than adult bay anchovy during the spawning season. Bay anchovy distribution and abundance has been shown to vary seasonally due to differences in freshwater flows (EA 1991; LMS 1992; Schmidt 1992). Periodic elevated discharge may influence adult bay anchovy distribution through the redistribution of abiotic and biotic resources over relatively fine spatial and temporal scales. Freshwater discharge during the 1996 and 1998 prespawn season was elevated compared to 1997 and may be partially responsible for lower observed densities of adult bay anchovy during the 1996 and 1998 prespawn season in lower river regions compared to 1997. Although salinity does not appear to limit larval bay anchovy in Indian Point, salinities below 8 ppt appear to decrease egg viability (ASA 1999). Unfavorable environmental conditions for egg survival may explain the low density of adults at Indian Point during spawning seasons, but larvae and juvenile bay anchovy abundance there suggests that other mechanisms function to increase young of year density in that region.

With the exception of the Indian Point region, adult abundance trends were consistent with larval abundance. However, at Indian Point trends in adult abundance ran counter to those for larval bay anchovy during the spawning season. In fact low abundance of adults in Indian Point during the spawning season was consistent over the course of this study in spite of the yearly variability in regional density observed elsewhere. Therefore, it is likely that some mechanism in addition to spawning location is responsible for the consistent elevated density of larvae there. The larval standing crop at Indian Point could benefit from conditions that translate into decreased larval mortality there; contribution of larvae through passive transport down river along shallow, less dense strata from northern regions; or directed transport up river along more dense deep strata from southern regions. However, the presence of three power plants utilizing the river for cooling in the vicinity of Indian Point may argue against lower larval mortality in this region.

Alternatively, total larval mortality rates may be higher downriver than in the Indian Point region due to gelatinous predators. Ctenophore predation on bay anchovy larvae has been shown to be a significant factor influencing the abundance and distribution of bay anchovy in Chesapeake Bay (Monteleone and Duguay 1988; Cowan and Houde 1993; Purcell et al. 1994). However ctenophores are restricted to higher salinity waters than are found at Indian Point. Therefore, although potential mortality through entrainment must be higher in the Indian Point region than elsewhere in the Hudson River due to the three power plants located therein drawing a combined 4.5 billion gallons per day at full capacity (http://riverkeeper.org/campaign.php/biodiversity/we_are_doing/190), total mortality on larval anchovy may be lower in Indian Point and areas upriver due to refuge from ctenophore predation.

Mechanisms other than population dynamics have also been put forth to explain larval bay anchovy distribution (Dovel 1981; EA 1991; LMS 1992; Schmidt 1992; PSE&G 1999; Schultz et al. 2003. Dovel (1981) attributed the distribution of bay anchovy larvae to hydraulic characteristics in the Hudson River Estuary. EA (1991), LMS (1992), and PSE&G (1999) suggest that vertical migration of larvae to deeper more saline water may facilitate upstream transport. Schmidt (1992) found Hudson River Estuary bay anchovy distribution to be directly related to salinity but found bay anchovy larvae in freshwater above the salt line, at distances too far to be explained by migration or passive transport alone, and concluded that salt front movement alone cannot explain larval distribution. More recently Schultz et al. (2003) found bay anchovy larvae in freshwater upstream of spawning and hypothesized that young bay anchovy have some means of directed transport in the Hudson River that may be related to strong nocturnal flood tides. However, directed transport of larvae would be difficult initially given larval bay anchovy's limited ability to affect their position, due to size and the fact that net transport in the Hudson River Estuary is south (Schmidt 1992), though larval development would increase mobility (Schultz et al. 2003). Schmidt (1992) cautioned that care be taken when inferring distributional processes of larval bay anchovy in the Hudson River Estuary because several processes may be occurring simultaneously including passive transport, active movement of larvae progressing with growth, and recruitment of newly hatched eggs. Indian Point remains an important nursery area for young of year bay anchovy due to relatively high density of larvae and juveniles there.

References

ASA (Applied Science Associates). 1999. 1998 Year class report for the Hudson River Estuary Monitoring Program. Prepared for Consolidated Edison Company of New York, New York.

Baird, D., and R. E. Ulanowicz. 1989. The seasonal dynamics of the Chesapeake Bay ecosystem. Ecological Monographs 59:329–364.

Barnthouse, L. W., R. J. Klauda, and D. S. Vaughn. 1988. Introduction to the monograph. Pages 1–8 *in* L. W. Barnthouse, R. J. Klauda, D. S. Vaughan, and R. L. Kendall, editors. Science, law, and Hudson River power: a case study in environmental impact assessment. American Fisheries Society, Monograph 4, Bethesda, Maryland.

Bassista, T. P. 2000. Reproductive biology and early life stage abundance and distribution of bay anchovy (*Anchoa mitchilli*) in the Hudson River Estuary, USA. Master's thesis. West Virginia University, Morgantown.

Bassista, T., and K. J. Hartman. 2005. Reproductive biology of bay anchovy in the Hudson River Estuary. Environmental Biology of Fishes 73:49–59.

Boreman, J., and C. P. Goodyear. 1988. Estimates of entrainment mortality for striped bass and other fish species inhabiting the Hudson River estuary. Pages 152–160 in L. W. Barnthouse, R. J. Klauda, D. S. Vaughan, and R. L. Kendall, editors. Science, law, and Hudson River power: a case study in environmental impact assessment. American Fisheries Society, Monograph 4, Bethesda, Maryland.

Brandt, S. B. 1992. Acoustic measurement of fish abundance in the Chesapeake Bay. Final Report to Maryland Department of Natural Resources, Power Plant Topical Research Program, [UMCEES]CBL 90–047, Annapolis, Maryland.

Brandt, S. B. 1996. Acoustic assessment of fish abundance and distribution. Pages 385–432 *in* B. R. Murphy and D. W. Willis, editors. Fisheries Techniques, 2nd edition. American Fisheries Society, Bethesda, Maryland.

Byrne, C. J. 1982. Anchovy. Pages 65–67 *in* M. D. Grosslein and T. R. Azarovitz, editors. Fish distribution. Mesa New York Bight Atlas Monograph 15. New York Sea Grant Institute, Albany.

Cowan, J. H., and E. D. Houde. 1993. Relative predation potentials of scyphomedusae, ctenophores, and planktivorous fish on ichthyoplankton in Chesapeake Bay. Marine Ecology Progress Series 95:55–65.

Dovel, W. L. 1981. Ichthyoplankton of the lower Hudson Estuary, New York. New York Fish and Game Journal 28:21–39.

Dunning, D. J., Q. E. Ross, M. T. Mattson, and D. G. Heimbuch. 2006. Distribution and abundance of bay anchovy eggs and larvae in the Hudson River and nearby waterways. Pages 215–226 *in* J.

R. Waldman, K. E. Limburg, and D. L. Strayer, editors. Hudson River fishes and their environment. American Fisheries Society, Symposium 51, Bethesda, Maryland.

EA (EA Engineering, Science, and Technology). 1991. Year class report for the Hudson River Estuary Monitoring Program. Prepared for Consolidated Edison Company of New York, New York Inc.

Foote, K. G. 1987. Fish target strengths for use in echo integration surveys. Journal of the Acoustical Society of America 82:981–987.

Foote, K. G., A. Aglen, and O. Nakken. 1986. Measure of fish target strength with a split-beam echo sounder. Journal of the Acoustical Society of America 80:612–621.

Friedland, K. D., G. C. Garman, A. J. Bedja, A. F. Studholme, and B. Olla. 1988. Interannual variation in diet and condition in juvenile bluefish during estuarine residency. Transactions of the American Fisheries Society 117:474–479.

Gallagher, R. P., M. F. Hirshfield, and E. S. Perry. 1983. Age, growth , and mortality estimates for larval bay anchovy (*Anchoa mitchilli*) in the Patuxent River. Report, Benedict Estuarine Research Laboratory, Benedict, Maryland.

Gardinier, M. N., and T. B. Hoff. 1982. Diet of striped bass in the Hudson River estuary. New York Fish and Game Journal 29(22):152–165.

Hartman, K. J., and S. B. Brandt. 1995. Trophic resource partitioning, diets, and growth of sympatric estuarine predators. Canadian Journal of Fish and Aquatic Science 124:520–537.

Hartman, K. J., and B. W. Nagy. 2006. Winter distribution and abundance of Hudson River fishes using hydroacoustics. Pages 175–196, *in* J. R. Waldman, K. E. Limburg, and D. L. Strayer, editors. Hudson River fishes and their environment. American Fisheries Society, Symposium 51, Bethesda, Maryland.

Hohn, M. E. 1988. Geostatistics and petroleum geology. Van Nostrand Reinhold, New York.

Horwitz, R. J. 1987. Fish. Pages 167–224 *in* K. L. Heck, editor. Ecological studies in the middle reach of Chesapeake Bay. Lecture notes on Coastal and Estuarine Studies 23, Springer-Verlag, Berlin.

Houde, E. D. 1996. Evaluating stage-specific survival during the early life of fish. Pages 51–66 *in* Y. Watanabe,Y. Yamashita, and Y. Oozeki, editors. Survival strategies in early life stages of marine resources. Balkema, Rotterdam, The Netherlands.

Houde, E. D., and C. E. Zastrow. 1991. Bay Anchovy. Pages. 8–1 to 8–14 *in* S. L. Funderburk, J. A. Mihursky, S. J. Jordan, and D. Riley, editors. Habitat requirements for Chesapeake Bay living resources. Chesapeake Research Consortium, Solomons, Maryland.

Jech, J. M., and J. Luo. 2000. Digital Echo Visualization and Information System (DEVIS) for processing spatially-explicit fisheries acoustic data. Journal of Fisheries Research 27:115–124.

Klebasko, M. J. 1991. Feeding ecology and daily ration of bay anchovy (*Anchoa mitchilli*) in the mid-Chesapeake Bay. Master's thesis. University of Maryland, College Park.

Limburg, K., and R. E. Schmidt. 1990. Patterns of fish spawning in Hudson River tributaries: response to an urban gradient? Ecology 71:1238–1245.

LMS (Lawler, Matusky, and Skelly Engineers). 1992. Year class report for the Hudson River Estuary Monitoring Program. Prepared for Consolidated Edison Company of New York, New York Inc.

Luo, J. 1993. Tidal transport of the bay anchovy, *Anchoa mitchilli*, in darkness. Journal of Fish Biology 42:531–539.

Luo, J., and S. B. Brandt. 1993. Bay anchovy, *Anchoa mitchilli*, production and consumption in mid-Chesapeake Bay based on a bioenergetics model and acoustic measures of fish abundance. Marine Ecology Progress Series 98:223–236.

Luo, J., and J. A. Musick. 1991. Reproductive biology of the bay anchovy in Chesapeake Bay. Transactions of the American Fisheries Society 120:701–710.

McGurk, M. D. 1986. Natural mortality of marine pelagic fish eggs and larvae: role of spatial patchiness. Marine Ecology Progress Series 34:227–242.

Merriner, J. V. 1975. Food habits of the weakfish, *Cynoscion regalis*, in North Carolina waters. Chesapeake Science 16:74–76.

Monteleone, D. M., and L. E. Duguay. 1988. Laboratory studies of predation by the ctenophore *Mnemiopsis leidyi* on the early stages in the life history of the bay anchovy, *Anchoa mitchilli*. Journal of Plankton Research 10:359–372.

Morgan, R. P., B. M. Baker, and J. H. Howard. 1995. Genetic structure of bay anchovy (*Anchoa mitchilli*) populations in Chesapeake Bay. Estuaries 18:482–493.

Newberger, T. A., and E. D. Houde. 1995. Population biology of bay anchovy *Anchoa mitchilli* in the mid Chesapeake Bay. Marine Ecology Progress Series 116:25–37.

PSE&G (Public Service Electric and Gas Company). 1999. Permit renewal application NJPDES permit no. NJ0005622. Salem Generating Station, Salem, New Jersey.

Purcell, J. E., D. A. Nemazie, S. E. Dorsey, E. D. Houde, and J. C. Gamble. 1994. Predation mortality of bay anchovy *Anchoa mitchilli* eggs and larvae due to scyphomedusae and ctenophores in Chesapeake Bay. Marine Ecology Progress Series 114:47–58.

Rivoirard, J., J. Simmonds, K. G. Foote, P. Fernandes, and N. Bez. 2000. Geostatistics for estimating fish abundance. Blackwell Scientific Publications Inc., Malden, Massachusetts.

SAS Institute. 1999. SAS/IML Software: usage and reference, Version 8.01, 1st edition. SAS Institute, Cary, North Carolina.

Schaefer, R. H. 1970. Feeding habits of striped bass from the surf waters of Long Island. New York Fish and Game Journal 17(1):1–17.

Schmidt, R. E. 1992. Temporal and spatial distribution of bay anchovy eggs through adults in the Hudson River Estuary. Pages 228–241 *in* C. L. Smith, editor. Estuarine research in the 1980s: The Hudson River Environmental Society Seventh Symposium on Hudson River Ecology. State University of New York Press, Albany.

Schultz, E. T., K. M. M. Lwiza, M. C. Fencil, and J. M. Martin. 2003. Mechanisms promoting upriver transport of larvae of two fishes in the Hudson River Estuary (USA). Marine Ecology Progress Series 251:263–277.

Schultz, E. T., K. M. M. Lwiza, J. R. Young, K. J. Hartman, and R. C. Tipton. 2006. The dynamics of bay anchovy in the Hudson River estuary: process-oriented studies and long-term changes. Pages 197–213 *in* J. R. Waldman, K. E. Limburg, and D. L. Strayer, editors. Hudson River Fishes and their Environment. American Fisheries Society, Symposium 51, Bethesda, Maryland.

Texas Instruments. 1976. Predation by bluefish in the lower Hudson River. Final Report to Consolidated Edison Company, New York.

Vouglitois, J. J., K. W. Able, R. J. Kurtz, and K. A. Tighe. 1987. Life history and population dynamics of the bay anchovy in New Jersey. Transactions of the American Fisheries Society 116:141–153.

Wang, S. B. 1992. Abundance, relative biomass, production and energy storage of bay anchovy, *Anchoa mitchilli*, in the Chesapeake Bay. Master's thesis. The University of Maryland, College Park.

Wang, S. B., and E. D. Houde. 1994. Energy storage and dynamics in bay anchovy *Anchoa mitchilli*. Marine Biology 121:219–227.

Wang, S. B., and E. D. Houde. 1995. Distribution, relative abundance, biomass and production of bay anchovy *Anchoa mitchilli* in the Chesapeake Bay. Marine Ecology Progress Series 121:27–38.

American Fisheries Society Symposium 51:251–272, 2006

Fine-Scale Distribution and Abundance of Pelagic Fish near Two Hudson River Power Plants

Ronald C. Tipton

United States Fish and Wildlife Service, Maryland Fishery Resources Office 177 Admiral Cochrance Drive, Annapolis, Maryland 21401 USA clif_tipton@fws.gov

Kyle J. Hartman

West Virginia University, Wildlife and Fisheries Resources Program 322 Percival Hall, Morgantown, West Virginia 26506-6125 USA

Abstract.—Pelagic fish abundance and distribution was estimated acoustically at Bowline and Indian Point power plants on the Hudson River, New York, during July, August, and September 1996, in a study designed to determine size-specific spatial and temporal fish abundance and distribution with respect to environmental variables (i.e., photoperiod, depth strata, field). August yielded the greatest ($P < 0.05$) mean density of fish at both power plants. Fish were concentrated in Bowline's near-field (lagoon) region, thus increasing the potential for deleterious impingement and entrainment effects there. For all size classes, mean fish density was greatest during darkness at both power plants. Generally, the two power plants exhibited similar trends in mean fish abundance with respect to water depth and photoperiod. During July and August mean fish density was significantly higher between 1.5 and 4.5 m (depth strata one) of depth at both power plants. Bowline's mean fish density was greater under near-field, shallow depth (depth strata one) darkness during July and August. Bowline's isolated and bathymetrically complex near-field lagoon concentrated pelagic fish compared to the river proper. Hydroacoustics were useful in providing a detailed map of fish distribution relative to each power plant over the course of a few days each month. Hydroacoustic monitoring could mitigate negative effects to pelagic fish at existing and proposed power plants, through sighting of power plant water intake structures or by providing a biological basis for modified production cycles.

Introduction

For several decades, there has been concern regarding deleterious anthropogenic effects on resident and transient fish populations within the Hudson River estuary. Of the myriad hazards to fish populations in the Hudson River impingement and entrainment losses have been at the forefront. Impingement (trapped against cooling water intake screens) may be an important source of mortality of fishes associated with power plant operations (Muessig et al. 1988; Barnthouse and Van Winkle 1988). Entrainment (passing through cooling water intake screens and cycling through the power plant)

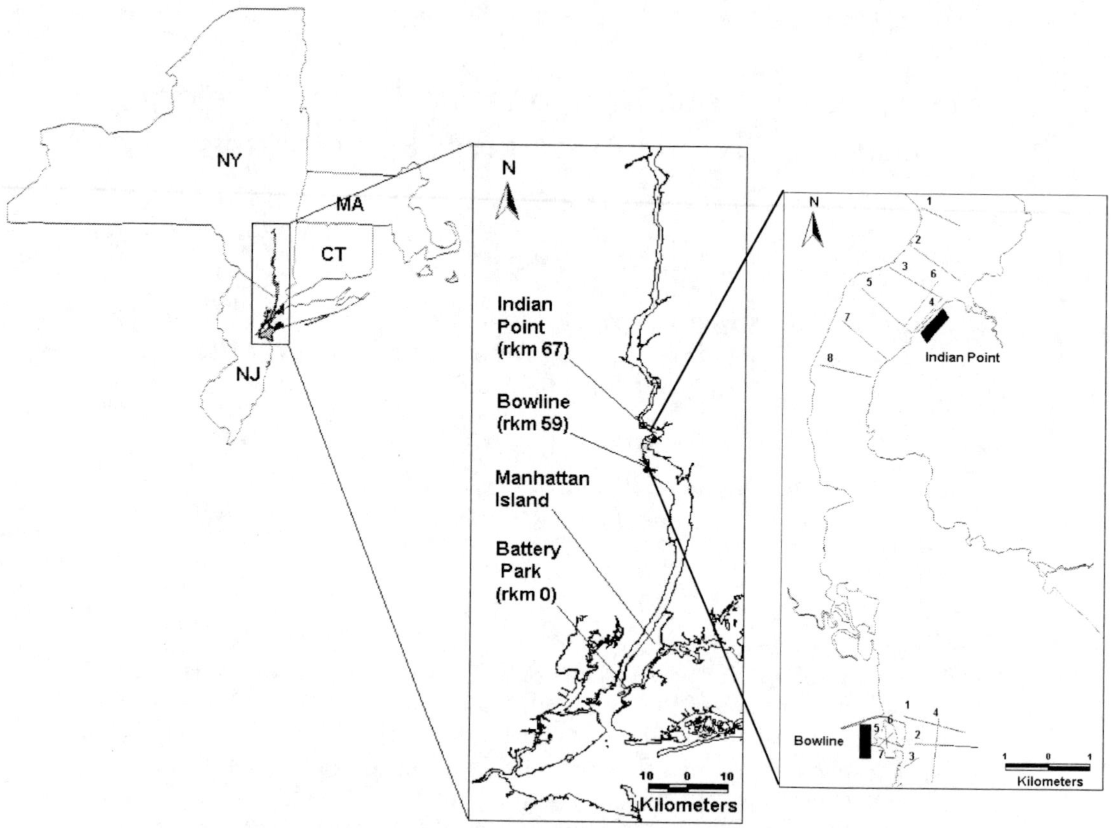

Figure 1. Map illustrating the location of the Hudson River and the position of Bowline and Indian Point power plants within the Hudson River. Bowline numbers 1, 2, 3, and 4 = far-field transects. Bowline numbers 5, 6, and 7 = near-field transects, rkm = river kilometer.

of fishes is also a potential source of bay anchovy *Anchoa mitchilli* mortality. The Polgar et al. (1988) model for Patuxent River bay anchovy assumed 100% mortality for entrained individuals. Of species historically known to the Hudson River, bay anchovy were estimated to suffer the highest conditional entrainment mortality (fractional population loss due to power plant operations) rate due to their pelagic nature and the length of time (June–October) they are in the vicinity of the power plants (Boreman and Goodyear 1988). Entrainment mortality at various life stages has been estimated to be as high as 45–80% at power plants (Boreman and Goodyear 1988).

Derived conditional mortality rates may be overestimating power plant impact due to underestimation of pelagic species' population size (through underrepresentation of population distributions) or nonuniform distribution in the vicinity of power stations. Spatial concentration or avoidance of areas near power plants may be a major factor in mortality experienced by pelagic species as a result of plant operations. Regardless of estimates of entrainment and impingement losses, the impact of power plant operations cannot be fully addressed without spatially and temporally fine scale investigation of vulnerable organisms (like small pelagic species) in the vicinity of power plant intakes. To address this need, we conducted monthly hydroacoustic surveys in the vicinity of Bowline (river km 59) and Indian Point (river km 67) power stations along the Hudson River, New York, during July,

August, and September 1996 (Figure 1). Hydroacoustic surveys can be used to provide cost effective and timely fine-scale information regarding size-specific fish distribution relative to intake structures, which would benefit conditional entrainment modeling efforts. Species composition may be obtained from impingement and entrainment data specific to intake structures, thus avoiding gear biases associated with other sampling methods and reducing cost.
The goals of this study were to:

1) Determine the timing of peaks in fish abundance over the course of this study, at Bowline and Indian Point;

2) Determine differences in fish distribution and abundance between transects inside Bowline lagoon and adjacent to intake structures (near field) and transects outside the lagoon and away from intake structures (far-field), representing spatial aspects impacting conditional impingement and entrainment estimates, at Bowline; and

3) Use fine scale spatial and temporal fish distribution and abundance information derived from this study to address implications of modified power plant operation to reduce impingement and entrainment effects on fish.

Methods

Study Area

Though geographically proximate, Bowline and Indian Point differ substantially with respect to bathymetry, depth, and situation relative to the Hudson River. Bowline's near-field is situated within an isolated 49 ha lagoon (Figure 1) The lagoon's bottom is complex, featuring a pronounced hump that occasionally breaks the water line during low tide. The hump is at least partially visible in each of the near-field transects (Figure 2, transects 5–7). Bowline's maximum depth was 12.00 m and averaged 7.07 m, over all surveys. Bowline's far-field is located outside the lagoon in the Hudson River proper. Bathymetry there lacks the seclusion and complexity of the near-field lagoon with depth gradually increasing without interruption from shore to channel (Figure 2, transects 1–4). Maximum depth in Bowline's far-field region was 14.00 m and averaged 8.36 m, over all samples.

Indian Point is located in a river bend North of Bowline (Figure 1). Depths at Indian Point were roughly twice those encountered at Bowline. Indian Point's maximum depth during the study was 28.00 m and averaged 17.31 m. Indian Point transects were deeper than those at Bowline, but exhibited bathymetric characteristics similar to those from Bowline's far-field (Figure 2, transects 1–4, Figure 3). In addition to greater depth, Indian Point also differs from Bowline with respect to bottom habitat. The bottom is neither complex nor irregular at Indian Point, compared to the Bowline lagoon. Indian Point also lacks isolation between intake and effluent areas that exist at Bowline. Substrate composition however, is consistent in the vicinity of both power stations, being comprised of sandy mud over intermittent gravel.

Hydroacoustics

Each month from July through September, 1996, two consecutive 24 h surveys were conducted at Indian Point and one 24 h survey was conducted at Bowline. These months were chosen to coincide with high densities of bay anchovy in the Hudson River, which has been confirmed in earlier studies (Dovel 1981; Byrne 1982; Beebe and Savidge 1988) and through our own

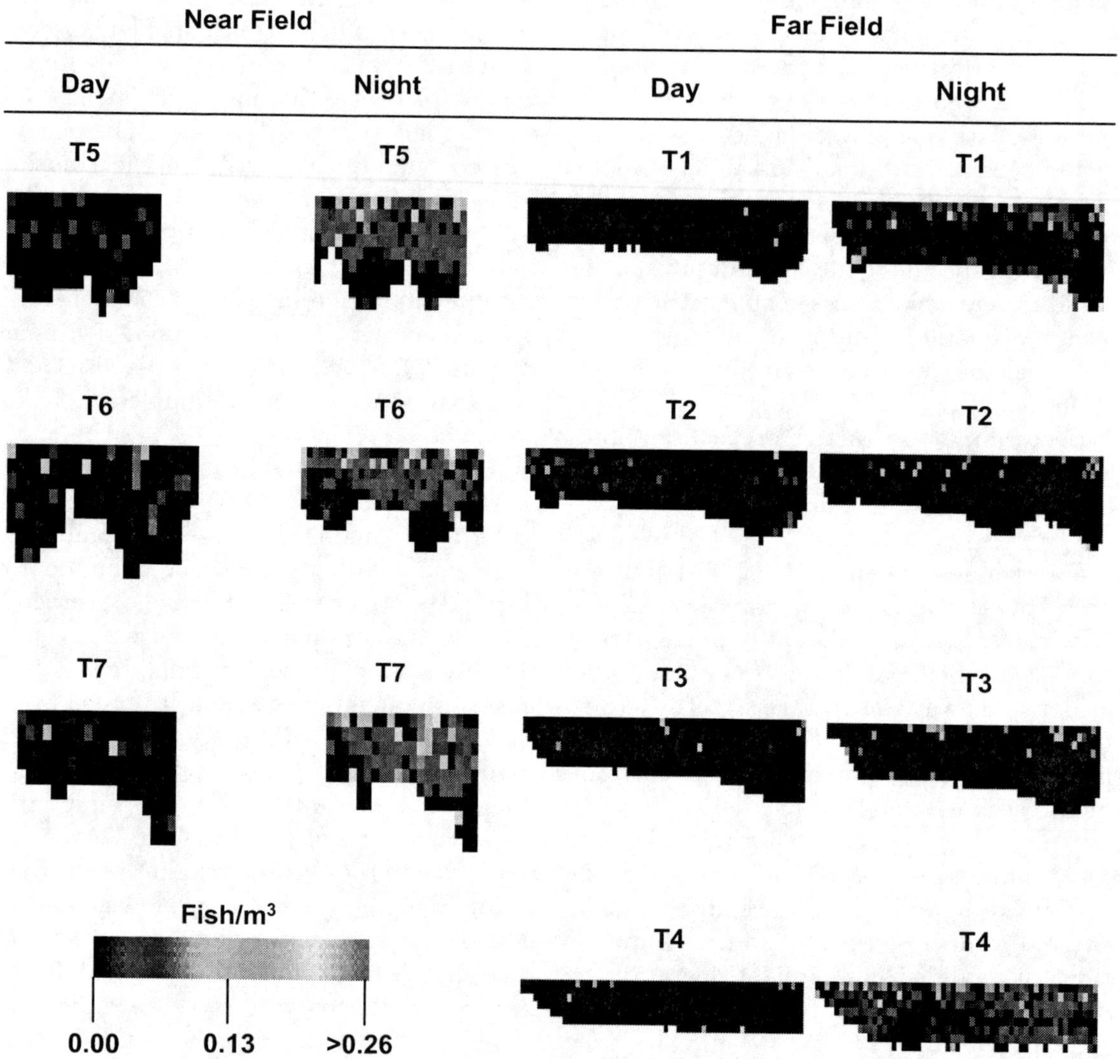

Figure 2. Corrected echogram from Bowline power plant illustrating fish density trends from near and far-fields during the day and night. Day echogram is taken from session ten (0854–0900 hours) on 22 September, 1996 sample. Night echogram is taken from session three (1903–1910 hours) on 21 September 1996. Echograms are not to scale with respect to transect length and depth. T = transect.

riverwide acoustic surveys (Tipton 2003). Monthly sampling consisted of a series of up to eight individual river transects in the vicinity of each of the power stations. At Bowline, a combination of near-field (transects 5–7) and far-field (transects 1–4) transects were surveyed hydroacoustically and repeated at 2-hour intervals. Eight individual transects were also surveyed and repeated at 2-hour intervals at Indian Point (Figure 2). Indian Point sampling was conducted over two consecutive days during each month of the study. To prevent bias associated with choosing data, mean fish densities were calculated using the entire Indian Point data set. A vessel breakdown

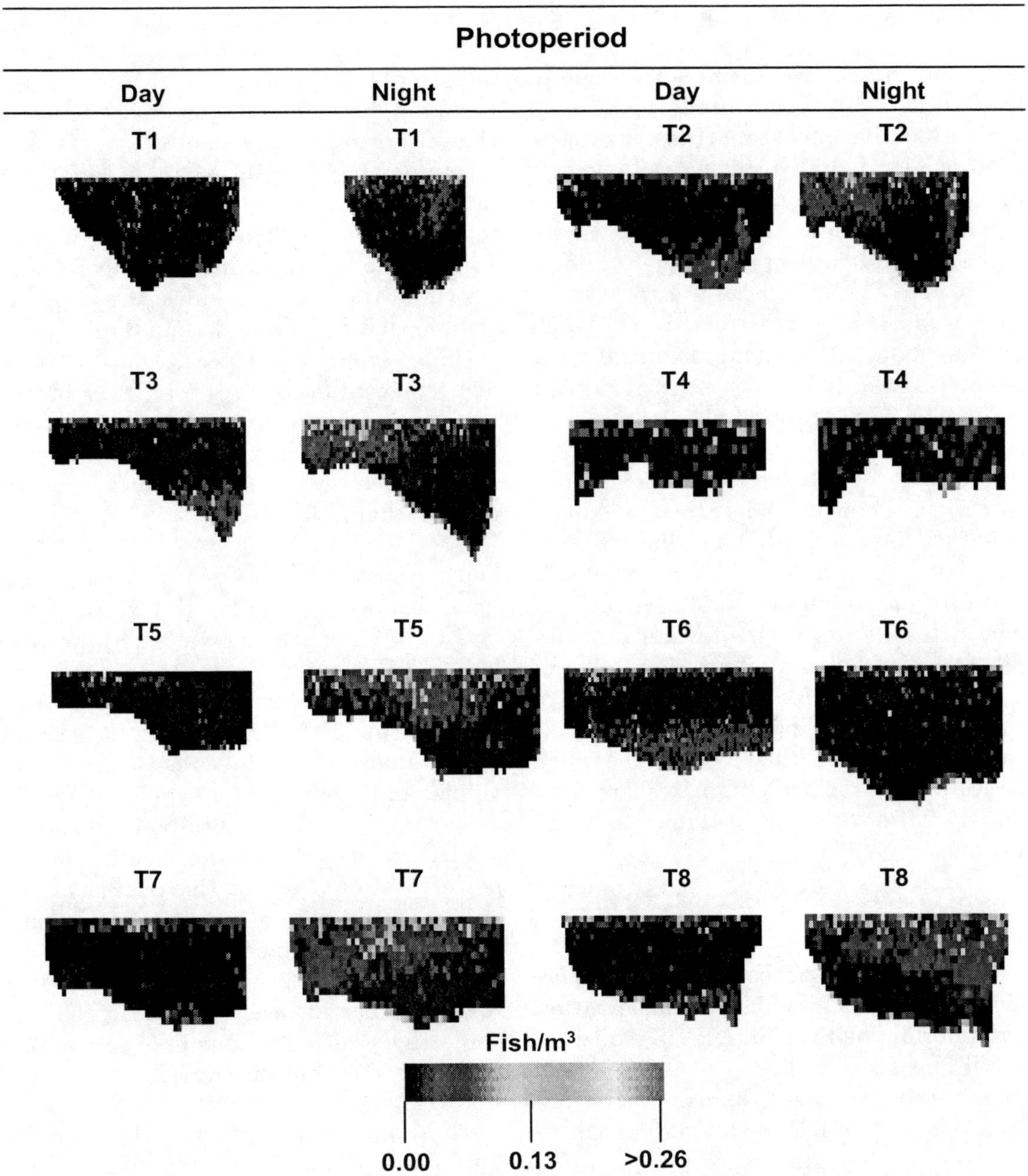

Figure 3. Corrected echogram from Indian Point power plant illustrating fish density trends during the day and night. Day echogram is taken from session three (1330–1338 hours) on 03 August 1996 and represents the "July" sample. Night echogram is taken from session ten (1903–1910 hours) on 04 August 1996 and represents the "July" sample. Echograms are not to scale with respect to transect length and depth. T = transect.

prevented completion of the Indian Point survey during mid July 1996. Thus, sampling scheduled for mid July was actually carried out on 02–03 August 1996. Therefore, references to the Indian Point July sample period herein effectively represent conditions during early August.

Acoustic surveys were conducted using a Simrad EY 500 echosounder equipped with a 120 kHz split-beam transducer. The echosounder digitally controls pulse length and pulse repetition rate. Time-varied gain is also controlled to correct for signal spreading and absorption through the water column. The transducer was towed near the surface via a vessel mounted boom and tow body. A blanking distance 1.0 m from the transducer face is necessary for acoustic beams from multiple pressure sources within the transducer to coalesce into a single sound wave. Because of this, targets in the near-field of the transducer could not be detected. Transducer blanking distance coupled with a tow body depth of approximately 0.5 m effectively eliminates the top 1.5 m of water for sampling. An off-bottom integration value of 0.5 m was set to prevent integration of the bottom during postprocessing of digital acoustic data. This combined with losses at the surface eliminates approximately 2.0 m of total available sampling volume.

Data Analysis

Post processing of acoustic signal was performed using Digital Echo Visualization and Information System (DEVIS) software (Jech and Luo 2000). Corrected acoustic data were processed using Statistical Analysis System (SAS) software (SAS Institute 1999). Analysis of variance (ANOVA) was used to determine statistical ($P < 0.05$) differences between mean fish density (number/m^3) within and across categorical environmental conditions. In all graphs the error bars represent 95% confidence intervals about mean values.

Fish densities were averaged over constant time intervals throughout the water column at 3-m depth intervals. Depth intervals were set vertically starting from the blanking distance plus tow body depth to the off-bottom integration depth. Depth-strata one includes less than or equal to 3 m of acoustic sample, depth-strata two includes greater than 3 to less than or equal to 6 m of acoustic sample, and depth-strata three includes the acoustic sample that was greater than 6 m. Due to the 1.5 m blanking distance, depth-strata one represents the 1.5–4.5 m layer of water, depth-strata two represents the water layer greater than 4.5–7.5 m, and depth-strata three represents the layer of water greater than 7.5 m.

During post processing, mean pelagic fish densities were calculated for four size classes of fish (size one = 10–21 mm total length [TL]; size two = 22–39 mm TL; size three = 40–89 mm TL; size four = 90–1000 mm TL). Partitioning of fish size in this manner largely reflects larvae, juvenile, and adult size bay anchovy as well as pelagic species generally larger than adult bay anchovy in the vicinity of the power stations during the course of the study. Our focus on bay anchovy results from their dominance (Dovel 1981; Byrne 1982; Beebe and Savidge 1988) from July to September in nearshore areas that include the position of Bowline and Indian point, on the Hudson River.

Each sampling period was also divided into photoperiods corresponding to day and night. We used general linear models (Proc GLM) and Tukey's studentized range test (PC SAS) to determine significant ($P < 0.05$) differences in fish density means among environmental categories (i.e., photoperiod, depth, field). Tide condition was categorized as either ebb or flood and used as a blocking factor during statistical analysis.

Results

Bowline

Bowline's highest mean density across all size classes of fish and environmental conditions (i.e., photoperiod, depth strata, field) for a given month occurred in August ($\overline{x} = 0.1561/\text{m}^3 \pm 0.0029$) and was significantly ($P < 0.05$) greater than means for September and July. The second highest densities were in September and were significantly greater than those from July (Figure 4; Table 1).

Fish density by size–class, over combined environmental conditions, was highest for size two and three size classes during July (Figure 5; Table 1). During August, abundance of all sizes had increased, with size-three fish being significantly more abundant than all others. By September there was a stair-stepped progression in density among size classes ranging from a high of $\overline{x} = 0.1073/\text{m}^3 \pm 0.0063$ for size one to a low of $\overline{x} = 0.0484/\text{m}^3 \pm 0.0019$ for size four fish (Figure 5; Table 1).

Overall, field (near-field [NF], far-field [FF]) played a significant role as a predictor of fish density at Bowline, with the near-field lagoon concentrating fish in two of three months (Figure 6). However, in July, there was no significant difference between NF and FF within individual size classes (Table 2). During August and September, near-field transects yielded significantly greater fish densities than far-field transects (Figure 6; Table 2). Trends exhibited for near-field versus far-field areas of Bowline were consistent, with the far-field yielding means 58%, 54%, and 52% lower than near-field means for July, August, and September.

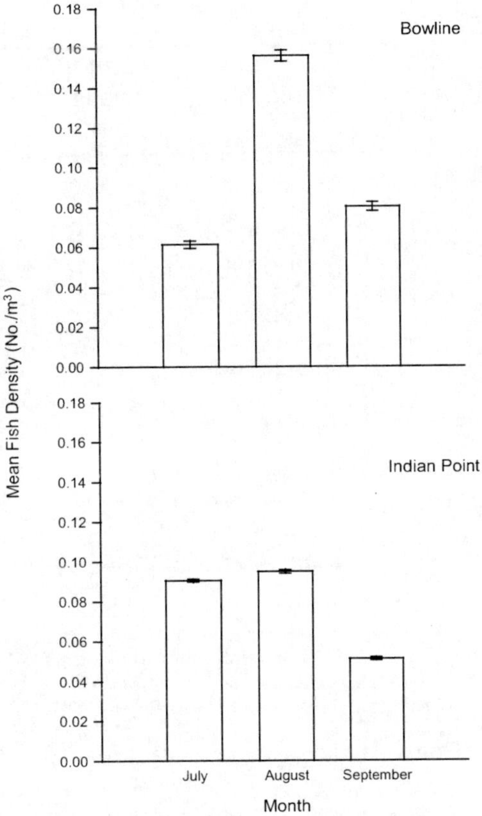

Figure 4. Mean pelagic fish density across all size classes and environmental groupings (i.e., photoperiod, depth, and field if applicable) at Bowline and Indian Point power plants.

Field effects seen over all size classes and environmental conditions generally persisted at the within size-class level for Bowline. July field results over combined size classes were consistent with results of field comparisons of means within size classes (Figure 6; Table 2). There was also no field effect for the size-four group during August and September at Bowline.

Photoperiod also affected fish density in the vicinity of Bowline power station (Figure 7). In all months, mean density of pelagic species was significantly ($P < 0.05$) higher during nighttime. Fish density means were 33%, 71%, and 44%, higher for nighttime versus

Table 1. Monthly mean pelagic fish density by size class for Bowline and Indian Point. CI = Confidence interval

Location	Month	Size Class	Mean (Fish/m³)	95% CI (+/–)
Bowline	July	1	0.0155	0.0017
		2	0.0928	0.0059
		3	0.0928	0.0040
		4	0.0446	0.0016
		1-4	0.0614	0.0019
	August	1	0.1280	0.0071
		2	0.1852	0.0065
		3	0.2362	0.0056
		4	0.0752	0.0021
		1-4	0.1561	0.0029
	September	1	0.1073	0.0063
		2	0.0879	0.0048
		3	0.0777	0.0032
		4	0.0484	0.0019
		1-4	0.0803	0.0022
Indian Point	July	1	0.0680	0.0018
		2	0.1022	0.0019
		3	0.1409	0.0015
		4	0.0508	0.0006
		1-4	0.0905	0.0008
	August	1	0.0929	0.0022
		2	0.1150	0.0022
		3	0.1282	0.0016
		4	0.0435	0.0006
		1-4	0.0949	0.0009
	September	1	0.0767	0.0023
		2	0.0543	0.0016
		3	0.0474	0.0011
		4	0.0261	0.0005
		1-4	0.0511	0.0008

daytime samples from July, August, and September samples (Figure 7; Table 3).

Analysis of depth-strata trends across all size classes and other environmental conditions within months consistently showed that depth-strata one contained greater densities of pelagic fish than depth-strata two and three, though the magnitude, and therefore significance of differences between depth-strata one and other strata varied between months. For July, depth-strata one density was significantly ($P < 0.05$) greater than depth-strata two and depth-strata three den-

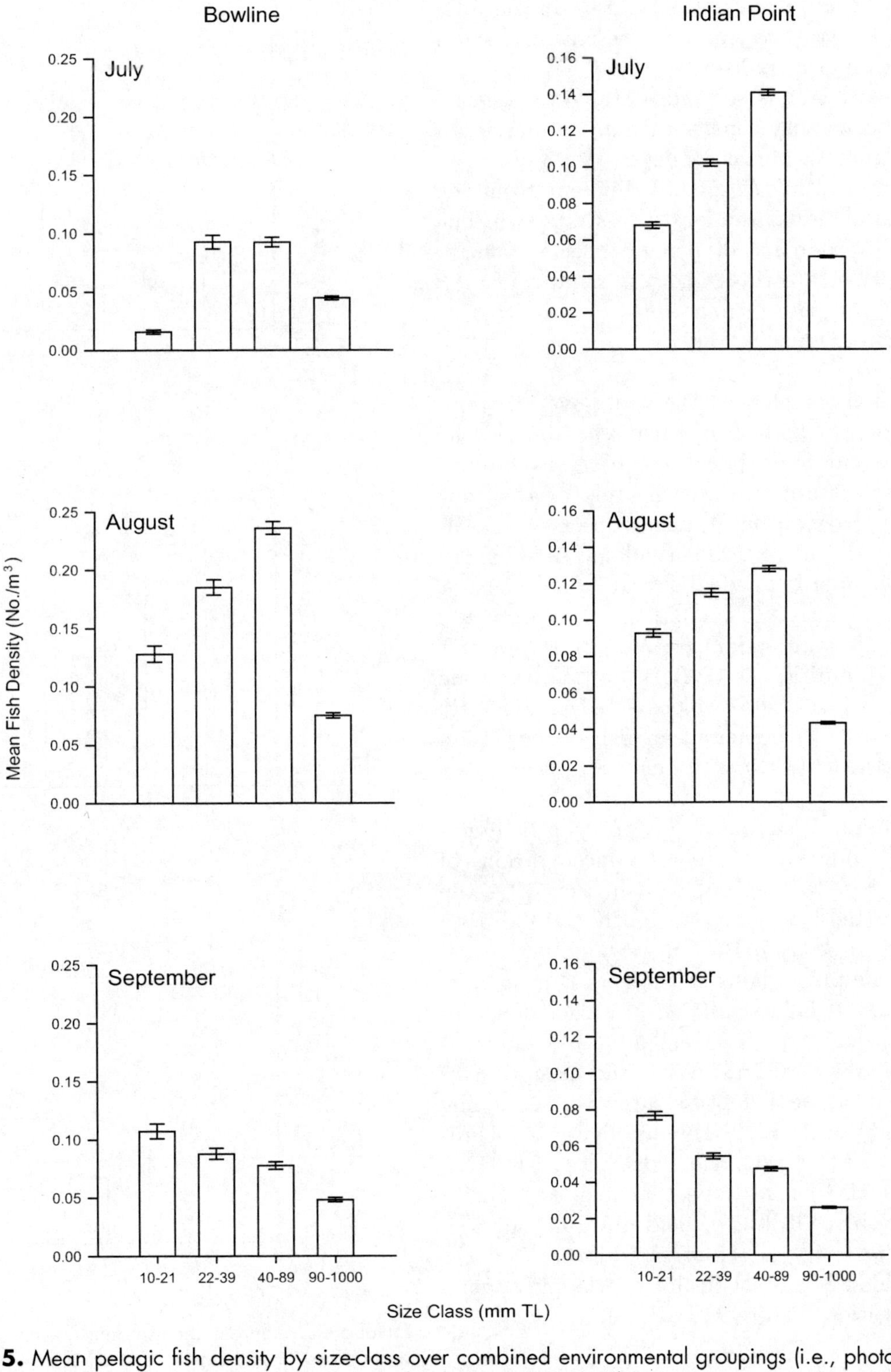

Figure 5. Mean pelagic fish density by size-class over combined environmental groupings (i.e., photoperiod, depth, and field if applicable), at Bowline and Indian Point power plants.

sity (Figure 8; Table 4). Also, during July depth-strata two and depth-strata three did not differ significantly ($P < 0.05$).
August trends showed a stair step decline in pelagic fish density with increasing depth (Figure 8). Means for depth-strata one were significantly ($P < 0.05$) different from the next highest mean for depth-strata two. The lowest density of pelagic species was in depth-strata three and was significantly ($P < 0.05$) lower than the depth-strata two mean (Figure 8; Table 4).

September showed the same overall trend in depth-strata fish density but lacked the sharp decline from depth-strata one to depth-strata two seen in July or the stair step progression of means observed in August (Figure 8). Mean pelagic fish density in September's depth-strata one was not significantly ($P < 0.05$) different from the second highest value for depth-strata two (Figure 8; Table 4). Both depth-strata one and depth strata-two possessed significantly ($P < 0.05$) higher mean fish density than depth-strata three (Figure 8; Table 4).

Overall, fine-scale trends of mean fish density showed the highest concentration, of combined size-class, of fish occupying the near-field at shallow depth (D1) during nighttime (NI), Bowline power plant acoustic samples (Figure 9; Table 5). During July, NFD1NI held significantly higher densities of pelagic species than all other environmental conditions. With the exception of the near-field depth-strata two nighttime condition (NFD2NI), combinations including depth-strata one (NFD1DA, FFD1NI, FFD1DA) held the second highest density of fish regardless of field and photoperiod. Otherwise, depth-strata two and three yielded low statistically overlapping mean densities (Figure 9; Table 5).

Trends in fish density at Bowline during July were also evident during August (Fig-

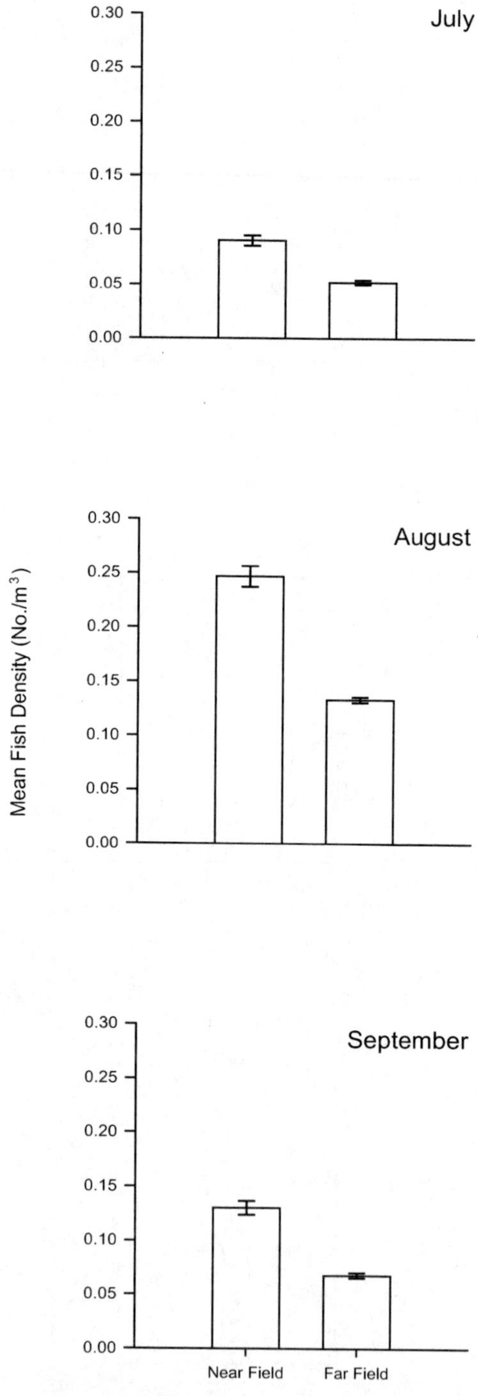

Figure 6. Mean pelagic fish density within field and across all size classes and other environmental groupings (i.e., photoperiod and depth) at Bowline power plant.

Table 2. Monthly mean pelagic fish density by field and size class for Bowline. NF = Near field, FF = Far-field. CI = Confidence interval.

			Mean (#/m³)	
Location	Month	Size Class	Near Field (95% CI)	Far-field (95%CI)
Bowline	July	1	0.0088 (0.0027)	0.0176 (0.0021)*
		2	0.0976 (0.0134)	0.0913 (0.0065)*
		3	0.1655 (0.0118)	0.0694 (0.0037)*
		4	0.0897 (0.0051)	0.0301 (0.0012)*
		1-4	0.0904 (0.0047)	0.0521 (0.0020)
	August	1	0.2140 (0.0242)	0.1059 (0.0064)
		2	0.3297 (0.0232)	0.1482 (0.0055)
		3	0.3525 (0.0170)	0.2064 (0.0054)
		4	0.0897 (0.0060)	0.0715 (0.0022)*
		1-4	0.2465 (0.0096)	0.1330 (0.0026)
	September	1	0.1943 (0.0177)	0.0852 (0.0065)
		2	0.1488 (0.0127)	0.0725 (0.0051)
		3	0.1208 (0.0105)	0.0667 (0.0029)
		4	0.0564 (0.0057)	0.0463 (0.0019)*
		1-4	0.1301 (0.0062)	0.0677 (0.0022)

* No significant difference

ure 9; Table 5). As in July, the greatest density of pelagic species occurred within NFD1NI environmental conditions. Near-field depth strata two at night (NFD2NI) also had high densities of pelagic species and was significantly ($P < 0.05$) higher than remaining environmental combinations. Fish densities in far-field depth-strata one at night (FFD1NI) and far-field depth strata-two at night (FFD2NI) were not statistically ($P < 0.05$) different but were significantly ($P < 0.05$) higher in density than remaining environmental conditions. Mean depth-strata one densities were consistently higher than depth-strata two or three regardless of field or photoperiod during the August survey. Fish density generally declined with increasing depth strata and was higher during nighttime (Figure 9; Table 5).

During the September Bowline survey, trends in fish density diverged from July and August trends (Figure 9; Table 5). The near-field depth-strata two nighttime condition (NFD2NI) held the greatest density of pelagic fish, during September. The near-field depth-strata one nighttime condition (NFD1NI) yielded the second highest fish density but was not significantly ($P < 0.05$) different than mean pelagic fish density from the near-field depth-strata one daytime condition (NFD1DA) (Figure 9; Table 5). The trend of generally decreasing fish density with increasing depth strata was seen only in the near-field and far-field daytime condition. Fish were more evenly distributed during nighttime in both near and far-fields. The higher densities in darkness (over light) was also found in the September survey (Figure 9; Table 5).

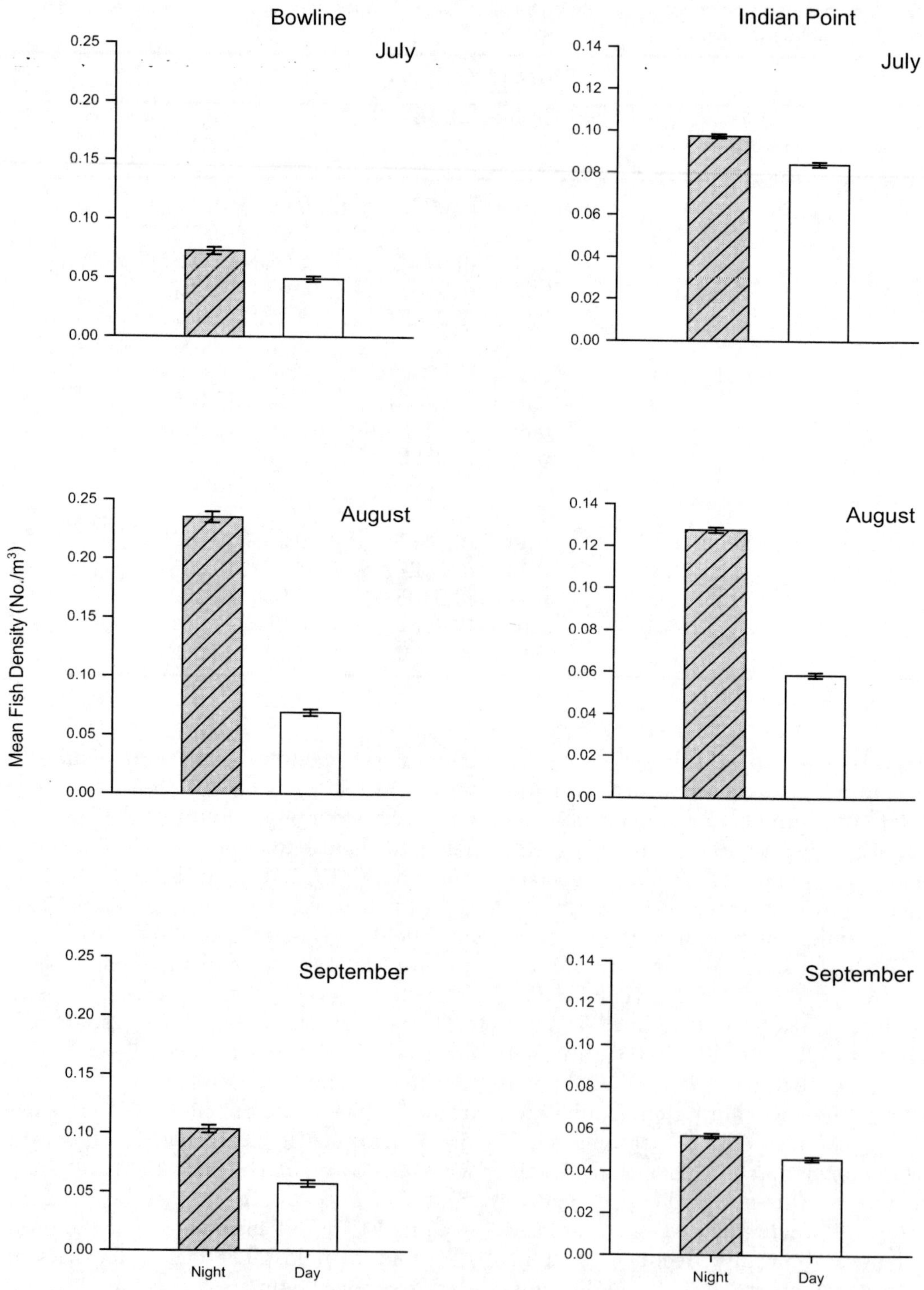

Figure 7. Mean pelagic fish density within photoperiod across all other environmental groupings (i.e., photoperiod, depth, and field if applicable) and size classes at Bowline and Indian Point power plants.

Table 3. Monthly mean pelagic fish density by photoperiod for Bowline and Indian Point. NI = Night, DA = Day. CI = Confidence interval

Location	Month	Photoperiod (Night/Day)	Mean (Fish/m^3)	95% CI (+/–)
Bowline	July	NI	0.0728	0.0030
		DA	0.0488	0.0022
	August	NI	0.2349	0.0047
		DA	0.0691	0.0028
	September	NI	0.1034	0.0033
		DA	0.0574	0.0029
Indian Point	July	NI	0.0975	0.0010
		DA	0.0841	0.0011
	August	NI	0.1275	0.0012
		DA	0.0586	0.0013
	September	NI	0.0567	0.0011
		DA	0.0457	0.0011

Indian Point

As observed for Bowline, the August Indian Point survey produced the highest mean density of fish ($0.0949/m^3 \pm 0.0009$) in comparisons that combine size classes and environmental conditions. July accounted for the second highest fish density at Indian Point (Figure 4; Table 1) and was significantly ($P < 0.05$) different than September densities. September's results exhibit the lowest density from Indian Point surveys ($0.0511/m^3 \pm 0.0008$), at slightly more than half the density seen for July and August (Figure 4).

Comparison of fish density differences between size classes, within a 24 h sample period and without regard to environmental conditions, showed consistent trends in size-class abundance for July and August and a decline in the density of smaller fish by September. Size-three fish dominated the July survey ($0.1409/m^3 \pm 0.0015$) and were significantly more abundant than other size classes. Size-two fish followed in abundance and were statistically ($P < 0.05$) greater than density for size-one and size-four fish (Figure 5; Table 1). The size-one group ranked third in abundance, but was not significantly greater than the mean of $0.0508/m^3 \pm 0.0006$ observed for the size-four group.

Size-class distribution at Indian Point was similar in July and August. Though trends were the same, the magnitude of differences between size classes was not as great in August as July (Figure 5). In August size-class, three was the most abundant ($0.1282/m^3 \pm 0.0016$), but not significantly ($P < 0.05$) greater than mean abundance for size-class two (Figure 5; Table 1). In contrast with the July survey, size-one fish in August were significantly ($P < 0.05$) more abundant than size-two and size-three fish.

By September, overall abundance of fish decreased at Indian Point. Size-class one fish were the most abundant (Figure 5; Table 1). Size-two and size-three fish were the next most abundant and did not differ statistically ($P < 0.05$). Densities of size-class four fish were below those observed for July and August and were the lowest for September (Figure 5; Table 1).

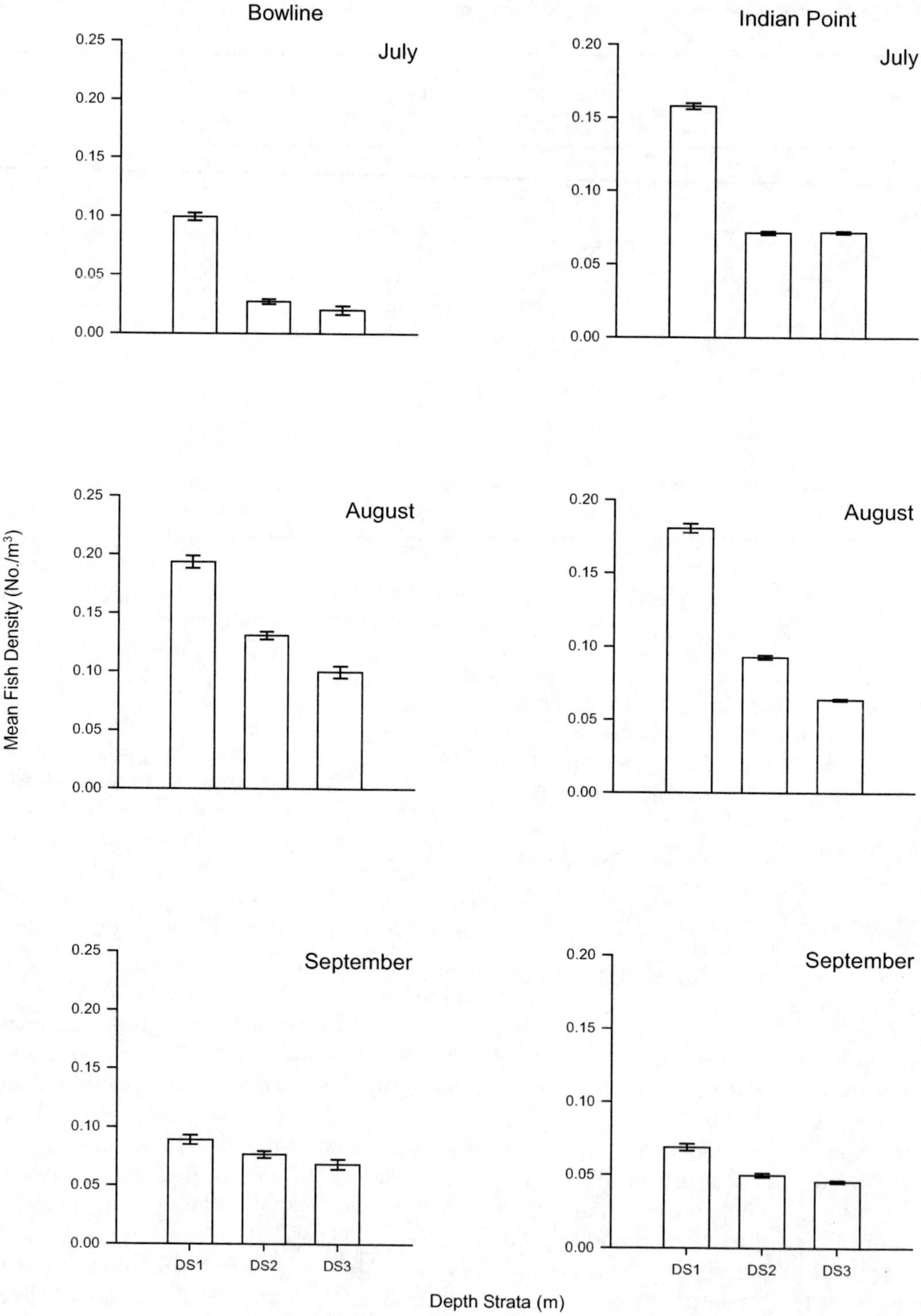

Figure 8. Mean pelagic fish density within depth strata across all other environmental groupings (i.e., photoperiod, depth, and field if applicable) and size classes at Bowline and Indian Point power plants. D1 = depth-strata one (1.5–4.5 m), D2 = depth-strata two (>4.5–7.5 m), D3 = depth-strata three (>7.5 m).

Table 4. Monthly mean pelagic fish density by depth strata for Bowline and Indian Point. Depth strata one = 1.5-4.5 m, depth strata two = > 4.5-7.5 m, depth strata three = > 7.5 m. CI = Confidence interval.

Location	Month	Depth Strata	Mean (Fish/m^3)	95% CI (+/–)
Bowline	July	1	0.0995	0.0034
		2	0.0272	0.0021
		3	0.0199	0.0037
	August	1	0.1935	0.0052
		2	0.1309	0.0033
		3	0.0998	0.0051
	September	1	0.0891	0.0040
		2	0.0763	0.0030
		3	0.0682	0.0043
Indian Point	July	1	0.1579	0.0021
		2	0.0715	0.0012
		3	0.0719	0.0010
	August	1	0.1806	0.0032
		2	0.0927	0.0015
		3	0.0637	0.0008
	September	1	0.0689	0.0024
		2	0.0496	0.0013
		3	0.0452	0.0009

The effect of photoperiod on fish density at Indian Point was similar to that of Bowline (Figure 7). The greatest difference in density between day and night samples at Indian Point was in August (NI-0.1275/m^3 $\pm$ 0.0012 versus DA-0.0586/m^3 $\pm$ 0.0013) (Figure 7; Table 3). Fish density in September continued to be greater for the nighttime group compared to the daytime group although the difference was less than in July and August.

Although mean fish abundance differed between Indian Point and Bowline, depth strata trends for July, August, and September were nearly identical between the two power stations (Figure 8; Table 4). Of notable difference, was the consistently higher density in depth-strata one for all months, including September, at Indian Point, while at Bowline there was no significant difference between depth-strata one and two. Indian Point's mean fish density from depth-strata one was significantly greater compared to lower depths within July, August, and September. During July, depth-strata two and three were not significantly ($P <$ 0.05) different at Indian Point. September samples at Indian Point also showed no statistical ($P < 0.05$) trend in fish density for depth-strata two compared to depth-strata three by pelagic species. However, during August at Indian Point, means from depth-strata two were significantly greater than means from depth-strata three (Figure 8; Table 4).

The interaction of depth strata and photoperiod at Indian Point showed the importance of shallow depth and darkness, but also indicated that shallow depth during daytime was important during July and

August (Figure 10; Table 6). In two of three months depth-strata one during conditions of darkness yielded significantly higher mean fish densities than other combinations. During July and August depth-strata one during daytime was the second highest group, with respect to mean fish density. As observed for other environmental comparisons during September, no clear trend in fish density was associated with any group (Figure 10; Table 6).

Discussion

Though many species of fish are at risk to impingement and entrainment effects in the Hudson River, the New York State Department of Environmental Conservation (NYSDEC) considers a few species the most susceptible. NYSDEC lists white perch *Morone americana*, bay anchovy, river herring *Alosa* spp., American shad *A. sapidissima*, and striped bass *Morone saxatilis* as species of concern for impingement and entrainment, due to their frequent presence in the vicinity of cooling water intake systems at the Bowline power generating station (Casutto 2001). These species were also included by Boreman and Goodyear (1988) in models estimating conditional entrainment mortality of fish in the vicinity of cooling water intake structures at Hudson River power plants (including Bowline and Indian Point).

Based upon a concurrent riverwide ichthyoplankton netting, trawling, and hydroacoustic survey, it is likely that most of the fish distribution patterns observed in the vicinity of Indian Point and Bowline were driven by the numerical dominance of life stages of bay anchovy. Bay anchovy represented 90% or more of the fishes collected in size categories one (Bassista 2000), two, and three (Tipton 2003) during 1996. Bay anchovy have also been shown to be numerically dominant species in fisheries re-

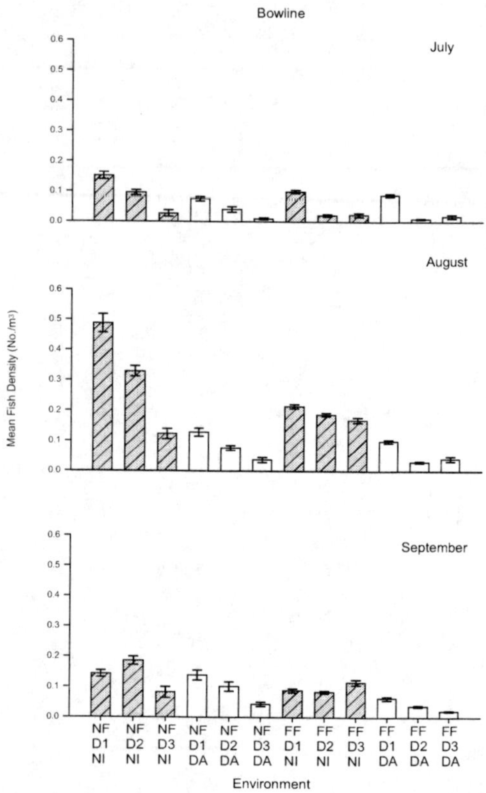

Figure 9. Mean pelagic fish density across size classes, within field, depth strata, and photoperiod, at Bowline power plant. NF = near field, FF = far-field, D1 = depth-strata one (1.5–4.5 m), D2 = depth-strata two (> 4.5–7.5 m), D3 = depth-strata three (> 7.5 m), NI = Night, and DA = Day.

search collections from the Hudson River Estuary (Beebe and Savidge 1988; Byrne 1982; Dovel 1981). Bay anchovy spawning was shown by Bassista and Hartman (2005) to have peaked in the Hudson River during July and August and ended by September 1996. This would make anchovy an abundant component of our first three size groupings. Boreman and Goodyear (1988) considered larval and juvenile bay anchovy to be the most susceptible to impingement and entrainment effects. Given the dominance of bay anchovy in fish communities along the U.S. Atlantic coast (Hildebrand and Schroeder 1928; Olney 1983; Morton 1989), inferences regarding patterns in fish distri-

Table 5. Monthly mean pelagic fish density across all size classes within field, depth strata, and photoperiod at Bowline. NF = Near field, FF = Far-field, D1 = Depth-strata one 91.5-4.5 m), D2 = Depth-strata two (>4.5-7.5 m), D3 = Depth-strata three (>7.5 m), NI = Night, DA = Day. CI = Confidence interval.

Location	Month	Environment	Mean (Fish/m^3)	95% CI (+/–)
Bowline	July	NFD1NI	0.1501	0.0122
		NFD2NI	0.0947	0.0089
		NFD3NI	0.0266	0.0106
		NFD1DA	0.0746	0.0073
		NFD2DA	0.0403	0.0090
		NFD3DA	0.0095	0.0029
		FFD1NI	0.0989	0.0056
		FFD2NI	0.0209	0.0034
		FFD3NI	0.0226	0.0065
		FFD1DA	0.0874	0.0047
		FFD2DA	0.0090	0.0016
		FFD3DA	0.0177	0.0057
	August	NFD1NI	0.4870	0.0306
		NFD2NI	0.3290	0.0169
		NFD3NI	0.1225	0.0172
		NFD1DA	0.1281	0.0136
		NFD2DA	0.0762	0.0080
		NFD3DA	0.0365	0.0082
		FFD1NI	0.2141	0.0070
		FFD2NI	0.1867	0.0056
		FFD3NI	0.1687	0.0081
		FFD1DA	0.0991	0.0055
		FFD2DA	0.0290	0.0026
		FFD3DA	0.0397	0.0079
Bowline	September	NFD1NI	0.1424	0.0114
		NFD2NI	0.1858	0.0138
		NFD3NI	0.0818	0.0181
		NFD1DA	0.1384	0.0164
		NFD2DA	0.1015	0.0154
		NFD3DA	0.0427	0.0068
		FFD1NI	0.0881	0.0064
		FFD2NI	0.0828	0.0045
		FFD3NI	0.1148	0.0089
		FFD1DA	0.0625	0.0057
		FFD2DA	0.0369	0.0028
		FFD3DA	0.0206	0.0020

butions in these Hudson River monthly 24 h surveys could be important in future studies associated with citing of new power plants or CWIS's.

Management Implications

Models estimating conditional entrainment mortality rate widely consider mortality of entrained organisms to be 100% (EPA 2004; Perry et al. 2002; Boreman and Goodyear 1988; Muessig et al. 1988; Polgar et al. 1988). In these models, conditional entrainment mortality is defined as the percent reduction in year-class strength of fish as a result of power plant entrainment mortality. Entrainment refers to the process by which smaller organisms (usually <50 mm; Vaughan 1988) are carried along with intake water through exclusion screens into the plant's cooling system. Impingement of species larger than 50 mm on intake screens may also be a significant source of mortality. Adult bay anchovy, for example, are fragile and mortality associated with their impingement would be expected to be nearly 100%. Boreman and Goodyear (1988) estimated that conditional entrainment mortality rates for bay anchovy in the Hudson River Estuary might range from 38 to 75% of the population. Boreman and Goodyear (1988) also caution that bay anchovy conditional entrainment mortality estimates from power plants may not be representative of the entire river population because available distributional data showed concentration of vulnerable life stages below river km 22. Because of this they felt that their conditional entrainment mortality estimates were only applicable to the portion of the bay anchovy population that occurred above river km 22 (Boreman and Goodyear 1988). Boreman and Goodyear's (1988) concerns about nonuniform fish distribution brought into focus the need for fine scale estimates of fish abundance in the vicinity of power plants. Such information would improve conditional entrainment model estimates, by providing time- (e.g., day versus night) and depth-specific (e.g., at the depth of cooling water intake) information on fish abundance. Fine scale abundance and distribution data would improve derived W-factors (i.e., the ratio of the density of organisms in power plant intake water to the average density of organisms nearby: Boreman and Goodyear 1988; Christensen and Englert 1988; Englert and Boreman 1988; Vaughan 1988) used in some entrainment models to estimate fish loss.

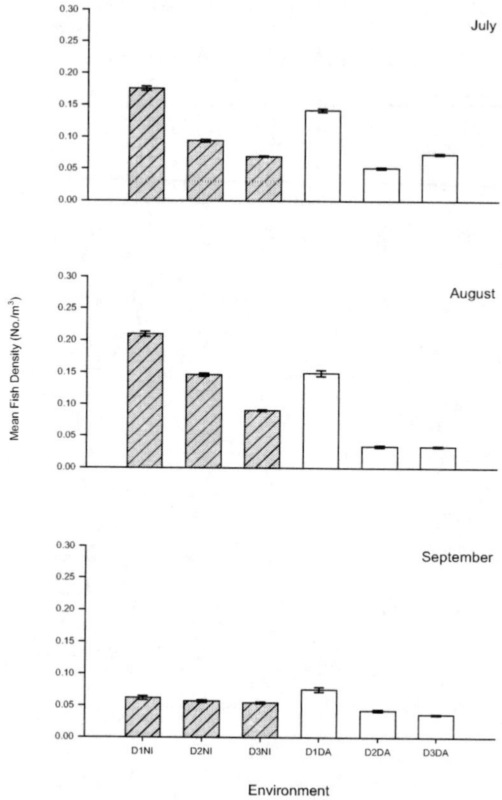

Figure 10. Mean pelagic fish density across size classes, within depth strata and photoperiod, at Indian Point power plant. D1 = depth-strata one (1.5–4.5 m), D2 = depth -strata two (>4.5–7.5 m), D3 = depth-strata three (>7.5 m), NI = Night, and DA = Day.

Information from this study suggests that substantial benefits to pelagic species are possible with modified power plant operations. Benefits to the ecologically important bay anchovy may be achieved if the timing of maximum water intake is modified dur-

Table 6. Monthly mean pelagic fish density across all size classes within depth strata, and photoperiod at Indian Point. D1 = Depth-strata one, D2 = Depth-strata two, D3 = Depth-strata three, NI = Night, DA = Day. CI = Confidence interval.

Location	Month	Environment	Mean (Fish/m^3)	95% CI (+/–)
Indian Point	July	D1NI	0.1757	0.0032
		D2NI	0.0943	0.0019
		D3NI	0.0698	0.0011
		D1DA	0.1422	0.0028
		D2DA	0.0514	0.0015
		D3DA	0.0739	0.0016
	August	D1NI	0.2096	0.0041
		D2NI	0.1462	0.0024
		D3NI	0.0905	0.0011
		D1DA	0.1488	0.0048
		D2DA	0.0342	0.0017
		D3DA	0.0336	0.0010
	September	D1NI	0.0620	0.0029
		D2NI	0.0570	0.0020
		D3NI	0.0546	0.0014
		D1DA	0.0755	0.0039
		D2DA	0.0425	0.0018
		D3DA	0.0359	0.0011

ing the relatively short period of time (July and August) bay anchovy spawn in the area. Unfortunately, spawning of bay anchovy coincides with peak electrical demands of summer. However, the timing of species concentrations in upper regions of the water column was consistently lower during daylight when energy and power plant cooling water demands are highest. Even Indian Point, where fish densities were greatest in shallow water during daytime, still had a significantly greater abundance of pelagic species at night during July and August. Therefore, reduced water intake at night could substantially decrease impingement and entrainment losses.

The potential for decreasing impingement and entrainment loss over the study period can be illustrated using simple calculations based on maximum water intake rates for Bowline and Indian Point, reported by Hutchison (1988) and fish abundance by fraction of day and night. For simplicity, we assumed 100% mortality of all individuals encountered acoustically regardless of their size and position relative to each power plant. We partitioned abundance estimates and water intake rates by percent day and night for each month and power plant and summarized the percent loss by day and night (Figure 11). With the exception of July, each power plant and month combination showed greater potential for impingement and entrainment loss at night. July results for Bowline and Indian Point were nearly equal with respect to day versus night, suggesting that a maximum of approximately half the loss due to impingement and entrainment could be avoided if intake

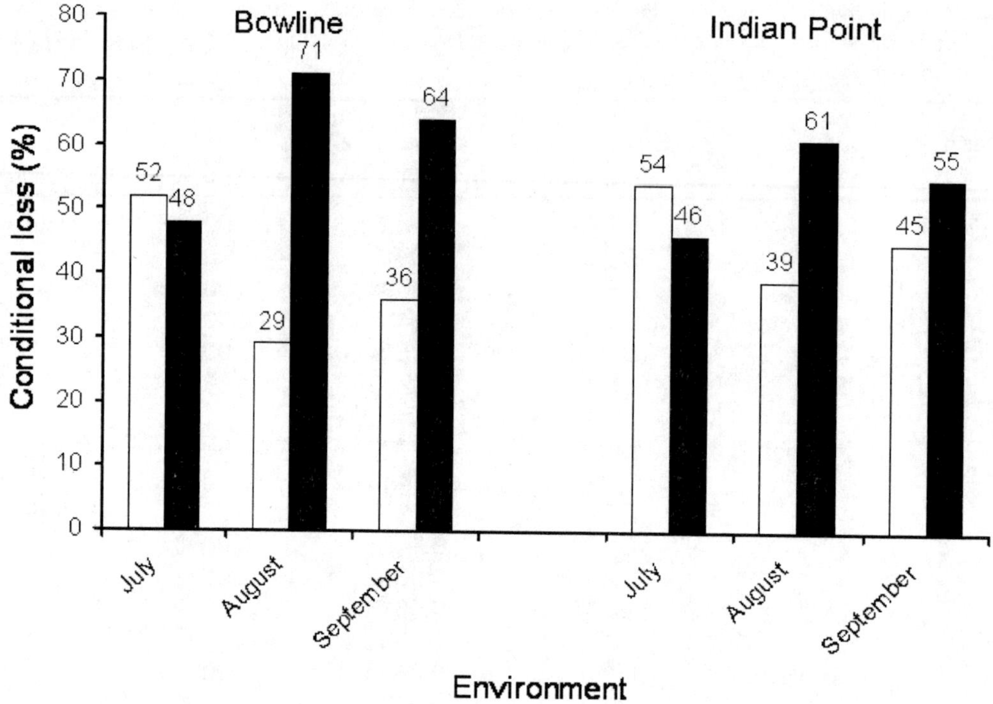

Figure 11. Conditional impingement and entrainment loss assuming 100% mortality of all size classes of fish at Bowline and Indian Point power plants. Filled bars and open bars represent estimates from densities during daylight and darkness.

did not occur at night. Conditional total loss estimates ranged from 194,291 fish/d (Indian, Point July) to 412,915 fish/d (Bowline, August) at night and from 167,844 fish/d (Bowline, August) to 231,278 fish/d (Indian, Point July) during the day. Though the proportionality of the trend is believed to be accurate, it is obvious that the assumptions applied are coarse and may yield values greater than what would actually be experienced. They are used here only for illustration of the potential benefit to fish when water intake is adapted to one environmental aspect under simplified conditions.

While a simpler evaluation of fish behavior (with regard to single environmental categories) is valuable, fish density relative to environmental interaction provides finer resolution of site specific fish behavior in the vicinity of power plants. Our study shows when, where, and at what depth fish, vulnerable to impingement and entrainment, occurred. Results from this study also show that hydroacoustic surveys can be a useful tool in processes involving power plants and fish. As illustrated above, scheduling periods of peak power production and therefore water withdrawal by power plants around periods when conditions favor elevated densities of pelagic fish would reduce impingement and entrainment potential. A priori fine-scale habitat and fish distribution assessments could aid in strategic placement of future intake and effluent structures. Physical conditions in the river vary stochastically through time and influence life history strategy and distribution of important pelagic species considerably over relatively short time spans.

Hydroacoustics can be used to determine size- and depth-specific fish distributions while environmental conditions are changing. Real time information regarding the position and distribution of entrainable fish could be used as a negative feedback system. Such a system could be used to modify water intake during periods in which a threshold density of entrainable targets are observed in the near-field of intake structures. Modification of water intake may include reduction of water withdrawal or switching temporarily to a stored source. The system would be cost effective compared to upgrading to more advanced cooling systems and could provide an interim solution during cooling technology upgrades. From a fisheries research standpoint, methods illustrated in this study may improve the efficiency of future surveys in areas where pelagic species may be at risk. This can be accomplished with even a cursory hydroacoustic determination of fish distribution and abundance prior to implementation of other fisheries gear. This could reduce cost and increase efficiency by focusing effort at locations in time that maximize the potential for success.

Differences in fish abundance and distribution were observed at Bowline and Indian Point over a period of time critical to recruitment success of bay anchovy in the area. Results with management implications include:

1) Elevated probability of impingement and entrainment losses during July and August at both power plants compared to September;

2) Increased density and therefore increased impingement and entrainment potential of fish in Bowline's near field;

3) Increased fish abundance in shallow water during nighttime at Bowline;

4) Increased fish abundance in shallow water during nighttime in July and August at Indian Point with substantial numbers of fish in shallow depths during the daytime; and

5) Overlap with regard to environmental interactions during September at Bowline and Indian Point, which is believed to result from emigration of bay anchovy from the area.

Assuming that trends observed in this study are applicable elsewhere in the region, modification of water withdrawal cycles could reduce the potential for impingement and entrainment losses at power facilities in the Lower Bay and Long Island Sound.

Acknowledgments

We thank P. Mazik, J. T. Petty, J. Luo, S. Welsh, and three anonymous reviewers for critical review and comments that improved this paper. We thank the Hudson River Foundation for funding of this study.

References

Barnthouse, L. W., and W. Van Winkle. 1988. Analysis of impingement impacts on Hudson River fish populations. Pages 182–190 *in* L. W. Barnthouse, R. J. Klauda, D. S. Vaughan, and R. L. Kendall, editors. Science, law, and Hudson River power: a case study in environmental impact assessment. American Fisheries Society, Monograph 4, Bethesda, Maryland.

Bassista, T. P. 2000. Reproductive biology and early life stage abundance and distribution of bay anchovy (*Anchoa mitchilli*) in the Hudson River Estuary, USA. Master's thesis. West Virginia University, Morgantown.

Bassista, T. P., and K. J. Hartman. 2005. Reproductive

biology and egg mortality of bay anchovy, *Anchoa mitchilli*, in the Hudson River Estuary. Environmental Biology of Fishes 73:49–59.

Beebe, C. A., and I. R. Savidge. 1988. Historical perspective on fish species composition and distribution in the Hudson River Estuary. Pages 25–36 *in* L. W. Barnthouse, R. J. Klauda, D. S. Vaughan, and R. L. Kendall, editors. Science, law, and Hudson River power: a case study in environmental impact assessment. American Fisheries Society, Monograph 4, Bethesda, Maryland.

Boreman, J., and C. P. Goodyear. 1988. Estimates of entrainment mortality for striped bass and other fish species inhabiting the Hudson River estuary. 152–160 *in* L. W. Barnthouse, R. J. Klauda, D. S. Vaughan, and R. L. Kendall, editors. Science, law, and Hudson River power: a case study in environmental impact assessment. American Fisheries Society, Monograph 4, Bethesda, Maryland.

Byrne, C. J. 1982. Anchovy. Pages 65–67 *in* M. D. Grosslein and T. R. Azarovitz, editors. Fish distribution. Mesa New York Bight Atlas Monograph 15. New York Sea Grant Institute, Albany.

Casutto, K. J. 2001. In the matter of Mirant Bowline LLC, state pollutant discharge elimination system permit hearing report and recommended decision. New York State Department of Environmental Conservation, Albany.

Christensen, S. W., and T. L. Englert. 1988. Historical development of entrainment models for Hudson River striped bass. Pages 133–142 *in* L. W. Barnthouse, R. J. Klauda, D. S. Vaughan, and R. L. Kendall, editors. Science, law, and Hudson River power: a case study in environmental impact assessment. American Fisheries Society, Monograph 4, Bethesda, Maryland.

Dovel, W. L. 1981. Ichthyoplankton of the lower Hudson Estuary, New York. New York Fish and Game Journal 28:21–39.

Englert, T. L., and J. Boreman. 1988. Historical review of entrainment impact estimates and the factors influencing them. Pages 143–151 *in* L. W. Barnthouse, R. J. Klauda, D. S. Vaughan, and R. L. Kendall, editors. Science, law, and Hudson River power: a case study in environmental impact assessment. American Fisheries Society, Monograph 4, Bethesda, Maryland.

EPA (U. S. Environmental Protection Agency). 2004. Cooling water intake structures-Section 316(b): Proposed Section 316(b) Phase II Existing Facilities Rule Economic Benefits Analysis; USEPA. Available: http://www.epa.gov/waterscience/316b/econbenefits/ (August 2004).

Hildebrand, S. F., and W. C. Schroeder. 1928. Fishes of Chesapeake Bay. Bulletin of the Bureau of Fisheries 53(1):1–388, Smithsonian Institute Press, Washington D.C.

Hutchison, J. B., Jr. 1988. Technical descriptions of Hudson River electricity generating stations. Pages 113–120 *in* L. W. Barnthouse, R. J. Klauda, D. S. Vaughan, and R. L. Kendall, editors. Science, law, and Hudson River power: a case study in environmental impact assessment. American Fisheries Society, Monograph 4, Bethesda, Maryland.

Jech, J. M., and J. Luo. 2000. Digital Echo Visualization and Information System (DEVIS) for processing spatially-explicit fisheries acoustic data. Journal of Fisheries Research 27:115–124.

Morton, T. 1989. Species profiles: life histories and environmental requirements of coastal fishes and invertebrates (mid-Atlantic). Bay anchovy. U. S. Fish and Wildlife Service Biological Report 82(11.97).

Muessig, P. H., J. R. Young, D. S. Vaughan, and B. A. Smith. 1988. Advances in field and analytical methods for estimating entrainment mortality factors. Pages 124–132 *in* L. W. Barnthouse, R. J. Klauda, D. S. Vaughan, and R. L. Kendall, editors. Science, law, and Hudson River power: a case study in environmental impact assessment. American Fisheries Society, Monograph 4, Bethesda, Maryland.

Olney, J. E. 1983. Eggs and early larvae of the bay anchovy, Anchoa mitchilli, and the weakfish, *Cynoscion regalis*, in lower Chesapeake Bay with notes on associated ichthyoplankton. Estuaries 6:20–35.

Perry, E., G. Seegert, J. Vondruska, T. Lohner, and R. Lewis. 2002. Modeling possible cooling-water intake system impacts on Ohio River fish populations. The Scientific World 2(S1):58–80.

Polgar, T. T., M. A. Turner, and J. K. Summers. 1988. Effect of power plant entrainment on the population dynamics of the bay anchovy (*Anchoa mitchilli*). Ecological Modelling 41:201–218.

SAS Institute. (1999). SAS/IML Software: usage and reference, Version 8.01, 1st edition. SAS Institute, Cary, North Carolina.

Tipton, R. C. 2003. Distributional ecology of bay anchovy (*Anchoa mitchilli*) in the Hudson River Estuary, USA. Dissertation. West Virginia University, Morgantown.

Vaughan, D. S. 1988. Entrainment and impingement impacts: Introduction American Fisheries Society Monograph 4:121–123.

American Fisheries Society Symposium 51:273–286, 2006

Transport of Striped Bass Larvae out of the Lower Hudson River Estuary

DENNIS J. DUNNING AND QUENTIN E. ROSS[1]
New York Power Authority, 123 Main Street, White Plains, New York 10601 USA
dennis.dunning@nypa.gov

ALAN F. BLUMBERG
Stevens Institute of Technology, Department of Civil, Environmental, and Ocean Engineering, Davidson Laboratory, Hoboken, New Jersey 07030 USA

DOUGLAS G. HEIMBUCH
AKRF, Inc., 7250 Parkway Drive, Suite 210, Hanover, Maryland 21076 USA

Abstract.—Our objectives were to determine if striped bass *Morone saxatilis* larvae were present in the East River and if so, could they have come from the Hudson River. To meet the first objective, we examined entrainment data collected at the Charles Poletti Power Plant (Poletti) during the years 1999 through 2002. To meet the second objective, we examined the simulated release of 168,000 neutrally buoyant, passive particles in the lower Hudson River Estuary, using a particle-tracking model that was linked to an estuarine circulation model. We also compared the abundance of striped bass post-yolk-sac larvae (PYSL) collected in the East River at Poletti with the abundance of striped bass PYSL collected in the Battery region of the lower Hudson River Estuary and the abundance of striped bass PYSL in the Battery region with freshwater flow in the estuary. Striped bass PYSL were collected by entrainment sampling in the East River at Poletti every year from 1999 through 2002. The striped bass PYSL in the East River probably came from the Hudson River Estuary because the median probability that neutrally buoyant, passive particles would be transported from the lower Hudson River Estuary to the upper East River and western Long Island Sound was 0.12, with a median transport time of 2 d, and because the mean density of striped bass PYSL was highest at Poletti and in the Battery region during the same year. The abundance of striped bass PYSL in the Battery region was higher when freshwater flow during May and early June was higher.

Introduction

The anadromous stock of striped bass *Morone saxatilis* that spawns in the Hudson River Estuary is a major contributor to Atlantic coastal fisheries (Richards and Rago 1999). Concern about the effects of water withdrawals on this stock arose in 1963 when an application was submitted to the Federal Power Commission for construction of a pump-storage power plant at Storm King Mountain and intensified when the Indian Point Unit No. 2 power plant became op-

[1]Current address: 348 County Highway 2, Mount Upton, New York 13809 USA

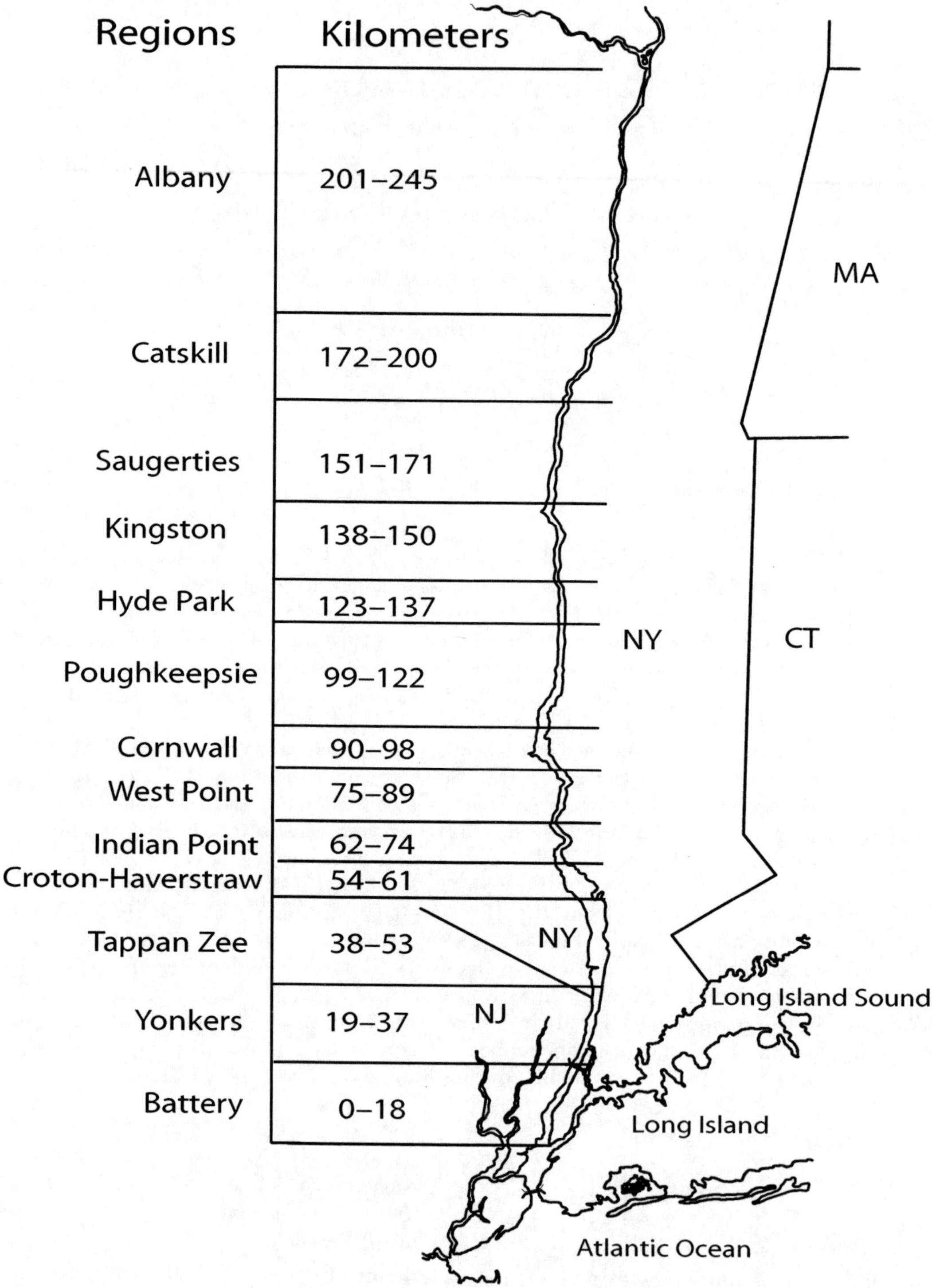

Figure 1. Regions of the Hudson River Estuary.

erational in 1973. Environmental advocacy groups claimed that the combined impacts of the Storm King Mountain and Indian Point power plants would pose catastrophic consequences for both the Hudson River and Atlantic coast stocks of striped bass (Cronin and Kennedy 1997).

In response to the dire predictions of environmental advocacy groups, annual ichthyoplankton sampling was started in 1974 to quantify the abundance and distribution of fish eggs and larvae in the Hudson River Estuary. From 1974 through 1988, annual ichthyoplankton sampling was done in the Albany through Yonkers regions (Figure 1) (Heimbuch et al. 1992; Metzger et al. 1992). Striped bass eggs and larvae were found in all of those river regions (Boreman and Klauda 1988). After 1988, sampling was extended into the Battery region, which ends at the mouth of the Hudson River Estuary. Striped bass larvae were consistently found in the Battery region(Central Hudson et al. 1999).Because ichthyoplankton sampling was not extended beyond the mouth of the Hudson River Estuary, the distribution of striped bass larvae outside the estuary was not determined.

Small striped bass (< 50 mm long) were collected during June 1999, 2000, and 2001 by a beach seine survey designed to monitor the abundance of young of the year (age-0) in western Long Island Sound (McKown and Brischler 2001a, 2001b, 2002); some were as small as 12 mm (K. McKown, New York State Department of Environmental Conservation, personal communication). Since striped bass transform into age-0 at 15 mm (Mansueti 1958), some of the small striped bass collected in western Long Island Sound during June could have been larvae; others may have been age-0 that recently transformed from larvae. The source of the small striped bass could have been the Hudson River Estuary; it could also have been a tributary of Long Island Sound. Historically, striped bass probably spawned in rivers flowing into Long Island Sound, but by 1941, spawning in those rivers seldom if ever occurred because Atlantic coast stocks of striped bass were overfished and striped bass spawning habitat was altered (Merriman 1941). Strict fishing regulations starting in 1986 eliminated overfishing and increased the size of Atlantic coast stocks of striped bass (Richards and Rago 1999). If the range of the Atlantic coast stocks of striped bass expanded as they recovered from overfishing, as seen in other species (Hilborn and Walters 1992), they may have begun to use rivers flowing into Long Island Sound for spawning.

If striped bass larvae from the Hudson River Estuary were the source of small striped bass collected in western Long Island Sound during June, they must have passed through the East River because it connects the estuary to western Long Island Sound (Figure 2). Serendipitously, entrainment sampling occurred at the Charles Poletti Power Plant (Poletti) during the same years when small striped bass were collected in western Long Island Sound during June. Poletti is located on the East River upriver of the confluence of the East River and the Harlem River, the only other connection between the Hudson River Estuary and western Long Island Sound. Therefore, entrainment sampling at Poletti provided an opportunity to detect the presence of striped bass larvae in the East River if they were going from the Hudson River Estuary to western Long Island Sound.

To complement entrainment sampling at Poletti, passive transport of striped bass larvae out of the lower Hudson River Estuary (km 0–km 29) to the upper East

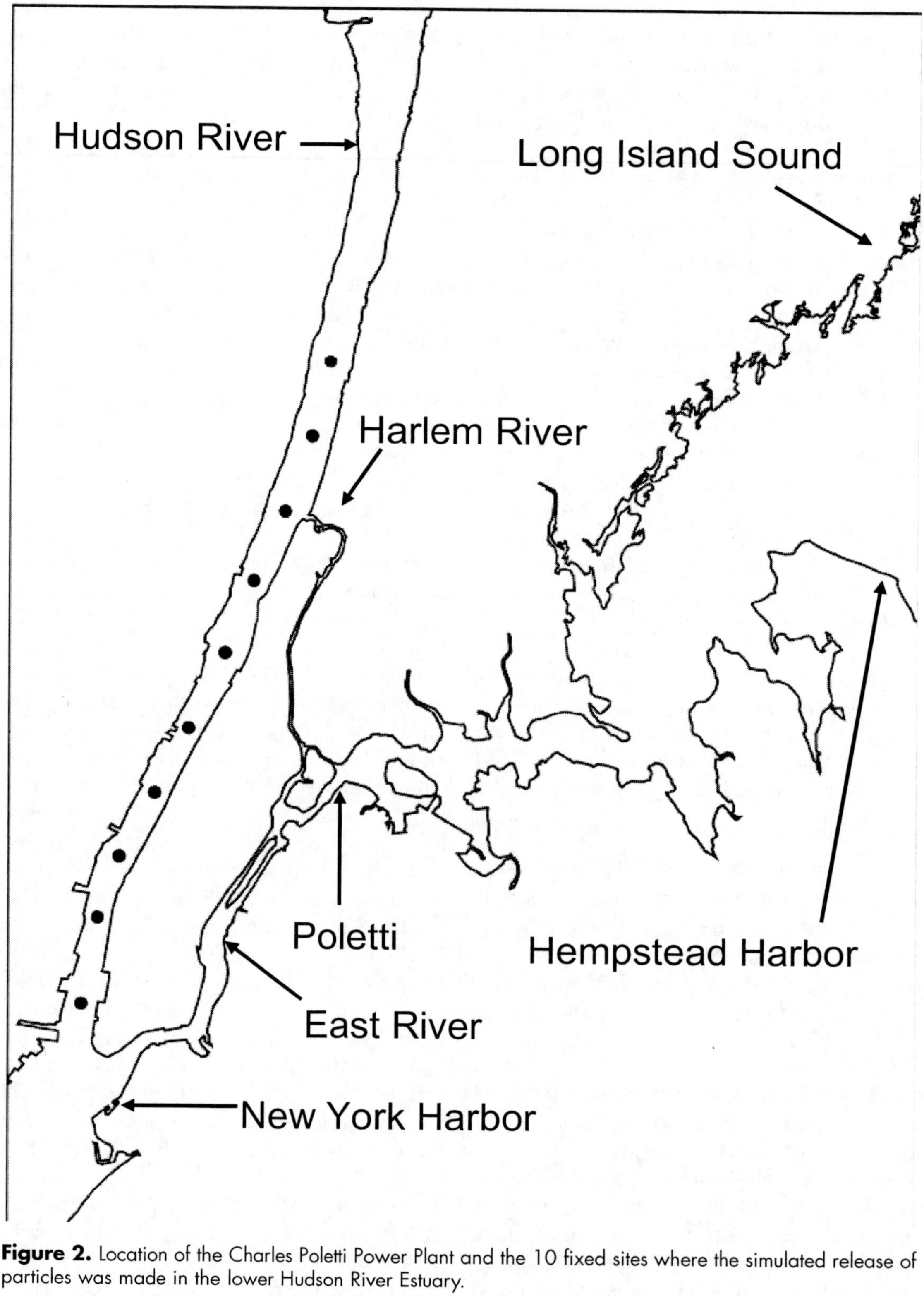

Figure 2. Location of the Charles Poletti Power Plant and the 10 fixed sites where the simulated release of particles was made in the lower Hudson River Estuary.

River and western Long Island Sound was simulated using a particle-tracking model coupled with a three-dimensional, estuarine circulation model. The estuarine circulation model simulated the hydrodynamics of the Hudson River Estuary, upper and lower bays of New York Harbor, the East River, and Long Island Sound. The particle-tracking model followed the movements of neutrally buoyant, passive particles.

If striped bass larvae are transported from the lower Hudson River Estuary to the East River, then the abundance of striped bass larvae in the East River would be directly related to the abundance of striped bass larvae in the lower estuary. Striped bass spawn in the middle of the Hudson River Estuary (Raney 1952) where freshwater flow has a profound effect on the distribution of neutrally buoyant, passive particles (Blumberg et al. 2004). Therefore, freshwater flow should strongly affect the abundance of striped bass larvae in the lower Hudson River Estuary.

Our objectives were to determine if striped bass larvae were present in the East River and if so, could they have come from the Hudson River Estuary. To meet the first objective, we examined entrainment data collected at Poletti during the years 1999 through 2002. To meet the second objective, we examined the simulated release of 168,000 neutrally buoyant, passive particles in the lower Hudson River Estuary. We also compared the abundance of striped bass PYSL collected in the East River at Poletti with the abundance of striped bass post-yolk-sac larvae (PYSL) collected in the Battery region of the lower Hudson River Estuary and the abundance of striped bass larvae in the Battery region with freshwater flow in the estuary.

Methods

Poletti Entrainment Sampling

Larvae were collected from the East River by entrainment sampling at Poletti, a fossil-fuel power plant with a maximum net generating capacity of 885 MW. Poletti has a once-through cooling water system with an intake and a discharge canal located along the shoreline. The Poletti intake withdraws water from the East River and is equipped with vertical traveling screens to prevent fish and debris from entering the cooling water system. The design flow rate at the intake is 1,634 m^3/min. However, the mean flow rate at the intake varies depending on the amount of electricity being generated at Poletti.

Entrainment samples were collected over one 24-h period weekly from January 1999 through December 2002. A single sample was generally collected every 6 h for a total of four samples during each 24-h sampling period. Collections were made using a 10.16-cm gasoline-powered trash pump connected to an aluminum pipe that extended into the discharge canal. Water from the pump passed into a barrel sampler containing a 500-μm mesh plankton net that was 1 m long and 0.76 m in diameter at the mouth. The barrel sampler was equipped with a primary drain system, which maintained the water level below the top of the tank. The net had a vented and removable cod end. A Signet flowmeter was used to determine the volume of water sampled, which typically ranged between 100 and 120 m^3 during each 2-h collection. After sampling, the contents of the plankton net were concentrated and preserved in 10% formalin and rose bengal. Preserved contents were examined and all larvae were identified to the lowest practical taxon and then counted.

The density (number/1,000 m^3) of larvae entrained was calculated for each 6-h period as the number caught divided by the sampling volume. The four 6-h density estimates within a 24-h period were averaged to produce a daily density. For comparisons with Hudson River ichthyoplankton collections, the life stage of PYSL was used because it lasts about 5 weeks, and therefore, it is relatively well sampled. The density of striped bass PYSL, for those weeks when striped bass PYSL were present in the Battery region of the Hudson River, were averaged to produce a density for each year, referred to as the mean density at Poletti in the remainder of this paper.

Hudson River Ichthyoplankton Sampling

From 1999 through 2002, ichthyoplankton sampling was done weekly in the Hudson River Estuary from March through June and biweekly from July through October in all river regions. The Battery region (km 0–18) lies within the lower Hudson River Estuary, as delineated for the particle-tracking model (Figure 1). Ichthyoplankton sampling was done with 1.0-m^2 Tucker trawls and 1.0-m^2 epibenthic sleds fitted with 500-μm mesh nets (Metzger et al. 1992). Both gear types were towed against the current for 5 min. A digital flowmeter was used to determine the volume of water sampled. After sampling, the contents of the plankton net were concentrated and preserved in 10% formalin. Preserved contents were examined and all fish eggs and larvae were identified to the lowest practical taxon and life stage unless a sample contained large numbers of eggs or larvae, in which case subsampling may have occurred; then they were counted (ASA 2000).

The weekly density of striped bass PYSL in the Battery region and the weekly standing crops of PYSL in the Battery, Yonkers, and Tappan Zee regions were estimated using approaches previously developed for the Hudson River (ASA 2000). The density of striped bass PYSL, for those weeks when striped bass PYSL were present in the Battery region, were averaged to produce a density of PYSL for the Battery region for each year, referred to as mean density in the Battery region in the remainder of this paper. The densities of striped bass PYSL in the Yonkers and Tappan Zee regions were used, along with the density in the Battery region, to assess the effect of freshwater flow.

Freshwater Flow

Freshwater flow for the Hudson River Estuary during 2001 and 2002 was measured by U.S. Geological Survey (USGS) at the federal dam gauge at Green Island, New York (USGS 2004). This gauge is typically used to represent freshwater flow into the Hudson River Estuary and measures the contributions of the upper Hudson River and the Mohawk River. However, data were not collected at that federal dam gauge during 1999 and 2000. So, freshwater flow into the Hudson River Estuary during 1999 and 2000 was estimated from USGS data collected at gauges on the upper Hudson River above Lock 1 and on the Mohawk River at Cohoes (USGS 2004). For the period 1 October 2000 through 30 September 2001, the combined flows from the upper Hudson River above Lock 1 and the Mohawk River at Cohoes differed from the flow at the federal dam by an average of about 3%. Abood et al. (1992) found a similar result when comparing the combined flows from the upper Hudson River above Lock 2 and the Mohawk River at Cohoes with the flow at the federal dam.

Particle-Tracking and Estuarine Circulation Models

The estuarine circulation model was a refinement of the System-Wide Eutrophication Model (SWEM) developed and cali-

Table 1. The density (number/1,000 m^3) of striped bass PYSL entrained at the Charles Poletti Power Plant, collected by trawling in the Battery region, and collected by trawling in all regions of the Hudson River Estuary during weeks when striped bass PYSL were collected in the Battery region from 1999 through 2002. NS means no sample was taken.

Year	Week of	PYSL density	
		Poletti	Battery
1999	May 24	0.0	1.4
	June 1	0.0	44.6
	June 7	3.4	3.1
	June 14	10.4	43.1
	June 21	0.0	0.0
	June 28	0.0	4.8
	Mean	2.3	16.2
2000	May 22	0.0	762.4
	May 30	62.0	1,130.8
	June 5	20.7	14,452.7
	June 12	285.3	1,827.2
	June 19	22.8	1,789.8
	June 26	11.7	808.5
	July 5	13.4	NS
	July 10	4.4	0.7
	Mean	58.1	2,967.4
2001	May 14	0.0	1.0
	May 21	0.0	0.0
	May 28	0.0	664.2
	June 4	11.1	1,148.2
	June 11	8.8	129.7
	June 18	6.4	23.0
	June 25	2.3	14.6
	July 1	0.0	3.0
	Mean	3.6	247.8
2002	May 13	0.0	1.8
	May 20	0.0	0.0
	May 28	0.0	0.0
	June 3	0.0	0.0
	June 10	0.0	1.2
	June 17	0.0	55.0
	June 24	2.0	125.0
	June 30	0.0	8.3
	Mean	0.2	23.9

brated by Blumberg et al. (1999). The first refinement was an expansion of the SWEM domain which encompassed the entire Hudson River Estuary, the New York Bight from Cape May, New Jersey to Nantucket shoals off the coast of Massachusetts, Long Island Sound, New York Harbor, and adjoining rivers in New York and New Jersey. The refined SWEM domain extended into four major tributaries of the East River: Newtown Creek, Bronx River, Westchester Creek, and Flushing Bay and one major tributary of western Long Island Sound, the Hutchinson River. The second refinement was an expansion of the model grid from 49×84 cells to 59×94 cells and an increase in the number of cells in the East River from 113 to 346.

The particle-tracking simulation involved releasing 168,000 particles at 10 fixed sites, about equally spaced apart, in the lower Hudson River Estuary and tracking them to a region that included the upper East River starting just east of Poletti and Long Island Sound starting just west of Hempstead Harbor (the upper East River/western Long Island Sound region) (Figure 2). At each site, five particles were released per h in each of 10 vertical grid cells over a 14-d period from May 21 through June 3, 2000. Each particle was tracked and its position determined and recorded every 10 s, the time step of the model. Tracking was done through June 5. As a result, the number of particles that spent t time steps in the upper East River/western Long Island Sound region could be counted. The probability that a particle would be transported to the upper East River/western Long Island Sound region was equal to the probability that it would be located there sometime during the simulation. The time it took for a particle to reach the upper East River/western Long Island Sound region was also recorded. The probability of transport and the time of transport to the upper East River/western Long Island Sound region were cumulated for all particles that were released in the estuary. Cumulative frequency distributions were used to calculate the median probability of transport and the median transport time to the upper East River/western Long Island Sound region and to determine the minimum and maximum values.

The model was run with meteorological and hydrodynamic data from the years 2000 and 2002 as the forcing functions. The results from both years were very similar and only those from the year 2000 are presented in this paper.

Results

Poletti Entrainment Sampling

Striped bass PYSL were collected from the East River by entrainment sampling at Poletti in every year: 176 in 2000, 13 in both 1999 and 2001, and 1 in 2002. No striped bass eggs or yolk sac larvae were collected. The density of striped bass PYSL collected at Poletti by entrainment sampling varied among weeks and years; the highest densities occurred during 2000 (Table 1). The weekly densities of striped bass PYSL from May 30 through July 5, 2000 were higher than the density from any other week during 1999, 2001, and 2002. The mean density of striped bass PYSL during 2000 was about 16 times higher than that during 2001, the next highest year.

Hudson River Ichthyoplankton Sampling

The density of striped bass PYSL collected by ichthyoplankton sampling in the Battery region varied considerably among weeks and years; the highest densities occurred during 2000 (Table 1). The densities of striped bass PYSL during the weeks from June 5 through June 19, 2000 were higher than those for any week during 1999, 2001, and 2002. Also, the densities of striped bass PYSL during the weeks from May 22 through June 26, 2000 were higher than those for any week during 1999, 2001, and 2002, except for 1 week during 2001.

The proportion of the riverwide standing crop that was found in the Battery region during four weeks in June 2000, based on ichthyoplankton sampling, ranged from 12% to 30% (Table 2). The highest proportion for any week during 1999, 2001, and 2002 was 2%. Additionally, the proportion of the riverwide standing crop found in the Yonkers and Tappan Zee, was higher during 2000 than during 1999, 2001, and 2002.

Freshwater Flow

Freshwater flow in the Hudson River Estuary was higher during May and the first week of June 2000 than that during May and the first week of June 1999 and 2001. Freshwater flow in the Hudson River Estuary was also higher during May 2002 than during May 1999 and 2001 but not as high or for as long as that during May 2000; it was considerably lower during June 2002 than during June 2000 (Figure 3).

Particle-Tracking Model

After 1 week, nearly all of the particles released in the lower Hudson River Estuary were transported out of the estuary. Most of the particles were transported through New York Harbor to the Atlantic Ocean and did not return by the end of the particle-tracking simulation. Most of the particles reaching the upper East River/western Long Island Sound region were transported through the Battery region; less than 1% of the particles were transported through Harlem River. The median probability of transport from the lower Hudson River Estuary to the upper East River/western Long Island Sound region was 0.12 with a minimum of 0.04 and a maximum of 0.19. The median transport time to the upper East River/western Long Island Sound region was 2 d with a minimum of 1 d and a maximum of 3 d.

Mean Densities of PYSL in the Battery Region and at Poletti

The mean densities of striped bass PYSL collected by ichthyoplankton sampling in the Battery region during 1999, 2001, and 2002 ranged from 16.2 to 247.8; the mean density was 2,967.4 during 2000. The mean densities of striped bass PYSL collected by entrainment sampling in the East River at Poletti during 1999, 2001, and 2002 ranged from 0.2 to 3.6; the mean density was 58.1 during 2000.

Discussion

Entrainment sampling at Poletti demonstrated that striped bass PYSL were present in the East River every year from 1999 through 2002. Based on results from the particle-tracking simulation, the striped bass PYSL in the East River probably came from the lower Hudson River Estuary. If striped bass PYSL behaved like neutrally buoyant, passive particles, 50% of the time 12% or more of the PYSL in the Battery region would have been transported by tidal currents to the upper East River/western Long Island Sound region. Since striped bass PYSL are capable of resisting water currents and migrating vertically within the water column (Collette and Klein-MacPhee 2002), selective tidal-stream transport, rather than passive transport, could explain the presence of striped bass PYSL in the East River. However, it seems unlikely that self-directed movements by striped bass PYSL would play a significant role in the transport of PYSL of out the lower Hudson River

Table 2. The standing crop (in thousands) of striped bass PYSL in all regions of the Hudson River Estuary and the percent of the standing crop in the Battery region and in the Yonkers and Tappan Zee regions combined from 1999 through 2002.

Year	Week of	Standing crop	Percent of standing crop	
			Battery region	Yonkers, and Tappan Zee regions
1999	May 10	29	0.0	0.0
	May 17	53,375	0.0	0.0
	May 24	16,813,362	0.0	0.2
	June 1	9,282,853	0.1	7.4
	June 7	2,738,146	0.0	7.7
	June 14	1,915,927	0.5	15.8
	June 21	350,243	0.0	2.0
	June 28	78,728	1.3	0.0
	July 12	422	0.0	0.0
	July 26	194	0.0	0.0
2000	May 8	413	0.0	0.0
	May 15	575,180	0.0	0.8
	May 22	34,109,735	0.5	80.9
	May 30	5,701,521	4.1	48.0
	June 5	9,999,620	30.2	49.3
	June 12	3,131,131	12.2	79.5
	June 19	2,397,616	15.6	70.0
	June 26	660,039	25.6	13.5
	July 5	141,791	0.1	10.3
	July 10	8,096	0.0	0.0
2001	May 14	1,131,330	0.0	0.0
	May 21	13,199,897	0.0	0.0
	May 28	15,429,611	0.9	7.2
	June 4	10,354,542	2.3	23.7
	June 11	5,619,794	0.5	20.6
	June 18	1,475,655	0.3	3.2
	June 25	527,020	0.6	31.1
	July 1	240,607	0.3	0.3
	July 17	2,350	0.0	0.0
	August 7	160	0.0	0.0
2002	May 6	36,432	0.0	1.4
	May 13	18,098	2.1	50.8
	May 20	28,009	0.2	59.9
	May 28	17,484	0.0	63.5
	June 3	292,823	0.0	8.7
	June 10	799,522	0.0	2.0
	June 17	2,779,210	0.4	0.4
	June 24	1,054,023	2.5	0.0
	June 30	297,623	0.6	0.0
	July 15	10,890	0.0	0.0

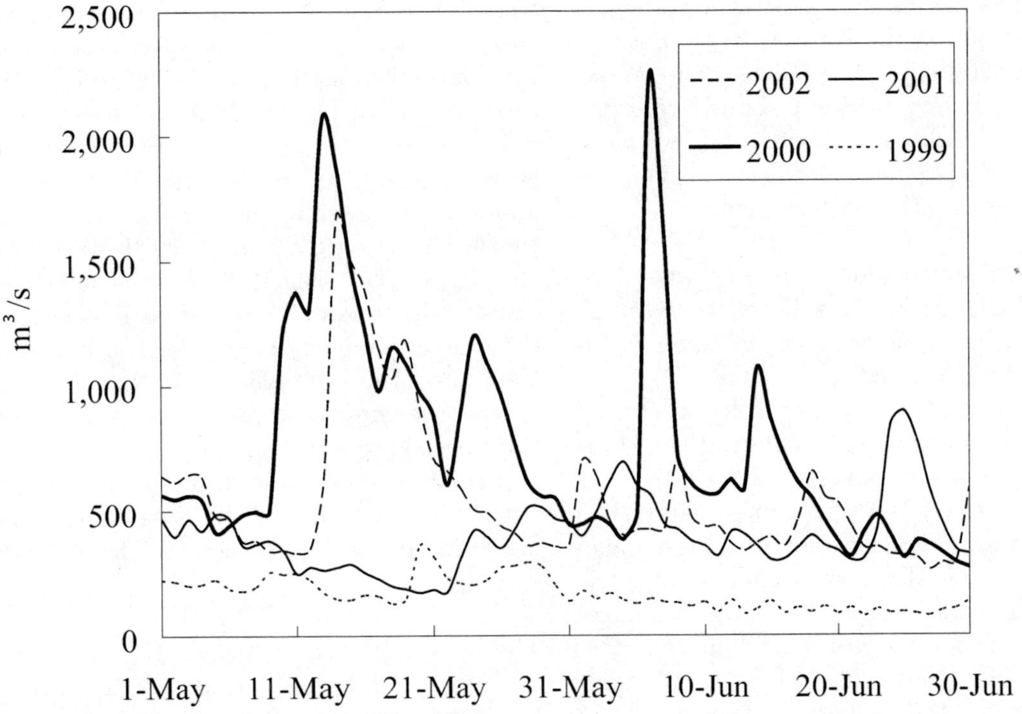

Figure 3. Freshwater flow in the Hudson River Estuary during May and June for the years 1999 through 2002.

Estuary to the East River because strong tidal currents and turbulent mixing occur in both the lower Hudson River Estuary (Nepf and Geyer 1996) and in the East River (Blumberg and Pritchard 1997). It also seems unlikely that self-directed movements by striped bass PYSL would be great enough to prevent their transport out of the lower Hudson River Estuary because of the strong tidal currents that occur there.

If passive transport of striped bass PYSL out of the Hudson River Estuary occurred, it is also likely that passive transport of striped bass PYSL downriver within the estuary occurred. Data from 2000 and 2002 indicate that passive transport of striped bass PYSL downriver within the Hudson River Estuary was greater when freshwater flow was higher. Unusually high freshwater flow during May and early June 2000 was associated with a 5 week period when the proportion of the standing crop of striped bass PYSL in the Battery region was higher than any year from 1999 through 2002. High freshwater flow during May 2002 was associated with a 3 week period when the proportion of the standing crop of striped bass PYSL in the Yonkers and Tappan Zee regions was comparable to that observed during 2000 and high relative to the proportions observed during 1999 and 2001. However, freshwater flow during early June 2002 was much lower than that during early June 2000 and the proportion of the standing crop of striped bass PYSL found in Battery region during 2002 was much lower than that observed during 2000. Thus, the low freshwater flow that occurred during early June 2002 apparently failed to substantially transport striped bass PYSL downriver from the Yonkers and Tappan Zee regions to the Battery region where they could be transported out of the lower Hudson River Estuary. Although we cannot rule out the possibility that selective tidal-stream transport occurred within the Hudson River Estuary and contributed to the downstream movement of striped bass PYSL, we can rule out the possibility that it was used by striped bass PYSL to prevent their movement downriver into the Battery region.

Additional support for the belief that striped bass PYSL in the East River were transported from the lower Hudson River Estuary is the fact that the mean density of striped bass PYSL was highest at Poletti and in the Battery region during the same year. The mean density of striped bass PYSL in the Battery region during 2000 was 12 times higher than the highest mean density in the Battery region among the years 1999, 2001, and 2002 while the mean density of striped bass PYSL in the East River at Poletti during 2000 was 16 times higher than the highest mean density in the East River among the years 1999, 2001, and 2002.

If the striped bass PYSL in the East River came from the Hudson River Estuary and they were transported passively, results from the particle-tracking simulation support the proposition that small striped bass caught in western Long Island Sound during June 1999, 2000, and 2001 originated from PYSL spawned in the estuary. The small striped bass were collected in bays that were within the boundaries of our upper East River/western Long Island Sound region, where the median probability of transport was 0.12. Furthermore, the median transport time to the bays was only 2 d.

Historically, the abundance of striped bass PYSL in the East River may have been affected by more than just the abundance of PYSL in the lower Hudson River Estuary. Although striped bass PYSL were collected every year from 1999 through 2002 by entrainment sampling at Poletti, none were collected by entrainment sampling at the Astoria Generating Station (Astoria), located on the East River adjacent to Poletti, during 1993 (LMS 1994) or at the Ravenswood Generating Station (Ravenswood), located on the East River, about 4 km closer to the Hudson River Estuary than Poletti, during 1992 (LMS 1993) and 1993 (NAI 1994). This is curious because striped bass PYSL were collected in the Battery region during both 1992 and 1993 (Con Edison 1996, 1997). In fact, the volume of water sampled at Astoria and Ravenswood was greater than that at Poletti and tidal currents should have transported striped bass PYSL into the East River during 1992 and 1993 as they did from 1999 through 2002.

The absence of striped bass PYSL in entrainment collections at Astoria and Ravenswood could reflect poorer survival of striped bass PYSL in the East River during 1992 and 1993. Many factors can affect survival of striped bass larvae, including water quality (Uphoff 1989). It is particularly interesting to note that universal secondary treatment was not in place at four major sewage treatment plants emptying into the East River until after 1993 (IEC 2001). These four sewage treatment plants: Newtown Creek, Wards Island, Bowery Bay, and Red Hook, have a maximum combined flow of 33.6 $m^3\ s^{-1}$ (Brosnan and O'Shea 1996). This represents about 67% of the annual mean near-surface flow rate in the East River toward Long Island Sound and about 13% of the annual rate near-bottom flow in the East River toward New York Harbor (Blumberg et al. 1999). Between 1993 and 1999, the secondary treatment systems at Newtown Creek, Wards Island, Bowery Bay, and Red Hook were either expanded or upgraded, which could account for higher water quality and higher survival during the period from 1999 through 2002.

Conclusions

Historically, annual monitoring of the spatial distribution of striped bass larvae spawned in the Hudson River Estuary has been restricted to the estuary itself, and therefore, the distribution of striped bass larvae outside the estuary was not known. The collection of striped bass PYSL at Poletti demonstrates that the distribution of striped bass PYSL extends beyond the Hudson River Estuary. Results from the particle tracking model demonstrate that passive transport is a reasonable explanation for the association between the abundance of striped bass PYSL in the lower Hudson River Estuary and in the East River at Poletti. It also accounts for the presence of small striped bass age-0 in western Long Island Sound during June. Improvements in water quality in the East River may have also contributed to the current distribution of striped bass PYSL outside of the Hudson River Estuary. The relationship between the abundance of striped bass PYSL transported outside the Hudson River Estuary and recruitment of striped bass to the Hudson River stock may have important management implications and merits additional study.

Acknowledgments

We thank Jim Reichle of ASA Analysis and Communications Inc. (ASA) for collecting entrainment samples, Bill Dey of ASA and Dennis Logan and John Seibel of Post Buckley, Schuh and Jernigan for compiling the Poletti data base, Paul Tartaglia and the staff at Poletti for logistical support associated with entrainment sampling, and the New York Power Authority for funding the Poletti entrainment program. We also thank Entergy Nuclear Northeast, Dynegy, and Mirant for sharing data from the Hudson River Monitoring Program and John Young and Nancy Decker of ASA for assembling and providing ichthyoplankton data for the Hudson River Estuary. We appreciate the dedication and professionalism of Quamrul Ahsan and Honghai Li of HydroQual Inc. in conducting the particle tracking simulations.

References

Abood, K. A., G. A. Apicella, and A. W. Wells. 1992. General evaluation of Hudson River freshwater flow trends. Pages 3–28 *in* C. L. Smith, editor. Estuarine Research in the 1980s. State University of New York Press, Albany.

ASA (ASA Analysis and Communication). 2000. 1996 year class report for the Hudson River Estuary Monitoring Program. Prepared for Central Hudson Gas and Electric Corporation, New York.

Blumberg, A. F., D. J. Dunning, H. Li, D. Heimbuch, and W. R. Geyer. 2004. Use of a particle-tracking model for predicting entrainment at power plants on the Hudson River. Estuaries 27:515–526.

Blumberg, A. F., L. A. Khan, and J. P. St. John. 1999. Three-dimensional hydrodynamic model of New York Harbor region. Journal of Hydraulic Engineering 125:799–816.

Blumberg, A. F., and D. W. Pritchard. 1997. Estimates of the transport through the East River, New York. Journal of Geophysical Research 102:5685–5703.

Boreman, J., and R. J. Klauda. 1988. Distributions of early life stages of striped bass in the Hudson River Estuary, 1974–1979. Pages 53-58 *in* L. W. Barnthouse, R. J. Klauda, D. S. Vaughan, and R. L. Kendall, editors. Science, Law, and Hudson River Power plants: a case study in Environmental Impact Assessment. American Fisheries Society, Monograph 4, Bethesda, Maryland.

Brosnan, T. M., and M. L. O'Shea. 1996. Sewage abatement and coliform bacteria trends in the lower Hudson-Raritan Estuary since passage of the Clean Water Act. Water Environment Research 68:25–35.

Central Hudson Gas & Electric Corp., Consolidated Edison Company of New York, Inc., New York Power Authority, and Southern Energy New York. 1999. Draft Environmental Impact Statement for State Pollutant Discharge Elimination System Permits for Bowline Point, Indian Point 2 & 3, and Roseton Steam Electric Generating Stations. New

York State Department of Environmental Conservation, Albany.

Collette, B. B. and G. Klein-MacPhee. 2002. Bigelow and Schroeder's Fishes of the Gulf of Maine. Smithsonian Institution Press, Washington, D.C.

Con Edison. 1996. 1992 Year Class Report for the Hudson River Estuary Monitoring Program. Submitted to the New York State Department of Environmental Conservation, Albany, New York.

Con Edison. 1997. 1993 year class report for the Hudson River Estuary Monitoring Program. Submitted to the New York State Department of Environmental Conservation, Albany, New York.

Cronin, J., and R. F. Kennedy, Jr. 1997. The Riverkeepers. Scribner, New York.

Heimbuch, D. G., D. J. Dunning, and J. R. Young. 1992. Post-yolk-sac larvae abundance as an index of year class strength of striped bass in the Hudson River. Pages 376–391 *in* C. L. Smith editor. Estuarine research in the 1980s. State University of New York Press, Albany.

Hilborn, R. and C. J. Walters. 1992. Quantitative fisheries stock assessment. Chapman and Hall, New York, New York.

IEC (Interstate Environmental Commission). 2001. 2000 Annual Report of the Interstate Environmental Commission. New York.

LMS (Lawler, Matusky and Skelly Engineers). 1993. Ravenswood Impingement and Entrainment Report September 1991-September 1992. Prepared for Consolidated Edison Company of New York, Inc., New York.

LMS (Lawler, Matusky and Skelly Engineers). 1994. Astoria Impingement and Entrainment Studies January 1993-December 1993. Prepared for Consolidated Edison Company of New York, Inc., New York.

Mansueti, R. J. 1958. Eggs, larvae, and young of striped bass, *Roccus saxatilis.* Maryland Department of Research and Education, Chesapeake Bay Laboratory No. 112, Solomons.

McKown, K. A., and J. M. Brischler. 2001a. A study of the striped bass in the marine district of New York VII. New York State Department of Environmental Conservation, Completion Report for Project AFC-24, Stony Brook.

McKown, K. A. and J. M. Brischler. 2001b. A study of the striped bass in the marine district of New York VII. New York State Department of Environmental Conservation, Completion Report for Project AFC-23, Stony Brook.

McKown, K. A., and J. M. Brischler. 2002. A study of the striped bass in the marine district of New York VII. New York State Department of Environmental Conservation, Completion Report for Project AFC-24, Stony Brook.

Merriman, D. 1941. Studies on the striped bass (*Roccus saxatilis*) of the Atlantic Coast. Fisheries Bulletin of the Fish and Wildlife Service 50:1–77.

Metzger, S. G., R. G. Keppel, P. Geoghegan, and A. W. Wells. 1992. Abundance of selected Hudson River fish species in previously unsampled regions: effects on standing crop estimates. Pages 348–375 *in* C. L. Smith editor. Estuarine research in the 1980s. State University of New York Press, Albany.

Nepf, H. M., and W. R. Geyer. 1996. Intratidal variations in stratification and mixing in the Hudson estuary. Journal of Geophysical Research 101:12,079- 12,086.

NAI (Normandeau Associates, Inc). 1994. Ravenswood Generating Station Impingement and Entrainment Report, February 1993 through January 1994. Prepared for Consolidated Edison Company of New York, Inc., New York.

Raney, E. C. 1952. The life history of the striped bass *Roccus saxatilis* (Walbaum). Pages 5–97 *in* E. C. Raney, E. F. Tresselt, E. H. Hollis, V. D. Vladykov, and D. H. Wallace. The striped bass *Roccus saxatilis.* Bulletin of the Bingham Oceanographic Collection Yale University 14(1).

Richards, R. A., and P. J. Rago. 1999. A case history of effective fishery management: Chesapeake Bay striped bass. North American Journal of Fisheries Management 19:356–375.

Uphoff, J. H., Jr. 1989. Environmental effects on survival of eggs, larvae, and juveniles of striped bass in the Choptank River, Maryland. Transactions of the American Fisheries Society 118:251–263.

USGS (United States Geological Survey). 2004. Daily stream flow statistics for New York. Available: http://nwis.waterdata.usgs.gov/ny/nwis (November 2004).

American Fisheries Society Symposium 51:287–294, 2006

Dispersal of Age-2+ Striped Bass Out of the Hudson River

DENNIS J. DUNNING
New York Power Authority
123 Main Street, White Plains, New York 10601 USA
dennis.dunning@nypa.com

JOHN R. WALDMAN
Queens College, City University of New York
65-30 Kissena Boulevard, Flushing, New York 11367 USA

QUENTIN E. ROSS
348 County Highway 2, Mount Upton, New York 13809 USA

MARK T. MATTSON
Normandeau Associates, Inc.
25 Nashua Road, Bedford, New Hampshire 03110 USA

Abstract.—Our objective was to determine if dispersal of age-2+ striped bass out of the Hudson River was affected by cohort abundance or stock size. We evaluated dispersal using the location of tagged age-2+ striped bass (in or outside of the Hudson River) and distance from the mouth of the Hudson River when they were recaptured by anglers. The proportion of age-2+ striped bass recaptured outside the Hudson River annually during May was significantly and positively correlated with cohort abundance ($r = 0.978$, $P < 0.001$), but not with stock size ($r = 0.739$, $P = 0.094$). The maximum distance at which an age-2+ striped bass was recaptured outside the Hudson River annually was also significantly and positively correlated with cohort abundance ($r = 0.869$, $P < 0.025$). Therefore, it appears that dispersal of age-2+ striped bass out of the Hudson River was density-dependent and due to cohort abundance.

Introduction

The Hudson River is a major source of striped bass *Morone saxatilis* in the Atlantic coastal fisheries (Richards and Rago 1999; Waldman and Fabrizio 1994). Seasonal movements of Hudson River striped bass along the Atlantic coast have been extensively studied by tagging mixed stocks of striped bass found in Long Island Sound and the New York Bight (Raney et al. 1952; Alperin 1966; Clark 1968; Kriete et al. 1978) and by tagging striped bass in the Hudson River (McLaren et al. 1981; Waldman et al. 1990). Based on these studies, it is believed that many striped bass enter the coastal fishery by leaving the Hudson River during late spring, dispersing along the Atlantic coast (principally north) during the summer, and migrating back to the Hudson River starting in the fall.

Using angler recaptures of striped bass tagged in the Hudson River, Waldman et al. (1990) concluded that the Hudson River stock was more widely dispersed during the period of 1984 through 1986 than at any other time it had been studied. During that period, three striped bass recaptured in Nova Scotia and one in Maine each exceeded the northern end of the range previously reported for striped bass tagged in the Hudson River, while two striped bass recaptured in Virginia and one in North Carolina each exceeded the southern range previously reported. Range expansion was attributed to a growing stock that was pressing outward from the Hudson River. Range expansion of striped bass from the Chesapeake Bay stock has also been reported during periods of growing stock size (Merriman 1941).

Growing stock size also appeared to increase the numbers of striped bass 200–399 mm long (a size range that includes age-2+ striped bass) that left the Hudson River as well as those that were larger. A higher percentage of Hudson River striped bass 200 mm to 399 mm total length were recaptured outside the Hudson River within 50 km of the mouth and beyond 199 km from the mouth during the period March 1986 through February 1987 compared with the years 1976 and 1977 combined (Waldman et al. 1990). Growing stock size was inferred from an increase in the abundance of post-yolk-sac larvae (PYSL) from 1977 through 1987. However, Waldman et al. (1990) examined range expansion for only two time periods (March 1986 through February 1987 and the years 1976 and 1977 combined), not enough to test for a trend.

Kriete et al. (1978) examined range expansion for striped bass based on recaptures from 17,595 fish tagged and released over 6 years in the James, York, and Rappahannock rivers in Virginia and concluded that the number of age-2+ striped bass joining the offshore migrating population was larger when cohort abundance was higher, not stock size. However, they did not test for a trend nor did they clearly identify the data that were used for determining cohort abundance. Furthermore, only 58 striped bass were recaptured outside of the James, York, and Rappahannock rivers over the 6 years, including 5 years when the number of fish recaptured was less than or equal to 10.

Our objective was to determine if dispersal of age-2+ striped bass out of the Hudson River was affected by cohort abundance or stock size. We met this objective by examining recaptures of age-2+ striped bass both in the Hudson River and outside of it by anglers from 1987 through 1992, a period that included 2 years with relatively high cohort abundance and 2 years with relatively low cohort abundance.

Methods

Striped bass were captured in trawls, marked with an internal anchor tag, and released in the lower Hudson River and in adjacent upper New York Harbor near Liberty Island from autumn through spring annually. Before the autumn of 1988, all striped bass greater than or equal to 200 mm total length (TL) and in good condition were tagged. After that, all striped bass greater than or equal to 150 mm TL and in good condition were tagged. Field procedures are outlined in Dunning et al. (1989) and Mattson et al. (1990). Scales removed from each striped bass at the time of tagging were used to estimate its age.

We evaluated dispersal using the location of age-2+ striped bass (in or outside of the Hudson River) and distance from the mouth of the Hudson River at recapture by anglers. The mouth of the Hudson River was defined as the southern tip of Manhattan. We restricted our evaluation to striped bass greater than or equal to 200 mm TL that were tagged as age-1+ and recaptured within 1 year following release. By limiting the recapture period to 1 year after release, location and distance traveled were not confounded by more than one annual migration. In assigning age, we assumed that striped bass were spawned starting in April (Smith 1985). Striped bass that survived until April of the year after being spawned were designated as age-1+ and of these, survivors captured after March of the next year, were designated as age-2+. Therefore, the age 2+ striped bass dispersal described in this paper represents movements occurring during the third year of life.

We used an annual index of PYSL abundance as a measure of stock size for Hudson River striped bass, as did Waldman et al. (1990), because there were no independent estimates of population size for the years 1987 through 1992 based on tagging and recapturing age-1+ and older striped bass. The PYSL index is based on ichthyoplankton sampling conducted throughout the estuarine portion of the Hudson River from spring through fall. It represents the average relative abundance of a weekly cohort from the time it enters the PYSL life stage to the time it leaves that life stage, expressed as the number of PYSL collected by ichthyoplankton sampling per 1,000 m^3 (Heimbuch et al. 1992). The PYSL index reflects the abundance of spawning females. Under the current fishing regulations, it is likely that stock size and abundance of spawning females are highly correlated. Thus, we used the PYSL index to test the hypothesis proposed by Waldman et al. (1990) about dispersal due to stock size.

We used the annual abundance of age-1+ striped bass, based on recaptures of fish caught by trawling in the lower Hudson River and in adjacent upper New York Harbor near Liberty Island during the fall and winter (Central Hudson et al. 1999), as the measure of cohort abundance at age-2+ because it represents abundance when the transition from age-1+ to age-2+ occurs. We did not use the abundance of age-2+ striped bass presented in Central Hudson et al. (1999) because it was estimated from recaptures of age-2+ fish caught by trawling in the lower Hudson River and in adjacent upper New York Harbor near Liberty Island beginning in November, i.e., six months after the transition from age-1+ to age-2+.

If cohort abundance influences the proportion of age-2+ striped bass recaptured outside of the Hudson River, this should be most apparent and detectable during a time of year when age-2+ striped bass are crowded and likely to leave the Hudson River. So, to test for the effect of cohort abundance, we first determined which month had the greatest number of age-2+ recaptures and the highest proportion of recaptures within the Hudson River on average across the years 1987 through 1992. For that month, we calculated the proportion of age-2+ striped bass recaptured outside the Hudson River each year. Then, we correlated the proportion of age-2+ striped bass recaptured outside the Hudson River with cohort abundance. To test for the effect of stock size on the range of age-2+ striped bass, we correlated the striped bass PYSL index for the years 1987 through 1992, with annual estimates of the proportion of age-2+ striped bass recaptured outside the Hudson River for the corresponding years.

Table 1. The proportion of age-2+ striped bass recaptured within the Hudson River monthly from 1987 through 1992 and the average and standard deviation (SD) across years. The total numbers of recaptures are in parentheses.

	Proportion of age-2+ striped bass (%)								
Year	Apr	May	Jun	Jul	Aug	Sep	Oct	Nov	Dec
1987	33	63	20	67	80	0	11	25	0
	(12)	(16)	(10)	(3)	(5)	(2)	(9)	(4)	(3)
1988	22	66	9	7	17	19	16	3	0
	(18)	(29)	(23)	(15)	(6)	(21)	(38)	(34)	(9)
1989	20	34	14	12	10	15	26	10	0
	(30)	(68)	(50)	(42)	(29)	(34)	(42)	(30)	(6)
1990	26	33	25	21	16	33	7	18	0
	(50)	(66)	(32)	(14)	(19)	(24)	(27)	(39)	(7)
1991	45	41	17	20	23	44	27	20	25
	(47)	(44)	(23)	(10)	(13)	(9)	(22)	(30)	(8)
1992	65	48	33	20	6	19	13	6	19
	(20)	(71)	(33)	(35)	(16)	(31)	(48)	(34)	(16)
Average	35	48	20	25	25	22	17	14	7
SD	17	14	8	22	27	15	8	9	12
	(177)	(294)	(171)	(119)	(88)	(121)	(186)	(171)	(49)

When the correlation between the proportion of age-2+ striped bass recaptured outside the Hudson River and cohort abundance or stock size was significant, we correlated the maximum distance at which an age-2+ striped bass was recaptured outside of the Hudson River with cohort abundance or stock size for that year. The maximum distance was the greatest distance, in km, from the mouth of the Hudson River to the location at which a striped bass was reported to have been recaptured. We used the maximum distance at which an age-2+ striped bass was recaptured because it was the measure of dispersal used by Waldman et al. (1990) to conclude that the range of the Hudson River stock had expanded.

Results

The proportion of age-2+ striped bass recaptured within the Hudson River, averaged over the years from 1987 through 1992, was highest during May and lowest during December (Table 1). Also, the average number of age-2+ striped bass recaptured in the Hudson River was highest during May.

The proportion of age-2+ striped bass recaptured outside the Hudson River annually during May was significantly and positively correlated with cohort abundance ($r = 0.978$, $P < 0.001$) (Figure 1) but not with stock size for the corresponding year ($r = 0.739$, $P = 0.094$) (Figure 2). The maximum distance at which an age-2+ striped bass was recaptured outside the Hudson River annually was also significantly and positively correlated with cohort abundance ($r = 0.869$, $P < 0.025$) (Figure 3).

Discussion

The high proportion and total number of age-2+ striped bass recaptured within the Hudson River during May, compared with

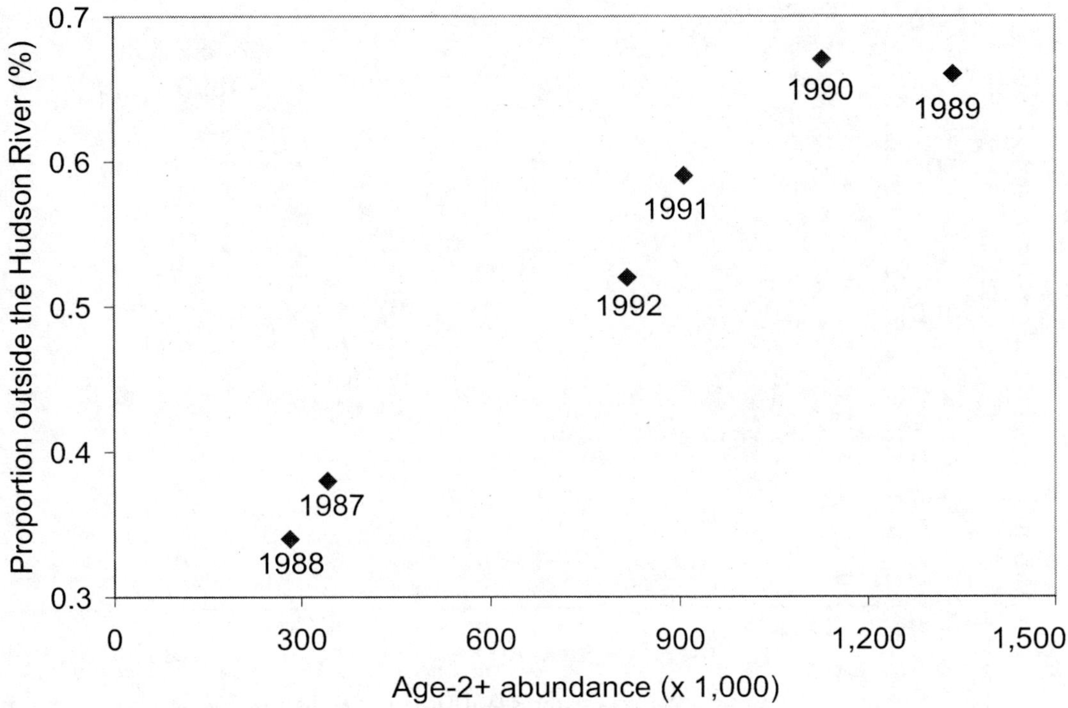

Figure 1. Proportion of age-2+ striped bass recaptured outside the Hudson River from 1987 through 1992 as a function of cohort abundance, represented by the number of age-2+ striped bass when the transition from age-1+ to age-2+ occurs.

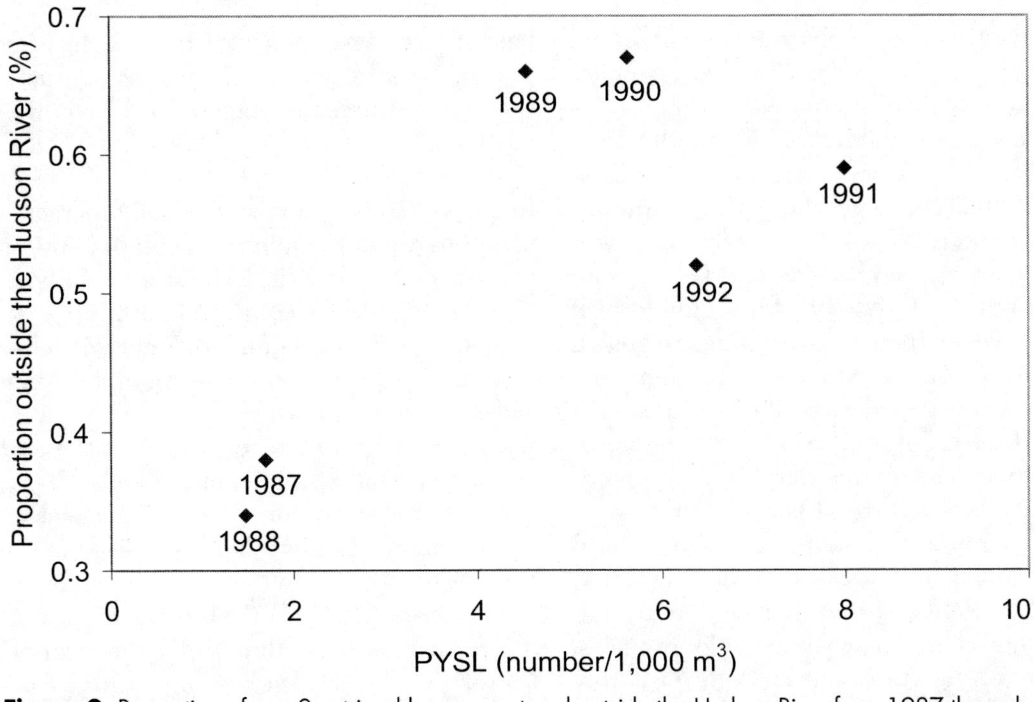

Figure 2. Proportion of age-2+ striped bass recaptured outside the Hudson River from 1987 through 1992 as a function of stock size during that year, represented by the number of striped bass PYSL per 1,000 m^3.

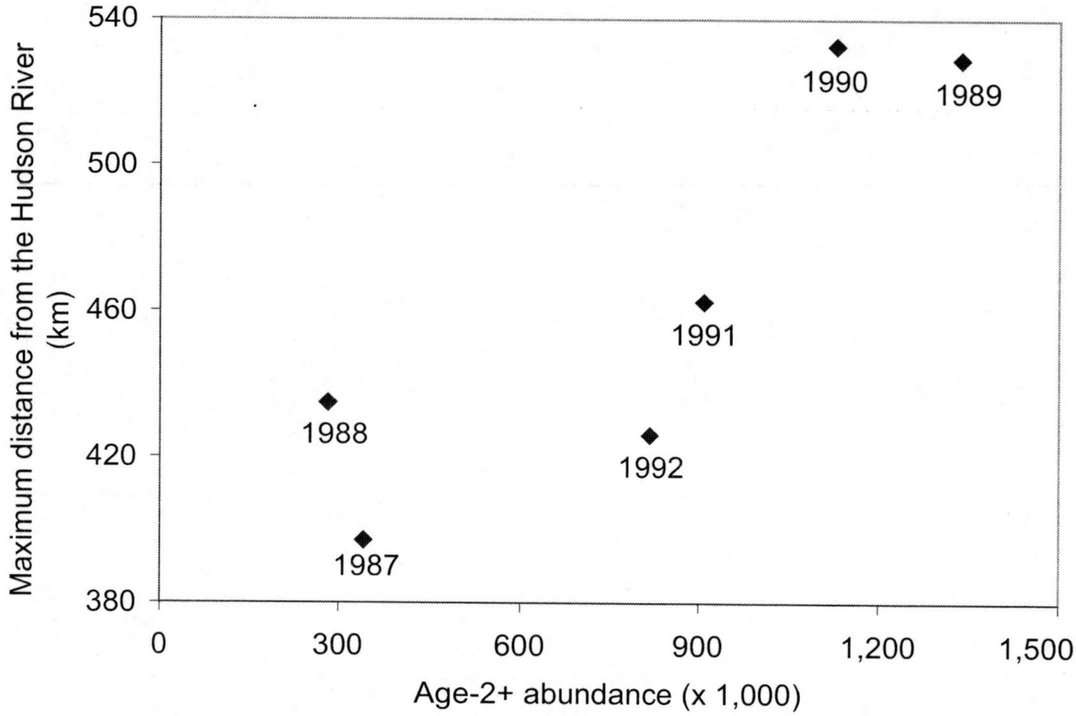

Figure 3. Maximum distance, in km, at which an age-2+ striped bass was recaptured outside the Hudson River annually as a function of cohort abundance, represented by the number of age-2+ striped bass when the transition from age-1+ to age-2+ occurs.

other months, probably reflects increased fishing effort for striped bass as well as higher abundance of, and increased feeding activity by, age-2+ striped bass. Seasonally, fishing effort for striped bass in the Hudson River is highest during the spring (Limburg et al. 2006; Euston et al., 2006. this volume), when striped bass ascend the river to spawn but feed readily. Although few, if any, age-2+ striped bass are likely to spawn (Hoff et al. 1988), relatively large numbers of immature striped bass have been caught in the Hudson River during the spring when their movements increase as water temperature increases (Dovel 1992). If the movements of immature striped bass were also accompanied by increased feeding activity, then the likelihood of being caught by anglers should have increased. Nonetheless, greater fishing effort during May probably mirrors the seasonal abundance of striped bass in the Hudson River. If so, then May was the month when intraspecific competition for resources among striped bass was most likely to occur.

Intraspecific competition normally increases in intensity as abundance increases and as intraspecific competition increases, emigration should increase (Murray 1979). Intraspecific competition should be more intense among age-2+ striped bass than between age-2+ striped bass and spawners because striped bass generally school with others of similar size and age (Raney 1952). If so, and our measures of dispersal accurately reflect emigration, we expected dispersal of age-2+ striped bass out of the Hudson River to increase during May when cohort abundance increased; it did. Both the proportion of age-2+ striped bass recaptured outside the Hudson River and the maximum distance at recapture for age-2+ striped bass increased when cohort abundance increased. Conversely, we expected that dispersal of age-2+ striped bass out of the Hudson River would not increase during May when stock size increased; it did not. In addition to the fact that the correlation between the proportion of age-2+ striped bass recaptured outside the Hudson River and stock size was not significant, the proportions of age-2+ striped bass recaptured outside the Hudson River during 1991 and 1992 (the 2 years when stock size was greater than the other 4 years) were lower than those during 1989 and 1990 (the years with the next two lower stock size values) when they should have been higher if stock size increased dispersal.

Intraspecific competition can occur for food or space. If intraspecific competition increased the dispersal of age-2+ striped bass from the Hudson River, it seems less likely that the limiting resource was space. Competition for space is generally associated with territorial behavior and spawning (Murray 1979). Striped bass are migratory and spawning does not involve establishing territories; in fact, spawning often involves a female accompanied by many males (Raney 1952). Furthermore, few, if any, age-2+ striped bass are likely to spawn (Hoff et al. 1988). Although it is also possible that aggression among age-2+ striped bass increased at higher levels of abundance, this seems unlikely because striped bass travel in schools and have been described as gregarious (Raney 1952).

Conclusions

It appears that dispersal of age-2+ striped bass out of the Hudson River was density-dependent and due to cohort abundance, as postulated by Kriete et al. (1978) for the James, York, and Rappahannock rivers in Virginia, rather than stock size, as postulated by Waldman et al. (1990). This density-dependent effect may be the result of competition among age-2+ striped bass for food, rather than space, which caused a larger proportion of age-2+ striped bass to disperse out of the Hudson River during years when their abundance was higher.

Acknowledgments

Central Hudson Gas & Electric Corporation, Con Edison of New York, Inc., the New York Power Authority, and Orange and Rockland Utilities funded the tagging of striped bass under terms of the Hudson River Cooling Tower Settlement Agreement. The field tagging program was performed by crews from Normandeau Associates, Inc., captained by Michael Ricci, Chris Sweeney, and Chris Burnett. Tag returns were handled by the Hudson River Foundation under contract with the New York Power Authority. We thank Kim McKown for reviewing this paper and offering extremely helpful and constructive comments.

References

Alperin, I. M. 1966. Dispersal, migration and origins of striped bass from Great South Bay, Long Island. New York Fish and Game Journal 13:79–112.

Central Hudson Gas and Electric Corporation, Consolidated Edison Company of New York, Inc., New York Power Authority, and Southern Energy New York. 1999. Draft Environmental Impact Statement for State Pollutant Discharge Elimination System Permits for Bowline Point, Indian Point 2 & 3, and Roseton Steam Electric Generating Stations. New York State Department of Environmental Conservation, Albany.

Clark, J. R. 1968. Seasonal movements of striped bass contingents of Long Island Sound and the New York Bight. Transactions of the American Fisheries Society 97:320–343.

Dovel, W. L. 1992. Movements of immature striped bass in the Hudson River Estuary. Pages 276–302 *in* C. L. Smith, editor. Estuarine Research in the 1980s.

State University of New York Press, Albany.

Dunning, D. J., Q. E. Ross, M. T. Mattson, P. Geoghegan, and J. R. Waldman. 1989. Reducing mortality of striped bass captured in seines and trawls. North American Journal of Fisheries Management 9:171–176.

Euston, E. T., S. A. Haney, K. A. Hattala, and A. W. Kahnle. 2006. Pages 295–315 in J. R. Waldman, K. E. Limburg, and D. L. Strayer, editors. Overview of Hudson River recreational fisheries, with an emphasis on striped bass. American Fisheries Society, Symposium 51, Bethesda, Maryland.

Heimbuch, D.G., D. J. Dunning, and J. R. Young. 1992. Post-yolk-sac larvae abundance as an index of year class strength of striped bass in the Hudson River. Pages 376–391 *in* C. L. Smith, editor. Estuarine research in the 1980s. State University of New York Press, Albany, New York.

Hoff, T. B., J. B. McLaren, and J. C. Cooper. 1988. Stock characteristics of Hudson River striped bass. Pages 59–68 in L. W. Barnthouse, R. J. Klauda, D. S. Vaughan, and R. L. Kendall, editors. Science, law, and Hudson River power: a case study in environmental impact assessment. American Fisheries Society, Monograph 4, Bethesda, Maryland.

Kriete, W. H., Jr., J. V. Merriner, and H. M. Austin. 1978. Movements of the 1970 yearclass striped bass between Virginia, New York, and New England. Proceedings of the Annual Conference Southeastern Association of Fish and Wildlife Agencies 32:692–696.

Limburg, K., J. Waldman, A. Kahnle, and K. Hattala. 2006. Fisheries of the Hudson River estuary. Pages 189–204 *in* J. S. Levinton and J. R. Waldman, editors. The Hudson River Estuary. Cambridge University Press, New York.

Mattson, M. T., J. R. Waldman, D. J. Dunning, and Q. E. Ross. 1990. Abrasion and protrusion of internal anchor tags in Hudson River striped bass. Pages 121–126 in N. C. Parker, editor. Fish-marking techniques. American Fisheries Society, Symposium 7, Bethesda, Maryland.

McLaren, J. B., J. C. Cooper, T. B. Hoff, and V. Lander. 1981. Movements of Hudson River striped bass. Transactions of the American Fisheries Society 110:158–167.

Merriman, D. 1941. Studies on the striped bass (*Roccus saxatilis*) of the Atlantic Coast. U.S. Fish and Wildlife Service Bulletin 35.

Murray, B. G. 1979. Population dynamics alternative models. Academic Press, New York.

Raney, E. C. 1952. The life history of the striped bass *Roccus saxatilis* (Walbaum). Pages 5–97 *in* E. C. Raney, E. F. Tresselt, E. H. Hollis, V. D. Vladykov, and D. H. Wallace, editors. The striped bass *Roccus saxatilis*. Bulletin of the Bingham Oceanographic Collection Yale University 14(1).

Richards, R. A., and P. J. Rago. 1999. A case history of effective fishery management: Chesapeake Bay striped bass. North American Journal of Fisheries Management 19:356–375.

Smith, C. L. 1985. The fishes of New York. New York State Department of Environmental Conservation, Albany.

Waldman, J. R., D. J. Dunning, Q. E. Ross, and M. T. Mattson. 1990. Range dynamics of Hudson River striped bass along the Atlantic coast. Transactions of the American Fisheries Society 119:910–919.

Waldman, J. R., and M. C. Fabrizio. 1994. Problems of stock definition in estimating relative contributions of Atlantic striped bass to the coastal fishery. Transactions of the American Fisheries Society 123:766–778.

American Fisheries Society Symposium 51:295–315, 2006

Overview of the Hudson River Recreational Fisheries, with an Emphasis on Striped Bass

E. Terry Euston and Susan A. Haney
Normandeau Associates Inc.
1921 River Road, Drumore, Pennsylvania 17518 USA
teuston@normandeau.com

Kathryn A. Hattala and Andrew W. Kahnle
New York State Department of Environmental Conservation, Hudson River Fisheries Unit
21 South Putt Corners Road, New Paltz, New York 12561 USA

Abstract.—Recreational fishing throughout the Hudson River estuary from the federal dam at Troy (river kilometer [rkm] 243) to the George Washington Bridge (rkm 19) was investigated during March 2001 through March 2002. Aerial counting surveys and angler interviews at nearly 200 access points were used to estimate fishing pressure, catch and harvest, catch rates, and various angler attributes. Fishing pressure for the mid-March through November period was estimated at 446,621 angler-hours. Most effort occurred in the late spring by anglers north of the Bear Mountain Bridge (rkm 74). Angling from boats comprised 72.6% of total effort. The total number of fish caught and harvested was estimated at 212,426 and 44,479 individuals, respectively, representing 31 species plus blue crab *Callinectes sapidus*. Most of the total catch was by boat anglers, although over the entire survey period shore anglers harvested the most fish. In sequence, striped bass *Morone saxatilis*, river herring *Alosa* spp., and white perch *M. americana* were the three most abundant species caught, whereas river herring, white perch, blue crab, and striped bass formed most of the harvest. Catch per unit effort (CPUE) and harvest per unit effort (HPUE) of shore anglers (0.69 fish/h and 0.22 fish/h) were higher than that of boat anglers (0.44 fish/h and 0.02 fish/h). Most anglers throughout spring sought striped bass, whereas during summer and fall boat anglers sought primarily black bass *Micropterus* spp., with much effort occurring during tournaments. Shore anglers were less focused and sought a broader variety of species. As a group, anglers fishing south of the Bear Mountain Bridge were less aware of fish consumption advisories due to contaminants than anglers fishing elsewhere in the estuary.

Introduction

The Hudson River Estuary is a unique resource that supports a variety of recreational fisheries. Anadromous fishes such as striped bass *Morone saxatilis*, American shad *Alosa sapidissima*, alewife *A. pseudoharengus*, and blueback herring *A. aestivalis* are pursued and may be harvested during spring spawning runs. Largemouth *Micropterus salmoides* and smallmouth *M. dolomieu* bass are sought extensively during summer and fall.

Abundant panfish species such as white perch *Morone americana*, sunfishes *Lepomis* spp., catfishes *Ictalurus* spp. and *Ameiurus* spp., crappies *Pomoxis* spp., and yellow perch *Perca flavescens* occur throughout much of the estuary. Marine species such as bluefish *Pomatomus saltatrix*, weakfish *Cynoscion regalis*, and Atlantic tomcod *Microgadus tomcod*, as well as blue crabs *Callinectes sapidus*, are sought seasonally in the more saline reaches of the lower estuary.

Such varied and extensive angling opportunities exist in proximity to major east coast population centers. Recent effort by licensed anglers in the Hudson River Estuary (Albany and Columbia counties and south) was estimated at 185,320 angler-days in 1996 (Connelly et al. 1997). This estimate is likely less than total estuary effort since no fishing license is required to fish either from shore or boat in the tidal Hudson River mainstem. A series of statewide mail surveys have documented continued growth of Hudson River sport fishery participation (Brown 1975; Kretser and Klatt 1981; Connelly et al. 1990, 1997). However, apparent steady growth of sport fishing effort has occurred without any accompanying quantitative estimates of catch or harvest. Thus, the effects of increased participation are largely undocumented.

Increased focus on Hudson River sport fisheries has also commanded attention due to several other factors. Foremost among them is the need for sport harvest data to comply with inter- and intrastate fisheries management plan objectives. Harvest data are particularly important for Hudson River species sought during spawning runs that are managed on a coastwide basis through New York's participation in the Atlantic States Marine Fisheries Commission (ASMFC). Management plans developed by ASMFC for striped bass, American shad, and river herring include a requirement to report harvest. In addition to the need for harvest data, an upstream contaminant source has resulted in fish consumption advisories throughout the estuary since 1976 (New York State Department of Health [DOH] 2000). The harvest of several Hudson River anadromous and resident species affected by the series of health advisories occurs at unknown levels throughout the estuary, particularly by ethnic groups that may rely on Hudson River fish as food (DOH 2000). As a result of these and other factors, a clear need existed to document overall recreational fishing effort, catch, and harvest, along with attendant fishery characteristics that included participant demographics.

Other than the series of mail surveys of licensed anglers, onsite contact surveys of Hudson River anglers are few and were limited spatially (Green and Jackson 1991), temporally (D. L. Peterson, Cornell University, unpublished manuscript), or by narrow focus (Barclay 1993; DOH 2000). The spring 1997 survey design and data reported by Peterson focused solely on striped bass and were intended to provide preliminary information to design the comprehensive recreational angling survey upon which this paper is based.

The full survey collected and reported data on the effort, catch, and harvest by recreational anglers throughout the Hudson River estuary for a 12-month period during 2001–2002, as well as demographics and expenditures of sport fishery participants. The focus of this paper, however, is on selected aspects of the primary open-water fishery that extended from March through November.

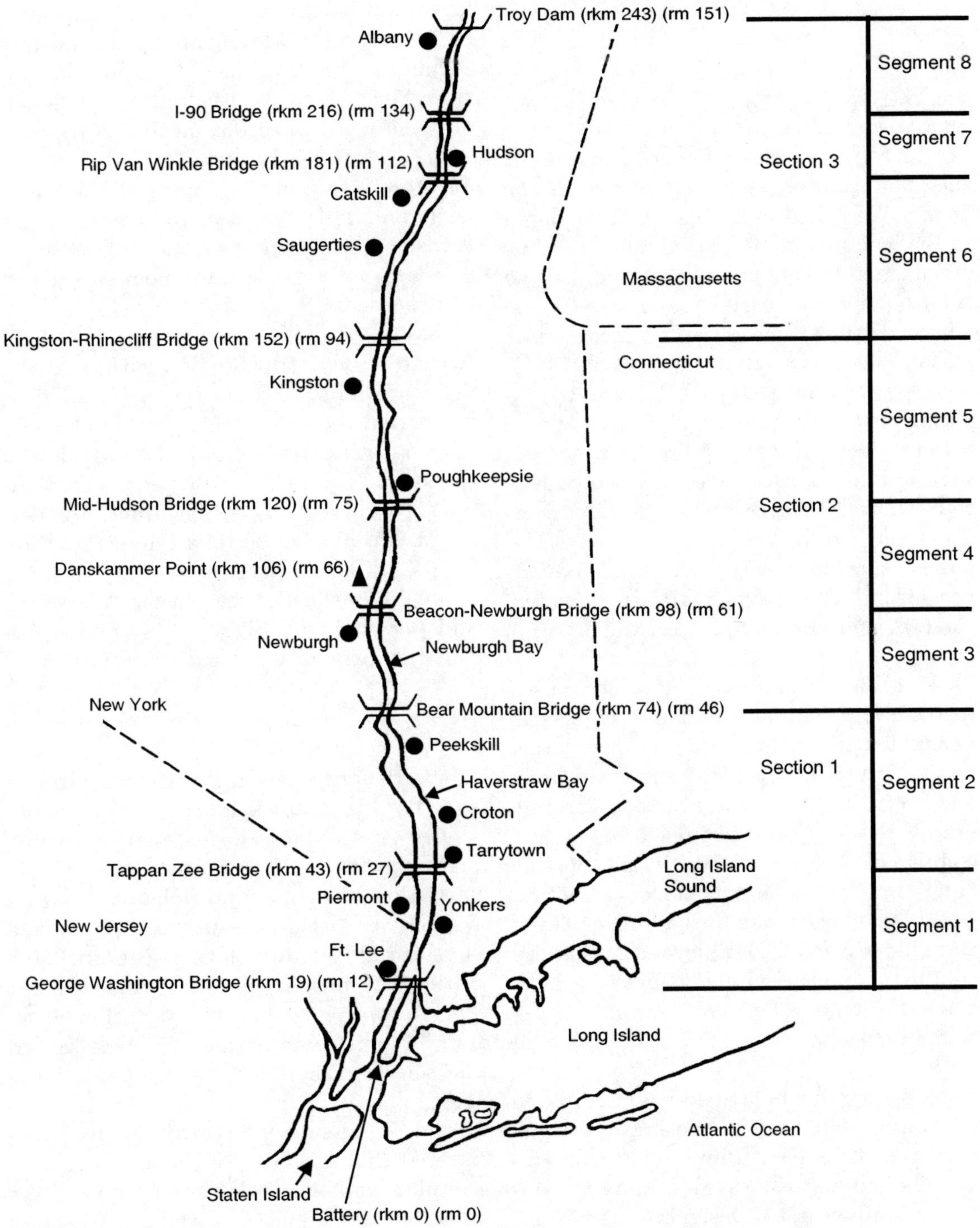

Figure 1. Hudson River recreational fisheries assessment study area (after Barnthouse et al. 1988). Landmark positions shown in river kilometers (rkm) and river miles (rm).

Methods

Study Area and Design

We investigated sport fishing throughout the tidal Hudson River estuary from the George Washington Bridge (GWB) north to the federal dam at Troy (Figure 1). This reach extends 224 km and also included the tidal portions of several major tributaries. Three distinct physiographic zones exist within the estuary (Barnthouse et al. 1988). The lower zone from GWB (rkm 19) to Haverstraw Bay (rkm 58) is shallow, up to 5.4 km wide, and typically brackish. A deep, narrower middle zone above Haverstraw Bay to Newburgh Bay (rkm 97) is usually freshwater but can be brackish seasonally. The largest zone upriver of Newburgh Bay to the federal dam (rkm 243) is freshwater, shallow, and less than 2.4 km wide.

Major bridges that cross the river were used to partition the study area into eight segments that were aggregated into three larger sections to facilitate overall survey logistics and reporting (Figure 1). Section 3 represented the longest study reach at 91 km, compared to 55 km of river within Section 1. The brackish zone is included within Section 1 but may extend upriver to Newburgh Bay (lower Section 2) or beyond during dry summers or otherwise low flow periods. Thus, the bulk of Section 2 and all of Section 3 is freshwater.

The opening day of striped bass fishing on 16 March 2001 and the closing of striped bass and black bass fishing on 30 November defined the time period of the open-water fishing survey. Designated "seasons" of approximately 6–9 weeks were established that facilitated temporal descriptions of fishing activity and also satisfied survey design considerations. The primary period of striped bass fishing during the spawning run was divided into two "seasons," early spring (16 March–30 April) and late spring (1 May–15 June). Opening day for New York black bass fishing is traditionally the third Saturday in June (16 June 2001). Thus, early summer (16 June–31 July), late summer (1 August–30 September), and fall (1 October–30 November) "seasons" encompassed the entire black bass season and the remainder of striped bass season.

A time of day stratification with separate morning (a.m.) and afternoon (p.m.) fishing periods was developed within the seasonal stratification, where the duration of the overall fishing day and the specific temporal periods were determined by day length. Fishing periods and time strata were adjusted to approximate day length throughout the fishing season. Weekdays and weekends or holidays (daytypes) provided additional temporal stratification.

Aerial Angler Counts

Low altitude flights in fixed-wing aircraft were used to count fishing boats and shore anglers as the basis for estimating fishing pressure. Proportionally more flights occurred on weekends and holidays than on weekdays. Aerial pressure counts took a substantial amount of time (up to 3.4 h during the busiest periods in spring) but were always less than the duration of an a.m. or p.m. time period. We randomized the selection of flight start time and the flight path within an a.m. or p.m. time period. Randomizing flight times and paths assured that all hours of the stratum were sampled and avoided sampling the same areas at the same time of day. Weather, mechanical problems, and severe flight restrictions following the World Trade Center attacks of September 11, 2001 prevented some flights, but eventually most were rescheduled.

Count flights covered the main river channel but also included passes over selected tributary mouths up to the first barrier (e.g., Eddyville Dam on Rondout Creek, about 5 km from the Hudson River) and backwaters that supported angling in some seasons. Counts of fishing boats and shore anglers were recorded by river segment. Only the number of fishing boats was counted; the number of boat anglers was later determined from ground intercepts.

Ground Surveys

Ground intercept surveys were conducted at access points that supported either shore or boat fishing, or sometimes both. However, since boat and shore angling were treated as independent strata, only one angler type was interviewed during a given survey. Temporal sampling intensity favored weekends, whereas spatial sampling was directed by where anglers were observed during aerial flights. Thus, access sites were visited in rough proportion to observed use; those sites receiving the most use were visited more often. Clerks waited at boat access sites to acquire completed-trip interviews, whereas shore fishing sites were visited by a roving clerk that conducted largely incomplete-trip interviews.

Data Analysis

Effort estimates for shore and boat anglers were instantaneous counts based on the aerial counts of both types of anglers. Fishing boat counts were expanded by the average number of anglers per boat per stratum to estimate boat angler counts. Instantaneous counts were expanded to estimate effort of the *i*th fishing day by:

$$\hat{e}_i = I_i \times T$$

where I_i = instantaneous count of anglers at time i,

T = length of the fishing period,

and the estimated effort for the stratum is:

$$\hat{E}_a = \sum_{i=1}^{n} (\hat{e}_i / \pi_i)$$

where $\pi_a = n_a/N_a$ for all of the sampling units in the *a*th stratum.

Total effort was the sum of the effort of each stratum where n_a = number of units sampled in stratum *a*, N_a = total number of units in stratum *a*.

The variance for each stratum and standard error is:

$$\hat{V}ar(E_a^n)\frac{N_a^2}{n_a - 1} = \sum_{i=1}^{n} (e_{ia} - \overline{e}_a)^2$$

$$\hat{S}E(\hat{E}_a) = \sqrt{Var(\hat{E}_a)}$$

The sum of the variances for each stratum was the variance for total effort. Precision of effort estimates was expressed as proportional standard error (PSE). Standard error was divided by the estimate to calculate PSE. A target PSE for survey estimates is 20% or less (Malvestuto 1983).

Angler trips were calculated by dividing the estimate of angler-hours for a stratum or fishing mode (shore/boat) by the mean trip length for the respective stratum/mode (Malvestuto 1983). Estimated total effort in angler trips was the aggregate of individual strata.

Catch per unit effort (CPUE) and harvest per unit effort (HPUE), expressed as fish/h, were calculated for all anglers (overall rates), as well as those targeting a specific fish species (directed rates). Overall rates were used to calculate total and individual fish catch and harvest. Directed rates were used to describe and compare angler success among targeted fish species. Rates for boat and shore anglers were calculated separately. Catch per unit effort and HPUE for boat anglers were determined using a ratio-of-means estimator, which is recommended when using completed trip interviews (Jones et al. 1995). Catch per unit effort and HPUE for shore anglers were determined using a mean of the ratios estimator, which is recommended when using incomplete trip interviews (Jones et al. 1995). The vast majority of boat interviews were completed trips, whereas most shore interviews were incomplete trips. Incomplete fishing trips of less than 0.5 h were discarded prior to calculating CPUE and HPUE to avoid extreme rates (Pollock et al. 1994).

Catch and harvest calculations for individual species and for all species combined estimated effort determined from aerial surveys with CPUE and HPUE determined from ground interviews (Pollock et al. 1994). Effort in angler-hours was multiplied by the respective rate within a stratum to estimate the number of fish caught, kept, or both. Stratum totals were summed to estimate catch and harvest by shore and boat angling for each season, river section, and the total estuary.

Angler characteristics such as residence location, license status, trip purpose, harvest disposition, and others were based on observed data only. Interview responses were sorted and summarized by river section, season, or species targeted as appropriate. All usable interviews were included for summaries of angler characteristics.

Results

Aerial fishing pressure counts occurred on 15–25 d per season. The fewest flights occurred in late spring and late summer. Bad flying weather prevented flights during successive weekends in late spring, including all of Memorial Day weekend. As a result, there were too many postponed flights that occurred late in the late spring–weekend stratum to make up. Likewise, flight restrictions following the September 11, 2001 World Trade Center attacks affected the number of flights completed in Late Summer. In other seasons 85% to 96% of planned flights occurred.

Seasonal and Spatial Trends in Angler Distribution

Angler distribution in the Hudson River Estuary during the spring mirrored the upstream movement of striped bass, commonly tracked by internet reports and the local outdoors press. Angling during the early spring season was focused in the lowermost Hudson River segments 1 and 2 (Figure 2). More than two-thirds (68%) of all shore anglers in early spring were counted in segments 1 and 2. Shore fishers in segments 1 and 2 were most frequently observed at bulkheads in Palisades Park (rkm 22–26) in New Jersey, at Piermont Pier (rkm 40), in Croton Point Park (rkm 55) and along the bank north of Croton Point (rkm 58, Haverstraw Bay), and at the Bear Mountain State Park fishing pier (rkm 74). Boat anglers in segment 2 represented 35% of all boats counted in early spring and were mostly associated with the Croton area and Haverstraw Bay. Little boat angling occurred in segment 1. By late April, boat angling was also common near Newburgh (rkm 97 in segment 3) and shore anglers frequented the west shore north of Marlboro (rkm 109) in segment 4.

Boat anglers were also observed on five occasions through April 28 in the power plant effluent below Danskammer Point (rkm 106) north of Newburgh. The most boats counted there were two.

During late spring most anglers fished from boats in the main river channel in study

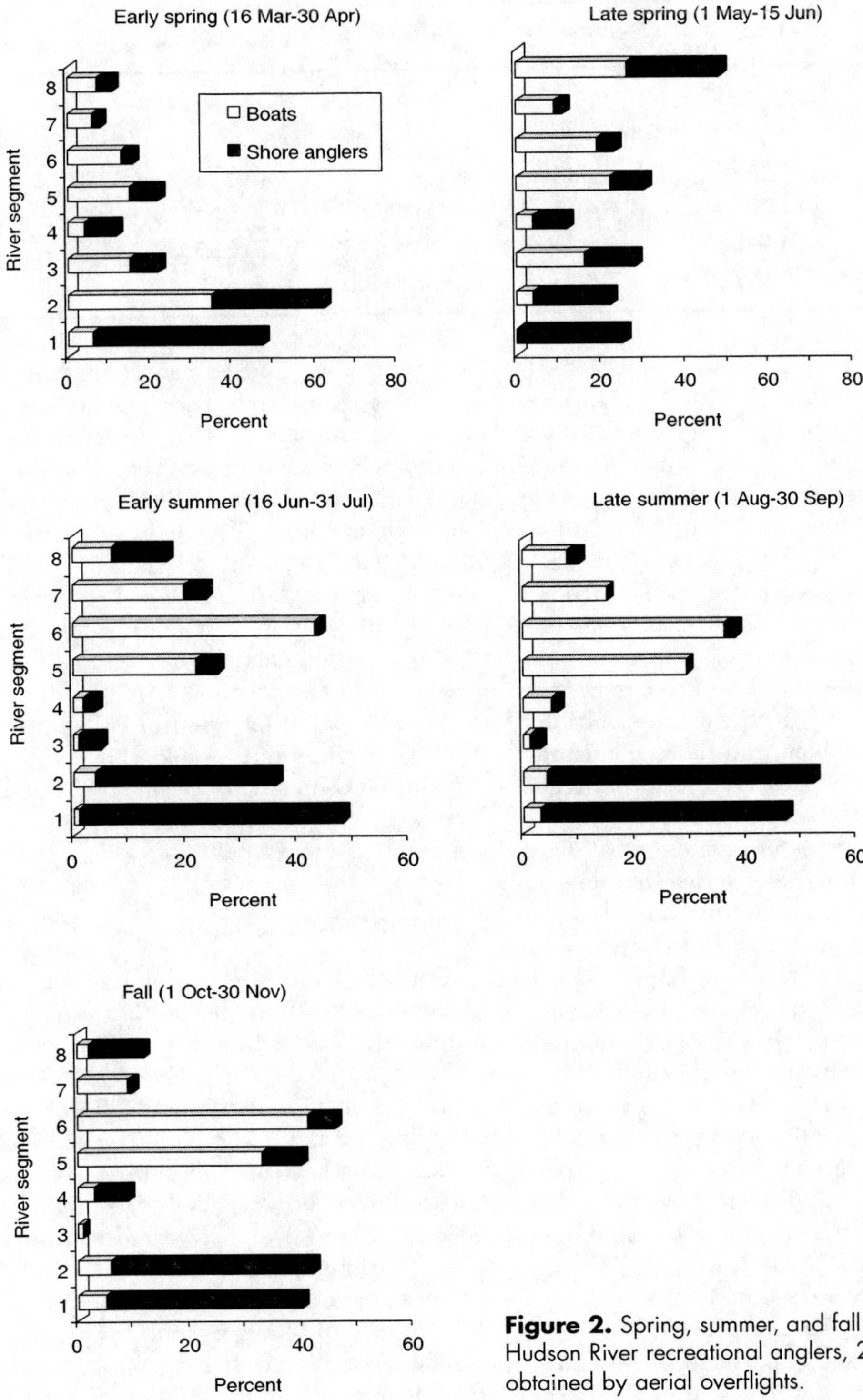

Figure 2. Spring, summer, and fall distribution of Hudson River recreational anglers, 2001. Data obtained by aerial overflights.

Table 1. Estimated recreational fishing effort (angler-hours [PSE[a]]) on the Hudson River, March-November 2001.

Season	Section 1	Section 2	Section 3	Hudson River
Early spring	48,626	27,477	18,460	94,563 (15)
Late spring	26,037	107,098	119,200	252,336 (29)
Early summer	17,452	8,066	22,785	48,303 (23)
Late summer	14,728	5,602	10,524	30,854 (18)
Fall	4,916	6,234	9,415	20,565 (28)
Total	111,760 (11)	154,477 (31)	180,384 (32)	446,621 (17)

segments 5 through 8 north of the mid-Hudson Bridge (rkm 120) at Poughkeepsie (Figure 2). The largest proportion (26%) of boats were counted in segment 8 (federal dam–Albany area), followed by segments 5 (Kingston, Esopus Meadows, Norrie Point) and 6 (Saugerties, Catskill, Germantown). Boat anglers also favored the reach from Cornwall to Newburgh (rkm 92–97) in segment 3. The peak period for boat angling occurred in May. By early June boat angling had declined substantially throughout the river, and more nonfishing boat activity was observed. Shore anglers in late spring were most abundant at the piers and bulkheads in segment 1 (25% of the seasonal total), along the shoreline in segment 8 (22%), primarily the result of anglers seeking black bass in the upper half of the estuary), or mesohaline species such as bluefish and blue crabs in the lower estuary. Boat angling during early summer occurred primarily in the main channel portions of segments 5 through 7 (Figure 2). Boats were typically scattered along the shorelines throughout these segments, particularly in segment 6 where 43% of all boats were counted. Shore anglers were abundant only in segments 1 and 2, where 79% of all shore anglers were counted. A lower level of shore angling continued throughout early summer in segment 8, particularly near Troy Dam and at the mouth of Vlomankill Creek (rkm 218) in Henry Hudson State Park.

By late summer and fall, shore angling was common only at a relatively few access sites in segments 1 and 2 (Figure 2). Sites in Palisades Park, Piermont Pier, Bear Mountain Pier, Charles Point Pier (rkm 69), and the bank north of Croton Point were the most heavily used. Substantial boat angling use in segments 5, 6, and 7 continued throughout late summer and fall, with the most boat-fishing activity in segment 6 each season. Additionally, boat fishing in the three principal tidal tributaries located in segment 5 (Rondout Creek, rkm 143) and segment 6 (Esopus Creek, rkm 164; Catskill Creek, rkm 182) increased throughout summer into fall.

Estimated Effort

Hudson River recreational anglers fished an estimated 446,621 angler-hours during the March 16 to November 30 fishing season (Table 1). Precision (PSE) of the overall estimate was 17%. Estimated effort in angler trips for the same period was 86,622. Section 3 furthest upriver supported 40.4% of total angler-hours, followed by mid-river section 2 (34.6%) and lowermost section 1 (25%). Fishing pressure estimated for the entire spring (346,899 angler-hours) was 77.7% of the full-season total. Spring (early and late combined) fishing effort was focused in sections 2 and 3, and within each section was remarkably similar in magni-

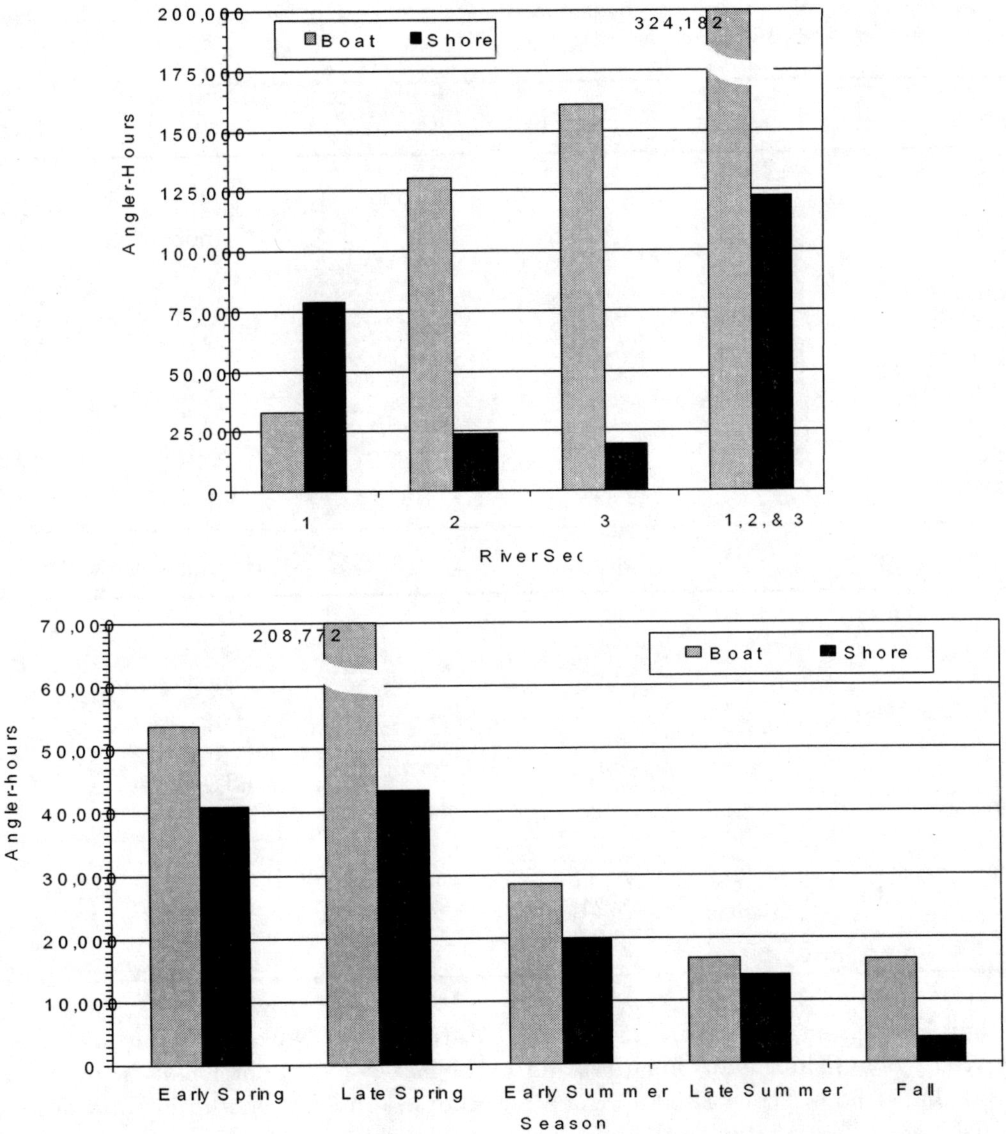

Figure 3. Boat and shore fishing effort (angler-hours) by Hudson River section (top) and season (bottom).

tude (134,575 and 137,660 angler-hours, respectively). Spring fishing effort accounted for 67% (Section 1) to 87% (section 2) of the total angler-hours estimated within each section. Section 3 also supported the most fishing effort throughout summer and fall, although the amount of section 3 effort after spring declined by 69%. Section 2 supported the least amount of fishing effort during the combined summer and fall seasons, declining by 85% from the amount of effort estimated for the combined spring seasons.

Boat fishing (324,182 angler-hours) represented 72.6% of the total amount of effort estimated for the Hudson River (Figure 3, top). Boat-based angling was intensively focused north of the Bear Mountain Bridge where it formed 84% and 89% of total an-

Table 2. Total numerical catch (all fish caught) and harvest (fish kept) of Hudson River fishes by recreational anglers, 16 March-30 November 2001.

Method	Section 1	Section 2	Section 3	Total
Boat				
Catch	16,010	31,920	94,761	152,691
Harvest	1,615	5,810	14,566	21,990
Shore				
Catch	31,611	19,703	18,421	69,734
Harvest	16,781	3,607	2,101	22,489
Total				
Catch	47,621	51,623	113,182	212,426
Harvest	18,396	9,417	16,667	44,479
	Spring		Summer–Fall	
Boat				
Catch	99,422		43,270	
Harvest	20,509		1,482	
Shore				
Catch	43,791		25,943	
Harvest	7,928		15,191	
Total				
Catch	143,213		69,212	
Harvest	27,806		16,672	

gler-hours estimated in sections 2 and 3, respectively. Boat fishing effort in late spring (208,772 angler-hours) comprised 64% of all boat fishing estimated during March–November (Figure 3, bottom). In contrast, shore-based fishing comprised 70.5% of total angler effort in section 1 (Figure 3, top). The amount of shore fishing effort exceeded boat fishing effort in section 1 during each season in contrast to upriver sections 2 and 3, where boat fishing prevailed each season.

Catch and Harvest

We estimated that Hudson River recreational anglers caught 212,426 fish (including blue crabs) and harvested 44,479 during the 16 March to 30 November season (Table 2). The largest (53%) portion of the overall catch occurred in upriver section 3, where more than twice as many fish were caught compared to sections 1 or 2. However, the largest fish harvest (18,396) came from the lower estuary in section 1.

Boat anglers caught twice as many fish as shore anglers, although the number of fish harvested by each method was similar (Table 2). The comparable harvest totals reflected that shore anglers kept 32.2% of their catch, more than twice the retention rate (15.4%) of boat anglers. Spatially, shore anglers in section 1 were responsible for 75% of the total shore angler harvest, retaining

Table 3. Estimated numerical recreational catch and harvest of selected Hudson River fishes by season and method, 16 March–30 November 2001.

		Spring		Summer–Fall		Total	
Species	Method	Catch	Harvest	Catch	Harvest	Catch	Harvest
Striped bass	Boat	31,832	6,050	381	0	32,213	6,050
	Shore	11,412	935	2,064	244	13,476	1,179
	Totals	43,244	6,985	2,445	244	45,689	7,229
River herring	Boat	31,203	10,189	1	0	31,204	10,189
	Shore	3,574	2,045	0	0	3,574	2,045
	Totals	34,777	12,235	1	0	34,778	12,235
White perch	Boat	5,434	0	5,992	564	11,426	564
	Shore	12,160	2,924	9,060	5,721	21,220	8,645
	Totals	17,594	2,924	15,052	6,286	32,645	9,210
Large-mouth bass	Boat	6,384	0	13,556	43	19,939	43
	Shore	0	0	0	0	0	0
	Totals	6,384	0	13,556	43	19,939	43
Ameri-can shad	Boat	13,034	1,047	6	0	13,040	1,047
	Shore	6,732	242	0	0	6,732	242
	Totals	19,766	1,289	6	0	19,772	1,289
Small-mouth bass	Boat	1,508	0	15,441	5	16,949	5
	Shore	627	0	654	6	1,282	6
	Totals	2,136	0	16,095	11	18,231	11

53% of their catch. In comparison, boat anglers in section 3 caught and harvested 66% of all fish caught from boats. For all boat anglers, as well as shore anglers other than in section 1, the retention rate varied from 10% to 18% of the respective catch.

Seasonally, more than two-thirds of all fish caught and 63% of fish harvested occurred during the spring (16 March to 15 June; Table 2), primarily the result of boat angling. Although shore anglers also caught more fish in the spring than later in the year, the number of fish harvested in summer through fall was more than twice the number harvested in the spring.

Hudson River recreational anglers caught at least 31 fish species plus blue crab. Striped bass ranked first among fishes caught and along with river herring (alewife and blueback) and white perch formed 53.3% of the total catch (Table 3). Black bass (combined species) and American shad formed 18.0% and 9.3% of the catch, respectively. River herring, white perch, blue crabs, and striped bass accounted for 82.3% of all fishes harvested.

A wide range in retention rates among species characterized the Hudson River sport fishery (Table 3). Blue crab and carp *Cyprinus carpio* were harvested at the highest rates (79% and 76%, respectively). White catfish *Ameiurus catus*, bluefish, river herring, and white perch were harvested at more

Table 3. continued

Species	Method	Spring Catch	Spring Harvest	Summer–Fall Catch	Summer–Fall Harvest	Total Catch	Total Harvest
Blue crab	Boat	0	0	808	783	808	783
	Shore	33	0	9,183	7,120	9,215	7,120
	Totals	33	0	9,931	7,904	10,024	7,904
Yellow perch	Boat	2,679	438	2,486	0	5,164	438
	Shore	1,389	75	84	0	1,473	75
	Totals	4,068	513	2,569	0	6,637	513
American eel	Boat	1,368	0	712	0	2,080	0
	Shore	2,827	390	980	396	3,807	786
	Totals	4,196	390	1,691	396	5,887	786
Common carp	Boat	2,784	2,784	616	0	3,400	2,784
	Shore	107	53	240	15	348	68
	Totals	2,892	2,837	856	15	3,748	2,852
White catfish	Boat	171	0	866	14	1,036	14
	Shore	404	327	626	585	1,030	912
	Totals	575	327	1,491	599	2,066	926
Bluefish	Boat	0	0	0	0	0	0
	Shore	0	0	1,924	870	1,924	870
	Totals	0	0	1,924	870	1,924	870

moderate rates of 45% to 28%. Striped bass were retained at an overall rate of 16%. The largemouth and smallmouth bass sport fishery was virtually all catch and release.

Seasonally, nearly the entire catch and harvest of river herring and American shad occurred in spring (Table 3). Most of the catch and harvest of striped bass also occurred in spring, but substantial numbers of striped bass were also caught (with only limited harvest) throughout summer and fall. Most of the carp catch and harvest also occurred in the spring, principally in early June as the result of archers targeting carp.

Largemouth and smallmouth bass, bluefish, and blue crab represented Hudson River species caught primarily in summer and fall (Table 3). Other fishes such as American eel *Anguilla rostrata*, white catfish, white perch, and yellow perch were caught throughout spring through fall.

Foremost among the less common fishes caught and with only limited harvest were several species of catfish. Bullheads *Ameiurus* spp. and channel catfish *Ictalurus punctatus* were each caught in numbers comparable to white catfish but fewer were retained.

Striped Bass Size and Age Data

Hudson River striped bass in 2001 were

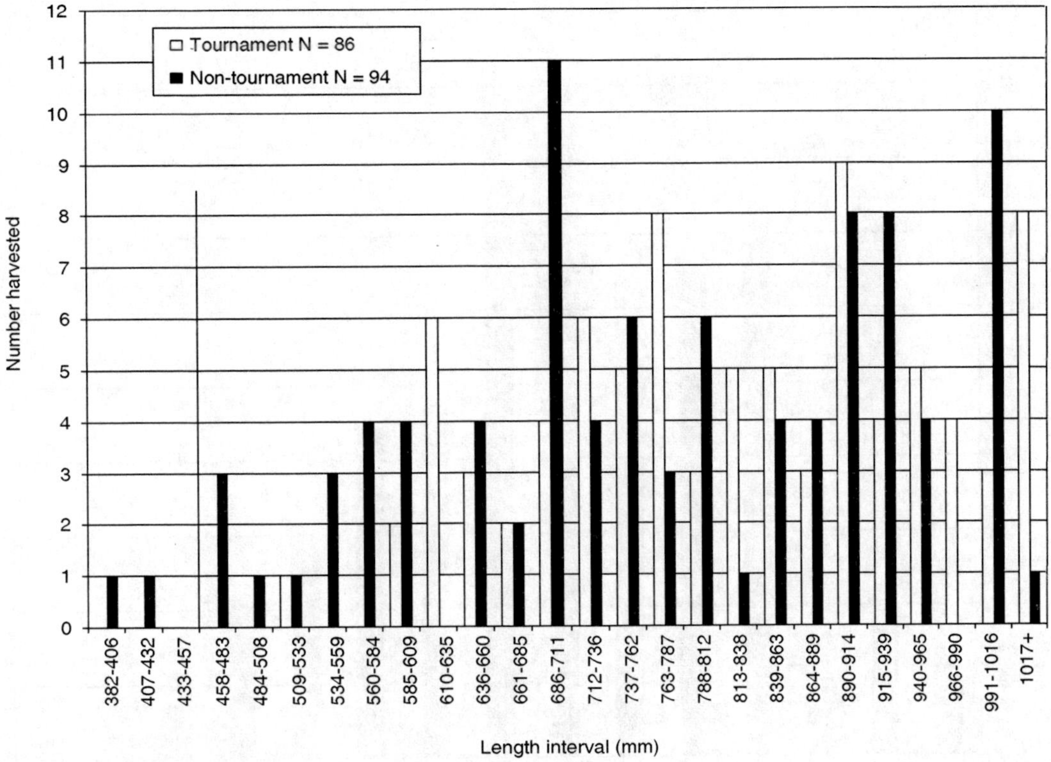

Figure 4. Length frequency of tournament (spring only) and nontournament (mostly spring) harvested striped bass.

managed with a 458-mm minimum size limit, although most anglers preferred to harvest larger fish at least 610 mm or longer (Figure 4). The numbers of harvested striped bass 610 mm or larger were generally comparable between tournament (n = 83) and nontournament (n = 76) groups, particularly for fish larger than 914 mm (36 in). However, nontournament anglers harvested substantially more striped bass less than 610 mm long (n = 18) than tournament anglers (n = 3), including some sublegal striped bass.

Age was determined for 169 harvested striped bass (sexes combined). Tournament caught fish (n = 80) were taken only during three weekends in May. Striped bass caught by nontournament anglers (n = 89) were harvested primarily in April and May (n = 81), with additional specimens in the sample from June (6) and October (2). The age distributions of tournament and nontournament harvested striped bass were judged sufficiently similar to combine and represent the overall 2001 striped bass harvest (Figure 5). Two modal striped bass age groups, Ages 5 and 6, and age 9, formed 28% and 18%, respectively, of the pooled sample. The oldest striped bass harvested was 16 years old (1985 year-class), but the largest striped bass harvested in our sample (1,120 mm) was 11 years old (1990 year-class). In addition, four 2-year old striped bass were harvested; two were the sublegal fish caught in late June.

Catch and Harvest Rates

Shore anglers were more successful than boat anglers (Table 4). The total CPUE of shore anglers (0.69 fish/h) exceeded that of boat anglers (0.44 fish/h) for the entire Hudson River, and was more than twice the CPUE

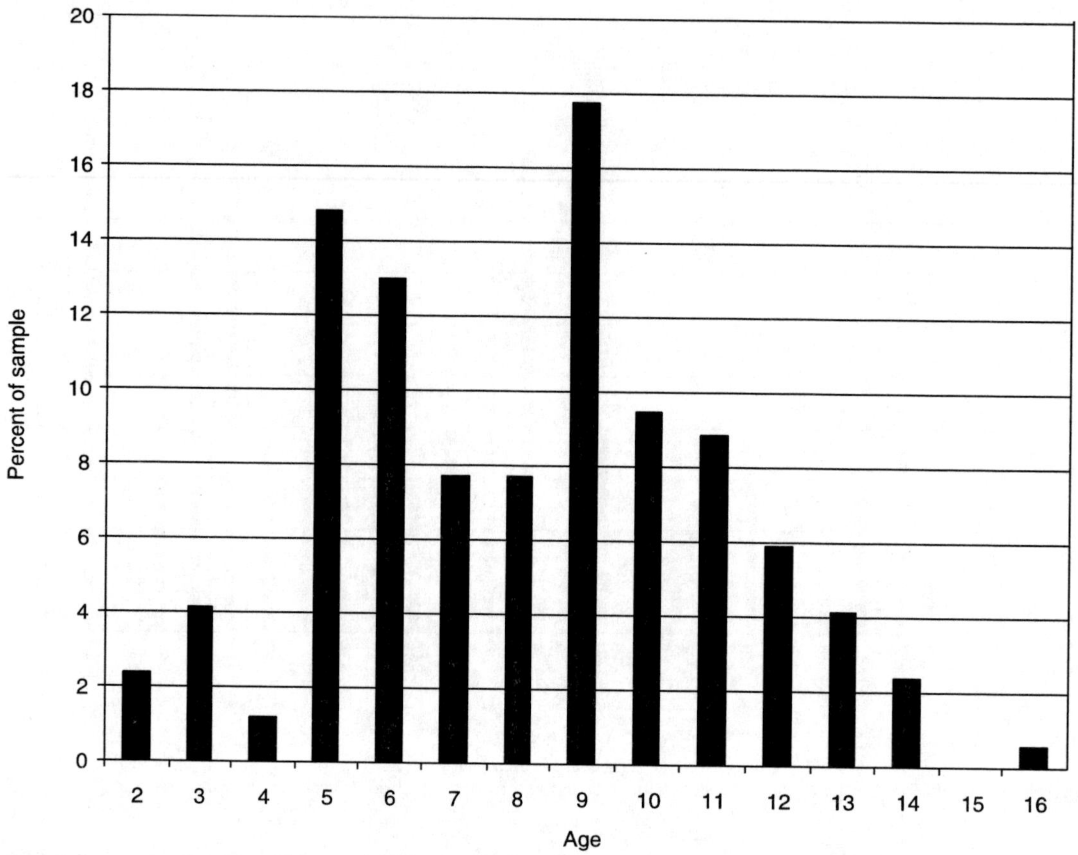

Figure 5. Combined age distribution (%) of tournament (*N* = 80) and nontournament (*N* = 89) striped bass harvest.

of boat anglers in sections 2 and 3. A large disparity also existed between boat and shore angler HPUE within sections 2 and 3. Although less disparity in HPUE existed between angler types in section 1, shore anglers harvested fish at more than twice the rate of boat anglers.

Seasonally, CPUE for shore anglers was highest during the late spring through late summer period when it ranged from 0.72 to 0.99 fish/h (Figure 6). The highest HPUE (0.44–0.46 fish/h) for shore anglers occurred during the summer seasons, compared to HPUE less than 0.10 fish/h during spring and fall. In contrast to the shore fishery, boat anglers experienced relatively high CPUE and HPUE (0.46 and 0.15 fish/h, respectively) in early spring which then declined through early summer. Boat angler CPUE subsequently increased to the peak seasonal value (0.51 fish/h) during fall, whereas the corresponding HPUE after late spring was negligible (< 0.01 fish/h).

Directed Fisheries

Hudson River recreational anglers targeted 19 species or species groups (Table 5). Either black bass or striped bass were sought by more than two-thirds of Hudson River anglers interviewed. Blue crab and catfishes (including white catfish, channel catfish, and bullheads) were each sought by approximately 3% of those interviewed. Bluefish, white perch, carp, and American shad com-

Table 4. Catch and harvest rates (fish/angler-hour) with associated precision (PSE), in parentheses, for Hudson River recreational anglers in 2001.

		Boat anglers	Shore anglers
Section 1	CPUE	0.32 (39.8)	0.52 (10.3)
	HPUE	0.11 (61.1)	0.28 (17.1)
Section 2	CPUE	0.37 (8.2)	0.88 (20.1)
	HPUE	0.03 (28.1)	0.16 (27.1)
Section 3	CPUE	0.49 (5.2)	1.04 (17.5)
	HPUE	0.01 (43.4)	0.09 (34.6)
Total river	CPUE	0.44 (4.5)	0.69 (8.7)
	HPUE	0.02 (23.8)	0.22 (14.5)

bined were targeted by 5% of anglers. More than 19% of anglers interviewed did not express a species preference.

Boat anglers clearly targeted either black bass or striped bass (Table 5). In contrast, the predominant response by queried shore anglers was "anything." Shore anglers that targeted certain species/groups sought primarily striped bass, or blue crab and catfishes. Pursuit of bluefish, white perch, carp, and American shad was also mainly by shore anglers.

The total of Hudson River open water fishing trips estimated for 16 March-30 November 2001 (86,622) was apportioned among target species based on the seasonal responses by interviewed anglers. Striped bass and black bass trips formed 72.7% of the total number of fishing trips, and trips for these species comprised nearly 90% of all trips by boat anglers (Table 5). Fishing trips targeting catfishes and American shad ranked third and fourth, respectively, among all anglers. Catfish trips were heavily shore-oriented, whereas the number and proportion of shad fishing trips was roughly comparable between angler types. Nearly 5% of the total number of shore trips targeted blue crab. The amount of untargeted fishing trips was highest among shore anglers who sought "anything" during more than one out of three trips.

Directed CPUE/HPUE

Fishers seeking American shad and blue crab, and archers seeking carp from boats had the highest directed CPUE (Table 6). However, HPUE was high only for blue crab and the limited number of carp archers. The directed striped bass CPUE for shore-based anglers (0.260 fish/h) exceeded that of boat anglers (0.148 fish/h). Conversely, boat anglers harvested striped bass at a higher rate than shore anglers (0.035 versus 0.028 fish/h, respectively). Catfish anglers in boats were comparatively rare but exhibited higher CPUE than shore catfish anglers. However, boat catfish anglers released their catch, whereas shore anglers harvested most of those caught. The largemouth bass CPUE by boat anglers was slightly higher than for smallmouth bass, but the HPUE for either species was negligible. Although the CPUE for shore-caught smallmouth bass was high, little shore angling for black bass occurred.

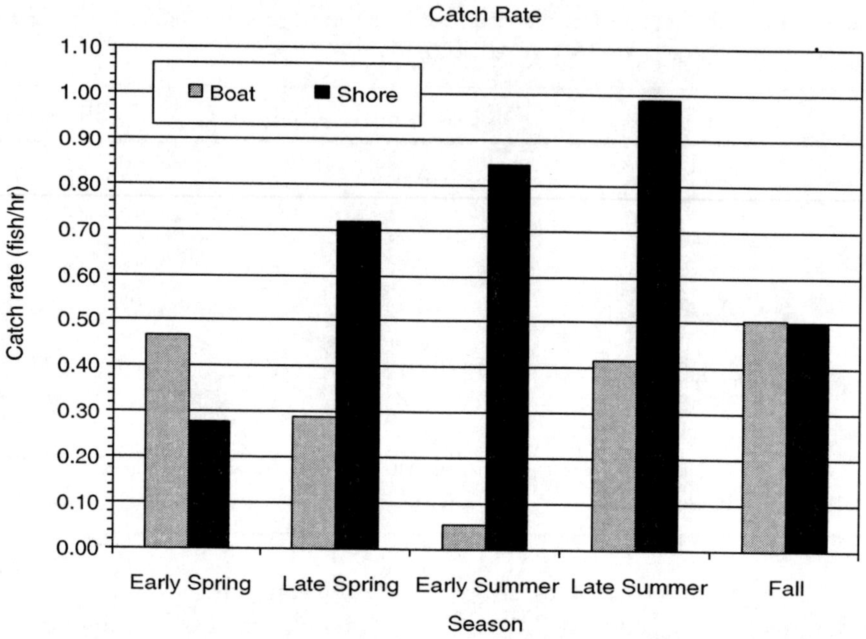

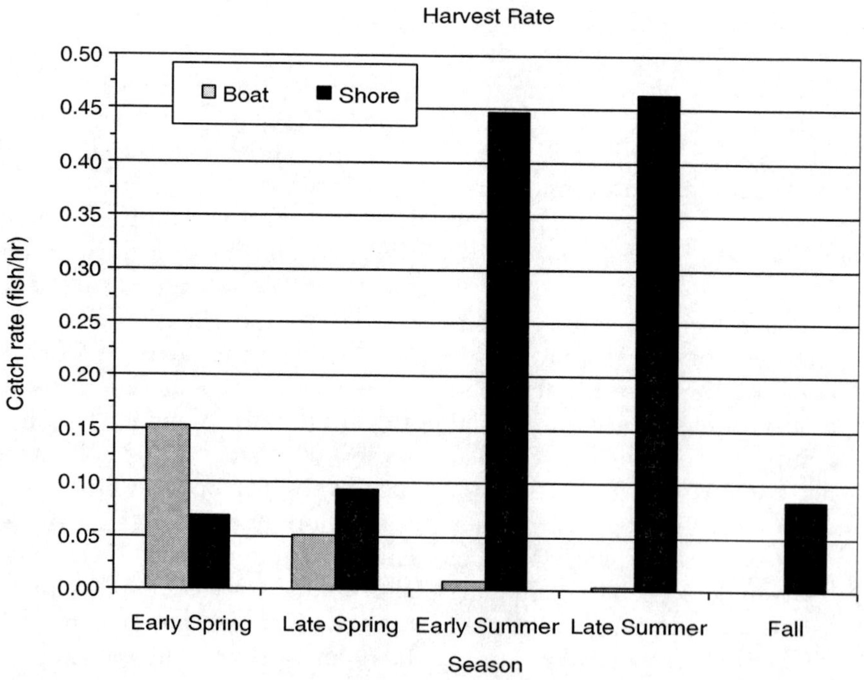

Figure 6. Total catch (CPUE) and harvest (HPUE) rates by boat and shore anglers fishing the Hudson River in 2001.

Table 5. Species targeted and number of Hudson River fishing trips by target species, 16 March–30 November 2001.

	Boat anglers		Shore anglers		All anglers	
Target species	Interviewed sample (%)	Estimated trips	Interviewed sample (%)	Estimated trips	Interviewed sample (%)	Estimated trips
Striped bass	28.4	41,139	34.5	12,849	31.1	53,988
Anything sought	4.3	3,310	38.5	11,166	19.6	14,476
Black bass	64.6	8,715	1.4	317	36.4	9,032
Catfishes	0.3	94	6.4	1,962	3.0	2,056
American shad	0.6	878	1.9	781	1.2	1,659
Blue crab	0.2	29	7.2	1,496	3.3	1,525
Common carp	0.2	220	2.7	626	1.3	845
White perch	0.2	29	2.7	780	1.4	809
Bluefish	0.2	24	2.9	570	1.4	594
All other species[a]	0.9	1,022	1.9	615	1.4	1,636
Totals		55,460		31,162		86,622

[a]Includes river herring, American eels, Atlantic tomcod, crappie, northern pike, walleye, Atlantic menhaden, sunfish, brook trout, weakfish, yellow perch.

Hudson River Angler Characteristics

Various characteristics of the interviewed angler sample such as residence, license status, ethnic and gender composition, fishing trip purpose and satisfaction, and intended harvest disposition were detailed in the comprehensive survey report. Herein, we discuss only sportfishing license data and the spatial variability in angler awareness of Hudson River fish consumption advisories, an ongoing concern for more than two decades (Barclay 1993; DOH 2000).

A New York sportfishing license was purchased by 59.3% of all interviewed anglers, but the licensed proportion among boat anglers (81.2%) was 2.5 times that of shore anglers (32.2%). A smaller proportion of both boat and shore anglers was licensed below Bear Mountain Bridge than upriver. Fewer than 50% of striped bass anglers interviewed were licensed.

Awareness of fish consumption advisories by Hudson River anglers varied spatially (Table 7). More than 95% of all anglers interviewed north of the Bear Mountain Bridge (sections 2 and 3) were aware of published health advisories, compared to 58.0% of anglers south of the Bear Mountain Bridge with the same knowledge. (Note: 58.0% represents a corrected percentage relative to 62.4% of aware anglers provided in the original source report). Further, lower estuary (section 1) anglers fishing during the summer seasons and residing in New York City boroughs were less aware of health advisories than those fishing in other seasons or residing elsewhere.

Table 6. Directed catch (CPUE) and harvest (HPUE) rates (fish/h) with associated precision (PSE) for the principal targeted Hudson River species, 2001.

	Boat				Shore			
Species sought	CPUE	PSE	HPUE	PSE	CPUE	PSE	HPUE	PSE
American shad	1.498	72	0.000	-	1.534	28	0.043	42
Common carp	1.380	97	1.380	97	0.102	37	0.019	86
Catfishes	0.236	14	0.000	-	0.148	49	0.097	71
Striped bass	0.148	19	0.035	14	0.260	17	0.028	19
Smallmouth bass	0.260	5	<0.001	83	1.571	52	0.053	137
Largemouth bass	0.298	5	<0.001	71	-	-	-	-
Blue crab	1.387	34	1.318	35	1.253	34	0.943	41

Fish consumption of certain species was common despite persistent health warnings. Anglers targeting and keeping catfish, blue crab, or striped bass, as well as anglers with no stated preference, intended to eat their catch.

Discussion

Striped bass fishing clearly provided the impetus for spring angling throughout the Hudson River estuary, and spring angling represented 78% of the amount of fishing effort (angler-hours) estimated for the 16 March through November period. Anglers also sought striped bass during summer and fall but primarily in discrete river areas as opposed to estuary-wide pursuit in spring. For example, renewed focus on striped bass by shore anglers in the fall occurred throughout the lower Hudson River (Section 1), and also, although more limited, below the Troy Dam.

Fishing effort built throughout early spring, and reached a peak in mid to late May. Although the seasonal and temporal fishing patterns of Hudson River anglers in spring 2001 were generally consistent with those described for spring 1997 (D. L. Peterson, Cornell University, unpublished manuscript), the amount of effort estimated in spring 2001 (346,899 angler-hours) was substantially less than the 619,000 angler-hours estimated during spring 1997. At least a portion of the disparity may be due to the difference in survey dates between years. The 1997 survey ran from 6 April–30 June (85 d), compared to the 16 March - 15 June (92 d) period designated "spring" in this survey. Although the 1997 survey frame was slightly shorter, it likely encompassed more consistent weather and, in late June, an open black bass season.

Further, early spring weather in 2001 was particularly unfavorable and likely delayed the spawning run of striped bass. The combined influences of poor weather and a delayed run of fish may have depressed the number of anglers seeking striped bass. Comparisons of raw aerial counts during comparable weekends in April and May revealed substantially fewer boats and shore anglers in 2001 than in 1997. Disparity between 1997 and 2001 aerial counts was most noticeable in section 1 south of the Bear Mountain Bridge, but also occurred in upriver reaches. Apparently a substantial reduction in angling pressure occurred during spring 2001.

Table 7. Summary of angler awareness of Hudson River fish consumption advisory.

	Section	Anglers interviewed		
		Total	Aware	Percent
	1	1,653	960	58.0
	2	1,098	1,089	99.2
	3	1,833	1,741	95.0
Total		4,584	3,790	82.7

Estimated fishing effort of 86,622 open water trips during 2001 was also less than total 1996 effort of 185,320 angler-days estimated by Connelly et al. (1997). Although trips and days appear to represent two comparable effort measures they are derived differently. The number of trips represents a specified calculation (estimated angler-hours/trip length), whereas angler-days were estimated by a mail survey of licensed anglers that depended upon trip recall.

Although estimated 2001 spring fishing effort was less than in 1997, likely the result of less effort for striped bass, HPUE of spring 2001 striped bass anglers was about the same as for spring 1997 shore anglers and slightly higher for boat anglers. The HPUE comparability yielded a 2001 spring striped bass harvest estimate (6,985 fish) that was approximately 49% of the striped bass harvest estimated in spring 1997. Since interyear striped bass harvest rate fluctuations may be dampened by the one-fish possession limit, any possible changes in 2001 striped bass catchability may be better indicated by CPUE. Spring striped bass CPUE in 2001 was higher for shore anglers in 1997 but substantially lower for the boat anglers that expended most of the effort. Thus, reduced success of striped bass boat anglers in 2001 coupled with poor weather may have reduced overall fishing effort.

Angler awareness of fish consumption advisories was high (>95%) in Hudson River study areas north of the Bear Mountain Bridge, but substantially lower (58%) among section 1 fishers. Advisory awareness in the lower estuary was especially lacking for those traveling from New York City boroughs to fish upriver in the summer. These anglers typically fished from shore at several public access sites, especially in the Croton area, and caught mostly white perch, blue crab, American eel, white catfish, bullheads, and bluefish. The retention rates among these species in section 1 were highest for white catfish (99%), blue crab (79%), and white perch (57% overall, but 65% for shore anglers that were responsible for 92% of the white perch harvest). Most of the harvest for each species occurred from shore during summer.

Several possible reasons exist for the disparate awareness of the health advisories between portions of the Hudson River estuary.

1) Fewer licensed anglers fished south of the Bear Mountain Bridge. Fishing License Guide booklets issued at license sale contain the most recent public health advisories published by DOH, but a New York license is not required to fish the tidal mainstem Hudson River. The guide booklet was judged more effective than other media for disseminating Hudson River health advisories by an earlier study (Barclay 1993).

2) Anglers residing or fishing in the upper estuary were closer to the contaminant source located above Troy Dam, and during 2001 may have been more cognizant of the publicity and controversy associated with this issue.

3) Many of the anglers interviewed in the upper estuary were black bass anglers. As a group, black bass anglers, even those from out of state, may be better informed or more aware of regional fisheries issues than other constituencies, either through publications received or membership in state federations that disseminate information that may affect their members.

4) Poor English comprehension may inhibit awareness of advisories either posted at fishing sites by DOH (DOH 2000) or printed in the Fishing License Guide booklet. Survey clerks working south of the Bear Mountain Bridge were much more likely to encounter anglers unable to complete interviews due to a language barrier.

Regardless, it may be erroneous to assume that reduced advisory awareness is the sole or even major factor in increased harvest and consumption of fish in the lower estuary. Further, Hudson River anglers are not unique in their apparent disregard for published health warnings (Connelly et al. 1996). Personal choice and risk perception are important roles (Velicer and Knuth 1994). Awareness of health advisories appears to have increased throughout most of the tidal Hudson River compared with previous surveys, although study methods differed (Barclay 1993; DOH 2000). Nonetheless, many anglers in the lower estuary continued to harvest blue crab and a variety of fish for consumption. White perch and, especially white catfish exceeded Food and Drug Administration (FDA) tolerance limits for commercial sale of these species and were among the principal fishes harvested among anglers surveyed in 1996 (DOH 2000). Angler preferences for harvest and consumption in the lower estuary appear not to have changed since 1996 despite enhanced awareness of the consequences.

Acknowledgments

This survey was funded through the NYSDEC Hudson River Estuary Program's Action Plan, supported by the New York State Environmental Protection Fund. All work related to this survey was performed under NYSDEC Contract No. C004005 with Normandeau Associates, Inc. The Hudson River Fisheries Unit (HRFU) of the Division of Marine Resources provided all technical guidance and administrative support.

References

Barclay, B. 1993. Hudson River Angler Survey. A report on the adherence to fish consumption health advisories among Hudson River anglers. Hudson River Sloop Clearwater, Inc. Poughkeepsie, New York.

Barnthouse, L. W., R. J. Klauda, and D. S. Vaughan. 1988. Introduction to the monograph. Pages 1–8 *in* L. W. Barnthouse, R. J. Klauda, D. S. Vaughan, and R. L. Kendall, editors. Science, law, and Hudson River power plants: a case study in environmental impact assessment. American Fisheries Society, Monograph 4, Bethesda, Maryland.

Brown, T. L. 1975. The 1973 New York statewide angler survey. Cornell University/NYSDEC.

Connelly, N. A., T. L. Brown, and B. A. Knuth. 1990. New York statewide angler survey 1988. New York Department of Environmental Conservation, Bureau of Fisheries. Albany.

Connelly, N. A., T. L. Brown, and B. A. Knuth. 1997. New York statewide angler survey 1996. Report 1: angler effort and expenditures. New York State Department of Environmental Conservation, Bureau of Fisheries, Albany.

Connelly, N. A., B. A. Knuth, and T. L. Brown. 1996. Sportfish consumption patterns of Lake Ontario anglers and the relationship of health advisories. North American Journal of Fisheries Management

16:90–101.

Green, D. M., and J. Jackson. 1991. Characterization of angler activity on the Hudson River estuary. Section VII, Pages 1–52 *in* E. A. Blair and J. R. Waldman, editors. Final Reports of the 1990 Tibor T. Polgar Fellowship Program. Hudson River Foundation, New York.

Jones, C. M., D. S. Robson, H. D. Lakkis, and J. Kressel. 1995. Properties of catch rates used in analysis of angler surveys. Transactions of the American Fisheries Society 124:911–928.

Kretser, W. A., and L. E. Klatt. 1981. 1976–77 New York angler survey final report. New York State Department of Environmental Conservation, Survey and Inventory Unit, Albany.

Malvestuto, S. P. 1983. Sampling the recreational fishery. Pages 397–419 *in* L. A. Nielsen and D. L. Johnson, editors. Fisheries techniques. American Fisheries Society, Bethesda, Maryland.

New York State Department of Health. 2000. Health consultation: 1996 survey of Hudson River anglers, Hudson Falls to Tappan Zee Bridge at Tarrytown, New York. Final Report. CERCLIS No. NYD980763841.

Pollock, K. H., C. M. Jones, and T. L. Brown. 1994. Angler survey methods and their application in fisheries management. American Fisheries Society Special Publication 25, Bethesda, Maryland.

Velicer, C. M., and B. A. Knuth. 1994. Communicating contaminant risks from sport-caught fish: the importance of target audience assessment. Risk Analysis 14:833–841.

American Fisheries Society Symposium 51:317–330, 2006

Hudson River Tributaries in the Lives of Fishes with Emphasis on the American Eel

Robert E. Schmidt and Rome Petersson
Simon's Rock College of Bard
84 Alford Road, Great Barrington, Massachusetts 01230 USA

Thomas R. Lake
New York State Department of Environmental Conservation, Hudson River Estuary Program
7 Steinhaus Lane, Wappingers Falls, New York 12590 USA

Abstract.—Hudson River tributaries serve as spawning areas for substantial numbers of anadromous and potamodromous fishes, and are critical habitat for at least alewife *Alosa pseudoharengus*, sea lamprey *Petromyzon marinus*, white sucker *Catostomus commersonii*, and smallmouth bass *Micropterus dolomieu*. Those species that are resident in tributary mouths are often common, widespread taxa in the Hudson River system but they are rarely present in large numbers. We document high densities of American eel *Anguilla rostrata* in one tributary mouth (>13,000 eels/ha; 15–30 g/m^2) and present observations supporting that this is typical of Hudson River tributaries. We suggest that the biology of American eel substantially determines the structure of the fish community and energy flow within these habitats. The presence of barriers and high density eel populations limit the import of marine and estuarine nutrients into Hudson River tributaries.

The tidal Hudson River has at least 79 small to large tributary streams, in addition to the upper Hudson and Mohawk Rivers, which enter the tidal Hudson at the federal dam at Troy, the upstream limit of tidal influence. There is also an unknown number of smaller, often ephemeral, streams that contribute water to the tidal Hudson River. All of the tributaries together (other than the Upper Hudson/Mohawk) contribute about 20% of the freshwater input to the Hudson River (Cooper et al. 1988). About half of these tributaries have dams or other anthropogenic obstructions on or near their mouths (Schmidt and Cooper 1996). Our observations over the past two decades have shown that there are exchanges of fishes between the tidal Hudson River and its tributaries in both directions that occur over several time scales and that vary from critical to the fishes and ecosystems to incidental. The many anthropogenic obstructions limit or prevent such exchanges from occurring over the full extent that may be possible. Our purposes in this paper are to summarize what is known about this exchange of fishes between tributaries and the tidal Hudson River and to present recent information about American eel *Anguilla rostrata*.

Table 1. Chronology, geography, and citations documenting sampling effort for adult migratory fishes in Hudson River tributaries. Mesh Size is the bar mesh in inches for 50 ft X 6 ft gill nets or 50 ft X 8 ft gill nets in Coote (2001) and "#" is the number of sets.

Tributary	Dates	Mesh Size	#	Citation
Canterbury Brook	3/30-6/7 '98	1.25	10	Schmidt & Lake 2000
Catskill Creek	10 '00-7 '01	0.5, 1.25 & 2.5	7	Coote 2001
Coxsackie Creek	3/31-6/7 '99	1.25 & 0.75	10	Schmidt & Lake 2000
Lattintown Brook	3/31-6/7 '99	1.25	10	Schmidt & Lake 2000
Moordener Kill	3/30-6/7 '98	1.25 & 0.75	10	Schmidt & Lake 2000
Pocantico River	4/8-5/19 '02	1.25	8	Schmidt & Lake 2002
Quassaic Creek	4/3-6/10 '96	1.25 & 0.75	30	Lake & Schmidt 1997
Quassaic Creek	3/28-6/10 '97	1.25 & 0.75	31	Lake & Schmidt 1998
Roeliff Jansen Kill	10 '00-7 '01	0.5, 1.25 & 2.5	7	Coote 2001
Rondout Creek	10 '00-7 '01	0.5, 1.25 & 2.5	7	Coote 2001
Saw Kill	~4/15&5/15 '01-04	1.25 & 0.75	2	Schmidt et al. 2001-04
Stockport Creek	10 '00-7 '01	0.5, 1.25 & 2.5	7	Coote 2001
Stony Creek	~4/15&5/15 '01-04	1.25 & 0.75	2	Schmidt et al. 2001-04

Methods

Migratory Fishes

Adult migratory fishes were sampled primarily with gill nets and secondarily with a backpack electroshocker in several Hudson River tributaries. Much of these data were subjectively summarized by Schmidt and Lake (2006) and detailed descriptions of methods and quantitative results are contained in unpublished reports (Table 1). In addition to the data in Table 1, single gill-net samples were taken in 31 Hudson River tributaries in 1998–1999 (Schmidt and Lake 2000).

Drift of eggs and larvae of migratory fishes has been quantitatively sampled (Table 2) using standard rectangular drift nets (330 μm mesh, 0.135 m^2 opening) or half-meter plankton nets (500 μm mesh, 0.5 m diameter opening) fished for about 20 min each sample. Drift nets were fished in triplicate in most studies or, in Stockport Creek, with six nets deployed simultaneously (Schmidt and Stillman 1994). Half-meter nets were used in lieu of drift nets when access to the stream bottom was difficult or hazardous (e.g., during spring floods). Drift was sampled after sundown except during two 24 h studies (Schmidt and Stillman 1994). In all cases, water velocities were measured in the mouth of each net with a Swoffer current meter or estimated by timing the drift of a floating object. Eggs and larvae were preserved in the field (4% formalin), sorted and identified (Auer 1982; Buynak and Mohr 1978; Fuiman 1979; Fuiman et al. 1983; Fuiman and Witman 1979; Gerlach 1983; Lippson and Moran 1974; Wang and Kernehan 1979) in the laboratory, and counted. Densities were calculated for each species and life stage (egg, yolk-sac larva, post-yolk-sac larvae, juvenile, other).

Young of Year American Eel

Study Area

Young of year (YOY) American eel were sampled in the mouths of two Hudson River tributaries, Hunters Brook and the Saw Kill (Figure 1). Both tributaries drain into the

Table 2. Chronology, geography, and citations documenting drift net sampling effort for early life stages of migratory fishes in Hudson River tributaries. Entries under Gear indicate the number and kind of nets deployed per sample and Effort indicates the number of samples in each study.

Tributary	Dates	Gear	Effort	Citation
16 different	3/14-6/20 '88	3 drift nets or 3 0.5 m nets	15 per trib	Schmidt & Limburg '89 Limburg & Schmidt '90
Stockport Crk.	5/6-7/30 '93	6 drift nets	60	Schmidt & Stillman '94 Schmidt & Stillman '98
Quassaic Crk.	4/23-6/10 '97	3 drift nets	10	Lake & Schmidt '98
Canterbury Brk.	3/30-6/7 '98	3 drift nets	9	Schmidt & Lake '00
Coxsackie Crk.	3/31-6/7 '99	" "	10	" "
Lattintown Crk.	3/31-6/7 '99	" "	9	" "
Moordener Kill	3/30-6/7 '98	" "	10	" "
Saw Kill	~4/15 & 5/15	3 drift nets	2	Schmidt et al. '01-'04
Stony Crk.	each year	" "	2	" "

eastern side of the tidal Hudson River and are about 50 km apart. These tributaries were chosen based on convenience, ease of access from the land, known presence of American eel, representation of different areas of the Hudson estuary, and similarity in size.

Hunters Brook empties into the tidal mouth of Wappinger Creek in Wappinger Falls, New York approximately 108 km north of the Battery (southern tip of Manhattan; Figure 1). Our sampling site was about 20 m upstream of the Creek Road bridge. This location is tidal and, although the Hudson River may occasionally have measurable salinity around the mouth of Wappinger Creek (probably no more than 1–2 ppt during drought events), Hunters Brook is completely freshwater.

The Saw Kill empties into the tidal Tivoli South Bay in Annandale, New York, approximately 158.5 km north of the Battery (Figure 1). Our sampling site was about 70 m west of the mouth of the Saw Kill adjacent to the north shore of the bay. This location is tidal but strictly freshwater and is located in a low marsh, where the marsh surface is exposed at low tide. There are several deeper channels winding through the marsh that are submerged even at low tide and our net was placed in one of them.

Field Procedure

Nets were Sheldon-type elver fykes (Young's Canvas Shop, Bangor, Maine). These nets were smaller than the commercial gear because we were setting them in restricted environments. Also, we could not sample the entire width of the streams because that would impede movement of migratory fishes. Mesh was about 1 mm throughout. Wings were 3.05 m long and 1.8 m deep. The fyke opening was 1.2 m in diameter and 2.4 m long with a single conical entrance. We modified the tip of the cone by sewing in a 2.5 cm inside diameter plastic ring and we tied the ring to the outer walls in three places, thus suspending the opening roughly in the middle of the trap. An excluder, 0.5 cm mesh, blocked entrance of large fishes into the net.

The bottom of the wings and opening were weighted with a chain (Saw Kill) or lead lines from gill nets (Hunters Brook) threaded through the bottom of the net. Addition-

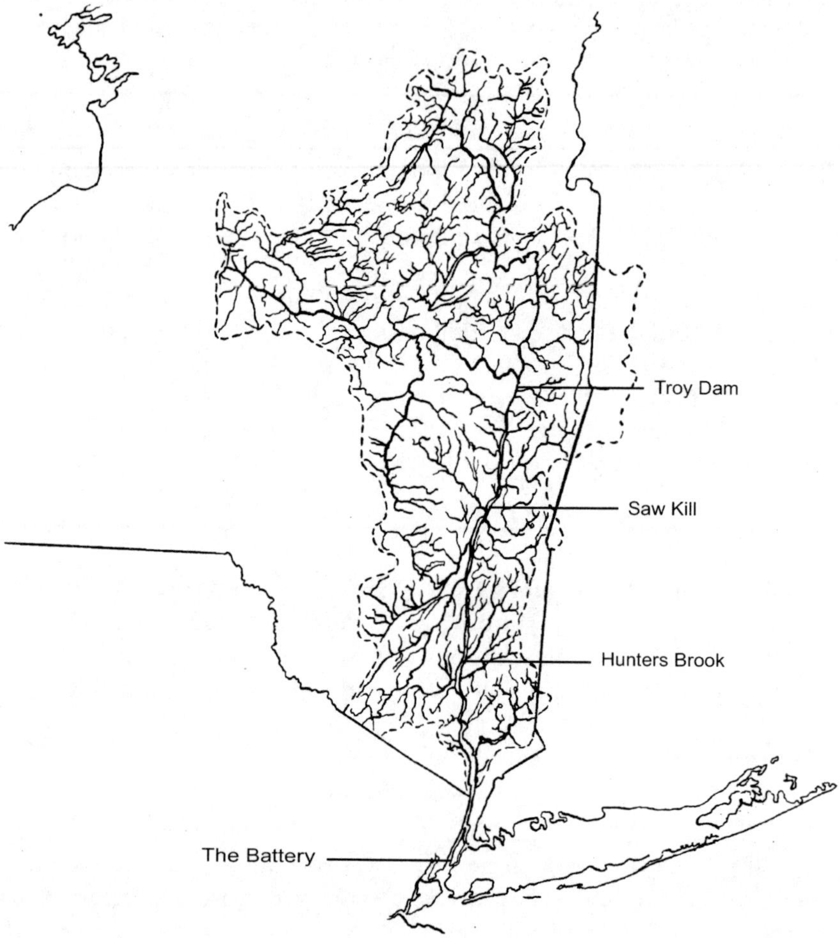

Figure 1. The Hudson River drainage basin and location of the two tributaries sampled for YOY American eel.

ally, rocks were placed on the lead line to assure a seal with the bottom of the stream. Corks on a line were likewise threaded through the top of the wings and net opening. The wings were staked at the end with metal garden stakes (Saw Kill) or 1.25 cm reinforcement rod (Hunters Brook) pounded into the streambed. In Hunters Brook, reinforcement rods were also placed along the wings to provide extra support in occasionally strong currents, which was not necessary in the generally slower flows in the Saw Kill. The cod end of the net was twisted to form a loop of netting which was then tied to cinder block(s) and the whole net stretched into fishing configuration.

Nets were deployed between March 15–17 in each year. Both nets were fished until June 7 in 2003 and May 31 in 2004. Nets were checked daily, generally around low tide for ease in handling. Information recorded was date, time, tidal stage, and air and water temperatures (hand held thermometer).

Laboratory Procedure

All fishes collected were identified and counted. Total weight of young of year eels was taken as well as total weight of larger eels. All fishes were returned to the stream. Young of year eels were returned to the stream well upstream of the net except on Tuesdays and Saturdays. On those 2 days a week, as many as 60 individuals were retained from each tributary and returned to the laboratory.

In the laboratory, young of year eels were anesthetized with clove oil, measured (total length, nearest mm), staged (Haro and Krueger 1988) under a dissecting microscope, and weighed (nearest 0.001 g) on an electronic balance. Approximately half of the specimens from each sample in each tributary were fixed in 70% ethanol and the other half in 10% formalin. These preserved specimens were catalogued into the New York State Museum fish collection.

Daily data on number of young of year American eel collected in the two tributaries in both years were smoothed (T4253H smoothing, SPSS) to reduce variation and delineate temporal patterns. Since we suspected that periods of higher than normal tides enhanced young of year eel migration, we plotted the smoothed data against maximum tide height (Columbia University website, data for New Hamburg and Tivoli) the previous night.

Yellow-Phase Eels

Study Area

There is a 4.3 m high waterfall (Schmidt and Cooper 1996) about 82 m upstream of the head of tide in the Saw Kill that restricts (but does not completely prohibit) the upstream movement of American eel. This part of the study concentrated on American eel inhabiting this short stretch of nontidal water.

Field Procedure

We collected American eel in the nontidal Saw Kill with a Coffelt Mark 10 backpack electroshocker. Eels were anesthetized with clove oil prior to handling. The total length of each eel collected was measured (TL cm) and all eels greater than 13 cm TL were fin clipped (lower caudal fin rays). Eels were then allowed to recover in clean water and released. We did not mark smaller eels because we determined that the mesh of our eel traps would not retain eels less than 13.0 cm long. During our first tagging session, all eels collected were weighed (nearest 0.1 g) with an electronic balance.

We calculated the population size of American eel in the nontidal portion of the Saw Kill using the Schnabel multiple mark and recapture technique. A length–weight regression was developed (Machut 2006). Total weight of eels was estimated by calculating the average weight of eels in each 0.5 cm size interval from the regression equation, multiplied by the observed number of eels in each size interval, multiplied by a proportion based on the Schnabel estimate, and summed. Length and five widths of the study area were measured with a tape and the area calculated.

We hypothesized that eels could be moving out of the Saw Kill to feed, so we deployed eel traps in Tivoli South Bay. We used 1-m long double-funnel eel pots (1.3 cm mesh) manufactured by Ron Ingold, the same traps used for a previous study on American eel in the tidal Hudson River (Morrison et al. 2002; Secor et al. 2002). We modified half of the traps to retain small eels by sewing 5 mm knotless nylon mesh around the outside and over the internal leads.

Traps were baited with frozen alewife *Alosa pseudoharengus*, one-quarter individual per trap. Traps were deployed in pairs, one large-mesh and one small-mesh trap per site, and within 250 m of the mouth of the Saw Kill in deeper areas where we knew traps would be submerged at low tide. Traps were checked and bait was changed daily. Captured eels were anesthetized in clove oil, examined for marks, measured (TL cm), and released.

Results

Fishes other than Eels

In order to organize our observations, we have categorized the fishes we have seen in tributaries as anadromous, catadromous, potamodromous, resident, and other species. These categories overlap and are not necessarily mutually exclusive. We have chosen to ignore the movements of fishes between the Upper Hudson/Mohawk River and the tidal Hudson River (through the locks on the Troy Dam) not because they aren't significant, but because we think these movements are complex and poorly documented.

Of the eight anadromous fishes in the Hudson River, only three are known to spawn in Hudson River tributaries; sea lamprey *Petromyzon marinus*, alewife, and blueback herring *Alosa aestivalis*. We have found sea lamprey ammocoetes in tributaries from Peekskill Hollow Brook to Stockport Creek, but have never seen them in large numbers. Sea lamprey in the Hudson River probably depend entirely on tributaries for spawning success.

In addition to tributary spawning, alewife and blueback herring spawn in the tidal Hudson River and the latter spawns in large numbers in the Mohawk River. Alewife is known to also spawn in most Hudson River tributaries, sometimes in large numbers. Blueback herring have been observed spawning in several tributaries, but in smaller numbers than alewife. Historically, rainbow smelt *Osmerus mordax* spawned in the mainstem Hudson and in tributaries and we collected small numbers of yolk-sac larvae (in Catskill Creek) in our studies, but this species is now virtually extinct in the Hudson River (Daniels et al. 2005).

There is only one catadromous species in the Hudson River, the American eel. That species is discussed in detail, below.

We have observed potamodromous behavior (living in the tidal Hudson River, but migrating into tributaries to spawn) in 11 fishes. White sucker *Catostomus commersonii* and smallmouth bass *Micropterus dolomieu* (Schmidt and Stillman 1998) depend on potamodromy to sustain their populations in the Hudson. Large numbers of white perch *Morone americana* and yellow perch *Perca flavescens* migrate into tributaries to spawn but also spawn elsewhere in the Hudson River. We think potamodromy in these species is significant to the Hudson River population, but we have not been able to quantify this hypothesis.

Three other cyprinoids show potamodromous behavior but tributary spawning in these species is probably not a significant fraction of spawning that occurs in tidal water; spottail shiner *Notropis hudsonius*, golden shiner *Notemigonus crysoleucas*, and shorthead redhorse *Moxostoma macrolepidotum*. We have observed four other species migrating into tributaries, but we are not sure that substantial spawning is occurring in them. "These include carp *Cyprinus carpio*, gizzard shad *Dorosoma cepedianum*, northern pike *Esox lucius*, and walleye *Sander vitreus*.

Although we have caught many species of fishes in tributary mouths, only nine were seen often enough for us to consider them as full-time residents in this habitat. One is the American eel (see below). Three are stream fishes that are often taken in tidal

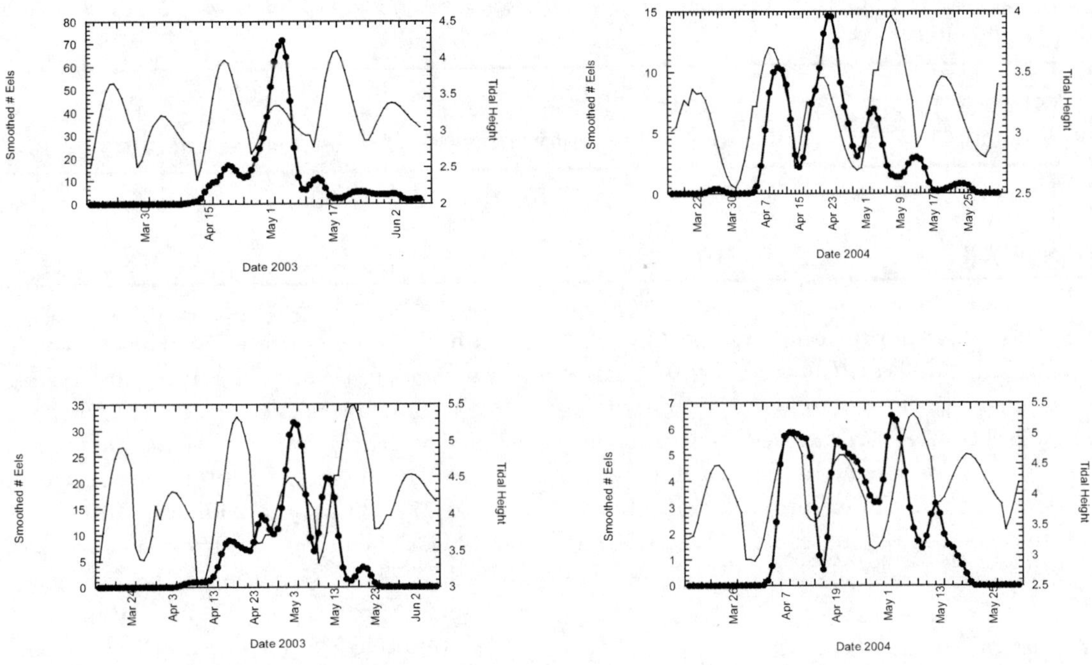

Figure 2. Daily catches of YOY American eel in two Hudson River tributaries, 2003 and 2004. Top two graphs are Hunters Brook and bottom two are the Saw Kill. Dark line with closed circles represents the smoothed daily catch and the light line represents the height above mean sea level of highest high tide on a given day.

tributary mouths; fathead minnow *Pimephales promelas*, fallfish *Semotilus corporalis*, and redfin pickerel *Esox americanus*. The banded killifish *Fundulus diaphanus* is a common and environmentally tolerant species that we frequently caught in tributaries. The three common sunfish species in the Hudson River are all taken regularly in tributaries and are often found in other habitats as well; these include redbreast sunfish *Lepomis auritus*, pumpkinseed *L. gibbosus*, and bluegill *L. macrochirus*. The final resident species is tessellated darter *Etheostoma olmstedi*, which also is abundant in many other Hudson River habitats.

We know of only one study that attempted to document the species composition of resident fishes in a Hudson River tributary mouth, in Quassaic Creek by Lake and Schmidt (1997). Their data and many incidental observations indicate that, in addition to a low species diversity, the numbers of individuals of resident species tends to be rather low.

Finally, we have observed species in tributary mouths that we consider transients between other habitats. They may be upland species that wander into the tributary mouths or species from the tidal river entering tributaries at random. With considerable sampling effort, every species known from the Hudson River would probably be collected in tributary mouths eventually. One set of these transients deserves special attention. Young of the year anadromous clupeids (including American shad *Alosa sapidissima*) have been collected in large numbers in several tributaries, especially those in the southern part of the estuary. Likewise, young of year striped bass *Morone saxatilis* and white perch have

Table 3. Total catch of YOY American eel in two Hudson River tributaries. Nets were checked daily and effort is days fished.

	2003		2004	
Tributary	Total #	Effort (days)	Total #	Effort (days)
Hunters Brook	877	82	305	77
Saw Kill	501	84	213	78

been found in large numbers, perhaps following the clupeids inshore. Yearling striped bass also appear to follow the schools of clupeids into the tributaries. It is not clear that the tributaries are a favored habitat for these life stages. Their presence in tributaries may be simply a result of their abundance in the Hudson River.

Young of Year Eels

A total of 1,896 young of year American eel were collected in this study (Table 3). Although specimens were collected in late March (Figure 2) into early June, the majority of young of year eels were taken between mid-April and mid-May.

The smoothed data showed periodicities in catches of American eel young of year (Figure 2) with a mean of 8.25 d (range 8–9 d) in the Saw Kill in 2003, 10.5 d (9–14 d, Hunters Brook 2003), 11.0 d (10–13 d, Saw Kill 2004), and 11.4 d (10–14 d, Hunters Brook 2004). On occasion, high catches occurred at periods of high tides similar to peak catches of glass eels in other anguillids (Tzeng 1985; Sugeha et al. 2001; Tesch 1977) but none of the periodicities seen in our catches were similar to the periodicities in tidal height (slightly greater than 14 d). We do not know which abiotic factor, if any, is correlated with the high catches of young of year American eel and therefore cannot explain why the 2003 Saw Kill data differ from the other three data sets.

Of all the eels collected in this study, 128 were measured, staged (Haro and Krueger 1988), and preserved from the Saw Kill (25.5% of the total catch) in 2003 and 39 (18.3%) in 2004. In Hunters Brook we preserved 277 (31.6%) in 2003 and 100 (32.8%) in 2004. These eels ranged from 48 to 70 mm TL. Mean size tended to decrease as the spring progressed (Figure 3) and as average pigmentation score increased (Figure 4), phenomena that are also seen in European eel *A. anguilla* (Tesch 1977). We caught glass eels (pigmentation score = 1) through the beginning of May in both tributaries. The latest capture of a glass eel was May 10, 2003 in Hunters Brook.

We estimated total numbers of young of year eels entering the tributaries by multiplying the number of eels collected by percent of the area sampled (Hunters Brook was 4,300 for the 2003 season). If our net in the Saw Kill intercepted 5% of the run (assumption of uniform distribution of eels across the mouth of the Saw Kill), the total run would be about 10,000 young of year eels. If we assume that the net intercepted 25% of the run (eels only moved along the channels and not across the marsh surface), the total run would be 2,000 eels.

In Hunters Brook, the bycatch was solely larger eels. We collected 150 specimens that averaged 1.46 g apiece. We collected about the same number of eels in the bycatch in the Saw Kill (144) but they were larger (av-

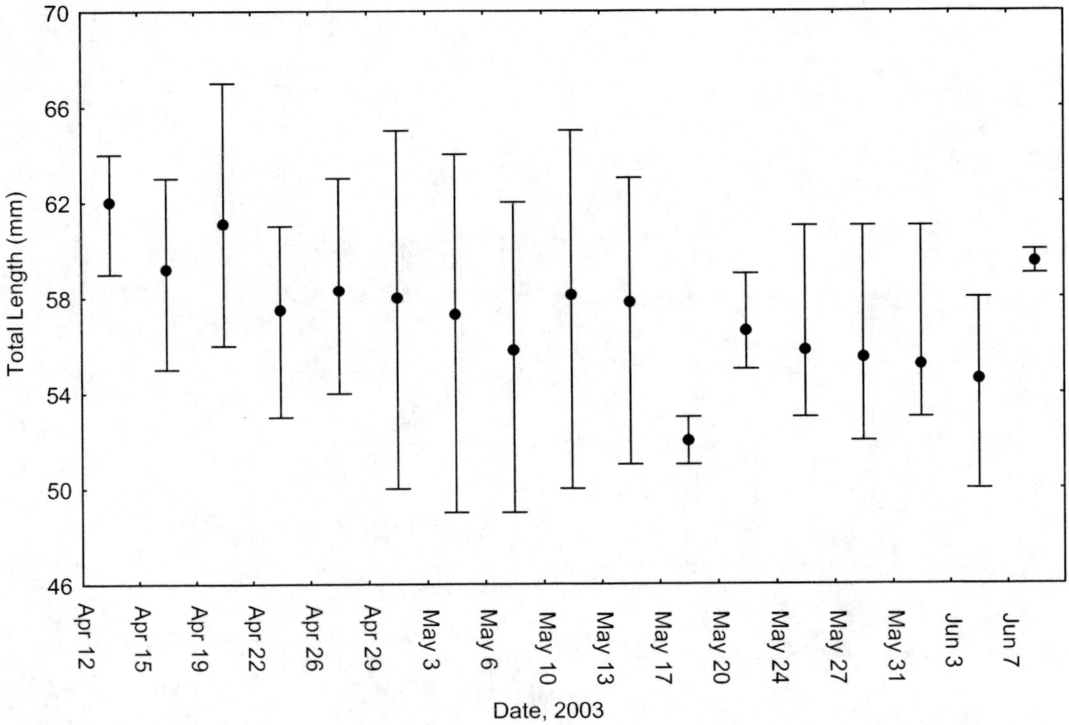

Figure 3. Mean (solid circle) and range (capped line) of total length of YOY American eel from Hunters Brook in 2003.

erage 2.61 g) than the eels in Hunters Brook. In addition we caught three other fishes in the Saw Kill; 76 fourspine stickleback *Apeltes quadracus*, 211 banded killifish, and 17 mummichog *Fundulus heteroclitus.*

Yellow Phase Eels

We marked eels in three trips, 17 June, 26 June, and 25 July 2003. We marked a total of 434 eels in the nontidal Saw Kill and recaptured a total of 30 eels in that area during marking. The Schnabel population estimate was 2234 eels in 82.5 m of stream for a total density of 13,100 eels/ha. The 95% confidence interval of our estimate was 1605–3296 eels greater than 13 cm TL.

As large as these numbers are, they are an underestimate. Examination of the length frequency of eels we collected (Figure 5) shows that the backpack electroshocker probably was biased towards sampling larger eels and that eels less than 13 cm TL were underrepresented. During electrofishing, the very small eels (5–7 cm TL) were caught mostly by chance. Thus, the absolute number of American eel in the study area could easily be two or three times our estimate.

The biomass of American eel (>13 cm TL) in the Saw Kill was 33.9 kg (24.3–49.9 kg; 95% confidence interval). The average eel weighed 15.2 g, therefore, the density of eels in this habitat was 19.9 g/m^2 (14.3–29.3 g/m^2). This density should also be an underestimate, but not by much since undersampled eels were very small.

Eel traps were fished beginning on 25 July through 4 September 2003. Traps were not fished continuously and our total effort was

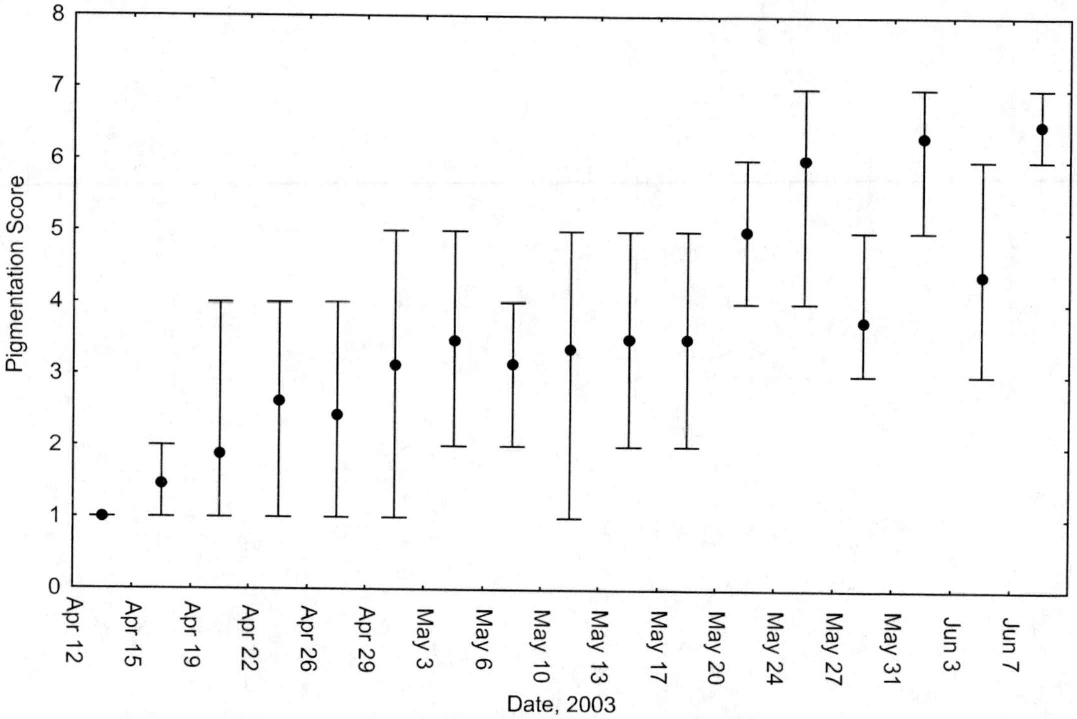

Figure 4. Mean (solid circle) and range (capped line) of pigmentation scores (Haro and Krueger 1988) of young of year American eel from Hunters Brook in 2003.

76 trap-nights. We captured 31 eels in the traps for a catch per unit effort of 0.41 eels/trap-night (0–9 eels per night). This catch rate seems low to us but we have no comparable data and the low catches may reflect the local environment.

Small mesh traps caught more eels (18) than large mesh traps (13). We collected no marked eels despite having marked 434 individuals within 250 m of our traps. In addition, the length frequency distribution of the trapped eels was very different compared to the length frequency distribution of the eels caught by electrofishing in the nontidal Saw Kill; eels were much larger (average 42.6 cm TL, 13.0–70.0 cm).

Discussion

We could not demonstrate that American eel move out of the Saw Kill to feed and, therefore, not only is there a high density of eels in the Saw Kill, but they are permanent residents. American eel can be considered the keystone species in the Saw Kill mouth and they have effects on several aspects of the ecology of this habitat.

American eel have been observed in other Hudson River tributaries: Quassaic Creek in Newburgh (Schmidt and Limburg 2003), Stony Creek in Tivoli (Schmidt et al. 2002), Hunters Brook in Wappingers Falls (Stainbrook 2004), and Black Creek in Esopus, Hannacrois Creek in New Baltimore, and Peekskill Hollow Brook in Peekskill (RES and K. Limburg, unpublished data). American eel densities in those tributary mouths are also high, some perhaps higher than in the Saw Kill.

The several thousand young of year American eel that we estimated entering the Saw

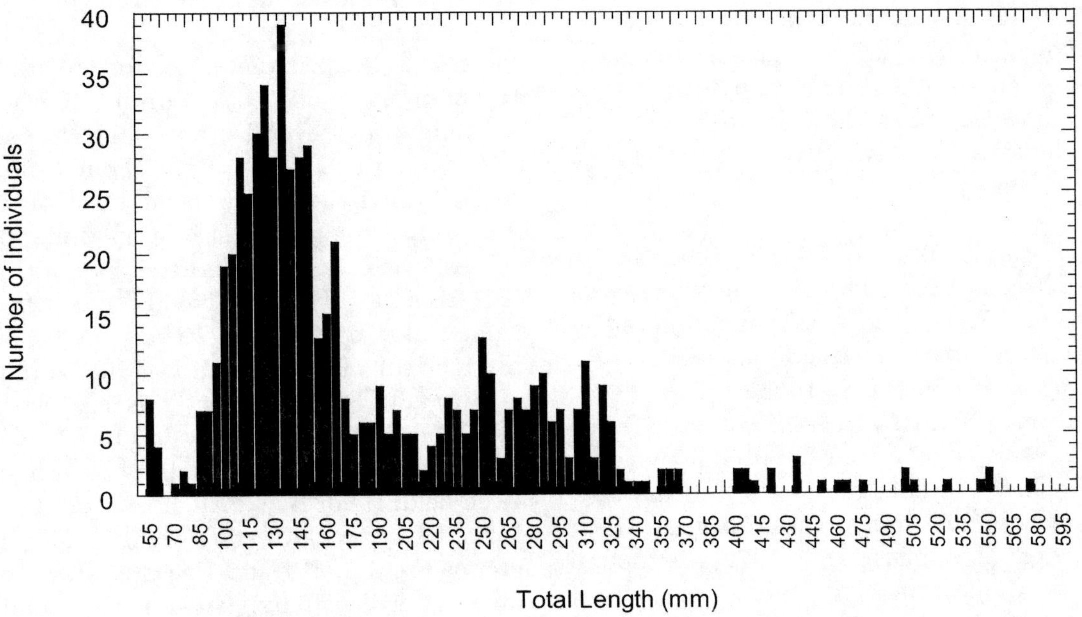

Figure 5. Length frequency of American eel from the Saw Kill collected with a backpack electroshocker.

Kill were not seen in our electrofishing effort. Electrofishing may not adequately sample these small eels but larger eels may be consuming them. Jessop (2000) attributed a substantial part of the greater than 99% mortality in young of year eels in Nova Scotia to predation by yellow eels.

American eel dominate the mouth of the Saw Kill and have a significant effect on structuring the species composition of this habitat. Additionally, energy flow is altered by American eel. Whether these tributary mouths are significant to the Hudson River eel population remains to be seen, but American eel are certainly significant residents in tributary mouths.

American eel feed on fishes, crayfishes, and aquatic insects (Facey and La Bar 1981; Ogden 1970). Eels in the Saw Kill could feed directly on smaller fishes and, given the high densities of eels, compete with other fishes for aquatic insects. We suspect that the combination of piscivory and competition from American eel explains the very small number of other fishes observed by Schmidt et al. (2002, 2004a, 2004b) in the mouth of the Saw Kill.

Anadromous fishes have been considered keystone species that import significant amounts of nutrients into freshwater ecosystems, like phosphorus, nitrogen, and carbon derived from marine sources (Durbin et al. 1979). Those statements were derived from alewife runs considered typical in the Northeast (e.g., Kissel 1974) wherein alewife ascend a stream, enter a lake or pond, and then suffer postspawning mortality. This scenario is not typical of Hudson River tributary spawning for any of the migratory fishes except perhaps, for sea lamprey and smallmouth bass, both nest builders. Our observations indicate that Hudson River migratory fishes enter tributaries on rising tides, spawn quickly, and leave, usually on the following ebbing tide. Drift of early life stages is predominantly eggs and yolk-sac larvae except for white sucker larvae which tend to stay longer in tributaries. There is only one Hudson River tributary, the Casperkill, south of Poughkeepsie, where migratory fishes can enter an impoundment and, other than documenting that there is an alewife run (Schmidt and Lake 2000), this tributary has not been studied.

Migratory behavior of fishes has been shown to affect nutrient dynamics in several ecosystems. Anadromous salmonids are a significant source of nutrients in western oligotrophic systems (e.g., Richey et al. 1975), Gulf menhaden *Brevoortia patronus* transport 5–10% of estuarine productivity to nearshore waters (Deegan 1993), and upstream migration in stream fishes can affect nutrient dynamics (Hall 1972). Although not documented, immigration of potamodromous fishes could potentially move significant amounts of estuarine nutrients into the tributaries in the Hudson River.

American eel transport nutrients upstream and recent data suggest that eels supplement stream but not pond habitats (Anderson and Schmidt 2006). Supplementing upland food webs with marine carbon by anadromous fishes or estuarine carbon by potamodromous fishes is minimal in Hudson River tributaries. This is due to the short residence time of the adults and larval stages in the tributaries and presumably relatively low mortality rates during residence. This short residence time, in turn, is due to the barriers present on the tributaries, lack of passage facilities in the Hudson Valley, and perhaps avoidance of predation by American eel.

We have observed fishes apparently feeding on eggs of alewife, predominantly spottail shiner and white perch, both migratory species. We suspect that American eel would feed on eggs and early larvae as well, but have not documented this behavior.

Acknowledgments

The research that went into documenting the anadromous and potamodromous fishes was funded by several grants to RES and TRL from the Hudson River Foundation and fellowships to TRL from the Polgar Fellowship Program. Research on young of year American eel was funded by a grant to RES from the Hudson River Estuary Program. Yellow eel research was supported by the Hudson River Foundation (Karin Limburg and RES). Tagging was supported by a Tibor T. Polgar Fellowship to RP. We thank the following individuals for their help in the field: Heidi Bock, Alida Cantor, Tom Coote, Nicholas DeGeyter, Thomas Diebel, Jeremy Frenzel, Bettina Hajagos, Tanessa Hartwig, Adlai Lang, Evan Leibu, John Licitra, Karin Limburg, Len Machut, Mayo Petersson, and Paz Petersson. We also thank Anne O'Dwyer for statistical help and David Strayer for ecological advice.

References

Anderson, J., and R. E. Schmidt. 2006. Significance of small impoundments to American eel (*Anguilla rostrata*). Pages 1–20 *in* W. C. Nieder and J. R. Waldman, editors. Final Reports of the Tibor T. Polgar Fellowship Program, 2005. Hudson River Foundation, New York, New York.

Auer, N. 1982. Identification of larval fishes of the Great Lakes Basin with emphasis on the Lake Michigan drainage. Great Lakes Fishery Commission, Special Publication #3, Ann Arbor, Michigan.

Buynak, G. L., and H. W. Mohr, Jr. 1978. Larval development of the redbreast sunfish (*Lepomis auritus*) from the Susquehanna River. Transactions of the American Fisheries Society 107:600–604.

Cooper, J. C., F. R. Cantelmo, and C. E. Newton. 1988. Overview of the Hudson River estuary. Pages 11–24 *in* L. W. Barnthouse, R. J. Klauda, D. S. Vaughan, and R. L. Kendall, editors. Science, law, and Hudson River power plants. American Fisheries Society, Monograph 4, Bethesda, Maryland.

Coote, T. 2001. Comparative analysis of fish assemblages in select Hudson River tributaries. Report to New York Sea Grant and the Hudson River National Estuarine Research Reserve, Annandale, New York.

Daniels, R. A., K. E. Limburg, R. E. Schmidt, D. L. Strayer, and C. Chambers. 2005. Changes in fish assemblages in the tidal Hudson River, New York. Pages 471–503 *in* J. N. Rinne, R. M. Hughes, and B. Calamusso, editors. Historical changes in large river fish assemblages of America. American Fisheries Society, Symposium 45, Bethesda, Maryland.

Deegan, L. A. 1993. Nutrient and energy transport between estuaries and coastal marine ecosystems by fish migration. Canadian Journal of Fisheries and Aquatic Sciences 50:74–79.

Durbin, A. G., S. W. Nixon, and C. A. Oviatt. 1979. Effects of the spawning migration of the alewife, *Alosa pseudoharengus*, on freshwater ecosystems. Ecology 60:8–17.

Facey, D. E., and G. W. La Bar. 1981. Biology of American eels in Lake Champlain, Vermont. Transactions of the American Fisheries Society 110:396–402.

Fuiman, L. A. 1979. Descriptions and comparisons of catostomid fish larvae: Northern Atlantic drainage species. Transactions of the American Fisheries Society 108:560–603.

Fuiman, L. A., J. V. Conner, B. F. Lathrop, G. L. Buynak, D. E. Snyder, and J. J. Loos. 1983. State of the art of identification for cyprinid fish larvae from eastern North America. Transactions of the American Fisheries Society 112:319–332.

Fuiman, L. A., and D. C. Witman. 1979. Descriptions and comparisons of catostomid fish larvae: *Catostomus catostomus* and *Moxostoma erythrurum*. Transactions of the American Fisheries Society 108:604–619.

Gerlach, J. M. 1983. Characters for distinguishing larvae of carp, *Cyprinus carpio*, and goldfish, *Carassius auratus*. Copeia 1983:116–121.

Hall, C. A. S. 1972. Migration and metabolism in a stream ecosystem. Ecology 53:585–604.

Haro, A. J., and W. H. Krueger. 1988. Pigmentation, size, and migration of elvers (*Anguilla rostrata* (Lesueur)) in a coastal Rhode Island stream. Canadian Journal of Zoology 66:2528–2533.

Jessop, B. 2000. Estimates of population size and instream mortality rate of American eel elvers in a Nova Scotia river. Transactions of the American Fisheries Society 129:514–526.

Kissel, G. W. 1974. Spawning of the anadromous alewife, *Alosa pseudoharengus*, in Bride Lake, Connecticut. Transactions of the American Fisheries Society 103:312–317.

Lake, T. R., and R. E. Schmidt. 1997. Seasonal presence and movement of fish populations in the tidal reach of Quassaic Creek, a Hudson River tributary (HRM 60): Documentation of potamodromy, anadromy, and residential components. Pages 1–36 *in* W. C. Nieder and J. R. Waldman, editors. Final Reports of the Tibor T. Polgar Fellowship Program, 1996. Hudson River Foundation, New York, New York.

Lake, T. R., and R. E. Schmidt. 1998. The relationship between fecundity of an alewife (*Alosa pseudoharengus*) spawning population and egg productivity in Quassaic Creek, a Hudson River tributary (HRM 60) in Orange County, New York. Pages 1–26 *in* W. C. Nieder and J. R. Waldman, editors. Final Reports of the Tibor T. Polgar Fellowship Program, 1997. Hudson River Foundation, New York, New York.

Limburg, K. E., and R. E. Schmidt. 1990. Patterns of fish spawning in Hudson River tributaries: response to an urban gradient? Ecology 71:1238–1245.

Lippson, A. J., and R. L. Moran. 1974. Manual for identification of early developmental stages of fishes of the Potomac River Estuary. Maryland Department of Natural Resources, PPSP-MP-13, Baltimore.

Morrison, W. E., D. H. Secor, and P. M. Piccoli. 2002. Estuarine habitat use by Hudson River American eels as determined by otolith Strontium:calcium ratios. Pages 87–116 in D. A. Dixon, editor. Biology, management, and protection of catadromous eels. American Fisheries Society, Symposium 33, Bethesda, Maryland.

Ogden, J. C. 1970. Relative abundance, food habits, and age of the American eel, *Anguilla rostrata* (Lesueur), in certain New Jersey streams. Transactions of the American Fisheries Society 99:54–59.

Richey, J. E., M. A. Perkins, and C. R. Goldman. 1975. Effects of kokanee salmon (*Onchorhynchus nerka*) decomposition on the ecology of a subalpine stream. Journal of the Fisheries Research Board of Canada 32:817–820.

Schmidt, R. E., N. Akpan, and J. Anderson. 2004a. Monitoring fish populations in the Tivoli Bays, Hudson River National Estuarine Research Reserve. Year Four. Report of Berkshire Environmental Research Center to the Hudson River National Estuarine Research Reserve, Annandale, New York.

Schmidt, R. E., and S. Cooper. 1996. A catalog of barriers to upstream movement of migratory fishes in Hudson River tributaries. Report of Hudsonia Limited to the Hudson River Foundation, New York.

Schmidt, R. E., E. G. Eck, and R. Petersson. 2002. Monitoring fish populations in the Tivoli Bays, Hudson River National Estuarine Research Reserve. Year Two. Report of Berkshire Environmental Research Center to the Hudson River National Estuarine Research Reserve, Annandale, New York.

Schmidt, R. E., E. Griffin-Noyes, and A. Lavine. 2001. Monitoring fish populations in the Tivoli Bays Hudson River National Estuarine Research Reserve: Year one. Report of Berkshire Environmental Research Center to the Hudson River National Estuarine Research Reserve, Annandale, New York.

Schmidt, R. E., and T. R. Lake. 2000. Alewives in Hudson River tributaries, two years of sampling. Report of Hudsonia Limited to Hudson River Foundation, New York, New York.

Schmidt, R. E., and T. R. Lake. 2002. Migratory fishes in the tidal Pocantico River. Report of Hudsonia Limited to Habitat Restoration Program, Hudson River National Estuarine Research Reserve, Annandale, New York.

Schmidt, R. E., and K. E. Limburg. 2003. Assessment of the current status of Quassaic Creek. Report of Hudsonia Limitied to the Hudson River Estuary Program, New Paltz, New York.

Schmidt, R. E., and T. R. Lake. 2006. The role of tributaries in the biology of Hudson River fishes. Pages 205–216 *in* J. S. Levinton and J. R. Waldman, editors. The Hudson River Estuary. Cambridge University Press, New York.

Schmidt, R. E., A. Lang, and E. Leibu. 2004b. Monitoring fish populations in the Tivoli Bays Hudson River National Estuarine Research Reserve: Year Three. Report of the Berkshire Environmental Research Center to the Hudson River National Estuarine Research Reserve, Annandale, New York.

Schmidt, R. E., and K. E. Limburg. 1989. Fishes spawning in non-tidal portions of Hudson River tributaries. Report of Hudsonia Limited to the Hudson River Foundation, New York, New York.

Schmidt, R. E., and T. Stillman. 1994. Drift of early life stages of fishes in Stockport Creek and significance of the phenomenon to the Hudson River Estuary. Report of Hudsonia Limited to the Hudson River Foundation, New York, New York.

Schmidt, R. E., and T. Stillman. 1998. Evidence of potamodromy in an estuarine population of smallmouth bass (*Micropterus dolomieu*). Journal of Freshwater Ecology 13:155–163.

Secor, D. H., J. E. Baker, W. E. Morrison, and J. C. Steinbacher. 2002. Ecology and contamination of the Hudson River American eel. Final Report of the Chesapeake Biological Laboratory to the Hudson River Foundation, New York.

Stainbrook, K. M. 2004. Using ecological indicators to detect environmental change in urbanizing watersheds: case study in Dutchess County, NY. Master's thesis. State University of New York, College of Environmental Science and Forestry, Syracuse.

Sugeha, H. Y., T. Arai, M. J. Miller, D. Limbong, and K. Tsukamoto. 2001. Inshore migration of the tropical eels *Anguilla* spp. recruiting to the Poigar River estuary on north Sulawesi Island. Marine Ecology Progress Series 221:233–243.

Tesch, E. W. 1977. The eel: biology and management of anguillid eels. Chapman and Hall, London.

Tzeng, W. 1985. Immigration timing and activity rhythms of the eel, *Anguilla japonica*, elvers in the estuary of northern Taiwan, with emphasis on environmental influences. Bulletin of the Japanese Society of Fisheries and Oceangraphy 47×48: 11–28.

Wang, J. C. S., and R. J. Kernehan. 1979. Fishes of the Delaware estuaries: a guide to the early life histories. Ecological Analysts Communications, Towson, Maryland.

American Fisheries Society Symposium 51:331–365, 2006

Atlantic Tomcod *Microgadus tomcod*: A Model Species for the Responses of Hudson River Fish to Toxicants

Isaac Wirgin
New York University, School of Medicine, Department of Environmental Medicine
57 Old Forge Road, Tuxedo, New York 10987 USA
wirgin@env.med.nyu.edu

R. Christopher Chambers
Howard Marine Sciences Laboratory, Northeast Fisheries Science Center, NOAA Fisheries Service
74 Magruder Road, Highlands, New Jersey 07732 USA

Abstract.—Despite recent successes in eliminating or reducing many point sources of chemical contaminants, sediments in the Hudson River Estuary are still highly contaminated with lipophilic and highly persistent polychlorinated biphenyls (PCBs), polychlorinated dibenzo-*p*-dioxins (PCDDs), polychlorinated dibenzofurans (PCDFs), and polycyclic aromatic hydrocarbons (PAHs). These have been shown to bioaccumulate to high levels in resource species and other key ecological components of the Hudson River food web. Resource managers and stewards must consider the possible toxic effects of these pollutants on the Hudson River biota, including its fish community; however, few studies have directly investigated these effects. A series of toxicological studies on Atlantic tomcod *Microgadus tomcod* from the Hudson River Estuary have demonstrated profound and broad-based changes in response to local contaminants. Levels of contaminants in the tissues of different life stages of tomcod from the Hudson River Estuary far exceed those in tomcod from other Atlantic Coast estuaries. More importantly, a combination of field and laboratory studies has demonstrated molecular to population level perturbations in tomcod from the Hudson River, all of which are consistent with chemical exposures and many of which appear to be mechanistically linked. These effects include induction of hepatic expression of cytochrome P4501A1 mRNA, high levels of hepatic DNA damage, somatic mutations at an oncogene locus critical to the initiation of chemical carcinogenesis, elevated prevalence of gross and histologically defined hepatic tumors, truncated age structure, and dramatic resistance at the molecular and organismal levels to halogenated aromatic hydrocarbons (HAHs). Resistance, a population level effect, was observed in the toxic responses of tomcod embryos and larvae to 2,3,7,8-tetrachlorodibenzo-*p*-dioxin (TCDD) and TCDD-like PCBs, but not PAHs. Because young of the year tomcod are a critical node in the Hudson River food chain, their evolved resistance to HAHs and high body burden of these and related contaminants has likely resulted in the trophic transfer of these contaminants to secondary and tertiary consumers of the Hudson River, including important resource species, and an elevated tissue burden of these contaminants in those consumers. In total, these studies are consistent with the hypothesis that exposure to Hudson River-borne contaminants has significantly altered its tomcod population and perhaps evoked broad change in the Hudson River fish community.

Introduction

The assessment of risk from chemical contamination to fish and other populations in the Hudson River Estuary is currently of great interest to New York State and United States federal resource stewards such as the National Oceanic and Atmospheric Administration (NOAA) Fisheries and the U.S. Fish and Wildlife Service. One responsibility of these agencies is to assess potential ecological damage to fish populations from exposure to xenobiotics and, if damage has occurred, to assign a monetary value to those perturbations. According to the Department of Interior Natural Resource Damage Assessment (NRDA) regulations, measures of reduced viability to fish populations could include mutations, liver and skin cancers, physiological abnormalities, skeletal deformities, histopathological lesions, fin erosion, impaired reproductive success, decreased ability to capture prey and avoid predators, and death [43 CFR 11.62(f)(1)(i)]. These biological endpoints could be evaluated by direct observations of affected fish in nature, by assays of in situ caged fish, or by toxicity testing of fish in the laboratory [43 CFR 11.62 (f)(2)(i-iv)]. In order to confidently link these changes in viability due to a xenobiotic of interest, a toxic response in free-ranging fish and in fish exposed under controlled laboratory conditions should be demonstrated. These last criteria, which combine field and laboratory studies, form a far more convincing case for ecological damage to natural populations from chemical exposure than implied by the demonstration of bioaccumulation of contaminants alone. In this review, we marshal evidence from a variety of levels of biological organization that support our view of Atlantic tomcod *Microgadus tomcod* from the Hudson River as a sensitive investigative model for identifying and evaluating the complex responses of fishes to environmental contaminants.

Chemical Contaminants of Concern in the Hudson River Estuary

The extent of risk to fish or other taxa from chemical pollutants depends on the identity of the specific chemical; the route, dose, and duration of exposure; and its toxicity in the target species. Uncertainties in assigning risks from exposure can be due to species, population, and individual differences in susceptibilities to contaminants. This may result from variation in the chemical's bioavailability, its uptake or efflux, or its toxicity at the target receptor. The fish community in the Hudson River Estuary has been chronically exposed for more than half a century to high and sometimes extreme levels of chemical contaminants (Wirgin and Waldman 1998). These chemical contaminants include organic and inorganic pollutants. Some of the organic pollutants of greatest concern for ecological health consequences include polychlorinated biphenyls (PCBs), polychlorinated dibenzo-*p*-dioxins (PCDDs), polychlorinated dibenzofurans (PCDFs), polycyclic aromatic hydrocarbons (PAHs), pesticides, and various estrogen-mimicking endocrine disruptors (Wirgin et al. 2006). Among the metal pollutants of greatest concern within the Estuary because of their concentrations or known

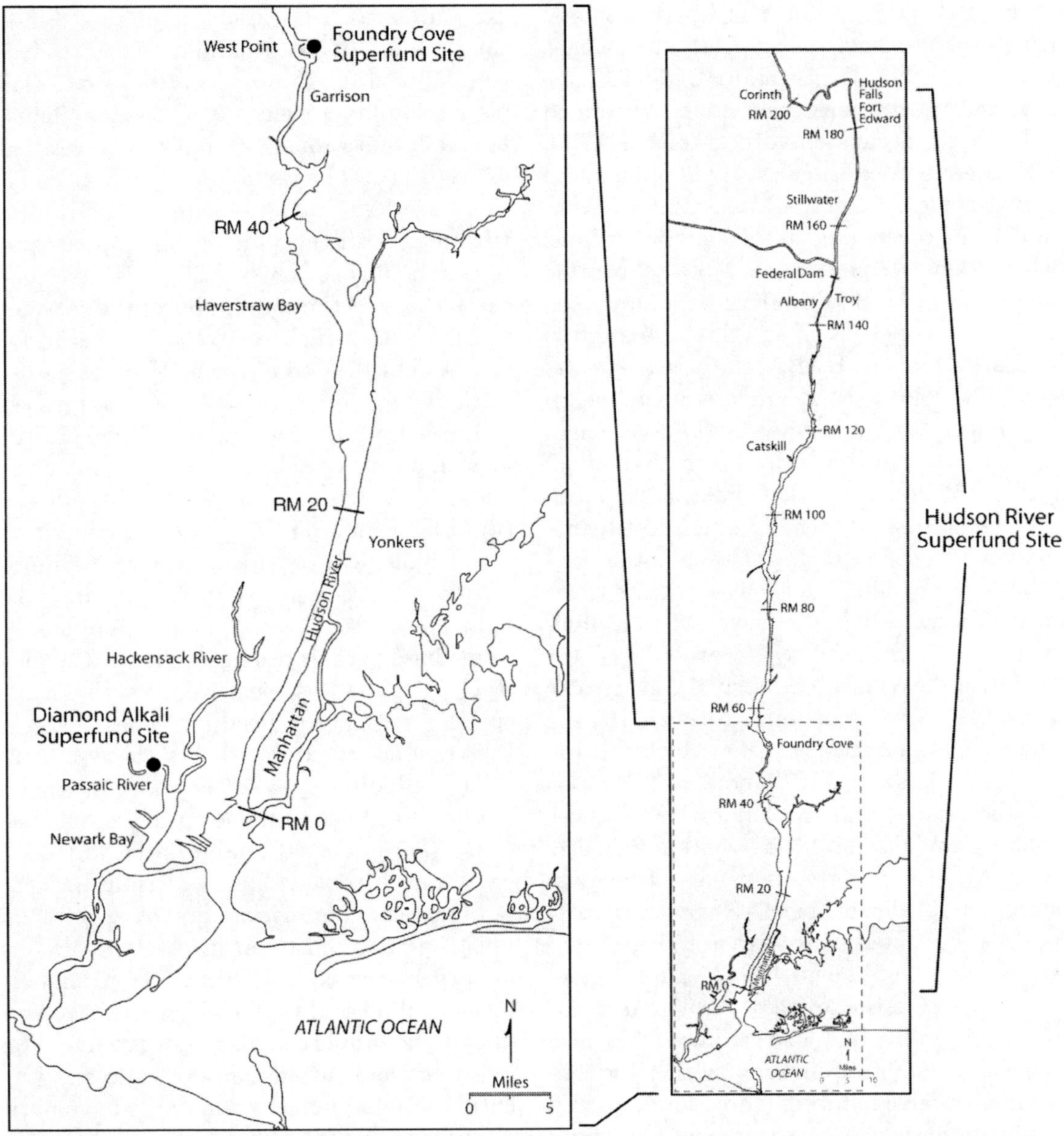

Figure 1. A map of the Hudson River Estuary showing the locations mentioned in the text and three U.S. federal Superfund sites.

toxicities are cadmium (Cd), mercury (Hg), chromium (Cr), nickel (Ni), and arsenic (As). The high sediment loads of these contaminants led to the designation of five U.S. federal Superfund sites in the estuary. The PCBs Superfund site, the longest such site in the nation, is the entire 200 mi of the Hudson River from Hudson Falls, New York to the Battery at the southern tip of Manhattan (Figure 1). The Diamond Alkali 2,3,7,8-tetrachlorodibenzo-*p*-dioxin (TCDD)Superfund site on the Passaic River resulted from chemical releases from a single herbicide manufacturing facility and their subsequent transport to much of downstream Newark Bay. In this paper, we focus on the biological consequences of exposure to PCBs, PCDD/Fs, and PAHs.

PCBs, PCDD/Fs, and PAHs all are aromatic (ring structure) hydrocarbon (AH) compounds, however, because PCBs and PCDD/Fs have chlorine substitutions, they are often termed halogenated aromatic hydrocarbons (HAHs). AHs are regarded as generally lipophilic and hydrophobic in character and tend to accumulate in the sediments of aquatic systems where they become bioavailable to benthic organisms. Chlorine dramatically alters the chemical properties of HAHs compared to PAHs by making HAHs relatively more resistant to biological degradation and hence more likely to biomagnify in the food chain and to persist in the environment. Individual PCBs and PCDD/Fs that differ in the number and position of their chlorine substituents are termed congeners. In fishes and other vertebrates, individual congeners usually differ in their toxicities due to their different abilities to bind the aryl hydrocarbon receptor (AHR) at the cellular level. Usually, the greater the affinity of a compound to the AHR, the greater is its toxicity. TCDD, which has the greatest affinity for AHR among all AHs, also has the greatest toxicity. Among PCBs, those that are coplanar are able to most efficiently bind AHR and are thus usually the most toxic. Knowledge of the HAH congener composition in contaminated environmental matrices (water, sediment, biota) enables more accurate evaluation of the toxicities of these matrices. Individual PAH compounds also differ in structure from one another by containing variable numbers (three or more) and configurations of benzene rings and sometimes by the presence of non-halogenated additions. These differences also result in different toxicities that are mediated by AHR binding.

Problems in Evaluating Toxicity of Chemical Contaminants to Natural Populations

Several issues complicate assessments of the damage to natural populations caused by exposures to chemical contaminants. First, the presence of a contaminant in an ecosystem is not synonymous with its bioavailability to any given species. Thus, bioavailability must be empirically verified for the species or tissue of interest. Second, there are large variations among and often within taxa in their biological responses to toxic chemicals. This reality also complicates the selection of an appropriate model species with which to evaluate these processes. Third, a wide range of responses—from the sublethal to lethal and expressed at levels from the molecular, cellular, histological, physiological, behavioral, organismal, and population to community endpoints—could be evaluated. Endpoints at lower levels of biological organization may be more sensitive, dose-responsive, specific to individual or classes of contaminants, and nearly immediate in their response. Those at higher levels may be less sensitive, less chemical-specific, with substantial lag times before a full response is exhibited. Yet these higher-order endpoints may be of more direct, if not greater, ecological relevance. Fourth, because there is usually large interindividual variation in toxic responses within natural populations, large numbers of specimens must be analyzed to ensure statistical confidence in the results. Fifth, chemical pollutants almost always co-occur in stressed urban or industrialized ecosystems. For many of the contaminants of concern, toxicities of individual contaminants have been defined in model fish species under controlled laboratory conditions. However, because of the typical co-occurrence of contaminants in the environment and in biological receptor species, predicting the toxicities of these contaminants in nature will be very difficult. These chemicals may exhibit additive toxicities, but often their combined toxicities are synergistic or antagonistic. Thus, using laboratory testing of single compounds as a benchmark to evaluate toxicity in fish exposed in nature may over- or underestimate realized toxic effects. Sixth, individuals from chronically and highly exposed environments may become resistant to the effects of chemical contaminants. This outcome not only makes it difficult to interpret the results from field surveys, but it also requires that some accounting be made for the costs associated with evolutionary resistance. While resistance may be beneficial in the short term to the affected population, it may hypersensitize individuals in the population to other stressors, negatively impact its viability after site remediation, and result in it serving as a means of transferring contamination to higher trophic levels. Because the endpoint is ecological rather than immediate, and because the link to ecosystem functions—including human health—is complex and indirect, long-term, well-funded investigations are necessary in order to fairly appraise these effects.

Tissue Burdens of Contaminants in Hudson River Fish Populations

Water and sediment concentrations of these contaminants are routinely monitored and their burdens in tissues have been determined in some Hudson River fish populations. Initial analysis focused on commercially or recreationally important resource species, most often striped bass *Morone saxatilis*. Because of the wide-ranging and seasonal movements of striped bass within the river and along the coast for spawning and feeding, especially in older age classes, the tissue burden of contaminants in this species provides a picture of spatially integrated contaminant levels from much of the tidal estuary and beyond (Sloan et al. 1995; Farley and Thomann 1998). Additionally, interest has also centered on species from locales within the Hudson River ecosystem with the greatest known concentrations of individual contaminants, such as brown bullhead *Ameiurus nebulosus*, goldfish *Carassius auratus*, pumpkinseed *Lepomis gibbosus*, and yellow perch *Perca flavescens* (Armstrong and Sloan 1988) from the main stem of the Hudson River at Stillwater, approximately 25 mi south of Ft. Edward, below the Federal Dam at Troy-Albany, and 85 mi south of Ft. Edward at Catskill, and in Atlantic tomcod (Yuan et al. 2001), striped bass, and bluefish *Pomatomus saltatrix* (Skinner et al. 1996, 1997) from the Newark Bay complex, which comprises the western estuary. Generally, tissue burdens of PCBs are high in fish from the upper Hudson River (above the Troy Dam) (Sloan et al. 2002). Often, but not always, the level of PCBs burdens in fish tissue corresponds directly with distance from the PCBs point sources with some punctuation associated with secondary inputs of PCBs in the New York City area (Farley and Thomann 1998). Long-term trends consist of a very rapid initial decline in PCBs burdens following the cessation of release of PCBs in 1977, followed by a much slower decrease over the past two decades with several rapid, short-term increases associated with episodic inputs from point sources (Sloan et al. 2005). Unfortunately, these analyses of PCBs were usually reported on an Aroclor (commercially available PCBs mixtures manufactured by Monsanto), not on a congener specific basis, so it is difficult to evaluate the potential toxicity of the specific tissue burdens to Hudson River populations.

Fish from Newark Bay and its associated tributaries have exhibited very high levels of total PCDD/Fs and PCBs. PCDD congener patterns were dominated by the most toxic congener TCDD. Levels of TCDD in fish and shellfish from Newark Bay were at times the highest ever reported in the literature (Rappe et al. 1991; Yuan et al. 2001). Most, if not all, of the TCDD in Newark Bay and its tributaries originated from a single herbicide manufacturing plant along the shores of the Passaic River, which produced chlorinated phenols from the late 1940s until 1969. It has been estimated that 4–8 kg of TCDD have been deposited in Newark Bay from releases from the Passaic River facility (Bopp et al. 1991).

Ecological Damage to Hudson River Fish Populations

Evidence of fish die-offs due to chemicals is rare but death due to any cause in the early life stages of fishes is rarely observed in nature. When such evidence does exist, it is certainly difficult to confidently ascribe it to a source. However, Cook et al. (2003) combined results of laboratory testing of young life stages of lake trout *Salvelinus namaycush* with retrospective analysis of TCDD-like chemicals in sediments and modeling of sediment bioaccumulation factors for TCDD-like chemicals in eggs, to conclude that environmental levels of TCDD from the 1930s through 1950s were sufficiently high in Lake Ontario to prohibit successful natural reproduction of lake trout and likely resulted in their extirpation from that lake by 1960.

There is no evidence of direct, chemically induced lethality to fish in the Hudson River yet experimental results demonstrating a genetic basis to resistance in some Hudson River ichthyofauna (discussed below) would implicate chemically mediated lethality in the past. Nevertheless, it is noteworthy that species diversity in the Hudson River is unusually high for a temperate estuary (Smith and Lake 1990), reflecting a combination of resident and seasonally marine migrant marine species from more southern climes as well as a large number of invasive species (Daniels et al. 2005). Extirpation of native Hudson River species is unknown with the probable exception of rainbow smelt *Osmerus mordax* which may be due to global warming (J. Waldman, Queens College, personal communication). Thus, evaluation of toxicities in nature is likely to require attention to more subtle sublethal effects.

Heightened interest in toxicity of chemicals in impacted ecosystems often follows reports of reduced abundance in economically or ecologically important species. In the mid-1970s, populations of striped bass along the Atlantic coast, including the Hudson River, declined to near historic low levels. Contaminants in the Hudson River were viewed as a factor that may have contributed to these radical declines in population size (Mehrle et al. 1982; Westin et al. 1983). Extensive evaluation of the toxic effects of a variety of chemicals, including PCBs, in young as well as adult striped bass, showed that although PCBs may reduce survivorship of larvae (Mehrle and Ludke 1983; Westin et al. 1985), they were not responsible for the population's decline. Instead, overfishing was implicated as the important mechanism for population reductions (Barnthouse et al. 2003) because severe limitations of harvests resulted in strong rebounds of striped bass adults along much of the Atlantic coast (Striped Bass Technical Committee 2003). Similarly, abundance of shortnose sturgeon *Acipenser brevirostrum* and Atlantic sturgeon *A. oxyrinchus* in the Estuary declined to historically low levels over the past 40 years, but overfishing rather than chemical pollution is believed to be the cause for these declines as well.

One manifestation of possible chemical-induced toxicity in Hudson River fish populations was the elevated prevalence of pathological aberrations in brown bullhead from the upper reaches of the Hudson River near Griffin Island and Stillwater at river mile (rm) 175 compared to those collected near Corinth (rm 200), just upstream of PCBs point sources (Kim et al. 1989; Bowser et al. 1990). Brown bullhead from contaminated sites exhibited 60-fold higher levels of total PCBs in skeletal muscle, significantly higher splenic and renal

A

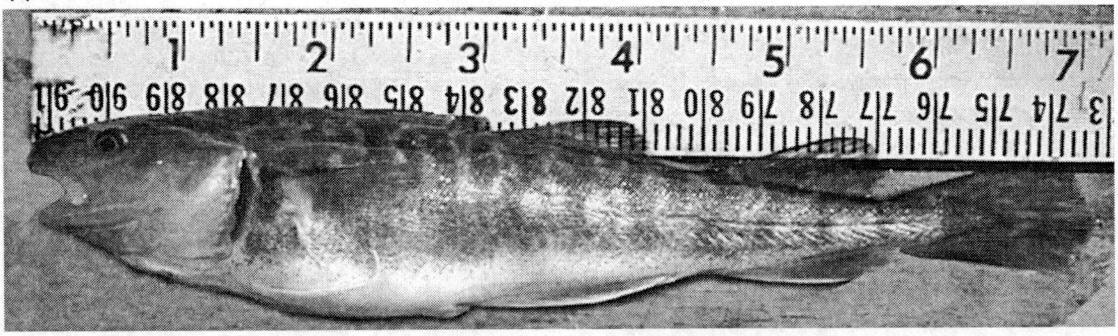

B

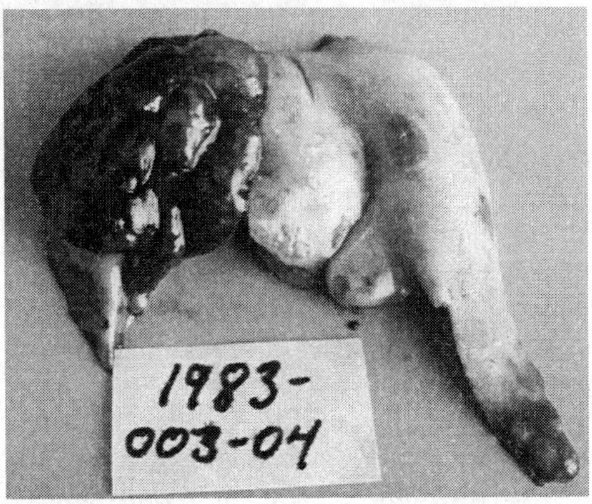

Figure 2. A: An adult one-year-old Atlantic tomcod from the Hudson River. B: Liver from an adult tomcod from the Hudson River with several large hepatocellular carcinomas and a tumor mass involving most of one of three liver lobes.

hemosiderous (pigmentation from hemoglobin degradation), and bile-duct hyperplasia than fish from the less impacted upriver site. The high incidence of these pathologies is consistent with the role of liver and kidney in detoxifying chemical contaminants. In view of the susceptibility of brown bullhead to liver and skin cancer at other contaminated sites (Baumann 1998), it is surprising that there are no reports of tumors in this species from the upper Hudson River. Furthermore, the proportion of female to male brown bullhead at the contaminated sites (0.75) far exceeded that at the less contaminated ones (0.55) suggesting the possible impact of endocrine disrupting chemicals. However, the impacts of these histological and demographic changes on population processes were not investigated further.

Epizootics of Neoplasia in North American Fish Populations

In the early 1980s, reports surfaced of epizootics of neoplasia in fish populations from contaminated urbanized and industrialized estuaries and rivers in different regions in the United States (reviewed in Wirgin and Waldman 1998). The most publicized of the epizootics and sites included English sole *Parophrys vetulus* from Puget Sound, Washington, winter flounder *Pseudo-pleuronectes americanus* from Boston Harbor, Massachusetts, brown bullhead from industrialized tributaries of Lake Ontario (Baumann et al. 1987) and the Anacostia River, Washington, D.C. (Pinkney et al. 2004), and Atlantic tomcod from the Hudson River (Dey et al. 1993). All of these affected taxa have benthic lifestyles as juveniles and adults. In almost all cases, liver or skin tumors were observed. Chemical etiologies were proposed for these epizootics because of the direct contact of skin with contaminated environmental matrices, the role of the liver in the detoxification of chemical contaminants, the lipophilicity of most cancer-causing xenobiotics, and the deposition of hydrophobic pollutants in the bottom stratum. This explanation is consistent with the finding of a significant decrease in the frequency of liver tumors in a brown bullhead population from the Black River, a tributary of Lake Erie, after the closing of a local coking facility resulted in reduced PAH pollution (Baumann and Harshbarger 1995).

Empirical determination of the etiology of environmentally induced tumors in fish populations is difficult. While associations between environmental or tissue contaminant levels and prevalence or severity of cancer or noncancer liver lesions have been made, most often for PAHs (Johnson et al. 1993) and sometimes PCBs and insecticides and their derivatives (Myers et al. 1994), controlled laboratory studies that experimentally attempt to induce tumors in fish from natural populations have been unsuccessful with few exceptions (Vethaak et al. 1996). This failure to induce tumors may have resulted from insufficient duration of exposure, exposure of inappropriate life stages, exposure to single chemicals or mixtures different than those present in nature, or combinations of these factors.

The complexity of the neoplastic process makes the experimental induction of tumors difficult. First, chemical carcinogenesis is known to be a multistep process including initiation (when DNA in a single cell is damaged by environmental mutagens), promotion (when cells with damaged DNA selectively outcompete neighboring cells with normal DNA), and progression (when cells with damaged DNA become genetically unstable and spontaneously acquire additional genetic alterations) (Lengauer et al. 1998). DNA damage to initiate this process probably must occur at specific genes, both oncogenes and tumor suppressor genes (Manam et al. 1992). Carcinogenesis often requires exposure to distinctly different classes of chemical carcinogens, those that initiate DNA damage and those that promote cells with damaged DNA. In natural environments such as the Hudson River, metabolites of PAHs probably serve as initiating agents by adducting to DNA (Kriek et al. 1998; Stein et al. 1994), and PCBs (Cogliano 1998) and PCDD/Fs (Huff et al. 1994) likely act as promoting agents. Second, early life stages are most sensitive to toxicity and carcinogenicity requiring that husbandry practices of vulnerable target species are sufficiently developed to allow for rearing of fish from the embryo/larvae through to the adult stage. This is uncharted territory for many species of marine and estuarine fish. Third, the development of tumors often requires that the rate of growth and sexual maturation in fish held in the laboratory closely reflects that in their natural environments, a difficult requirement for many taxa of concern.

Atlantic Tomcod from the Hudson River: A Model for Carcinogenesis?

In the 1970s and early 1980s, Atlantic tomcod (Figure 2A) from the Hudson River exhibited one of the highest prevalences of tumors ever seen in natural populations of any organism. Tumors were initially observed incidentally in tomcod collected from the Hudson River in routine biomonitoring programs conducted by the utility companies sited along the river. Smith et al. (1979) originally reported that the livers of 25% of adult tomcod collected in 1978 between rm 25 and rm 71 during their spawning season (late December to February) exhibited gross neoplastic nodules and hepatocellular carcinomas (Figure 2B). In later studies, the presence of gross and histologically identified liver lesions was related to the age and size of the fish. The prevalences of gross hepatic lesions in adult tomcod from the Hudson River ranged from 32% for age 1 fish to 78% for age 2 individuals, while such lesions were absent in adult tomcod from

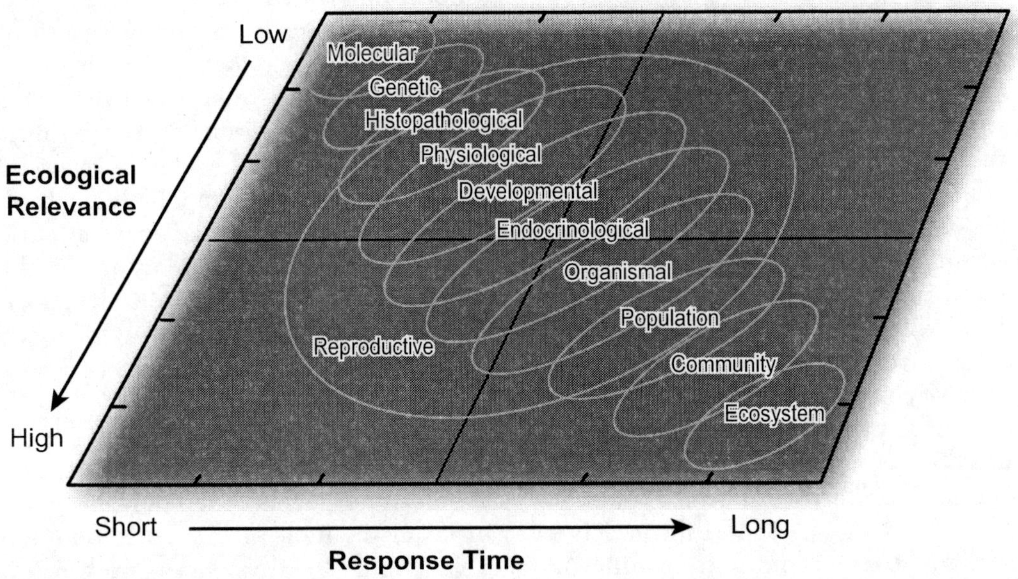

Figure 3. The distribution of hierarchical levels of biological organization and how the response time and ecological relevancy of biomarkers covary across this hierarchy.

the Pawcatuck River, a less contaminated waterway along the Connecticut–Rhode Island border (Dey et al. 1993). Histopathological examination revealed hepatomas (neoplastic nodules and hepatocellular carcinomas) in 44% of age-1 and 93% of age-2 Hudson River tomcod. In comparison, the incidence of histologically defined liver tumors in age-1 and age-2 tomcod from the Pawcatuck River was 3% and 10%, respectively. Furthermore, histopathological evaluation failed to reveal preneoplastic or neoplastic lesions in tomcod from five estuaries of varying environmental quality in the southern Gulf of St. Lawrence (Couillard et al. 1999).

The lesions observed in the Hudson River tomcod covered a continuum from basophilic foci with little cellular alteration to highly invasive hepatocellular carcinomas affecting the entire liver. The lesion morphology in tomcod was similar to chemically induced liver lesions observed in rodents used as toxicological models. A positive association was detected between the prevalence of liver lesions and the size of fish within an age-group, indicating that high growth rate may enhance tumorigenesis. Gross hepatic lesions and hepatomas were absent in juvenile young of the year (age-0) tomcod collected in summers from the lower Hudson River Estuary (Dey et al. 1993) suggesting that duration of exposure, source of contaminants, or processes associated with sexual maturation in the fall months might play a role in the neoplastic process in this population. Environmental factors operating on a broad scale in the Hudson River ecosystem may contribute to the propensity of tomcod to exhibit chemically induced tumors.

The presence of preneoplastic and neoplastic lesions in Hudson River tomcod has also been associated with other abnormalities and pathologies of the liver, including significantly elevated hepatic lipid levels, perhaps indicative of increased detoxification activities in livers of Hudson River tomcod (Cormier et al. 1989). Lipophilic contaminants such as PAHs and PCBs would be expected to occur at higher levels in these fish during growth spurts, one of which occurs in the first autumn of age-0 fish as they approach sexual maturation.

A short-term laboratory study did not detect liver cancers in juvenile age-0 Hudson River tomcod (Cormier and Racine 1990). Juvenile tomcod were collected in June from the Hudson River and Sheepscot River, Maine, and reared until December in clean laboratory water. The expectation was that the Hudson River tomcod, four to six months old at time of collection, would develop liver cancer because of early life stage environmental exposures to contaminants and that tomcod from the Sheepscot River would not. The lack of tumors in the Hudson River tomcod might have been due to insufficient exposures through feeding on contaminated prey or, more likely, to reduced growth and an absence of sexual maturation in these laboratory-reared fish.

Use of Biomarkers to Determine the Etiology of Disease and Health of Ecosystems

Biomarkers and bioindicators have been used as alternatives to controlled laboratory studies to investigate the etiology of cancers or other perturbations in natural populations. Biomarkers in fishes have also been used to evaluate the overall health of impacted aquatic ecosystems. Because the presence of contaminants in the environment does not necessarily mean that they are bioavailable to potentially impacted populations, the use of biomarkers and bioindicators confirms bioavailability and usually does so in the milieu of the natural environment. Biomarkers have been defined as functional measures of exposure to environmental stressors and are often measured at the suborganismal level of biological organization (Adams 2002). In contrast, bioindicators can be viewed as measurable responses at higher levels of biological organization from the organismal through the community levels. Thus, it has been suggested that biomarkers can be viewed as measures of exposure and bioindicators as measures of effects. It is our contention, however, that any response at or above the molecular level is indicative of a biological effect. Thus, we will use the term 'biomarker' in this chapter for all biological responses to exposures.

Biomarker responses can be quantified at a hierarchy of levels of biological organization from the molecular to the community levels (Figure 3). Responses at lower levels of complexity (molecular) are immediate, sensitive, and specific to toxicant exposure relative to responses at higher levels of organization (population, community). Conversely, responses at higher levels have more ecological relevancy but are delayed, diffuse, and difficult to assign to specific stressors relative to responses at lower levels of organization. Current research efforts are focused on identifying mechanistic linkages between responses at different levels of biological organization (Johnson and Collier 2002). Success in these efforts would include the capability to predict population and community effects from perturbations at lower levels of organization. Biomarkers can be evaluated in natural populations, in situ enclosures, or laboratory settings using single or multiple contaminants or complex contaminated environmental matrices. Over the past decade, biomarker studies in tomcod have employed a variety of investigative methods and have been implemented at multiple levels of biological organization in order to determine the etiology of tumors in the Hudson River population and to evaluate the effects of pollutants on tomcod and the Hudson River fish community.

Life History Characteristics of Atlantic Tomcod

Tomcod lend themselves well to biomarker studies for several reasons. Tomcod are usually very common in estuaries of the northeastern United States and Atlantic Canada. The Hudson River Estuary supports the southernmost spawning population of tomcod (Klauda et al. 1988). The egg stage of tomcod is benthic and prolonged (at least 1 month), larvae at hatching have a large yolk complement from their mothers, and juveniles and adults are bottom-dwelling omnivorous feeders on benthic invertebrates. Tomcod thus have the potential to bioaccumulate high levels of lipophilic contaminants through direct contact with sediments, maternal transfer, and consumption of benthic prey. They have a lipid-rich liver that further serves to accumulate these pollutants. Tomcod are resident in the Hudson River year round, although they do undergo seasonal movements upstream to the salt front for spawning (Klauda et al. 1988). The tissue burdens and biological responses of tomcod, therefore, reflect exposure histories that are specific to, and perhaps integrated across, much of their natal estuaries. This is in contrast to species such as mummichog (also known as Atlantic killifish) *Fundulus heteroclitus* which have a more limited range during their life history and therefore have tissue burdens and biomarker responses that reflect more localized signals of contaminant levels. Additionally, recent advances in tomcod husbandry now allow for multi-generational, controlled laboratory studies of the responses of various life stages of tomcod to contaminants. Furthermore, the timing of events in the tomcod life history (e.g., the winter spawning of tomcod is unique in the Hudson River) results in the young life stages of tomcod being a key prey for major resource and other ecologically important species within the lower estuary in spring and early summer. The wide distribution and high abundance of juvenile tomcod throughout the tidal estuary combined with its role as prey for common freshwater, estuarine, and marine species, place tomcod at a critical node in the Hudson River food web (Chambers and Witting, 2006, unpublished).

Tissue Burdens of HAHs in Tomcod

Chemical carcinogenesis is a multistage process and in natural populations may require high levels of chronic exposure to DNA-damaging PAHs and to tumor promoting PCBs and PCDD/Fs. Tomcod from the Hudson River Estuary bioaccumulate much higher levels of PCBs and PCDD/Fs and are exposed to significantly higher levels of PAHs than tomcod from elsewhere.

Initial studies of tissue burdens in tomcod compared hepatic levels of PCBs and PCDD/Fs on a congener-and-sex specific basis among adult tomcod collected from the Hudson River at Garrison, New York (rm 51), the Hackensack River, New Jersey, (one of two major tributaries to Newark Bay in the western Hudson River Estuary; Figure 1), and the Miramichi and Margaree rivers in the southern Gulf of St. Lawrence, Canada (Courtenay et al. 1999). Hepatic PCDD burdens in tomcod from the Hackensack River were dominated by TCDD, which provided greater than 80% of total TCDD toxic equivalency quotients (hereafter, "TEQ," which is a summary measure of the total toxicity of PCDDs, PCDFs, and coplanar PCBs in environmental samples scaled to the toxicity of TCDD and in fishes is usually assessed by early life stage toxicities, gene induction, or both [Van den Berg et al. 1998]). Levels of hepatic TCDD in tomcod from the Hackensack River were among the highest ever reported in a natural population. TEQs derived from PCDD/Fs were seven- to eightfold higher in tomcod from the Hackensack River than in those from the Hudson River and 250-fold higher than in tomcod from the Margaree River. Surprisingly, total coplanar PCB levels were very similar between tomcod from the Hackensack River and the Hudson River and 40 and 90-fold higher than in tomcod from the Miramichi River and Margaree River, respectively. The contribution of PCB77, expressed as lipid-normalized/body weight (ng/kg wet weight), comprised about 82% of the total coplanar PCBs in Hudson River tomcod. Hepatic levels of all of these HAH contaminants were much higher in male than female tomcod (2.4-to 4.6-fold), suggesting that a sizeable amount of these contaminants is shunted into tomcod eggs and, importantly, potentially transferred to and affecting early life stages of offspring.

Subsequent studies of tissue burdens have examined hepatic, congener-specific levels of HAHs in juvenile age-0 tomcod (5–9 months of age) from multiple sites within the Hudson River Estuary complex and in Atlantic Canada. Age-0 tomcod were collected from 19 locales in the Hudson River Estuary (between RM1 and RM107), Newark Bay, the Hackensack River, the Miramichi River, and age-matched, laboratory-spawned F_1 juvenile tomcod of Hudson River descent (Yuan et al. 2001; Yuan 2003; Fernandez et al. 2004) (Table 1). Juvenile fish from the Hackensack River and Newark Bay exhibited 930 times higher total hepatic TEQs than similarly aged tomcod from the Miramichi River. Mean total TEQs in juvenile tomcod collected from the Hudson River were 62-fold higher than in tomcod from the Miramichi River and sevenfold higher than in the laboratory-reared F_1 Hudson River reared under clean conditions. Coplanar PCB burdens expressed as total TEQs using fish toxic equivalency factors (TEFs) (a relative measure of toxicity of individual PCDDs, PCDFs, and coplanar PCBs congeners compared to that of TCDD which is assigned a value of 1) were 58- and 19-fold higher in tomcod from the Hackensack River/Newark Bay than in juveniles from the Miramichi River and laboratory-reared F_1 Hudson River controls, respectively. Similarly, mean TEQs from coplanar PCBs were 30- and nine-fold higher in age-0 tomcod from the Hudson River compared to age-0 juveniles from the Miramichi River or laboratory-reared F_1 Hudson River fish, respectively. However, coplanar PCBs only contributed 2% of the total TEQs in tomcod from the Hackensack River/Newark Bay compared to a mean of 25.6% in tomcod from the Hudson River. The contribution of coplanar PCBs to total TEQs among locales in the Hudson River varied greatly; the lowest contribution (1.5%) was at Yonkers (rm 17), where TCDD contributions were high, and the highest contribution (57%) was in fish from Haverstraw Bay (rm 37). Surprisingly, there were no trends in total hepatic (coplanar and noncoplanar) PCBs levels with river miles in the Hudson River. Variation in PCBs levels in tomcod among collection sites was great, but the levels did not decrease with distance downstream and away from the historic point sources of PCBs release into the upper Hudson River.

Furthermore, a direct mixing model (Sather et al. 2001) was used to estimate the relative contributions of three commercially available PCBs mixtures (Aroclor 1242 [A1242], A1254, A1260) to the hepatic PCBs profiles in age-0 tomcod from multiple Hudson River sites (Fernandez et al. 2004). The last two digits in the numerical designation of Aroclors denote the percentage of chlorine that they contain, such that A1242 contains fewer chlorine substituents than

Table 1. Wet-weight concentrations of halogenated aromatic hydrocarbons (HAHs) expressed as TCDD Toxic Equivalency Quotients (TCDD-TEQs)[d] in livers of juvenile Atlantic tomcod collected from 1994 to 1998 from various rivers miles (rm) on the main-stem Hudson River and from the Hackensack River and upper Newark Bay in the western Hudson River Estuary.

	Newark Bay	Hackensack River	Hudson River (RM) 1	8	10	17	37	58	67	77	82
PCDDs and PCDFs											
TCDD	655	897	26	28	14	27	9	4	0	1	0
PCDDs[a]	671	921	31	36	14	35	16	8	3	4	0
PCDFs[b]	50	103	13	20	12	31	27	13	9	7	1
Coplanar PCBs											
PCB77	3	4	1	2	1	3	3	2	2	1	1
PCB81	1	1	0	0	0	1	1	1	0	0	0
PCB126	11	15	4	7	3	9	11	6	4	4	3
PCB169	0	0	0	0	0	0	0	0	0	0	0
Mono-ortho PCBs[c]	6	8	3	5	3	6	10	6	4	4	2
Total TCDD-TEQs	741	1052	52	70	33	85	68	36	22	20	7

[a] Sum of PCDDs from tetra-to octa-chlorinated dibenzo-p-dioxins-

[b] Sum of PCDFs from tetra-to octa-chlorinated dibenzo-furans

[c] Sum of all 46 mono-ortho polychlorinated biphenyls including tetra-, penta-, hexa-, and hepta-chlorinated biphenyls

[d] Fish-specific Toxic Equivalency Factors (TEFs) for each HAH were from Van den Berg et al. (1998)

A1254 or A1260. A1242 was the predominant mixture released at the upstream GE manufacturing facilities, whereas A1254 and A1260 predominate in NYC municipal and industrial effluents. As expected, a linear increase in the A1242 contribution with river mile was observed in age-0 tomcod collected between rm 0 and rm 80, a result which was consistent with its upriver sources. Unexpectedly, age-0 tomcod collected upriver of rm 80 exhibited a PCBs congener pattern with decreasing A1242 characteristics. Because this shift in pattern occurred near the zone of seasonal transition from brackish to freshwater, it was hypothesized that this was due to increased depuration or decreased uptake by tomcod of lower chlorinated PCBs congeners in freshwater.

Similar concentrations of PCDD/PCDFs were observed in age-0 juveniles as reported for adults from the same locales (Yuan et al. 2001, Yuan 2003; Fernandez et al. 2004). Juveniles from the Hackensack River/Newark Bay exhibited slightly higher levels of PCDD/Fs than adults, indicating that exposure occurs early in their life cycles. Approximately 90% of the total TEQs in tomcod from these two sites was contributed by one congener, TCDD, most likely originating from the single industrial point source on the Passaic River (Figure 1; Bopp et al. 1991). Total TEQs were highest in age-0 juveniles from the Hackensack River (1052 ng/kg) compared to a mean of 62 ng/kg TEQs in the 19 samples from the Hudson River. There were some TEQ "hotspots" in the Hudson River with fish from Yonkers (rm 17) exhibiting 420 ng/kg total TEQs. The TCDD contribution to total TEQs was much smaller in the Hudson River than in the Hackensack River/Newark Bay, ranging from 42% at rm 17 in Yonkers to 8% at rm 107.

Interestingly, PCBs congener patterns differed between age-0 juvenile and adult tomcod collected from the same sites within the Hudson River Estuary (Fernandez et al. 2004). The di- to tetra substituted PCB congeners dominated the PCBs composition in age-0 juvenile tomcod, whereas the penta- to nonachlorinated substituted congeners, particularly those with di-ortho substitutions, predominated in adults. This difference in patterns between life stages may have been due to dietary or metabolic variation between age-0 juvenile and adult tomcod and may have implications for differences in PCBs toxicity in young versus older life stages and for those consumers of this resource.

Because of their long-term persistence within tissues, ratios of total PCDDs to PCDFs can provide information on sources of contaminants and the movement of fish within and between estuaries. The ratios of total PCDDs:PCDFs expressed on a wet weight basis were very different in juvenile tomcod from the Hackensack River/Newark Bay compared to those from the Hudson River. In tomcod from the Hackensack River/Newark Bay the ratio of PCDDs:PCDFs varied between 1.5 and 1.9, while PCDFs concentrations exceeded those of PCDDs in eight of nine collections from the Hudson River (Yuan et al. 2001; Yuan 2003; Fernandez et al. 2004). These results suggest that tomcod from Newark Bay and the main-stem Hudson River are exposed to different levels and sources of PCDDs/Fs and, given the persistence of these contaminants, that exchange of tomcod between Newark Bay and the Hudson River is rare. This suggests that spawning populations of tomcod in the tributaries of Newark Bay are probably distinct from those in the Hudson River.

Hepatic burdens of PCDDs and PCDFs, expressed as TEQs, were very similar in tomcod from most Hudson River sites. On a wet weight basis, however, the contribution of TCDD to total PCDDs among

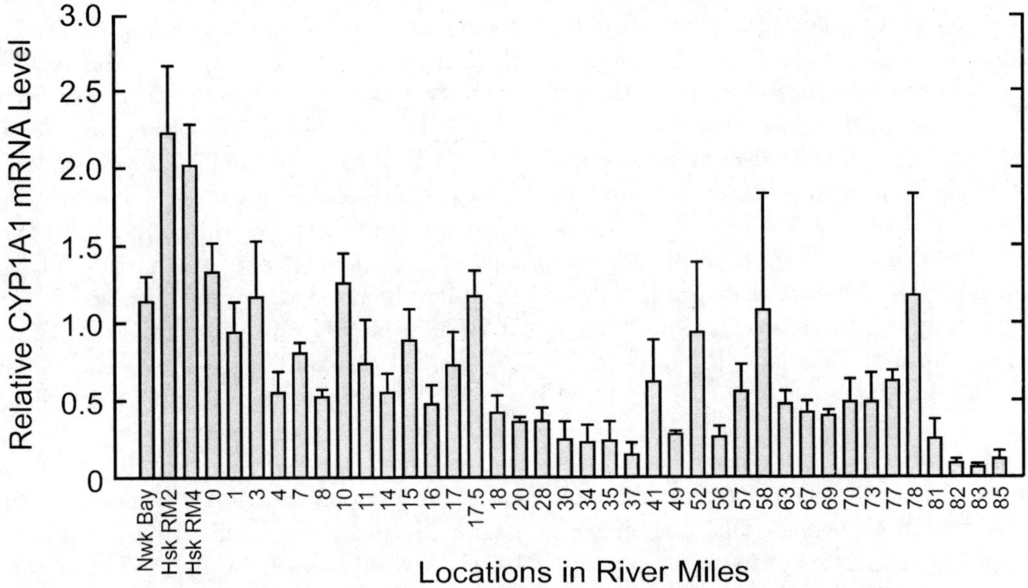

Figure 4. Relative levels of hepatic cytochrome P4501A1 (CYP1A1) mRNA expression in age-0 juvenile Atlantic tomcod collected from 1994 to 1998 and in 2000 from 40 sites in the Hudson River Estuary complex including two locales in the Hackensack River and one in Newark Bay (see Figure 1). The numerals along the x-axis refer to location (in river miles) in the Hudson River upstream of the southern tip of Manhattan Island. The mean (± 95% CI) relative CYP1A1 mRNA concentrations are displayed.

Hudson River collections decreased with distance upstream from lower Manhattan. This pattern supports the assessment of an increased bioavailability of upstream sources of PCDDs other than TCDD, most likely octachlorodibenzo-*p*-dioxins (Bopp et al. 1998). In summary, these studies of tissue burdens in juvenile tomcod demonstrate that after only 5 to 9 months of environmental exposure, juvenile tomcod from the Hudson River Estuary exhibit contaminant levels that are comparable to those found in adults.

Levels of PCDD/Fs and PCBs in unfertilized eggs of tomcod collected from the Hudson River (RM50) and from the lower Miramichi River (Loggieville, New Brunswick) were analyzed and compared to the hepatic levels of their mothers (Roy et al. 2001). Total TEQs from PCDD/Fs were 8- to 15-fold higher in Hudson River tomcod eggs compared to those from the Miramichi River, while coplanar PCBs were 73- to 79-fold higher in eggs from Hudson River tomcod than those from the Miramichi River. Lipid-adjusted TEQs from total PCDD/Fs and total coplanar PCBs were very similar in the unfertilized eggs and in the livers of their mothers. This similarity suggests that the maternal transfer of these persistent lipophilic contaminants to developing embryos is substantial and could contribute to various early life stage toxicities in tomcod of Hudson River origin.

Bile Metabolites of PAHs in Tomcod

PAHs are very difficult to measure in tissues of fish because they are metabolized very rapidly (Varanasi and Stein 1991). A surrogate measure of exposure has been developed, termed "fluorescent aromatic compounds" (FACs), that quantifies bile metabolites of PAHs that are comprised of four to six aromatic rings (including benzo[a]pyrene (B[a]P)) by using high performance liquid chromatography with detection at an excitation/emission wavelength pair of 380/430 nm (Krahn et al. 1986). Levels of FACs were significantly higher in adult tomcod from the Hudson River compared to those from three other cleaner rivers (Wirgin et al. 1994). For example, FACs levels were approximately eightfold higher in tomcod from the Hudson River than in the Miramichi River.

Although Hudson River tomcod have been exposed to higher levels of PAHs compared to tomcod from elsewhere, this does necessarily mean that these compounds are biologically active. Parent PAH compounds are not inherently mutagenic and carcinogenic. They become mutagenic once they are metabolically converted to forms that are highly reactive with cellular biomolecules, at which time they can initiate cancer and cause other perturbations by damaging DNA (Stein et al. 1994). Vertebrates, including fishes, have a system of enzymes (Phase I enzymes) that oxidize PAHs and some other lipophilic compounds to form more polar metabolites (Stegeman and Hahn 1994), which are then conjugated by other enzymes (Phase II enzymes) to carrier molecules that facilitate the excretion of these conjugated metabolites from the body (George 1994). Sometimes, however, the metabolites of Phase I enzymes escape the conjugating activity of Phase II enzymes, which can result in damage to DNA.

Cytochrome P4501A1 Expression in Environmentally Exposed Tomcod

Cytochrome P4501A1 (CYP1A1) is a major Phase I enzyme gene whose products metabolize PAHs and other compounds to reactants that can form adducts and damage DNA. Levels of CYP1A1 expression in fishes can be dramatically increased (induced) at the mRNA, protein, and enzyme activity levels by exposure to many forms of HAHs and PAHs (Stegeman and Hahn 1994). Levels of CYP1A1 expression in fishes, therefore, are widely used to quantify exposure to these compounds and their early biological effect. Induced CYP1A1 activity is needed to metabolize PAHs; HAHs are largely refractory to CYP1A1 metabolism. Persistent HAH-induced CYP1A1 activity, however, can result in increased DNA damage. This process is known to generate reactive oxygen species that can cause mutations in DNA by modifying its bases (Schlezinger et al. 1999). Oxidized bases have also been implicated as a cause of cancer in fish populations from contaminated locales (Malins et al. 1990).

AHs differ in their efficacies in inducing CYP1A1 due to their differing abilities to bind AHR. PCDDs, the most active AHs, often elicit induction at low parts per trillion levels, whereas PAHs are the least active AHs (induction occurs at ppm levels). It is thought that the AHR pathway mediates most, if not all, biological responses to AH contaminants including transcription of CYP1A1 and a battery of other xenobiotic metabolism genes. As a result, expression levels of CYP1A1 are believed to be predictive of most toxic responses to these contaminants. Since TCDD best binds AHR, other molecules that are structurally similar to TCDD share an ability to bind AHR and share potency in eliciting AHR-mediated toxicities. A numeric system based on empirically derived toxicity and CYP1A1 expression data in mammals, birds, and fishes has been developed to compare the toxicities of individual PCDDs, PCDFs, and coplanar PCBs congeners (Van den Berg et al. 1998). Because it is most toxic, TCDD is given a Toxic Equivalency Factor (TEF) of 1. Toxicities of all other PCDD/Fs and

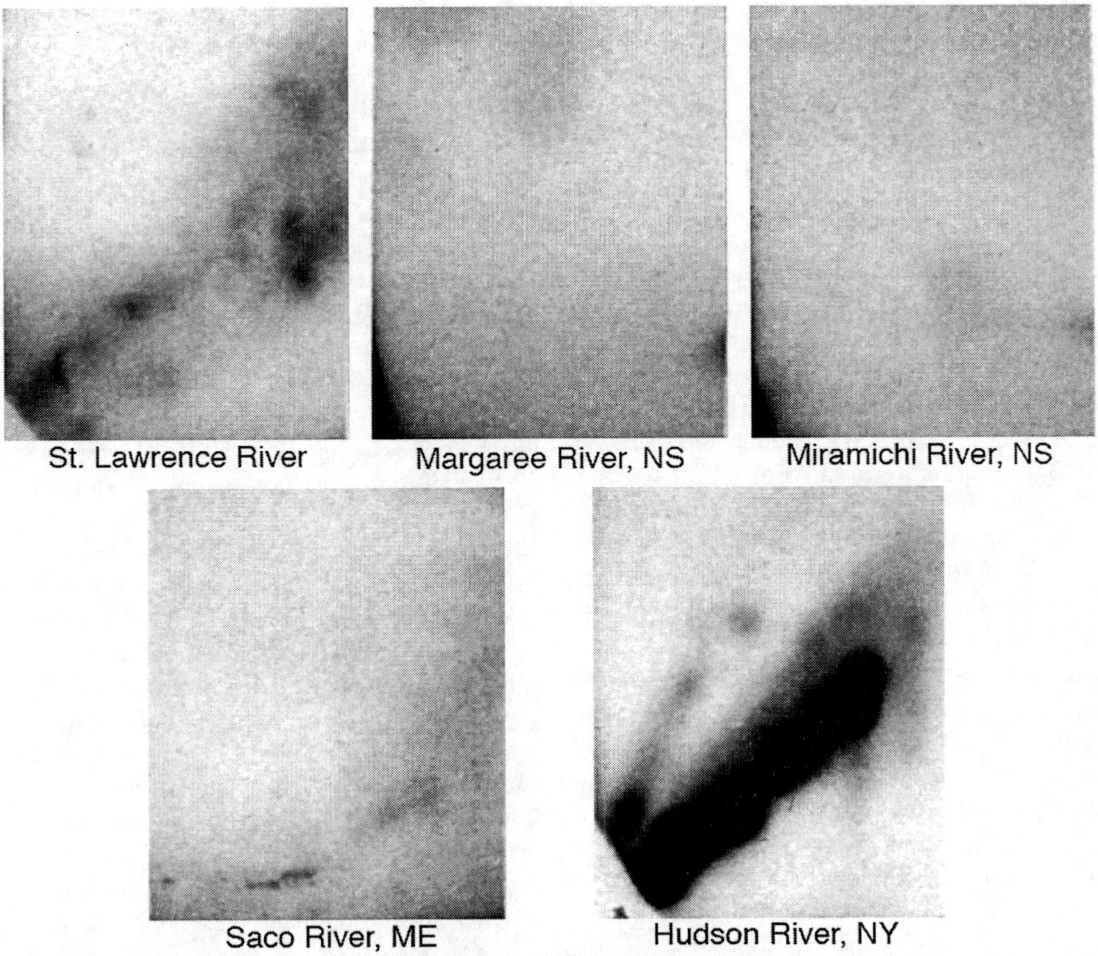

Figure 5. Autoradiograms of thin-layer chromatograms of hepatic DNA adducts detected by ^{32}P postlabeling analysis in Atlantic tomcod collected from 1987 to 1993 in five Atlantic coast estuaries. Locations of tomcod collections are (north to south) St. Lawrence River, Quebec; Miramichi River, New Brunswick; Margaree River, Nova Scotia; Saco River, Maine; and Hudson River, New York. DNA adducts were visualized by storage phosphor imaging (Reproduced from Wirgin et al. [1994] with permission from *Environmental Health Perspectives*).

coplanar PCBs congeners are scaled to that exhibited by TCDD and have values less than 1. Noncoplanar PCBs also have been shown to exert toxicities, most likely through a mechanism independent of the AHR pathway and therefore are not assigned TEF values.

Levels of hepatic CYP1A1 mRNA expression have been found to vary among tomcod from different estuaries (Kreamer et al. 1991). Wirgin et al. (1994) compared adult tomcod collected from five Atlantic coast rivers and found, not unexpectedly, that CYP1A1 mRNA expression was highest by a considerable degree in tomcod from the Hudson River. Levels were 16- and 28-fold higher in Hudson River tomcod than in tomcod from the cleaner Saco River (Maine) and the Margaree River (Nova Scotia), respectively. These results confirmed that AHs are bioavailable to tomcod from the Hudson River and that AHs elicit a molecular-level effect that can be mechanistically linked to perturbations at higher levels of biological organization, in this case at the level of DNA damage.

Levels of hepatic CYP1A1 mRNA in tomcod also varied among locations within an ecosystem although no significant trend between CYP1A1 mRNA levels and location in the Hudson River was evident. CYPIA1 expression was quantified for age-0 juvenile tomcod collected from 40 sites in the Hudson River (RM1 to RM107) and from the Hackensack River and Newark Bay (Yuan et al. 2001; Yuan 2003). Levels of CYP1A1 mRNA gene expression were found to vary by as much as 23- to 34-fold among sites within the estuary (Figure 4). As expected, levels of CYP1A1 and HAH contaminants were highest in tomcod collected from the Hackensack River and Newark Bay. In tomcod from the Hudson River, CYP1A1 was highest at the most downriver sites and displayed secondary peaks just downstream of municipalities such as Newburgh, New York (rm 61) and Kingston, New York (rm 80). The significant and consistent site-to-site differences in the levels of the CYP1A1 biomarker support the utility of this and other biomarkers as indicators of the microgeographic distribution of bioavailable contaminants in this large system. Although levels of hepatic HAHs (expressed as TEQs) and hepatic CYP1A1 were highest in juvenile age-0 tomcod from Newark Bay/ Hackensack, there was not, however, a significant among-site relationship between levels of CYP1A1 mRNA and hepatic burdens of HAHs in tomcod from the Hudson River (Yuan et al. 2001). This may have been due to exposure to environmental inducers other than HAHs (such as PAHs), resistance to CYP1A1 mRNA inducibility or a combination of these and other factors.

The age at which induced CYP1A1 mRNA expression occurs in environmentally exposed Hudson River tomcod is yet to be fully determined. CYP1A1 mRNA expression was not induced in environmentally exposed Hudson River tomcod larvae (1–2 months old) relative to matched, similarly aged, and untreated laboratory-reared F_1 larvae of Hudson River descent (Roy et al. 2002). CYP1A1 mRNA expression in these laboratory-reared tomcod was highly inducible, however, by waterborne exposure to B[a]P. In view of the high levels of CYP1A1 mRNA and HAH burdens in wild 5- to 9-month old juvenile tomcod from the Hudson River (Yuan et al. 2001; Yuan 2003), these results suggest that 1) the HAHs in young larvae that might have been transferred to the egg and onto the yolk sac larvae from the mother have been diluted by growth and depuration to levels that do not induce CYP1A1 expression, and 2) bioaccumulation of contaminants has yet to occur in larvae although at 2 months of age these fish would have been consuming prey, albeit pelagic zooplankton, for most of this period. An increase in tissue burdens of HAHs and CYP1A1 mRNA may be associated with the habitat and diet shift of tomcod associated with their transition from pelagic larvae to the benthic associated juvenile life stage.

DNA Damage in Tomcod from the Hudson River

Although these studies indicate that AHs accumulate and are biologically active in Hudson River tomcod, they do not demonstrate injury to individual fish nor evidence of historical effects on the population. Use of the ^{32}P- postlabeling technique has indicated that tomcod from the Hudson River suffered higher levels of hepatic DNA damage than tomcod from cleaner rivers. The ^{32}P-postlabeling technique is a very sensitive assay for quantifying levels of hepatic DNA adducts in environmentally exposed fish (~ 1 adduct in 10^{9} to 10^{10} nucleotides)

(Stein et al. 1994). DNA adducts are DNA bases to which the reactive metabolites of PAHs or other compounds have covalently bound. Presumably the presence of these bulky DNA adducts results in errors during DNA replication and therefore the introduction of mutations (Wirgin and Theodorakis 2002). Among Hudson River contaminants of concern, ^{32}P postlabeling is sensitive to PAHs but not to PCDD/Fs or PCBs exposure because the latter compounds are not usually biotransformed to reactive metabolites (Wirgin et al. 1994). Nevertheless, it is difficult to determine to which individual PAHs the organism has been exposed because the ^{32}P postlabeling assay detects exposure to many different DNA-bound PAH metabolites.

DNA adducts in fishes provide a cumulative, lifetime measure of exposure to these compounds because fishes are inefficient at repairing damaged DNA compared to, for example, rodents (Stein et al. 1993). Moreover, DNA adducts have proven to be an informative tool for biomonitoring because they are unambiguous indicators of genotoxicity and have been clearly associated with health effects such as neoplasia, tissue damage, and early life stage toxicities (Stein et al. 1994; Kriek et al. 1998, Wirgin and Theodorakis 2002). Importantly, the formation of DNA adducts, if in carcinogenically important genes such as oncogenes or tumor suppressor genes, is probably a necessary first step in the initiation of chemical-caused neoplasia. Additionally, the levels of hepatic CYP1A1 and hepatic DNA adducts should be correlated because the generation of DNA adducts is mechanistically linked to prior activity of CYP1A1 in order to metabolically transform PAHs.

The levels of hepatic DNA adducts were significantly higher in adult tomcod from the Hudson River than in adult tomcod from four cleaner rivers (Wirgin et al. 1994) (Figure 5). Hudson River tomcod exhibited 40-fold higher levels of hepatic DNA adducts than tomcod from the least impacted locale, the Margaree River. Similarly, levels of DNA adducts were significantly higher in tomcod from the AH-contaminated St. Lawrence River than in all of the other rivers except the Hudson River. This confirms that adult tomcod from the Hudson River have been exposed to much higher levels of PAHs than tomcod from other locales and that these almost certainly have had detrimental biological effects such as the initiation of neoplasia or early life stage toxicities. In later studies (Yuan 2003), levels of hepatic DNA adducts were compared in juvenile tomcod collected in the year 2000 from 10 sites within the Hudson River (rm 1 to rm 107). Although hot spots were identified and there were fourfold differences in levels of DNA adducts among sites, no geographic pattern among river locales was observed. Interestingly, overall levels of hepatic DNA adducts in juvenile fish were approximately an order of magnitude less in 2000 than that previously reported for adult Hudson River tomcod collected between 1989 and 1993 (Wirgin et al. 1994). This result is consistent with cumulative lifetime exposure and inefficient DNA repair in fish, but it is also consistent with a possible decrease in PAH levels during the 7- to 11-years period between collections.

While high levels of overall DNA damage are almost certainly deleterious to a population, it is the genomic location of the damage that is most predictive of its carcinogenic potential. Oncogenes and tumor suppressor genes are known targets of PAH-DNA adducts in mammals and fish (reviewed in Wirgin and Theodorakis 2002) and their mutagenic activation and inactivation, respectively, is necessary for initiation of chemical carcinogenesis. In fact, the specific sensitive parts of these genes and patterns of mutation are often identical between mammals and fish (Bailey et al. 1996). For example, codons 12–13 and 62–63 of the K-*ras* oncogene in rodents, humans, and rainbow trout *Oncorhynchus mykiss* are vulnerable to PAH-induced mutations. DNA from liver tumors of tomcod from the Hudson River was found to have an activated K-*ras* oncogene, whereas there was no evidence of K-*ras* activation in DNA from normal livers of tomcod from the Hudson River or Saco River (Wirgin et al. 1989). In total, results from these studies are consistent with a chemical basis to the initiation of the high prevalence of tumors in tomcod and the sensitivity of the Hudson River population to toxicity from these contaminants.

CYP1A1 Expression in Chemically Treated Tomcod

Experimental studies were conducted in order to calibrate the responses from field-exposed fish and to determine the role of different classes of AHs in eliciting responses, particularly CYP1A1 induction. In these studies, adult tomcod from the Hudson River and Miramichi River were returned to the laboratory, depurated in clean water for extended periods (usually 20, but up to 300 d), and treated under controlled experimental conditions with AHs of concern. Although this period was undoubtedly insufficient to depurate HAHs to background levels, CYP1A1 mRNA were decreased to basal levels (Kreamer et al. 1991). As expected, hepatic CYP1A1 mRNA expression in adult tomcod from both populations was dose-responsive and highly inducible (up to 500-fold) by exposure to PAHs (B[a]P and beta-naphthoflavone [B-NF]). Tomcod from the two populations, however, differed dramatically in their sensitivities to CYP1A1 mRNA induction by coplanar PCBs or TCDD; tomcod from the Miramichi River were highly sensitive, whereas tomcod from the Hudson River were insensitive. In these initial studies with limited doses of PCB77 and TCDD, there was no evidence of CYP1A1 induction by these chemicals in tomcod from the Hudson River (Wirgin et al. 1992; Courtenay et al. 1999).

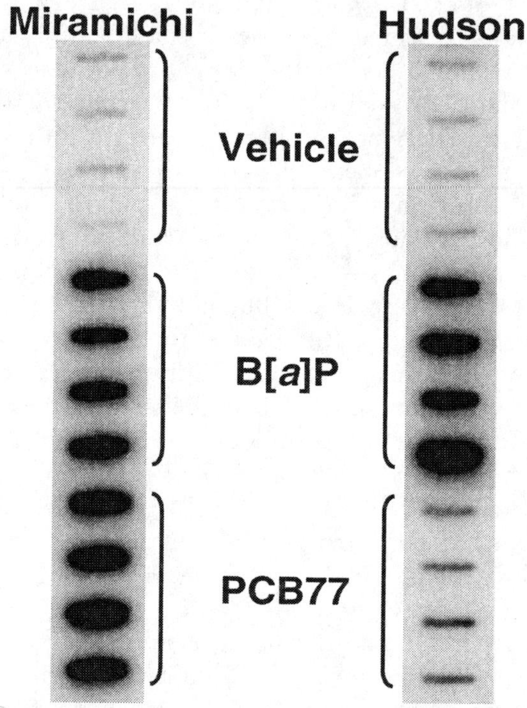

Figure 6a. A slot blot hybridization of cytochrome P4501A1 (CYP1A1) mRNA expression in laboratory-reared, F_1 early yolk-sac tomcod larvae from Miramichi River and Hudson River parents. Larvae were waterborne-exposed to 1 ppm B[a]P for 48 hr, 1 ppm PCB77 for 144 hr, or acetone vehicle for 144 hr and immediately sacrificed. CYP1A1 mRNA expression is depicted for four larvae for each treatment group in each population

In more extensive dose–response studies, significant differences in the sensitivity of hepatic CYP1A1 mRNA inducibility was observed between adult tomcod from the two aforementioned populations for three different coplanar PCBs congeners (PCB77, PCB81, and PCB126) and TCDD. When

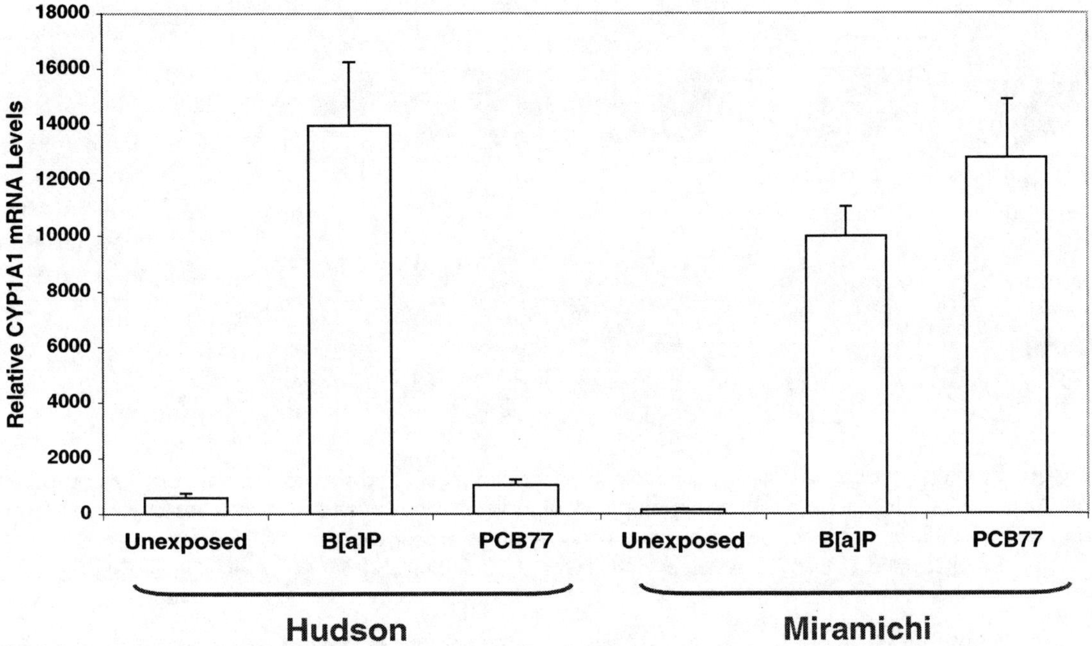

Figure 6b. A comparison of relative cytochrome P4501A1 (CYP1A1) mRNA expression in 1 ppm B[a]P, 1 ppm PCB77, and acetone vehicle waterborne-exposed laboratory-reared, F_1 early yolk-sac tomcod larvae from Miramichi River and Hudson River parents (n=8 larvae/treatment group in each population). The mean (± 95% CI) CYP1A1 mRNA concentrations are displayed.

higher doses of these chemicals were used, significant hepatic CYP1A1 induction was observed in Hudson River tomcod, albeit requiring 100-fold higher doses than that needed to significantly induce CYP1A1 in Miramichi River tomcod (Yuan 2003). Additionally, the extra-hepatic inducibility of CYP1A1 mRNA was compared in brain, heart, gills, gonads, intestine, kidney, and spleen in adult and juvenile tomcod from the two populations. For all tissues tested, tomcod from the Hudson River were approximately 100-fold less sensitive to CYPIA1 mRNA induction than tomcod from the Miramichi River (Yuan 2003). Thus, tomcod from the Hudson River exhibited significantly less sensitivity for all HAHs tested in all tissues tested. Furthermore, this difference between populations in sensitivity to coplanar PCBs, but not B[a]P-induced CYP1A1 induction, was also observed for PCB77 in waterborne exposed embryos, early yolk sac larvae, and late yolk sac larvae (Wirgin unpublished data). Thus, tomcod from the Hudson River exhibited significantly less sensitivity for all HAHs, in all tissues, and all life stages tested.

Yuan (2003) addressed the issue of whether resistance to CYP1A1 inductibility by HAHs, but not PAHs occurs throughout the Hudson River tomcod population by collecting juvenile age-0 tomcod from seven sites (rm 1 to rm 107) in the Hudson River and one site in the Miramichi River, depurating them in clean laboratory water, and treating them with PCB77 or B[a]P. Fish from all Hudson River sites showed strong induction of CYP1A1 mRNA after exposure to B[a]P, but not after exposure to PCB77. Juvenile age-0 tomcod from the Miramichi River showed strong induction

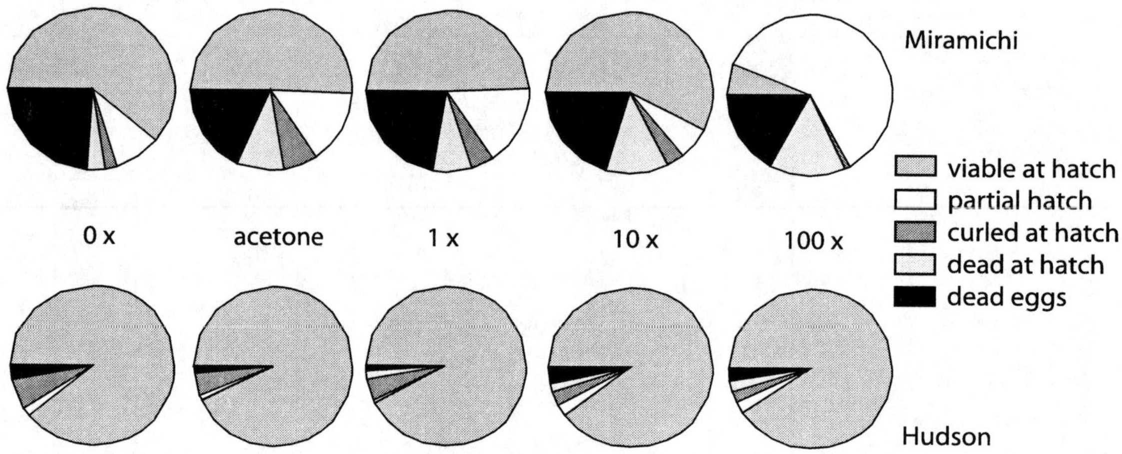

Figure 7. Effects of dose of a mixture of coplanar PCBs congeners and two controls (water (0 x groups), and acetone (contaminant vehicle)) on development of replicate populations of F_2 embryos of Hudson River and Miramichi River ancestry. The x represents concentrations of four coplanar PCB congeners (PCB77, PCB81, PCB126, and PCB169) measured in the livers of adult tomcod from the Hudson River Estuary.

for both chemicals. This demonstrates that juvenile tomcod from throughout the Hudson River population exhibit impaired CYP1A1 inducibility and further strengthens the argument that this is due to a population-wide genetic adaptation.

Researchers working on mummichog (also known as Atlantic killifish) from highly polluted Atlantic Coast estuaries have made similar, but not identical, observations. Killifish from the Hudson River (Elskus et al. 1999), Newark Bay (Prince and Cooper 1995b), New Bedford Harbor, Massachusetts (Nacci et al. 1999; Bello et al. 2001), and the Elizabeth River, Virginia (Van Veld and Westbrook 1995) exhibited reduced CYP1A protein or EROD inducibility compared to conspecifics from reference sites for AHs. Killifish from these highly polluted locales also were resistant to overt adult lethality or early life stage toxicities induced by these and other AH compounds (Prince and Cooper 1995a; Nacci et al. 1999; Meyer et al. 2002). In most cases, in contrast to tomcod, resistance to gene induction and toxicities was observed for both PAHs and HAHs in these killifish populations.

Resistance as a Genetic Adaptation or a Physiological Acclimation?

Resistance in tomcod and other fish species probably resulted from chronic exposures to highly toxic AHs, other contaminants, or combinations of these. Resistance to toxicants can result from genetic adaptations or physiological acclimations. If genetic, resistance is a population-level effect that is transmitted across generations and that persists long after site remediation. Development of genetic adaptations requires that variant resistant individuals exist, albeit at low frequencies, in the population prior to chronic exposure to chemical pollutants (Nacci et al. 1999). Chemical exposure results in the selection of individuals with resistant phenotypes and an increase in their frequency over time in the population. Studies showed that inter-individual variation in CYP1A1 inducibility was high within tomcod populations (Courtenay et al. 1994) as was variation in the structure of the CYP1A1 gene (Roy et al. 1995). If resistance has a physiological basis, it will quickly be lost as exposures to the operative toxicant diminish in frequency and intensity.

In either case, it is thought that resistance is associated with a cost, such as an increased sensitivity to other stressors or reduced performance (e.g., fecundity, growth, etc.) relative to individuals lacking resistance when both groups are in environments free of the toxicant. For example, killifish from the AH-resistant Elizabeth River population are significantly more susceptible than those from a nearby clean reference site to acute phototoxicity from a PAH, ambient UV light, and low oxygen conditions (Meyer and Di Giulio 2003). Additionally, a cadmium (Cd)-resistant population of the oligochaete worm *Limnodrilus hoffmeisteri* inhabited a highly contaminated locale (Cd, Ni, cobalt) in the Hudson River (Foundry Cove [Figure 1]). Following remediation efforts, the genetically adapted resistant phenotype in the population was rapidly displaced by Cd-sensitive worms that were either at low background levels prior to the clean-up or rapidly immigrated to the site suggesting that selective pressure against the resistant phenotype was intense in the remediated environment (Levinton et al. 2003).

We conducted experiments to determine whether resistance in tomcod has a genetic or physiological basis. Adult tomcod were collected from the Hudson and Miramichi rivers and F_1 and F_2 pure Hudson River, pure Miramichi River, and F_1 hybrid offspring were produced and evaluated for CYP1A1 mRNA inducibility as embryos or larvae (Roy et al. 2001; Chambers and Wirgin unpublished data). In initial studies with a single dose of each chemical, CYP1A1 mRNA was significantly inducible in pure F_1 larvae from both populations with B[a]P, however, gene expression was significantly higher in larvae of Miramichi River descent. For PCB 77, sig-

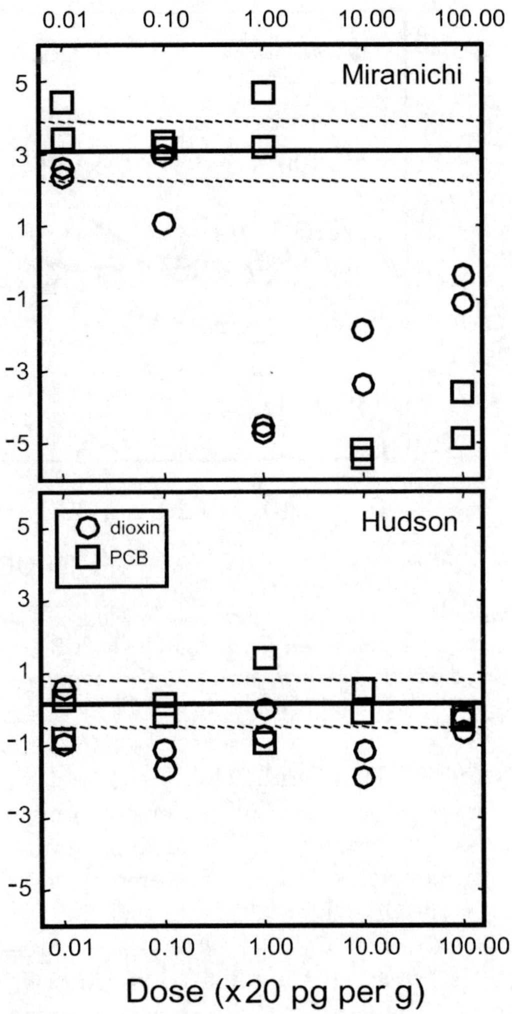

Figure 8. Effects of population source (Hudson River vs. Miramichi River), contaminant (PCBs and TCDD), and dose (0.01 to 100 times the concentrations of PCBs or TCDD measured in the liver of adult tomcod from the Hudson River Estuary) on morphometric variables of 1-d old F_1 tomcod larvae as summarized by the scores on the first principal component axis (PCA 1). The controls (water and acetone (the contaminant vehicle)) did not differ in their effects on PCA 1 values and are represented by the solid horizontal lines. The sensitivity of tomcod from the Miramichi River, and the resistance of Hudson River tomcod to dose were best reflected in PCA 1, a composite of whose values increases with increasing fish length, head and eye size, and yolk-sac depth. PCA 2 values (not shown) best separated the two source populations from one another.

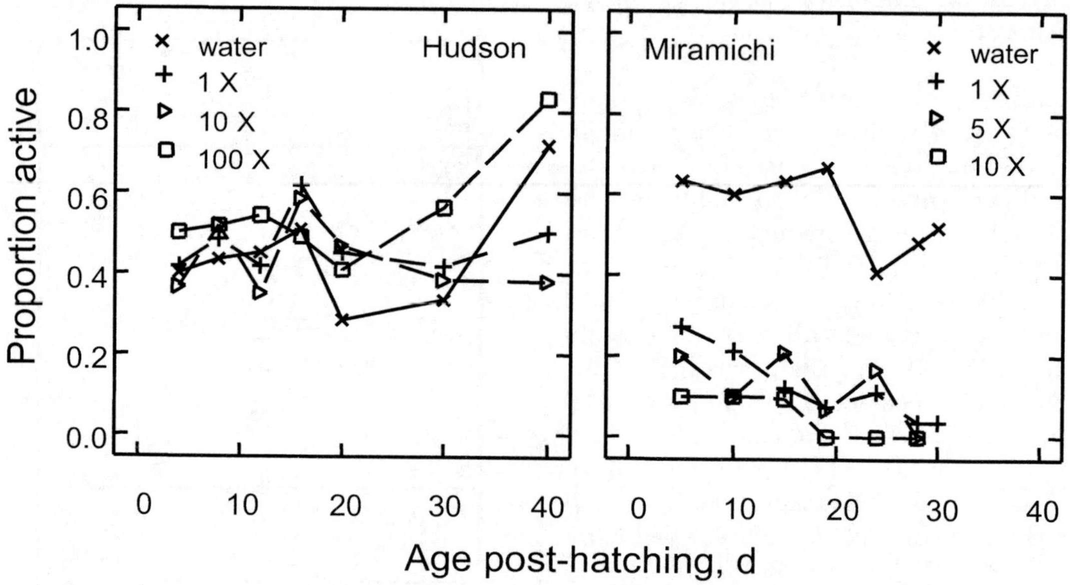

Figure 9. Effect of population source (Hudson River vs. Miramichi River) and dose of PCBs mixture of four coplanar congeners (PCB77, PCB81, PCB126, and PCB169) on activity level of unfed F_2 larvae during their first 40 d after hatching. Dose represents 0 to 100 times 'x', the concentration of these congeners measured in the livers of adult tomcod from the Hudson River Estuary.

nificant CYP1A1 induction was observed in larvae of both populations, although levels of gene expression were significantly less in larvae of Hudson River compared to Miramichi River descent. Hybrids exhibited intermediate levels of gene expression. In later, more extensive experiments, F_1 and F_2 embryos, early yolk sac larvae, and late yolk sac larvae from both populations were exposed to more doses of B[a]P and PCB77 and evaluated for CYP1A1 mRNA inducibility. Responses in the F_1 and F_2 offspring were nearly identical in all three early life stages. CYP1A1 mRNA was highly inducible by treatment with PCB77 or B[a]P in both generations of Miramichi River offspring. In both generations of Hudson River offspring, CYP1A1 mRNA was highly inducible by B[a]P, but only slightly by PCB77 (Figures 6a, 6b). Statistical analysis indicated that CYP1A1 mRNA inducibility was significantly more sensitive to PCB77, but not for B[a]P treatment in tomcod from the Miramichi River than in those from the Hudson River. These results indicate that genetic adaptation has occurred in the Hudson River tomcod population to HAHs, but not PAHs. This also suggests that more than one molecular pathway – likely an AHR-dependent and an AHR-independent pathway – may mediate CYP1A1 induction and perhaps higher-level biological responses in this population. Although levels of both PAHs and HAHs are unusually high in the Hudson River environment, it is possible that CYP1A1 activity in detoxifying PAHs is critical to organismic survival and population viability, but because HAHs induce but are poorly metabolized and generate reactive oxygen species, the second alternative pathway is shut down in the Hudson River population.

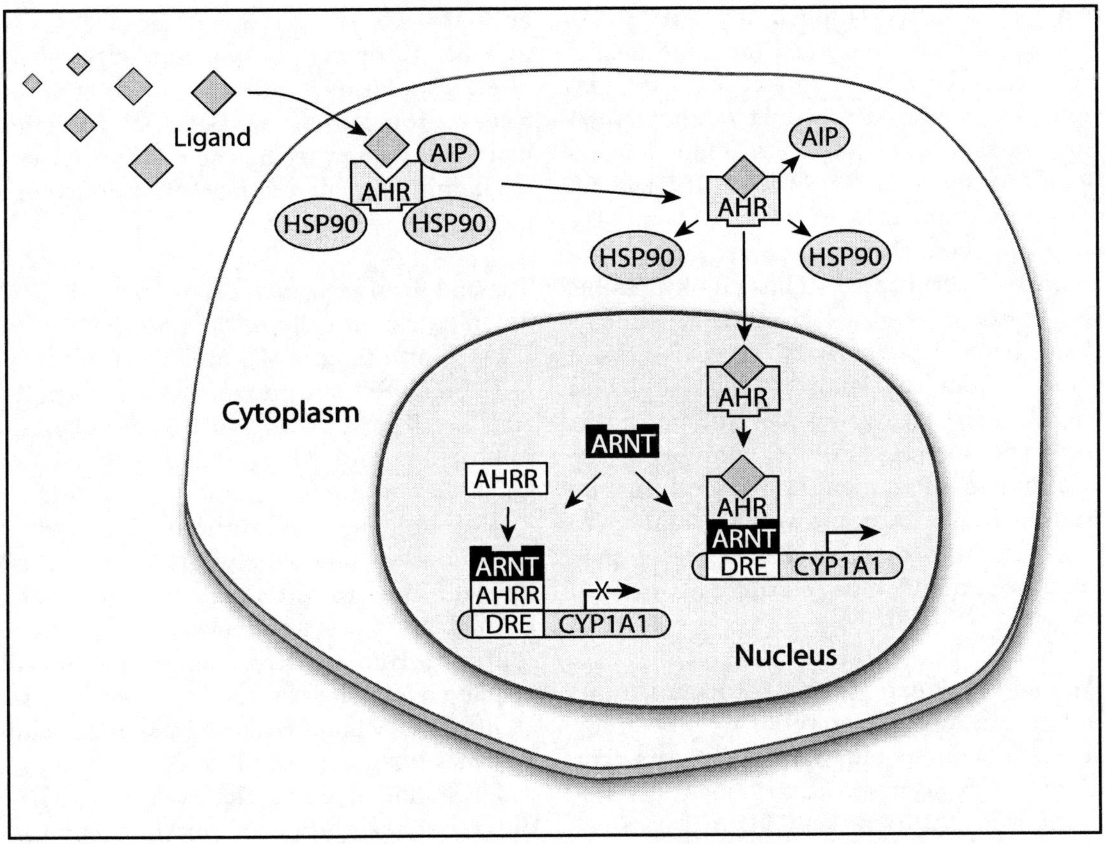

Figure 10. A schematic of a cell depicting molecules that are part of the aryl hydrocarbon receptor (AHR) pathway and whose expression was quantified and compared between Atlantic tomcod from the Miramichi River and Hudson River. Abbreviations: AIP,-aryl hydrocarbon receptor interacting protein; AHRR,-aryl hydrocarbon receptor repressor; ARNT, -aryl hydrocarbon receptor nuclear translocator; CYP1A1, -cytochrome P4501A1; DRE, -dioxin response element; HSP90, -heat shock protein 90.

Early Life stage Toxicities in PCBs and TCDD Exposed Tomcod

While CYP1A1 induction is predictive of many toxic responses to PCDD/Fs, PCBs, and PAHs because it is AHR mediated, no empirical evidence of resistance at higher levels of biological organization had been observed in tomcod from the Hudson River population. It is known that early life stage toxicities, including craniofacial malformations, multifocal hemorrhages, reduced growth, yolk sac edema, pericardial edema, and reduced blood flow (Hornung et al. 1996, 1999) are very sensitive biological responses to TCDD and dioxin-like PCBs in some, but not all species of fish (Elonen et al. 1998). It is thought that the early life stage toxic effects of individual PCB congeners covary with CYP1A1 inducibility and are mediated by the AHR pathway. Thus, it is likely that early life stage toxic effects of individual PCB congeners can be predicted based on their structural similarity to TCDD. These effects are also expected to be additive and species-specific. In controlled laboratory experiments, early life stages of lake trout are exquisitely sensitive to TCDD (Walker et al. 1991; Walker et al. 1994) and dioxin-like PCBs (Walker et al. 1996), and much more sensitive (16–180 fold) than other fish species tested (Elonen et al. 1998). As a result, despite continued introductions of captive-reared fish to restore wild populations, it is believed that lake trout populations in the Great Lakes have failed to rebound to historical levels of abundance because environmental exposures of young life stages to these chemicals has severely compromised recruitment. Thus, it is possible that chronic exposure to these chemicals may have affected early life stage success in tomcod from the Hudson River and this may consistently restrict recruitment in this population as well. In fact, analyses of the abundance of spawning tomcod in the Hudson River show an overall decline over the past twenty years (M. Mattson, personal communication, Normandeau Associates).

We undertook experimental studies of tomcod as an investigative bridge between organismal and population-level effects. The effects of exposure to environmentally relevant concentrations and identities of coplanar PCBs and TCDD were experimentally investigated in young life stages of laboratory-reared tomcod from the Hudson River and the Miramichi River. In two consecutive years, F_1 and F_2 embryos of Hudson River and Miramichi River ancestry were exposed as embryos to graded doses of a waterborne mixture of coplanar PCBs (PCB77, PCB81, PCB126, and PCB169) or TCDD. Doses were selected based on tissue burdens and congener patterns previously measured in wild tomcod from the Hudson River (Courtenay et al. 1999; Yuan et al. 2001, Roy et al. 2001). Measures of viability of embryos, larvae, and yolk sac larvae, and morphometric and behavioral variables of yolk sac larvae were scored. The morphometric variables, including measures of curvature of the larvae, yolk sac volume, and the sizes of eyes, head, and jaws, were reduced by principle component analysis (PCA) and the resulting PCA scores were evaluated via analysis of variance (ANOVA) in order to detect the impacts of toxicant, dose, population, and their interactive effects on tomcod larvae. Behavior was summarized by quantifying the relative activity (swimming versus resting) of variously exposed yolk sac larvae.

Tomcod from the two source populations differed dramatically in their sensitivities to HAH contaminants. F_1 and F_2 embryos of Miramichi River tomcod showed a graded response to PCBs dose with lower viability to hatching and greater levels of partial or failed hatch-outs at higher doses (Figure 7). Miramichi River embryos, exposed to PCBs doses at and above baseline ('1 x') that survived to hatching, did not survive through their first six weeks of larval life. In contrast, Hudson River tomcod embryos displayed a consistently high level of viability to hatching with minimal abnormalities at all ages regardless of PCBs dose, which included doses that were 10 to 100 times greater those measured in environmentally exposed tomcod from the Hudson River.

These differences between population sources in their responses to dose were also expressed in the morphology of survivors. In this case, a parallel response occurred to PCBs and TCDD (Figure 8). Over 80% of the experiment-wide variation in morphometric features was captured in the first two principal component (PC) axes. Total length, head dimensions (including eye and jaw size), and yolk amount contributed most to the first PC axis. Miramichi River yolk sac larvae showed dramatic change in their morphology at doses at and above the nominal baseline dose. In contrast, Hudson River individuals showed no significant effect of doses of either PCBs or TCDD.

These experimental studies were also designed to test for more subtle and sublethal behavioral responses. Vertical position in the water column is critical to the success of the nearly passive earliest larval stages of estuarine fishes. Because the Hudson River is a vertically stratified estuary of overlying freshwater flowing downstream and deeper seawater flowing upstream, vertical position within the water column will affect the spatial distribution of larvae and the location in the benthic habitat of recently transformed juvenile fish. In the laboratory, untreated larval tomcod from both populations exhibit characteristic regular vertical movements in which they ascend from the bottom to near the surface of holding vessels. Presumably, this behavior is reflective of similar movements in the natural environment. Vertical movements during the first 2–40 d of larval life for PCBs exposed F_2 Miramichi and Hudson River tomcod were monitored while they resided in 400 mL beakers. Activity of Miramichi River larvae was sensitive to all doses of PCBs exposure (doses at 1–10 times those measured in tomcod from the Hudson River), whereas tomcod larvae of Hudson River descent were unaffected by all doses of PCBs mix (up to 100 x) (Figure 9). In total, these studies indicate that tomcod from the Hudson River population have undergone significant evolutionary change at gene(s) that mediate responses at organismal and population levels to HAHs, perhaps due to chronic exposure to these contaminants.

Studies of the Mechanistic Basis of Resistance in Hudson River Tomcod

Because of its importance in mediating many toxic responses to HAHs and PAHs, it is possible that the mechanistic bases of resistance resides in the structure (amino acid sequence) or expression of molecules in the AHR pathway (Figure 10). Briefly, in unexposed cells, AHR resides in the cytoplasm where it is bound to two molecules of heat shock protein 90 (HSP90) and a single molecule of AHR interacting protein (AIP, also known as ZAP2 and Ara9). When lipophilic AH contaminants (ligands) diffuse into the cell, they bind AHR in the cytoplasm, HSP90s and AIP are released, and the ligand-AHR complex is translocated into the nucleus by an unknown mechanism. Within the nucleus, ligand-AHR binds the aryl hydrocarbon nuclear translocator (ARNT) and this three-part complex binds enhancer elements (dioxin response elements [DREs] also known as XREs and AHREs) upstream of AH responsive genes such as CYP1A1. Binding of the ligand-AHR-ARNT complex to these enhancer elements activates induced transcription of these genes. The aryl hydrocarbon receptor repressor (AHRR) serves to down-regulate activation of the AHR pathway by binding with ARNT and the AHRR-ARNT complex competing with the ligand-AHR-ARNT complex for DRE binding sites. AHRR-ARNT binding to DREs does not activate transcription of CYP1A1.

To investigate the possible role of altered AHR pathway molecules in conferring resistance, AHR, ARNT and AHRR were cloned from tomcod and their DNA sequences determined (Roy and Wirgin 1997; Roy and Wirgin 2002). For AHR, sequences were compared between tomcod from the Hudson River and three populations in Atlantic Canada. All tomcod from the Hudson River shared AHR genotypes not seen in tomcod from elsewhere. All tomcod from the Hudson River showed a two amino acid deletion in the AHR that was absent in tomcod from the three Canadian rivers (Yuan 2003). The significance of this genetic polymorphism has yet to be assessed in functional assays.

Cloning and characterization of these genes allowed for development of polymerase chain reaction (PCR)-based assays to quantify and compare the expression of AHR, ARNT, and AHRR at the RNA level among multiple life stages (embryo, larval, juvenile, and adult) and tissues of tomcod from the Hudson River and Miramichi River populations (Roy et al. 2002). The effects of PCBs or TCDD treatment on the expression of these genes were then evaluated. It was hypothesized that resistance of the Hudson River population results from decreased basal or induced expression of AHR and ARNT or increased expression of AHRR. Levels of AHR and ARNT expression were similar between tomcod from the two rivers, except that AHR expression was significantly higher in environmentally exposed tomcod from the Hudson River compared to those from the St. Lawrence or Miramichi Rivers. Exposure to coplanar PCB77 or TCDD under controlled laboratory did not affect levels of AHR in any life stage or tissue of tomcod from either population. As expected, AHRR expression was equally inducible in tomcod from the two populations by B[a]P, but more so in tomcod from the Miramichi River than the Hudson River by PCB77. Levels of hepatic AHRR expression also differed among environmentally exposed adult tomcod from the Hudson, St. Lawrence, and Miramichi Rivers, with highest levels of gene expression in tomcod from the Hudson River. Thus, levels of AHRR expression paralleled that of CYP1A1 in both controlled laboratory experiments and environmental exposures.

Several important conclusions can be reached from these studies of the mechanistic basis of resistance. First, differential levels of expression of molecules in the AHR pathway at the RNA level did not seem to contribute to the resistant phenotype in tomcod from the Hudson River. If anything, AHR was higher and AHRR expression lower in tomcod from the Hudson River compared to those from more sensitive populations. Second, levels of AHRR expression paralleled those of CYP1A1 in these populations in that it was higher in environmentally exposed tomcod from the Hudson River than elsewhere and it was less inducible by PCB77 in tomcod from the Hudson River than in those from the Miramichi River. This similarity in patterns of expression between AHRR and CYP1A1 in these populations may be due to both genes being similarly regulated at the transcriptional level. Both genes are induced by exposure to AH compounds and binding of AHR-ARNT to enhancer elements in their regulatory regions. Third, AHR expression was not inducible or down-regulated by AH treatment in tomcod. These results contrast with those in resistant killifish from the Elizabeth River (Meyer et al. 2003), but are in agreement with those in resistant killifish from New Bedford Harbor (Powell et al. 2000). Thus, it does not seem that chronic exposure of tomcod in the Hudson River to AHs has down-regulated AHR expression and thereby impaired inducibility of CYP1A1. One caveat is that these analyses were at the RNA level, and it is expression of the proteins encoded by these genes that are functionally important. Studies using rodents have shown that exposure to HAHs destabilizes AHR and results in down-regulation of its expression (Pollenz et al. 1998). Studies of AHR pathway proteins of tomcod must be conducted to truly understand the role of this pathway in resistant phenotypes.

Summary and Conclusions

Surprisingly few examples of toxic effects of chemical contaminants on the Hudson River fish community have been reported. Compared to neighboring large estuaries, the Hudson River hosts at least as rich an ichthyofaunal community. In recent decades, abundance of glamour species in the Hudson River such as shortnose sturgeon and striped bass are near historically high levels. However, one species, Atlantic tomcod exhibits manifestations of toxic damage consistent with exposure to AHs at a hierarchy of mechanistically linked biological endpoints, including altered gene expression, DNA damage, liver cancer, a truncated age structure, and reduced adult abundance. Linking the higher-level perturbations in a natural population to exposure to single or even groups of contaminants, however, is tenuous in the absence of empirical verification. Controlled laboratory exposure experiments are needed to evaluate the sensitivity of naïve tomcod to environmentally relevant mixtures of PAHs, PCDD/Fs, and PCBs in inducing hepatic neoplasms or preneoplastic lesions and to determine the effects of tumors on longevity of individuals and the overall population age structure.

Despite extensive evidence of toxic effects, the Hudson River tomcod population is remarkably resistant to biological effects of coplanar PCBs and TCDD at a variety of molecular and organismal endpoints. These altered biological responses are mechanistically linked and suggest a role for the AHR pathway in conferring resistance. Resistance seems to have largely a genetic basis, although shorter-term acclimation also contributes to modulated gene expression in this population. Most of the known components of the AHR pathway have been characterized in tomcod. However, their expression at the mRNA level does not seem to explain the resistance phenomenon. Studies are needed to characterize the expression of these genes at the protein level because it is here where the phenotype is determined.

Given their high levels of chronic exposure to and bioaccumulation of AHs, it is not surprising that resistance has developed in the Hudson River tomcod population. In fact, it has been suggested that resistance may be common in fish populations that are chronically exposed to high levels of AHs (Wirgin and Waldman 2004). Tissue burdens of HAHs in environmentally exposed tomcod from the Hudson River are at or exceed those that induce acute toxicity in tomcod from other populations that we have examined. On the surface, the development of resistance seems inconsistent with increased DNA damage, elevated prevalence of tumors, and a truncated age structure in this population. It is possible that resistance has only very recently developed and that the prevalence of tumors today is less than that in the early 1980s. Targeted field studies are needed to document the prevalence of tumors and preneoplastic lesions in the current Hudson River tomcod population. Alternatively, as has been suggested in the mammalian literature, the development of liver tumors may serve as a strategy of resistance to toxicants. In tomcod from the Hudson River, it is likely that at least one spawning occurs in age-1 fish before any cancer-associated mortality, thus negating any selective load of cancer on the population. Liver tumors may allow individuals to survive in the face of chronic exposure to hepatotoxicants (Farber 1990; Farber and Rubin 1991; Yusuf et al. 1999). In fish and in mammals, hepatic tumors and preneoplastic lesions often exhibit decreased expression of Phase I and increased expression of Phase II enzymes compared to normal tissues. These strategies may enhance detoxification of environmental hepatotoxicants in tumors compared to normal tissues and provide them a selective advantage.

While resistance does provide short-term benefits to populations in the face of chronic exposure to toxicants, it does not come without costs at the population and community levels. Fish that are resistant to one chemical may be more sensitive to the effects of a second stressor than fish that are not resistant. Investigations into interactive effects of multiple chemical stressors (AH and metals) in adult and young life stages of tomcod from the Hudson River are underway and suggest that tomcod from the Hudson River may be more sensitive to metals than those from the Miramichi River (Sorrentino et al. 2004; Sorrentino 2004). Furthermore, tomcod in the Hudson River are at the very southern extreme of the species' geographic distribution and the additional effects of thermal stress during the warmer months are of special concern and warrant investigation.

After the removal of a strong selective agent (e.g., through site remediation), the performance of resistant populations may be diminished compared to that of sensitive populations. This reduction in performance may be due to the allocation of a disproportionate share of a limited energy budget towards resistance. Thus, fecundity, growth rate, and survival may all be reduced in a population that has been challenged by and adapted to contaminants. These and other demographic parameters can be empirically tested in controlled laboratory experiments with tomcod as has been done for killifish from the Elizabeth River.

The evolution of resistance permits the survival of populations in habitats that are chronically challenged by high levels of contaminants, but given the demonstrated importance of young-of-the-year tomcod as prey to resource and other key species throughout the lower Hudson River Estuary, resistance by tomcod almost certainly results in elevated transfer of contaminants up the food chain. Several aspects of the rate of consumption of tomcod by striped bass and other fishes are currently being investigated by us, but the full impact of resistance in tomcod on its immediate and secondary consumers warrants full evaluation. Ultimately, however, resistant populations are an aberration. Their persistence, as well as repercussions throughout the community and ecosystem in which they are embedded, are best considered as toxic responses to chemical contaminants in their environment.

Acknowledgments

We acknowledge support of the Hudson River Foundation, SBRP Grant ES10344, NIEHS Center Grant ES00260, and the Northeast Fisheries Center (NOAA Fisheries Service) for the studies described in this review.

References

Adams, S. M. 2002. Biological indicators of aquatic ecosystem stress. American Fisheries Society, Bethesda, Maryland.

Armstrong, R. W., and R. J. Sloan. 1988. PCB patterns in Hudson River fish: I. resident freshwater species. Pages 304–324 *in* C. L. Smith, editor. Fisheries research in the Hudson River. The Hudson River Environmental Society, State University of New York Press, Albany.

Bailey, G. S., D. E. Williams, and J. D. Hendricks. 1996. Fish models of environmental carcinogenesis: the rainbow trout. Environmental Health Perspectives. 104 (Supplement 1):5–21.

Barnthouse, L. W., D. Glaser, and J. Young. 2003. Effects of historic PCB exposures on the reproductive success of the Hudson River striped bass population. Environmental Science and Technology 37:223–228.

Baumann, P. C. 1998. Epizootics of cancer in fish associated with genotoxins in sediment and water. Mutation Research 411:227–233.

Baumann, P. C., and J. C. Harshbarger. 1995. Decline in liver neoplasms in wild brown bullhead catfish after coking plant closes and environmental PAHs plummet. Environmental Health Perspectives 103:168–170.

Baumann, P. C., W. D. Smith, and W. K. Parland. 1987. Tumor frequencies and contaminant concentrations in brown bullheads from an industrialized river and a recreational lake. Transactions of the American Fisheries Society 116:79–86.

Bello, S. M., D. G. Franks, J. J. Stegeman, and M. E.

Hahn. 2001. Acquired resistance to AH receptor agonists in a population of Atlantic killifish (*Fundulus heteroclitus*) inhabiting a marine Superfund site: *in vivo* and *in vitro* studies on the inducibility of xenobiotic metabolizing enzymes. Toxicological Sciences 60:77–91.

Bopp, R. F., S. N. Chillrud, E. L. Shuster, H. J. Simpson, and F.D. Estabrooks. 1998. Trends in chlorinated hydrocarbon levels in Hudson River basin sediments. Environmental Health Perspectives 106 (Supplement 4):1075–1081.

Bopp, R. F., M. L. Gross, H. Tong, H. J. Simpson, S. J. Monson, B. L. Deck, and F. C. Moser. 1991. A major incident of dioxin contamination: sediments of New Jersey estuaries. Environmental Science and Technology 25:951–956.

Bowser, P. R., D. Martineau, R. Sloan, M. Brown, and C. Carusone. 1990. Prevalence of liver lesions in brown bullheads (*Ictalurus nebulosus*) collected from a polluted and "non-polluted" site on the Hudson River, New York. Journal of Aquatic Animal Health 2:177–181.

Cogliano, V. J. 1998. Assessing the cancer risks from environmental PCBs. Environmental Health Perspectives 106:317–323.

Cook, P. M., J. A. Robbins, D. D. Endicott, K. B. Lodge, P. D. Guiney, M. K. Walker, E. W. Zabel, and R. D. Peterson. 2003. Effects of aryl hydrocarbon receptor-mediated early life stage toxicity on lake trout populations in Lake Ontario during the 20th century. Environmental Science and Technology 37:3864–3877.

Cormier, S. M. and R. N. Racine. 1990. Histopathology of Atlantic tomcod: a possible monitor of xenobiotics in northeast tidal rivers and estuaries. Pages 59–71 *in* J. F. McCarthy and L. R. Shugart, editors. Biomarkers of environmental contamination. Lewis Publishers, Boca Raton, Florida.

Cormier, S. M., R. N. Racine, C. E. Smith, W. P. Dey, and T. H. Peck. 1989. Hepatocellular carcinoma and fatty infiltration in the Atlantic tomcod, *Microgadus tomcod* (Walbaum). Journal of Fish Diseases 12:105–116.

Couillard, C. M., P.J. Williams, S. C. Courtenay, and G. P. Rawn. 1999. Histopathological evaluation of Atlantic tomcod (*Microgadus tomcod*) collected at estuarine sites receiving pulp and paper mill effluent. Aquatic Toxicology 44:263–278.

Courtenay, S. C., C. M. Grunwald, G.-L. Kreamer, W. L. Fairchild, J. T. Arsenault, M. Ikonomou, and I. I. Wirgin. 1999. A comparison of the dose and time response of CYP1A1 mRNA induction in chemically treated Atlantic tomcod from two populations. Aquatic Toxicology 47:43–69.

Courtenay, S., P. J. Williams, C. Grunwald, T.-L. Ong, B. Konkle, and I. I. Wirgin. 1994. An assessment of within group variation in CYP1A1 mRNA inducibility in Atlantic tomcod. Environmental Health Perspectives 102(Supplement 12):85–90.

Daniels, R.A., K. E. Limburg, R. E. Schmidt, D. L. Strayer, and R. C. Chambers. 2005. Changes in fish assemblages in the tidal Hudson River, New York. Pages 471–503 *in* J. N. Rinne, R. M. Hughes, and B. Calamusso. Historical changes in large river fish assemblages of the Americas. American Fisheries Society, Symposium 45, Bethesda, Maryland.

Dey, W. P., T. H. Peck, C. E. Smith, and G.-L. Kreamer. 1993. Epizoology of hepatic neoplasia in Atlantic tomcod (*Microgadus tomcod*) from the Hudson River estuary. Canadian Journal of Fisheries and Aquatic Sciences 50:1897–1907.

Elonen, G. E., R. L. Spehar, G. W. Holcombe, R. D. Johnson, J. D. Fernandez, R. J. Erickson, J. E. Tietge, and P. M. Cook. 1998. Comparative toxicity of 2,3,7,8-tetrachlorodibenzo-*p*-dioxin to seven freshwater fish species during early life-stage development. Environmental Toxicology and Chemistry 17:472–483.

Elskus, A. A., E. Monosson, A. E. McElroy, J. J. Stegeman, and D. S. Woltering. 1999. Altered CYP1A expression in *Fundulus heteroclitus* adults and larvae: a sign of pollutant resistance. Aquatic Toxicology 45:99–113.

Farber, E. 1990. Clonal adaptation during carcinogenesis. Biochemical Pharmacology 39:1837–1846.

Farber, E., and H. Rubin. 1991. Cellular adaptation in the origin and development of cancer. Cancer Research 51:2751–2761.

Farley, K. J., and R. V. Thomann. 1998. Fate and bioaccumulation of PCBs in aquatic environments. Pages 1581–1593 *in* W. N. Rom, editor. Environmental and occupational medicine (3rd Edition). Lippincott-Raven, Philadelphia.

Fernandez, M. P., M. G. Ikonomou, S. C. Courtenay, and I. I. Wirgin. 2004. Spatial variation and source prediction of PCBs and PCDD/Fs among young-of-the-year and adult tomcod (*Microgadus tomcod*) in the Hudson River Estuary. Environmental Science and Technology 38:976–983.

George, S. G. 1994. Enzymology and molecular biology of Phase II xenobiotic conjugating enzymes in fish. Pages 37–85 *in* D. C. Malins and G. K. Ostrander, editors. Aquatic Toxicology: Molecular,

Biochemical, and Cellular Perspectives. Lewis Publishers, Boca Raton, Florida.

Hornung, M. W., J. M. Spitsbergen, and R. E. Peterson. 1999. 2,3,7,8-tetrachlorodibenzo-p-dioxin alters cardiovascular and craniofacial development and function in sac fry of rainbow trout (*Oncorhynchus mykiss*). Toxicological Sciences 47:40–51.

Hornung, M. W., E. W. Zabel, and R. E. Peterson. 1996. Toxic equivalency factors of polybrominated dibenzo-*p*-dioxin, dibenzofuran, biphenyl, and polyhalogenated diphenyl ether congeners based on rainbow trout early life stage mortality. Toxicology and Applied Pharmacology 140:227–234.

Huff, J., G. Lucier, and A. Tritscher. 1994. Carcinogenicity of TCDD: experimental, mechanistic, and epidemiologic evidence. Annual Review of Pharmacology and Toxicology 34:343–372.

Johnson, L., and T. K. Collier. 2002. Assessing contaminant-induced stress across levels of biological organization. Pages 533–563 *in* S. M. Adams, editor. Biological indicators of aquatic ecosystem stress. American Fisheries Society, Bethesda, Maryland.

Johnson, L., C. M. Stehr, O. P. Olson, M.S. Myers, S. M. Pierce, C. A. Wigren, B. B. McCain, and U. Varanasi. 1993. Chemical contaminants and hepatic lesions in winter flounder (*Pleuronectes americanus*) from the Northeast coast of the United States. Environmental Science and Technology 27:2759–2771.

Kim, J. C. S., E. S. Chao, M. P. Brown, and R. Sloan. 1989. Pathology of brown bullhead, *Ictalurus nebulosus* from highly contaminated and relatively clean sections of the Hudson River. Bulletin of Environmental Contamination and Toxicology 43:144–150.

Klauda, R. J., R. E. Moos, and R.E. Schmidt. 1988. Life history of Atlantic tomcod, *Microgadus tomcod*, in the Hudson River Estuary, with emphasis on spatio-temporal distribution and movements. Pages 219–215 *in* C. L. Smith, editor. Fisheries research in the Hudson River. State University of New York Press, Albany, New York

Krahn, M. M., L. D. Rhodes, M. S. Meyers, L. K. Moore, W. D. MacLeod Jr., and D. C. Malins. 1986. Associations between metabolites of aromatic compounds in bile and the occurrence of hepatic lesions in English sole (*Parophrys vetulus*) from Puget Sound, Washington. Archives of Environmental Contamination and Toxicology 15:61–67.

Kreamer, G.-L., K. Squibb, D. Gioeli, S. J. Garte, and I. Wirgin. 1991. Cytochrome P450I mRNA expression in feral Hudson River tomcod. Environmental Research 55:64–78.

Kriek, E., M. Rojas, K. Alexandrov, and H. Bartsch. 1998. Polycyclic aromatic hydrocarbon-DNA adducts in humans: relevance as biomarkers for exposure and cancer risk. Mutation Research 400:215–231.

Lengauer, C., K. W. Kinzler, and B. Vogelstein. 1998. Genetic instability in human cancers. Nature (London) 396:643–649.

Levinton, J. S., E. Suatoni, W. Wallace, R. Junkins, B. Kelaher, and B. Allen. 2003. Rapid loss of genetically based resistance to metals after the cleanup of a Superfund site. Proceedings of the National Academy of Science 100:9889–9891.

Malins, D. C., G. K. Ostrander, R. Haimanot, and P. Williams. 1990. A novel DNA lesion in neoplastic livers of feral fish: 2,6-diamino-4-hydroxy-5-formamidopyrimidine. Carcinogenesis 11:1045–1047.

Manam, S., R. D. Storer, S. Prahalada, K. R. Leander, A. R. Kraynak, B. J. Ledwith, M. J. van Zwieten, M. O. Bradley, and W. W. Nichols. 1992. Activation of the Ha-, Ki- and N-*ras* genes in chemically induced liver tumors from CD-1 mice. Cancer Research 52:3347–3352.

Mehrle, P. M., T. A. Haines. S. Hamilton, J. L. Ludke, F. L. Mayer, and M. A. Ribick. 1982. Relationship between body contaminants and bone development in East-Coast striped bass. Transactions of the American Fisheries Society 111:231–241.

Mehrle, P. and L. Ludke. 1983. Impacts of contaminants on early life stages of striped bass. Progress Report 1980–1983. Columbia National Fisheries Research Laboratory, Fish and Wildlife Service, U.S. Department of Interior, Columbia, Missouri.

Meyer, J. N., and R. T. Di Giulio. 2003. Heritable adaptation and fitness costs in killifish (*Fundulus heteroclitus*) inhabiting a polluted estuary. Ecological Applications 13:490–503.

Meyer, J. N., D. E. Nacci, and R. T. Di Giulio. 2002. Cytochrome P4501A in killifish (*Fundulus heteroclitus*): heritability of altered expression and relationship to survival in contaminated sediments. Toxicological Sciences 68:69–81.

Meyer, J. N., D. A. Wassenberg, S. I. Karchner, M. E. Hahn, and R. T. Di Giulio. 2003. Expression and inducibility of aryl hydrocarbon receptor pathway genes in wild-caught killifish (*Fundulus heteroclitus*) with different contaminant-exposure histories. Environmental Toxicology and Chemistry

22:2337–2343.

Myers, M. S., C. M. Stehr, O. P. Olson, L. L. Johnson, B. B. McCain, S.-L. Chan, and U. Varanasi. 1994. Relationships between toxicopathic hepatic lesions and exposure to chemical contaminants in English sole (*Pleuronectes vetulus*), starry flounder (*Platichthys stellatus*), and white croaker (*Genyonemus lineatus*) from selected marine sites on the Pacific Coast, USA. Environmental Health Perspectives 102:200–215.

Nacci, D., L. Coiro, D. Champlin, S. Jayaraman, R. McKinney, T. R. Gleason, W. R. Munns, Jr., J. L. Specker, and K. R. Cooper. 1999. Adaptations of wild populations of the estuarine fish *Fundulus heteroclitus* to persistent environmental contaminants. Marine Biology 134:9–17.

Pinkney, A. E., J. C. Harshbarger, E. B. May, and W. L Reichert. 2004. Tumor prevalence and biomarkers of exposure and response in brown bullheads (*Ameiurus nebulosus*) from the Anacostia River, Washington, DC and Tuckahoe River, Maryland, USA. Environmental Toxicology and Chemistry 23:638–647.

Pollenz, R. S., M. J. Santostefano, E. Klett, V. M. Richardson, B. Necela, and L. S. Birnbaum. 1998. Female Sprague-Dawley rats exposed to a single oral dose of 2,3,7,8-tetrachlorodibenzo-*p*-dioxin exhibit sustained depletion of aryl hydrocarbon receptor protein in liver, spleen, thymus, and lung. Toxicological Sciences 42:117–128.

Powell, W. H., R. Bright, S. Bello, and M. E. Hahn. 2000. Developmental and tissue-specific expression of AHR1, AHR2, and ARNT2 *in* dioxin-sensitive and -resistant populations of the marine fish *Fundulus heteroclitus*. Toxicological Sciences 57:229–239.

Prince, R., and K. R. Cooper. 1995a. Comparisons of the effects of 2,3,7,8-tetrachlordibenzo-*p*-dioxin on chemically impacted and nonimpacted subpopulations of *Fundulus heteroclitus*. I. TCDD toxicity. Environmental Toxicology and Chemistry 14:579–588.

Prince, R., and K. R. Cooper. 1995b. Comparisons of the effects of 2,3,7,8-tetrachlorodibenzo-*p*-dioxin on chemically impacted and nonimpacted subpopulations of *Fundulus heteroclitus*. II. Metabolic considerations. Environmental Toxicology and Chemistry 14:589–596.

Rappe, C., P.-A. Bergqvist, L.-O. Kjeller, and S. Swanson. 1991. Levels and patterns of PCDD and PCDF contamination in fish, crabs, and lobsters from Newark Bay and the New York Bight. Chemosphere 22:239–266.

Roy, N. K., S. Courtenay, G. Maxwell, Z. Yuan, R. C. Chambers, and I. Wirgin. 2002. Cytochrome P4501A1 is induced by PCB 77 and benzo[a]pyrene treatment but not by exposure to the Hudson River environment in Atlantic tomcod (*Microgadus tomcod*) post-yolk sac larvae. Biomarkers 7:162–173.

Roy, N. K., S. Courtenay, Z. Yuan, M. Ikonomou, and I. Wirgin. 2001. An evaluation of the etiology of reduced CYP1A1 messenger RNA expression in the Atlantic tomcod from the Hudson River, New York, USA, using reverse transcriptase polymerase chain reaction analysis. Environmental Toxicology and Chemistry 20:1022–1030.

Roy, N. K., J. Duffy, M. Zarzecka, and I. Wirgin. 2002. Comparison of AHR, ARNT, and AHRR expression in chemically-treated and environmentally exposed Atlantic tomcod from two populations. SETAC 23rd Annual Meeting in North America. 16–20 November 2002, Salt Lake City, Utah.

Roy, N. K., G.-L. Kreamer, B. Konkle, C. Grunwald, and I. Wirgin. 1995. Characterization and prevalence of a polymorphism in the 3[prime] untranslated region of cytochrome P4501A1 in cancer-prone Atlantic tomcod. Archives of Biochemistry and Biophysics 322:204–213.

Roy, N. K. and I. Wirgin. 1997. Characterization of the aromatic hydrocarbon receptor gene and its expression in Atlantic tomcod. Archives of Biochemistry and Biophysics 344:373–386.

Roy, N. K. and I. Wirgin. 2002. Characterization of ARNT and AHRR in Atlantic tomcod. SETAC 23rd Annual Meeting in North America. 16–20 November 2002, Salt Lake City, Utah.

Sather, P., M. G. Ikonomou, R. F. Addison, T. He, P. Ross, and B. Fowler. 2001. Similarity of an Aroclor-based and a full congener-based method in determining total PCBs and a modeling approach to estimate Aroclor speciation from congener-specific PCB data. Environmental Science and Technology 35:4874–4880.

Schlezinger, J. J., R. D. White, and J. J. Stegeman. 1999. Oxidative inactivation of cytochrome P4501A stimulated by 3, 3′, 4, 4′-tetrachlorobiphenyl: production of reactive oxygen by vertebrate CYP1As. Molecular Pharmacology 56:588–597.

Skinner, L. C., S. J. Jackling, G. Kimber, J. Waldman, J. Shastay, Jr., and A. J. Newell. 1996. Chemical residues in fish, bivalves, crustaceans and a cephalopod from the New York-New Jersey Harbor estuary: PCB, organochlorine and mercury. New York

State Department of Environmental Conservation, Division of Fish, Wildlife, and Marine Resources, Albany, New York.

Skinner, L. C., R. Prince, J. Waldman, A. J. Newell, and J. Shastay, Jr. 1997. Chemical residues in fish, bivalves, crustaceans and a cephalopod from the New York-New Jersey Harbor estuary: Dioxins and furans. New York State Department of Environmental Conservation, Division of Fish, Wildlife and Marine Resources, Albany, New York.

Sloan, R. J., M. W. Kane, and L. C. Skinner. 2002. 1999 as a special spatial year for PCBs in Hudson River fish. Bureau of Habitat, Division of Fish, Wildlife and Marine Resources, New York State Department of Environmental Conservation, Albany, New York.

Sloan, R. J., M.W. Kane, and L. C. Skinner. 2005. Of Time, PCBs and the Fish of the Hudson River. Bureau of Habitat, Division of Fish, Wildlife and Marine Resources, New York State Department of Environmental Conservation, Albany, New York.

Sloan, R. J., B. Young, and K. Hattala. 1995. PCB Paradigms for Striped Bass in New York State. Division of Fish and Wildlife, Division of Marine Resources, New York State Department of Environmental Conservation, Albany, New York. Technical Report Rep. 95–1 (BEP).

Smith, C. E., T. H. Peck, R. J. Klauda, and J. B. McLaren. 1979. Hepatomas in Atlantic tomcod *Microgadus tomcod* (Walbaum) collected in the Hudson River estuary in New York. Journal of Fish Diseases 2:313–319.

Smith, C. L. and T. R. Lake. 1990. Documentation of Hudson River fish fauna. American Museum Novitates 2981.

Sorrentino, C. 2004. Impact of exposure to metals on Gene Expression Induced by Aromatic Hydrocarbons and its Ecotoxicological Implications. Doctoral dissertation. Department of Environmental Medicine, New York University, New York.

Sorrentino, C., N. K. Roy, R. C. Chambers, S. C. Courtenay, and I. Wirgin. 2004. B[a]P-DNA binding in early life-stages of Atlantic tomcod: Population differences and chromium modulation. Marine Environmental Research 53:383–388.

Stegeman, J. J., and M. E. Hahn. 1994. Biochemistry and molecular biology of monooxygenases: current perspectives on forms, functions, and regulation of cytochrome P450 in aquatic species. Pages 87–206 *in* D. C. Malins and G. K. Ostrander, editors. Aquatic Toxicology: Molecular, Biochemical, and Cellular Perspectives. Lewis Publishers, Boca Raton, Florida.

Stein, J. E., W. L. Reichert, B. French, and U. Varanasi. 1993. ^{32}P postlabeling analysis of DNA adduct formation and persistence in English sole (*Pleuronectes vetulus*) exposed to benzo[a]pyrene and 7H-dibenzo[*c,g*] carbazole. Chemical and Biological Interactions 88:55–69.

Stein, J. E., W. L. Reichert, and U. Varanasi. 1994. Molecular epizootiology: assessment of exposure to genotoxic compounds in teleosts. Environmental Health Perspectives 102(Supplement 12):19–23.

Striped Bass Technical Committee. 2003. 2003 Stock Assessment Report for Atlantic Striped Bass: Catch-at-age based VPA & Tag Release/Recovery based Survival Estimation. Report #SBTC-2003–3. Atlantic Striped Bass Management Board, Atlantic States Marine Fisheries Commission.

Van den Berg, M., L. Birbaum, A. T. C. Bosveld, B. Brunstrom, P. Cook, M. Feeley, J. P. Giesy, A. Handberg, R. Hasegawa, S. W. Kennedy, T. Kubiak, J. C. Larsen, F. X. Rolaf van Leeuwen, A. K. Djien Liem, C. Nolt, R. E. Peterson, L. Poellinger, S. Safe, D. Schrenk, D. Tillitt, M. Tysklind, M. Younes, F. Waern, and T. Zacharewski. 1998. Toxic equivalency factors (TEFs) for PCBs, PCDDs, PCDDs, PCDFs for humans and wildlife. Environmental Health Perspectives 106:775–792.

Van Veld, P., and D. Westbrook. 1995. Evidence for depression of cytochrome P4501A in a population of chemically resistant mummichog (*Fundulus heteroclitus*). Environmental Sciences 3/ 4:221–234.

Varanasi, U., and J. E. Stein. 1991. Disposition of xenobiotic chemicals and metabolites in marine organisms. Environmental Health Perspectives 90:93–100.

Vethaak, A. D., J. G. Jol, A. Meijboom, M. L. Eggens, T. ap Rheinallt, P. W. Wester, T. van de Zande, A. Bergman, N. Dankers, F. Ariese, R. A. Baan, J. M. Everts, A. Opperhuizen, and J. M. Marquenie. 1996. Skin and liver diseases induced in flounder (*Platichthys flesus*) after long-term exposure to contaminated sediments in large-scale mesocosms. Environmental Health Perspectives 104:1218–1229.

Walker, M. K., P. M. Cook, A. R. Batterman, B. C. Butterworth, C. Berini, J. J. Libal, L. C. Hufnagle, and R. E. Peterson. 1994. Translocation of 2,3,7,8-tetrachlorodibenzo-*p*-dioxin from adult female lake trout (*Salvelinus namaycush*) to oocytes: effects on early life stage development and sac fry survival.

Canadian Journal of Fisheries and Aquatic Sciences 51:1410–1419.

Walker, M. K., P. M. Cook, B. C. Butterworth, E. W. Zabel, and R. E. Peterson. 1996. Potency of a complex mixture of polychlorinated dibenzo-*p*-dioxin, dibenzofuran, and biphenyl congeners compared to 2,3,7,8-tetrachlorodibenzo-*p*-dioxin in causing fish early life stage mortality. Fundamental and Applied Toxicology 30:178–186.

Walker, M. K., J. M. Spitsbergen, J. R. Olsen, and R. E. Peterson. 1991. 2,3,7,8-tetrachlorodibenzo-*p*-dioxin (TCDD) toxicity during early life stage development of lake trout (*Salvelinus namaycush*). Canadian Journal of Fisheries and Aquatic Sciences 48:875–883.

Westin, D. T., C. E. Olney, and B. A. Rogers. 1983. Effects of parental and dietary PCBs on survival, growth, and body burdens of larval striped bass. Bulletin of Environmental Contamination and Toxicology 30:50–57.

Westin, D. T., C. E. Olney, and B. A. Rogers. 1985. Effects of parental and dietary organochlorines on survival and body burdens of striped bass larvae. Transactions of the American Fisheries Society 114:125–136.

Wirgin, I. I., D. C. Currie, and S. J. Garte. 1989. Activation of the K-*ras* oncogene in liver tumors of Hudson River tomcod. Carcinogenesis 10:2311–2315.

Wirgin, I. I., C. Grunwald, S. Courtenay, G.-L. Kreamer, W. L. Reichert, and J. E. Stein. 1994. A biomarker approach to assessing xenobiotic exposure in Atlantic tomcod from the North American Atlantic coast. Environmental Health Perspectives 102:764–770.

Wirgin, I. I., G.-L. Kreamer, C. Grunwald, K. Squibb, S. J. Garte, and S. Courtenay. 1992. Effects of prior exposure history on cytochrome P450IA mRNA induction by PCB congener 77 **in** Atlantic tomcod. Marine Environmental Research 34:103–108.

Wirgin, I., and C. W. Theodorakis. 2002. Molecular biomarkers in aquatic organisms: DNA damage and RNA expression. Pages 43–110 *in* S. M. Adams, editor. Biological Indicators of Aquatic Ecosystem Stress. American Fisheries Society, Bethesda, Maryland.

Wirgin, I., and J. R. Waldman. 1998. Altered gene expression and genetic damage in North American fish populations. Mutation Research 399:193–219.

Wirgin, I., and J. R. Waldman. 2004. Resistance to contaminants in North American fish populations. Mutation Research 552:73–100.

Wirgin, I., J. S. Weis, and A. E. McElroy. 2006. Physiological and genetic aspects of toxicity in Hudson River species. Pages 441–464 *in* J. S. Levinton and J. R. Waldman, editors. The Hudson River estuary. Cambridge University Press, Cambridge, New York.

Yuan, A. 2003. Resistance of cytochrome P4501A1 mRNA to induction by halogenated aryl hydrocarbons in Hudson River Atlantic tomcod: its prevalence and mechanistic basis. Doctoral dissertation, Department of Environmental Medicine, New York University School of Medicine, Tuxedo, New York.

Yuan, Z., M. Wirgin, S. Courtenay, M. Ikonomou, and I. Wirgin. 2001. Is hepatic cytochrome P4501A1 expression predictive of dioxins, furans, and PCB burdens in Atlantic tomcod from the Hudson River estuary? Aquatic Toxicology 54:217–230.

Yusuf, A., P. M. Rao, S. Rajalakshmi, and D. S. Sarma. 1999. Development of resistance during the early stages of experimental liver carcinogenesis. Carcinogenesis 20:1641–1644.